Rat Hybridomas and Rat Monoclonal Antibodies

Editor

Hervé Bazin
Professor
Experimental Immunology Unit
Catholic University of Louvain
Brussels, Belgium
and
Commission of the European Communities
Directorate-General for Science Research and Development
Brussels, Belgium

CRC Press, Inc.
Boca Raton, Florida

Library of Congress Cataloging in Publication Data

Bazin, H.
Rat hybridomas and rat monoclonal antibodies/author, Hervé Bazin
p. cm.
Includes bibliographies and index.
ISBN 0-8493-5438-2
1. Hybridomas. 2. Antibodies, Monoclonal. 3. Rats — Immunology.
I. Title.
[DNLM: 1. Antibodies, Monoclonal. 2. Hybridomas. 3. Rats. QW 575 B363r]
QR185.8.H93B39 1990
599'.0293— dc19 89-854

This book represents information obtained from authentic and highly regarded sources. Reprinted material is quoted with permission, and sources are indicated. A wide variety of references are listed. Every reasonable effort has been made to give reliable data and information, but the author and the publisher cannot assume responsibility for the validity of all materials or for the consequences of their use.

Direct all inquiries to CRC Press, Inc., 2000 Corporate Blvd., N.W., Boca Raton, Florida, 33431.

International Standard Book Number 0-8493-5438-2

Library of Congress Card Number 89-854
Printed in the United States

PREFACE

Professor Joseph Maisin, to whose memory this book is dedicated, was a flamboyant personality. With the help of wealthy patients and other patrons, as early as 1925 he had built a large, modern Cancer Institute in the Louvain Medical School.

Three floors of the building were occupied by wards, operating theatres, examination rooms, and radiotherapy equipment, including a unique radium bomb donated by the Union Minière du Haut-Katanga, the main mining company in the Belgian Congo. Laboratories and offices filled the ground floor. The greater part of the basement housed thousands of mice and rats. With ventilation being what it was in those days, this was not an ideal situation from the point of view of the patients, but Maisin saw the inconvenience as their personal contribution to the research aimed at improving their lot. I lived in this Institute during three war years, serving as "full-time" clinical assistant to the boss, yet enjoying enough freedom to earn a master's degree in chemistry at the university. It is typical of Maisin's broadmindedness and vision that he approved of this strange arrangement. Joining the resources of medicine and chemistry would produce what he felt would be the science of the future and, in his eyes, was a sufficient reason for supporting my undertaking.

Maisin was a distinguished pathologist and an outstanding clinician, who took care of his patients with remarkable competence and dedication. Patients flocked to him from all over the country and elsewhere, yet I never saw him cutting an examination short or even failing to give comfort and hope for lack of time. His enormous vitality made up the gaps and allowed him to devote his energies to many extra-clinical activities. He was one of the founders of the International Union Against Cancer, of which he was successively Secretary General, President, and Honorary President, and edited its Acta for many years. He was involved in many other organizations, national and international, and traveled a great deal. He had a heavy teaching load and interrogated hundreds of students every year.

He was an eloquent and charismatic speaker and used the teaching of histopathology as a means of conveying the latest discoveries of medical science to his students. Even so, he still found time for research, of all his activities the closest to his heart.

Every evening, after finishing the rounds of his patients, he would go down to the basement to inspect his beloved mice and rats. I confess that I was not much attracted to his brand of research. Groups of animals were put on different diets or given various dietary supplements or injections, chosen with a sometimes dismaying eclecticism. The mice had a patch of skin painted with benzopyrene or methylcholanthrene, whereas the rats were fed the liver carcinogen 4-dimethylamino-azobenzene (butter yellow). Maisin followed with care the frequency, aspect, and development of precancerous and cancerous lesions in the various groups, in the hope of discovering "problastic" or "antiblastic" substances. The approach was too empirical for my taste. However, in the light of today's epidemiological investigations on the relation between diet and cancer, Maisin now appears to be an authentic pioneer.

It was in this celebrated basement that the ancestral couple of the LOU rats first arrived, presumably from the U.S. some time between 1937 and 1941. As related by Hervé Bazin in the second chapter of this book, they turned out to be the bearers of a remarkable form of immunocytoma, which was first discovered by Maisin and his co-workers in the early 1950s. The LOU rats have since produced a huge progeny, and with them an impressive body of research, which includes the development of hybridomas and of monoclonal antibodies. They have thereby earned the rat a respectable place in immunology, a discipline known to be somewhat mouse-biased.

When Dr. Bazin asked me to write a foreword to this book, my first reaction was to say no. Even in the most charitable context, I could not possibly qualify as an immunologist nor as an expert on monoclonal antibodies. He kindly insisted and I finally accepted his invitation, seeing in it an opportunity to recall the memory of a remarkable man for whom I had great respect, affection, and gratitude. No doubt, Joseph Maisin would have welcomed with enthusiasm the publication of this book, the topic of which, through the strange and devious ways of science, he unknowingly initiated some 50 years ago.

Christian de Duve
Nobel Prize 1974

ACKNOWLEDGMENTS

We are indebted to a large number of individuals: B. de Séjournet, F. Flemal, M. Godichelle, B. Libert, D. Bourgois, M. Calier, J. De Mets, L. De Clercq, B. De Keyser, T. Goffart, P. Grognet, D. Jourdain, R. Meykens, C. Negel, J. Relles, and L. M. Xhurdebise who worked more or less a long time in the Experimental Immunology Unit. They all have largely contributed to the development of the LOUVAIN rat model in collaboration with those who have written sections of this book and with many distinguished colleagues who have participated to parts of these studies, in particular the late Prof. J. F. Heremans, C. Milstein, A. Beckers, H. Bennich, J. D. Capra, D. Carson, P. Carter, J. M. Davie, C. Deckers, F. Dessy, A. Froese, M. Fougereau, H. H. Fudenberg, G. Fust, J. Gergely, I. C. M. MacLennan, G. A. Medgyesi, H. Metzger, M. Moriamé, R. Pauwels, I. Schechter, D. Secher, A. H. Sehon, H. L. Spiegelberg, V. Starace, J. Urbain, J. P. Vaerman, and A. C. Wang.

The idea for a book especially devoted to "Rat Hybridomas and Rat Monoclonal Antibodies" was conceived during a course on that subject given in 1986 at the Beijing Institute for Cancer Research by four members of the Experimental Immunology Unit of the Catholic University of Louvain, under the auspices of a collaboration between these Institutes and the National Center for Biotechnology of the People's Republic of China and the Commission of the European Communities, Division for Science, Research and Development. We particularly thank Professors Yonghui Liu, Zhi-Wei Dong, Yong-Su Zhen, P. Fasella, G. Valentini, F. Van Hoeck, S. Finzi, G. Gerber, and D. de Nettancourt for their help and encouragement in the implementation of this collaboration which we hope will continue for years to come and contribute to developing scientific cooperation and friendship between the People's Republic of China and the European Communities.

Studies on rat immunocytomas began with the financial support of the "Fonds Cancérologique de la Caisse Générale d'Epargne et de Retraite" Belgium to whom we are particularly indebted. Their support was determining at the beginning of these studies and very useful throughout their development. We also received funds from the "Fonds National de la Recherche Scientifique" (Belgium), the "Fonds de la Recherche Médicale" (Belgium), the Radiation Protection Research Programme of the CEC (Publication 2459), the "National Institute of Health" (U.S.A.), the "Institut National de la Santé et de la Recherche Médicale" (France), the "Délégation Générale à la Recherche et aux Sciences et Techniques" (France), the Institut pour l'Encouragement de la Recherche Scientifique dans l'Industrie et l'Agriculture" (Belgium), the "Service de Programmation de la Politique Scientifique" (Belgium), and the Bekales Foundation (Liechtenstein). They have all contributed to the development of the LOUVAIN rat model and also the Directorate-General Division for Science, Research and Development of the Commission of the European Communities of which I have been a scientific staff member for 20 years.

I am especially indebted to F. Bolle, secretary of the unit, who has tirelessly prepared the manuscript of this book and improved the final draft. I wish to acknowledge N. Gardiner for supervising the English text. My thanks are also to L. De Greef, who often alone assumed the secretarial tasks of the unit during the last months of preparation of this manuscript and finally to M. Brancart, who has paid meticulous attention to the unit finances.

Hervé Bazin

DEDICATION

To the late Professor Joseph Maisin, who brought the ancestors of LOUVAIN rats to the Catholic University of Louvain and worked many years with them. This small rat nucleus has now developed into a large breed which lives throughout the world.

CONTRIBUTORS

Fabienne Ackermans
Assistant Professor
Experimental Immunology Unit
Faculty of Medicine
University of Louvain
Brussels, Belgium

Hervé Bazin
Professor, Head
Experimental Immunology Unit
Faculty of Medicine
University of Louvain
Brussels, Belgium

Blanche Bellon
Research Fellow
Pathologie Rénale et Vasculaire
INSERM U28
Hôpital Broussais
Paris, France

Monique Bodeus
Research Assistant
Microbiology Unit
Faculty of Medicine
University of Louvain
Brussels, Belgium

Erik Briers
General Director
N. V. Eco-Bio
Genk, Belgium

Yveline Bruinen
Executive Secretary of the European Node
CERDIC
Faculté de Médecine
Chemin de Vallombrose
Nice, France

Guy Burtonboy
Associate Professor
Microbiology Unit
Faculty of Medicine
University of Louvain
Brussels, Belgium

André Capron
Professor, Head
Institut Pasteur de Lille
Centre d'Immunologie et de Biologie Parasitaire
Unité Mixte INSERM
Lille, France

Monique Capron
Associate Professor
Centre d'Immunologie et de Biologie Parasitaire
Unité Mixte INSERM 167 — CNRS 624
Lille, France

Andreas Cerny
Research Assistant
Institute of Pathology
Experimental Pathology
University Hospital
Zurich, Switzerland

Danielle Chassoux
Chargé de Recherches
Institut de Recherches Scientifiques sur le Cancer
CNRS ER 278
Villejuif, France

Marcela-Viviana Chazev
Experimental Immunology Unit
Faculty of Medicine
University of Louvain
Brussels, Belgium

Françoise Cormont
Research Assistant
Experimental Immunology Unit
Faculty of Medicine
University of Louvain
Brussels, Belgium

Guy Cornelis
Associate Professor
Microbiology Unit
Faculty of Medicine
University of Louvain
Brussels, Belgium

Anne Cornet
Technician
Haematology Unit
University of Louvain
Brussels, Belgium

Marc De Bruyère
Professor, Head,
Immuno-Haematology Unit
Faculty of Medicine
University of Louvain
Brussels, Belgium

Alain De Cremer
Technician
Experimental Immunology Unit
Faculty of Medicine
University of Louvain
Brussels, Belgium

Thierry Delaunay
Research Associate
Hoechst-Behring (France)
Experimental Immunology Unit
Faculty of Medicine
University of Louvain
Brussels, Belgium

Nicole Delferrière
Research Associate
Microbiology Unit
Faculty of Medicine
University of Louvain
Brussels, Belgium

Colette Digneffe
Research Associate
Experimental Immunology Unit
Faculty of Medicine
University of Louvain
Brussels, Belgium

Philippe Druet
Professor, Head,
Pathologie Rénale et Vasculaire
INSERM U28
Hôpital Broussais
Paris, France

Claire Durieux
Institut Pasteur de Lille
SERLIA
Lille, France

Béatrice Dutertre
Scientific Attaché of the European Node
CERDIC
Faculté de Médecine
Chemin de Vallombrose
Nice, France

Jia-Chun Fang
Research Assistant
Beijing Institute for Cancer Resarch
Beijing, People's Republic of China

Catherine Fievet
Chargé de Recherches à l'INSERM
Institut Pasteur de Lille
SERLIA
Lille, France

Jean-Charles Fruchart
Professor, Director
SERLIA
Institut Pasteur de Lille
Lille, France

Christine Genart
Technician
Experimental Immunology Unit
Faculty of Medicine
University of Louvain
Brussels, Belgium

Agnès Goris
Research Associate
N.V. Eco-Bio
Genk, Belgium

Jean-Marie Grzych
Research Assistant
Institut Pasteur de Lille
Centre d'Immunologie et de Biologie Parasitaire
Unité Mixte INSERM 167 — CNRS 624
Lille, France

Jean-Charles Guery
Graduate Student
Pathologie Rénale et Vasculaire
INSERM U28
Hôpital Broussais
Paris, France

François Hirsch
Research Fellow
Pathologie Rénale et Vasculaire
INSERM U28
Hôpital Broussais
Paris, France

Josèe Hutschemackers
Research Associate
Phytopathology Unit
Faculty of Agronomic Sciences
University of Louvain
Louvain-La-Neuve, Belgium

Michèle Janssens
Research Associate
Microbiology Unit
Faculty of Medicine
University of Louvain
Brussels, Belgium

Min Jiang
Research Associate
Institute of Medicinal Biotechnology
Chinese Academy of Medical Sciences
Tiantan
Beijing, People's Republic of China

Jean-Pierre Kints
Animal House Manager
Experimental Immunology Unit
Faculty of Medicine
University of Louvain
Brussels, Belgium

George Klein
Professor
Department of Tumor Biology
Karolinska Institute
Stockholm, Sweden

Monique Koffigan
Institut Pasteur de Lille
SERLIA
Lille, France

Dominque Latinne
Immuno-Haematology Unit
Haematology Unit
Faculty of Medicine
University of Louvain
Brussels, Belgium

Marc Lefebvre
Research Assistant
Experimental Immunology Unit
Faculty of Medicine
University of Louvain
Brussels, Belgium

Anne-Marie Lebacq-Verheyden
Haematology Unit
Faculty of Medicine
University of Louvain
Brussels, Belgium

L. G. Linares-Cruz
Research Assistant
Institut de Recherches Scientifiques sur le Cancer
CNRS ER 278
Villejuif, France

Jean-Marie Malache
Experimental Immunology Unit
Faculty of Medicine
University of Louvain
Brussels, Belgium

Patrick Manouvriez
Research Associate
Experimental Immunology Unit
Faculty of Medicine
University of Louvain
Brussels, Belgium

G. M. Marchal
Laboratory of Pathology and Microbiology
Thuin, Belgium

Ludo Meulemans
Development Manager
N.V. Eco-Bio
Genk, Belgium

Paul C. Montgomery
Professor, Head
Department of Immunology and Microbiology
Wayne State University School of Medicine
Detroit, Michigan

Benoît Mousset
Research Assistant
Microbiology Unit
Faculty of Medicine
University of Louvain
Brussels, Belgium

E. Mrena
Laboratory of Pathology and Microbiology
Thuin, Belgium

Anne Neirynck
Bachelor of Science
Haematology Unit
Faculty of Medicine
University of Louvain
Brussels, Belgium

Jacques Ninane
Consultant
Haematology Unit
Faculty of Medicine
University of Louvain
Brussels, Belgium

Françoise Nisol
Technician
Experimental Immunology Unit
Faculty of Medicine
University of Louvain
Brussels, Belgium

W. S. Pear
Department of Tumor Biology
Karolinska Institute
Stockholm, Sweden

Jane V. Peppard
Assistant Professor
Department of Immunology and Microbiology
Wayne State University School of Medicine
Detroit, Michigan

Bernadette Platteau
Research Associate
Experimental Immunology Unit
Faculty of Medicine
University of Louvain
Brussels, Belgium

Pierre Querinjean
Research Associate
Genetics Unit
Faculty of Sciences
University of Louvain
Louvain-La-Neuve, Belgium

Anne-Marie Ravoet
Research Associate
Haematology Unit
Faculty of Medicine
University of Louvain
Brussels, Belgium

Michel Rits
A.Z. Leiden
Nierziektendienst C3P
Leiden, The Netherlands

Jean Rousseaux
Professor of Biochemistry
Unité INSERM CNRS 409
Institut de Recherches sur le Cancer
Lille, France

Roselyne Rousseaux-Prévost
Chargee de Recherches
Unité CNRS 409
Institut de Recherches sur le Cancer
Lille, France

A. Rovayo
Graduate Student
Microbiology Unit
Faculty of Medicine
University of Louvain
Brussels, Belgium

Jacques Seghers
Haematology Unit
Faculty of Medicine
University of Louvain
Brussels, Belgium

Maryam Sekhavat
Bachelor of Science
Haematology Unit
Faculty of Medicine
University of Louvain
Brussels, Belgium

Gerard Sokal
Professor, Head
Haematology Unit
Faculty of Medicine
University of Louvain
Brussels, Belgium

Marie-Paule Sory
Graduate Student
Microbiology Unit
Faculty of Medicine
University of Louvain
Brussels, Belgium

Dirk Stynen
Research Manager
N.V. Eco-Bio
Genk, Belgium

Janos Sümegi
Associate Professor
Department of Tumor Biology
Karolinska Institute
Stockholm, Sweden

Su-Lian Sun
Associate Professor
Department of Immunology
Beijing Institute for Cancer Research
Beijing, People's Republic of China

Michèle Surleraux
Graduate Student
Microbiology Unit
Faculty of Medicine
University of Louvain
Brussels, Belgium

A. Tacquet
Service de Médecine Générale et Néphrologie
Centre Régional d'Hemodialyse
Hôpital A. Calmette
Lille, France

Vincent Vercruysse
Graduate Student
Microbiology Unit
Faculty of Medicine
University of Louvain
Brussels, Belgium

M. Verhoyen
Professor, Head
Phytopathology Unit
Faculty of Agronomic Sciences
University of Louvain
Louvain-La-Neuve, Belgium

C. Verwaerde
Research Assistant
Centre d'Immunologie et de Biologie Parasitaire
Unité Mixte INSERM 167 — CNRS 624
Institut Pasteur de Lille
Lille, France

Christian Vincenzotto
Research Associate
Experimental Immunology Unit
Faculty of Medicine
University of Louvain
Brussels, Belgium

P. J. Volle
Director of Development and Research
Hoechst-Behring
Rueil-Malmaison, France

Dominique Wauters
Technician
Experimental Immunology Unit
Faculty of Medicine
University of Louvain
Brussels, Belgium

Georges Wauters
Professor
Microbiology Unit
Faculty of Medicine
University of Louvain
Brussels, Belgium

Han-Zhang Xia
Research Associate
Department of Immunology
Cancer Institute
Chinese Academy of Medical Sciences
Beijing, People's Republic of China

Qiang Zhao
Department of Immunology
Beijing Institute for Cancer Research
Beijing, People's Republic of China

Xio-Xia Zhang
Department of Immunology
Beijing Institute for Cancer Research
Beijing, People's Republic of China

Pi Zhiming
Department of Immunology
Beijing Institute for Cancer Research
Beijing, People's Republic of China

Xu Zuoling
Vice Director
Department of Immunology
Beijing Institute for Cancer Research
Beijing, People's Republic of China

TABLE OF CONTENTS

INTRODUCTION

The first model of monoclonal antibody (MAb) secreting hybridomas developed by Köhler and Milstein,[1] was the mouse model. It can almost be considered perfect. Its efficiency and extreme stability probably account for its success. However, there are several reasons for developing another rodent model, particularly a rat model.

The first reason is certainly linked to the rat antibody repertoire which is different from that of the mouse. This is true for the mouse xenoantigens which can be studied with rat but not with mouse MAbs. Likewise, rat alloantigens can be better studied with rat than with mouse MAbs. At last, some antigens from origins other than mouse or rat species can induce much stronger immune responses in rats than in mice and evidently the reciprocal could be true for other antigens.

The second reason relates to physicochemical and biological properties of rat immunoglobulins which do not seem to exist in the mouse species. Rat antibodies of the IgG1, IgG2a, and IgG2b isotypes can easily fix the human or the rabbit complement. Likewise, the rat IgG2b antibodies can be efficiently used by human K cells to kill their target cells.

The third reason is linked to the *in vivo* production of rat MAbs. Rats are approximately ten times bigger than mice, but both species have gestation times of about 3 weeks and they are adults at the same age. The litter size of the LOU/C rats, calculated on 183 gestations, has been found to be 7.67 and by comparison, that of the BALB/c mice is equal to 5.2 according to Crispens.[2] All these data explain why the price of an inbred rat is approximately twice the cost of an inbred mouse. But a LOU/C rat could give a mean production of 50 to 150 mg of MAbs per animal for labor of inoculation and bleeding, which is approximately the same as that given by a mouse. Hybridoma development is about as expensive in rats as in mice, but the antibody production is much easier and cheaper in the rat species. LOU/C rats are easy to handle with a minimum of experience and are certainly no more difficult to handle than BALB/c mice. Although the future of the production of MAbs most probably lies in *in vitro* culture, at the present time cheap quantities of MAbs mean *in vivo* production in rats.

The fourth reason concerns the purification methods which have been set up for rat MAbs; they are easy and rapid to use, highly efficient, and inexpensive. An *in vitro* culture supernatant[3] and an *in vivo* ascitic fluid method[4] have been developed for purification of rat MAbs by immunoaffinity chromatography. However, conventional purification techniques of rat immunoglobulins can also be used, but these techniques are based on properties of heterogenous molecule populations which are rarely found in a given MAb preparation. Techniques of DEAE chromatography, gel filtration, or preparative electrophoresis can be used, but the degree of purity and the percentage of recovery achieved by them are limited. In most cases, affinity chromatography is preferable.

The fifth reason is the absence of viral particles in rat hybridomas and therefore in rat MAbs; this is an important fact to take into account in the case of human *in vivo* therapeutic use.

However, obtaining rat-rat hybridomas still seems to be considered difficult. In sentences like, "At the time of writing, successful production of rat-rat hybridomas have been limited to a small number of laboratories. Several experienced investigators have had difficulty in keeping rat-rat hybrids alive for more than 2 to 3 weeks...",[5] the same author in the second edition of his book[6] adds to the same paragraph, "The reason for such problems is not yet clear. The hybrids seem to grow initially, and then die. Perhaps rat cells are unduly susceptible to 'natural killer' cells, or some rat colonies have unusually high numbers of natural killer cells in the spleen." Similar sentences are frequently written or heard in conferences and seminars. Already we have obtained, and so have scientists around us, thousands and thousands of rat-rat hybridomas with no more difficulty than if we are making mouse-mouse hybridomas. We do not believe that we are especially clever, but rather we believe it is just a question of experiments which can be communicated to interested colleagues.

Moreover, many scientists are not familiar with rat immunoglobulins and rat MAbs. There are no books and no reviews devoted either to the development of rat hybridomas or to the use of rat MAbs. The purpose of this book is to fill this gap. However, it should be made clear that we have not tried by any means to write an exhaustive review on hybridomas in general as many excellent books have already been written on this subject, although they generally concern mouse[5-8] or human hybridomas[9,10] which are now too numerous to all be described or on rat hybridomas. We have attempted rather to write on our own experiments or on those of friend laboratories which have used similar techniques against many different kinds of antigens.

We hope that this book will further the development of the use of the rat hybridomas and the rat monoclonal antibodies.

REFERENCES

1. **Köhler, G. and Milstein, C.,** Continuous cultures of fused cells secreting antibody of predefined specificity, *Nature,* 256, 495, 1975.
2. **Crispens, C. G.,** Reproduction and growth characteristics: mouse, in *Inbred and Genetically Defined Strains of Laboratory Animals — Part I. Mouse and Rat,* Altman, Ph. and Katz, D. D., Eds., Federation of American Societies for Experimental Biology, Bethesda, MD, 1979, 45.
3. **Bazin, H., Xhurdebise, L. M., Burtonboy, G., Lebacq, A. M., De Clercq, and Cormont, F.,** Rat monoclonal antibodies. I. Rapid purification from *in vitro* culture supernatants, *J. Immunol. Methods,* 66, 261, 1984.
4. **Bazin, H., Cormont, F., and De Clercq, L.,** Rat moncolonal antibodies. II. Rapid and efficient method of purification from ascitic fluid or serum, *J. Immunol. Methods,* 71, 9, 1984.
5. **Goding, J. W.,** *Monoclonal Antibodies: Principle and Practice,* 1st ed., Academic Press, London, 1983, 276.
6. **Goding, J. W.,** *Monoclonal Antibodies: Principle and Practice,* 2nd ed, Academic Press, London, 1986, 315.
7. **Hurrell, J. G. R.,** *Monoclonal Hybridoma Antibodies: Techniques and Applications,* CRC Press, Boca Raton, FL, 1985, 240.
8. **Bartal, H. and Hirshaut, Y.,** *Methods of Hybridoma Formation,* Humana Press, Clifton, NJ, 1987.
9. **Engleman, E. G., Foung, S. K. H., Larrick, J., and Raubitschek, A.,** *Human Hybridomas and Monoclonal Antibodies,* Plenum Press, New York, 1985.
10. **Strelkauskas, A. J.,** *Human Hybridomas - Diagnostic and Therapeutic Applications,* Marcel Dekker, Inc., New York, 1987.

Part I
Immunocytomas in LOUVAIN (LOU) Rats

Chapter 1

RAT IMMUNOGLOBULINS

H. Bazin, J. Rousseaux, R. Rousseaux-Prévost, B. Platteau, P. Querinjean, J. M. Malache, and T. Delaunay

TABLE OF CONTENTS

I. HISTORICAL

The first data concerning the rat serum proteins were obtained by Grabar and Courcon[1] and Escribano and Grabar,[2] using the immunoelectrophoretic analysis. They identified two lines of precipitation in the gammaglobulin area. Later, Arnason et al.[3,4] found three proteins with antibody activity in rat serum and called them IgM, IgX, and IgG. IgG were further divided into three subclasses with the IgA class clearly set apart. Then, a fourth IgG subclass, as well as the IgE and the IgD were found.

A. CHARACTERIZATION OF RAT IMMUNOGLOBULIN M

Arnason et al.[3,4] identified the isotype first by its susceptibility to cysteine and then by its molecular weight.

B. CHARACTERIZATION OF RAT IMMUNOGLOBULIN G

Two isotypes were identified by Arnason et al.[3] At first, they were called IgX and IgG, and later IgA and IgG.[4] Some time later, Binaghi and Sarando de Merlo[5] were able to identify two

IgG subclasses, which they named IgGa and IgGb. Both these subclasses were of low anodal mobility and were probably the equivalents of the present IgG2a and IgG2b subclasses. Three years later, two IgG subclasses called gamma1 and gamma2 were identified by Jones[6] from purified anti-DNP rat polyclonal antibodies. These two rat IgG subclasses probably corresponded to the present IgG1 and "IgG2a + IgG2b" subclasses. Quite clearly, the same serum component was called IgA by some authors and gamma1 by others. However, Bazin et al.[7] and Orlans[8] obtained antigenic cross-reactions by using anti-human IgG sera as antibodies and LOUVAIN IgG1 monoclonal or polyclonal normal rat IgG1 as antigens. Moreover, isotype-specific anti-mouse IgG1 serum from rabbit or sheep always strongly cross-reacted with purified rat monoclonal IgG1.[9] These cross-reacting antigens, which belonged either to human IgG and mouse IgG1 or to rat "IgG1 or IgA", clearly supported the IgG nature of the rat component. The last rat IgG subclass, called IgG2c, was identified by Bazin et al.,[7] who used LOUVAIN monoclonal immunoglobulin secreted by IR tumors.

C. CHARACTERIZATION OF RAT IMMUNOGLOBULIN A

The characterization of rat IgA has been difficult to establish. For a long time, IgA and IgG subclasses have been confused. Arnason et al.[3] identified two isotypes, IgX and IgG. IgX was later called IgA, as molecules of this isotype were found in saliva and in limited quantities in the sera.[4] However, Nash et al.[10] identified a component of the rat milk with a rabbit anti-mouse IgA. Antiserum raised against this rat component, using the cross-reaction existing between it and a rabbit anti-mouse IgA, showed the presence of a large number of IgA containing cells in the *lamina propria* of the rat intestine. These authors named this class IgA and showed that its anodal mobility was faster than that of the class Arnason et al.[4] had previously termed "IgA", which was in fact the IgG1 isotype. Orlans and Feinstein[11] later demonstrated a cross-reaction between human and rat IgA.

D. CHARACTERIZATION OF RAT IMMUNOGLOBULIN E

Reaginic antibodies were known to exist in rat species[12,13] and to have properties similar to those of man.[14,15] Stechschulte et al.[16] and Jones and Edwards[17] published the identification they had made that a rat IgE was equivalent to human IgE. Their results were confirmed by Bazin et al.,[18] who used rat IgE monoclonal proteins.

E. CHARACTERIZATION OF RAT IMMUNOGLOBULIN D

The first demonstration of the presence of a rat IgD equivalent to the human was given by Ruddick and Leslie[19] in 1977; they used a chicken antiserum to human IgD which was cross-reacting with a rat lymphocyte membrane component. This work was fully confirmed by Bazin et al.[20] in 1978 in their research on LOUVAIN IR monoclonal proteins determined on that isotype.

II. NOMENCLATURE OF RAT IMMUNOGLOBULIN ISOTYPES

Old usage of immunoglobulin nomenclature has now disappeared. Historical designation in the case of rat immunoglobulin isotypes is given in Table 1, and it is sometimes useful to know when reading old literature dealing with studies concerning rat immunoglobulins. Table 2 gives this correspondence. According to the recommendations of the committee of immunoglobulin nomenclature of the World Health Organization,[27] the two initial letters "Ig" must be adopted for all immunoglobulin isotypes and for all animal species. They also prescribed that all the letters be written on the same line, thus IgG1 and not IgG_1. The division of IgG isotype into subclasses on the basis of their electrophoretic mobility, can be maintained at the present stage of knowledge. This system has also been adopted for most animal species in the absence of more advanced knowledge on the chemical and biological structures of these proteins.

TABLE 1
Historical Designation of the Rat Immunoglobulin Isotypes

Ref. 7, 20	IgM	IgD	IgG1	IgG2a	IgG2b	IgG2c	IgE	IgA
3	IgM	—	IgX	IgG		—	—	—
4	IgM	—	IgA	IgG		—	—	—
21	γM	—	γ1	γ2		—	U2(?)	U1(?)
22	IgM	—	IgA	IgG2a	IgG2b	—	—	—
10	—	—	—	—	—	—	—	IgA
23, 24	—	—	—	—	—	—	—	IgA
6	γM	—	γ1	γ2		—	—	—
25	—	—	—	—	—	—	IgE	—
17	—	—	—	—	—	—	IgE	—
26	IgM	—	IgG1	IgG2a	IgG2b	IgG2c	IgE	IgA
19	—	IgD	—	—	—	—	—	—

TABLE 2
Nomenclature of Rat Immunoglobulin Isotypes

Ref. 20, 26	Isotypes							
	IgM	IgD	IgG1	IgG2a	IgG2b	IgG2c	IgE	IgA
Old usage	γM	γD		γG			γE	γA
	γ1M			7Sγ				
	β2M			γSS				
Transitory usage	—	—	IgX	γ2		—	—	—
			IgA	IgGa	IgGb			
			γ1					

Rat IgM, IgD, IgG, IgE, and IgA isotypes seem to have properties quite analogous to human homologues. It now seems possible to correlate the mouse and the rat IgG subclasses (Figure 1). Rat light chains were characterized by Querinjean et al.,[33,34] who established the presence of kappa and lambda chains as described for other species.[35]

III. ALLOTYPES OF RAT IMMUNOGLOBULINS

Molecules of an immunoglobulin class or subclass do not always have the same amino acid sequences and thus do not always bear identical antigenic determinants. They can present two or more antigenic forms which are called allotypes according to the Oudin terminology.[36] In this case, individuals from the same species could be divided into two or more allotypic groups, each one bearing its own allotype. Allotypic specificities are supported by structural differences determined by genes which are inherited as Mendelian character. Immunoglobulin genes are co-dominant and always phenotypically expressed in the heterozygote. However, each immunoglobulin-producing cell expresses only one of the allelic genes, according to the rule of allelic exclusion. Up to now, four allotypies have been identified in rat immunoglobulins.

A. KAPPA LIGHT CHAINS

Immunizing "black and white hooded rats" with Wistar immunoglobulins, Barabas and Kelus[37] in 1967 found the first allotypic difference in rat immunoglobulins. Soon afterwards Wistar[38] demonstrated the presence of an allotypic marker on rat light chains. These data were confirmed by several authors[39-41] using various rat strains. Rohklin et al.[42] suggested that this

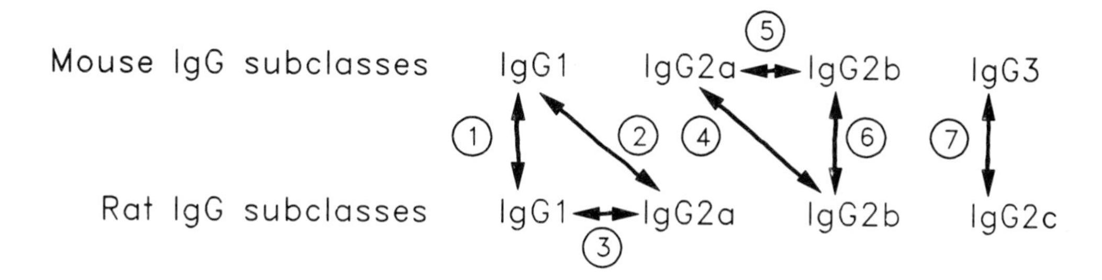

FIGURE 1. Correlation between mouse and rat IgG subclasses. The various homologies are respectively supported: (1) Antigenic cross-reactions[9] and sequence homology;[28] (2) Homocytotropic antibodies and sequence homology;[28] (3) Antigenic cross-reactions (Bazin, unpublished data) and sequence homology;[28] (4) Sequence homology;[28] (5) Sequence homology;[29] (6) Sequence homology;[28] (7) Homologies based on physicochemical and biological properties.[30-32]

allotypic difference was located on the kappa light chains. This suggestion was later entirely confirmed by Beckers et al.,[43] who used kappa Bence Jones proteins from LOUVAIN immunocytomas and by Nezlin et al.,[44] who localized the allotypic marker on the constant part of polyclonal kappa light chains. By agreement between the scientists working in the field, a common nomenclature has been adopted;[45] it considers the allotype called IgK-1, which has two alleles, IgK-1a (reference strain, LOU/C) and IgK-1b (reference strain, DA).

The differences between the two alleles were described by Gutman et al.,[46] who found one sequence gap and many amino acid substitutions between the kappa chains of the DA and LEW rat strains. The rat kappa allotype is complex and suggests that the genes coding for it are not real alleles. Moreover, there are two amino acid differences between the LOUVAIN kappa Bence Jones and the LEWIS kappa polyclonal light chains, both coming from the IgK-1a allele and suggesting isotypic variation.

B. ALPHA HEAVY CHAINS

Bazin et al.[47] in 1974 described an allotype located on the rat alpha heavy chains of immunoglobulins. This allotype was called Iα-1, which is now known as IgH-1[45] with two alleles, IgH-1a (reference strain, LOU/C) and IgH-non1a (reference strain, OKA). The antigenic determinant(s) corresponding to this allotype was found in 100% of the alpha chains from LOU/C rats, homozygous for the IgH-1 allotype.

C. GAMMA 2B HEAVY CHAINS

Beckers and Bazin[48] in 1975 described another difference for the rat gamma2b heavy chains with two alleles, Iγ2b-1a, now called IgH-2a (reference strain, LOU/C) and Iγ2b-1b, now called IgH-2b (reference strain, OKA).[45]

D. GAMMA 2C HEAVY CHAINS

Leslie[49] detected an allotype designated IgH-3 associated with the rat IgG2c subclass. Two alleles were found; IgH-3a on the heavy chain of IgG2c from COP, F344, BN, DA rats and IgH-3b on the heavy chain of IgG2c molecules from WISTAR/Fu, LOU/M, SHR rats.

E. LOCATION OF RAT ALLOTYPE GENES

IgH-1 and IgH-2 loci are linked to each other,[48] but they are not linked to the IgK-1 locus.[43] The heavy chain genes segregate independently of those concerned with sex, coat pattern and color, and eye color.[47] The IgK-1 locus is not linked to the major histocompatibility complex, Rt-1 of the rat, three coat color genes,[40] or a gene controlling the immunological responsiveness to the lactate dehydrogenase antigen.[39]

TABLE 3
Normal and Congenic Strains of Rat for Allotypic Markers

Rat strains	Genetic background	Number of backcrosses	Allotypic markers IgK-1	IgH-1	IgH-2
LOU/C	LOU/C	—	a	a	a
LOU/C[a]·IgK-1b(OKA)	LOU/C	12	b	a	a
LOU/C·IgK-2b(OKA)	LOU/C	12	a	b	b
OKA	OKA	—	b	b	b
PVG·1a[b]	PVG/c	—	a		
PVG·1[b]	PVG/c	15	b		
DA	DA	—	b		

[a] From Bazin, H. and Beckers, A., *Molecular and Biological Aspects of the Acute Allergic Reaction,* 1976, 125. With permission.
[b] Hunt and Gowans, personal communication.

The IgH-1 and IgH-2 genes, which are linked to each other,[48] have been placed on the linkage group VIII[50] and located on chromosome 6.[51] The kappa gene locus is located on chromosome 4[52] and the lambda on chromosome 11 (Pear, W. S. and Wahlström, personal communication).

F. RAT CONGENIC STRAINS FOR ALLOTYPE MARKERS

Different strains of rat, congenic to inbred strains, have been created (Table 3). They can be used to follow immunoglobulin of the IgA[54] or IgG2b, and especially the kappa allotype makes it possible to follow immunoglobulins from any class.[55-57] Likewise the kappa allotype can be used for easy purification of all monoclonal immunoglobulins or monoclonal antibodies.[58]

IV. PHYSICOCHEMICAL PROPERTIES

A. GENERALITIES

Studies on physicochemical properties of rat immunoglobulins have been greatly helped by the monoclonal proteins synthesized by the LOUVAIN rat immunocytomas. Most published data (Table 4) correspond to monoclonal immunoglobulins synthesized by LOU immunocytomas, as it was nearly impossible to purify rat polyclonal Ig subclass(es) before the appearance of the affinity chromatography using monoclonal antibodies anti-rat immunoglobulin class or subclass.

B. ISOELECTRIC POINT — ELECTROPHORETIC MOBILITY

Immunoelectrophoresis analysis of normal rat serum (Figure 2), as well as zone electrophoresis in agar, agarose, or cellulose acetate of LOUVAIN monoclonal immunoglobulins, have shown a large diversity of electrophoretic mobilities. IgG1 and IgE generally have a fast anodic mobility. An isoelectric point of 5.9 has been reported for the rat IR162 IgE.[65] IgG2a and IgG2b monoclonal immunoglobulins have electrophoretic mobilities from fast gamma1 to slow gamma2. Such differences of charge among proteins which have the same constant part must be due to the sequence of the variable regions of heavy and light chains. IgG2c monoclonal immunoglobulins have the slowest electrophoretic migration. IgG2c and IgM monoclonal proteins can, owing to their euglobulin properties, precipitate in the gel and create an artifact. IgA monoclonal proteins often show more than one component having various isoelectric points (Figure 3).

C. MOLECULAR WEIGHTS OF THE LIGHT CHAINS

To determine the molecular weights of the kappa and lambda light chains of rat immuno-

TABLE 4
Physicochemical Properties of Rat Immunoglobulin Isotypes

Designation	Ref.	IgM	IgD	IgG1	IgG2a	IgG2b	IgG2c	IgE	IgA
Heavy chains	18	μ	δ	γ1	γ2a	γ2b	γ2c	ε	α
Light chains		κ or γ	κ or γ	κ or γ	κ or γ	κ or γ	κ or γ	κ or γ	κ or γ
Molecular formula	18, 20, 59	(2H+2L)5	(2H+2L)	(2H+2L)	(2H+2L)	(2H+2L)	(2H+2L)	(2H+2L)	(2H+2L)n[a]
Presence of the J chain	60	Yes	—	—	—	—	—	—	Yes (in polymeric)
Presence of secretory component	61	Sometimes in secretion	—	—	—	—	—	—	Yes (in secretion)
Sedimentation coefficient	16, 18, 20, 23, 62,63	17—19S	—	6.7S	6.4S	6.5S	6.7S	7.2S	7—19S
Molecular weight[b] (kDa)	20,64	900	570	156	156	156	156	183—198	163
Molecular weight of heavy chain (kDa)[c]	64	72	60	55	50—52	55	50—52	72—75	ND

[a] n = 1 or more.
[b] Determined by SDS-polyacrilamide gel electrophoresis (5.7 or 12% acrylamide).
[c] Determined by SDS-polyacrilamide gel electrophoresis in reducing conditions (5 to 15% acrylamide).

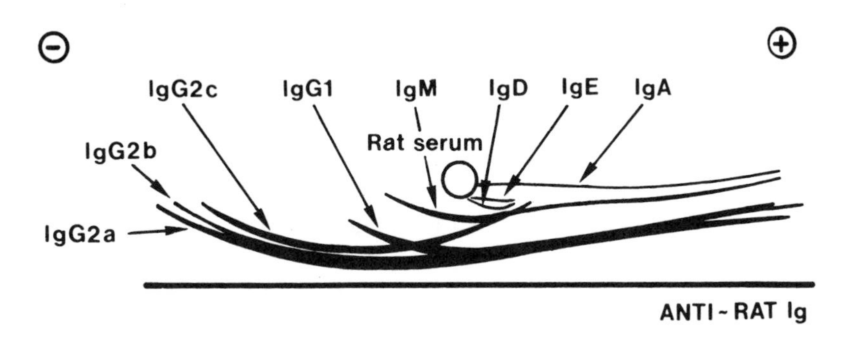

FIGURE 2. Immunoelectrophoretic pattern of normal rat serum developed with an anti-rat immunoglobulin serum.

globulin classes and subclasses, gradient pore SDS polyacrylamide gel electrophoresis (7.5 to 15% acrylamide) was carried out. Highly purified rat immunoglobulins from rat immunocytomas or rat hybridomas were electrophoretically separated, in reducing conditions (1% SDS, 5% 2,beta-mercaptoethanol) at constant voltage (200 V) for 45 min. Molecular weight of kappa and lambda light chains were then determined according to the migration of SDS-PAGE molecular weight standards (Table 5).

D. SOLUBILITY

Euglobulin properties — In most cases rat immunoglobulins, either polyclonal or monoclonal, have a normal solubility of pseudoglobulins at 0.15 *M* ionic strength. However, rat IgM and IgG2c generally behave as euglobulins and precipitate in solutions of low ionic strength. Their redissolution in physiological saline is not always possible.

Ammonium sulfate precipitation — Rat IgE differs from rat IgM and IgG by its solubility in 40% saturated ammonium sulfate.[66] For more information, see Chapter 15.

Rivanol precipitation — See Chapter 15.

Caprylic acid precipitation — See Chapter 15.

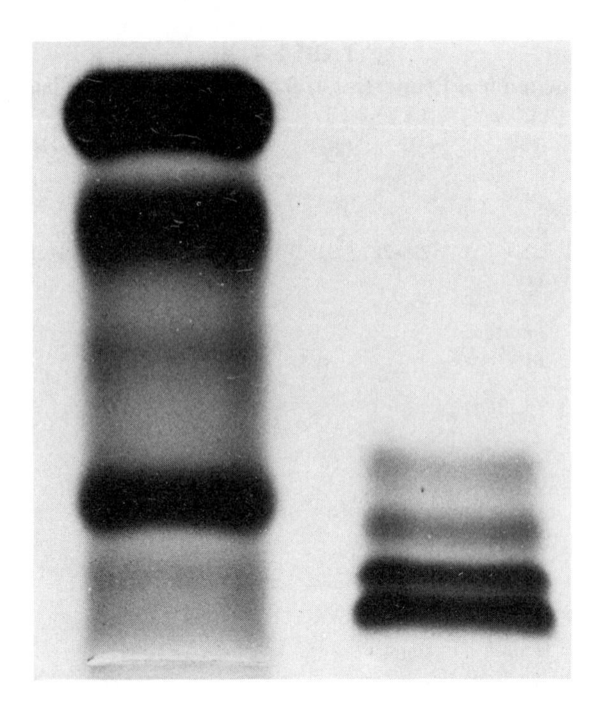

FIGURE 3. Agarose electrophoresis of a normal rat serum (left) and purified rat monoclonal IgA from the IR699 immunocytoma (right). Top: anode, bottom: cathode. The purified protein shows various electrophoretic mobilities.

TABLE 5
Molecular Weights of Rat Kappa and Lambda Light Chains of Immunoglobulins

Rat immunoglobulins from immunocytomas and hybridomas		Molecular weight of rat immunoglobulins light chains (Kd)	
		kappa	lambda
IR202	(IgM)	24.5	
IR731	(IgD)	22.8	
IR31	(IgG1)		27.5
IR418	(IgG2a)	23.3	
IR863	(IgG2b)	24.1	
RH58	(IgG2b)[a]		25.7
RH68	(IgG2b)[a]		24.0
IR304	(IgG2c)	25.0	
IR162	(IgE)	21.2	
IR1060	(IgA)	21.2	

[a] From Burtonboy, G., University of Louvain. With permission.

E. ABSORPTION COEFFICIENT AT 280 NM

It is possible to carry out quantitative estimation of various substances by measuring their absorption in the ultraviolet. The specific absorption of proteins depends on their contents in certain aromatic aminoacids, especially tyrosine and tryptophane. The maximum absorption is around 280 nm and allows quantitative measurement of protein concentration.

TABLE 6
Molar Absorption Coefficient (Epsilon) at 280 nm of Various Preparations of Rat Ig (1/mol × cm)

Immunocytomas or LOU/C rat hybridomas	Isotype	Molar absorption coefficient (epsilon)	
		Bradford method	G method
IR202	IgM-kappa	1.09	—
IR473	IgM-kappa	1.12	1.07
IR731	IgD-kappa	1.10	—
IR162	IgE-kappa	1.07	—
IR1060	IgA-kappa	1.34	—
IR27	IgG1-kappa	—	1.29
IR595	IgG1-kappa	1.26	1.32
IR871	IgG1-kappa	1.04	—
IR418	IgG2a-kappa	1.24	—
IR452	IgG2a-kappa	1.10	1.30
IR863	IgG2b-kappa	1.05	—
LO-DNP-11	IgG2b-kappa	1.27	—
IR304	IgG2c-kappa	1.05	—
IR1148	IgG2c-kappa	1.07	—

We have compared two techniques of protein quantification and calculated the absorption coefficient of various rat monoclonal immunoglobulins at 280 nm.

1. The First Method (Bradford)

Materials — Coomassie brillant blue G250 (Biorad 161-0406): 100 mg dissolved in 50 ml ethanol at 95%. Add 100 ml of 85% (weight/volume) of phosphoric acid. Complete to 1 l with distilled water.

Methods — A protein sample from 10 to 100 mg in 0.1 ml solvant (like saline or PBS) is added to 5 ml Coomassie solution; 5 min later, read the optical density at 595 nm against the pure Coomassie solution. The standard curve is given by a commercial preparation of gammaglobulin.

2. The Second Method (Gravimetric)

Materials — PBS: NaCl 8.0 g/l; KCl: 0.2 g/l; Na_2HPO_4 $2H_2O$: 1.4 g/l; KH_2PO4: 0.2 g/l.

Methods — After an extensive dialysis of the immunoglobulin preparation in PBS, lyophilize samples of 1 ml of the immunoglobulin preparation and of 1 ml of PBS in weighted flasks. Determine the quantities of immunoglobulin by comparison of the different flasks. Adjust the Ig preparation to 1.0 mg/ml and read the absorption at 2800 Ä. Results are given in Table 6.

Quite clearly, there are fluctuations between the values obtained by the two methods and also among immunoglobulins of the same isotype.

The method of quantification of MAbs by their absorption at 2800 Ä is certainly useful, but its value is limited (Table 6) and needs a correct standard of the same isotype, or even better of the same immunoglobulin.

F. SENSITIVITY TO REDUCTION OF RAT MONOCLONAL IgG SUBCLASSES AND IgE

1. IgG Subclasses

Sensitivity to reduction of rat IgG subclasses has been studied on several monoclonal IgG from each subclass: IgG1, IgG2a, IgG2b, IgG2c.[68] The results have shown a clear-cut difference between rat IgG2a and the other subclasses. Indeed, in the case of IgG2a the main pathway for reduction is

$$H_2L_2 \rightarrow 2HL \rightarrow 2H + 2L$$

while for the other subclasses, all the following pathways are observed:

$$\begin{array}{ccccc} & \nearrow & H_2L + L \rightarrow & H_2 + 2L & \\ & & \nearrow & \downarrow & \\ H_2L_2 & \longrightarrow & 2HL \longrightarrow & 2H + 2L & \end{array}$$

The characteristic pattern of rat IgG2a is similar to human IgG4 subclass,[69] but different from mouse IgG2a for which H_2L and H_2 intermediates are present.[70] Whether this peculiar sensitivity to reduction is due to a characteristic structure of the hinge region or to a particular quaternary structure, has not been determined.

2. IgE

Reductive cleavage of rat monoclonal IgE has been studied in relation to its biological activity.[71] The results have shown patterns of reduction strikingly different from those of human IgE.[72] Two inter-heavy chain disulfide bonds located in the C epsilon 2 domain were found to be the most sensitive to reduction (reduced with 1 m*M* dithiothreitol, DTT), while homologous disulfide bonds in human IgE are the most resistant. Furthermore, the intra-epsilon chain labile disulfide bond of human IgE (located in the C epsilon 1 domain) was found absent in rat IgE. Inter-heavy-light chain disulfide bridges were found to be the most resistant to reduction (reduction with 10 m*M* DTT), while these disulfide bonds are the most labile in the case of human IgE. The main pattern of reduction may be summarized as follows:

$$H_2L_2 \xrightarrow{1\ mM\ DTT} 2HL \xrightarrow{10\ mM\ DTT} 2H + 2L$$

$$H_2L_2 \xrightarrow{1\ mM\ DTT} H_2 + 2L \xrightarrow{10\ mM\ DTT} 2H + 2L$$

While the patterns of reduction are very different for rat and human IgE, the effects of reduction on biological activity were found identical: absence of effect with 1 m*M* DTT and decrease of affinity for mast cells with 10 m*M* DTT. We have suggested that reduction of all inter-chain disulfide bridges induces a modification of the IgE quaternary structure; a steric hindrance by the Fab part could be exerted on the effector sites (binding sites to mast cells) located within Fc.[71] This hypothesis is favored by the fact that a noncovalent dimer of epsilon chains has a better biological activity than IgE in which all inter-chain disulfide bonds have been reduced.

G. PROTEOLYSIS OF RAT IgG SUBCLASSES AND IgE

1. IgG Subclasses

Proteolytic cleavage of monoclonal rat IgG (IgG1, IgG2a, IgG2b, IgG2c) has been studied with various enzymes: papain, pepsin,[73] trypsin,[74] *Staphylococcus aureus* V8 proteinase.[75] These studies have been performed, first as an approach to a chemical typing of IgG subclasses, secondly in order to define the optimal conditions for preparation of different fragments: Fab, Fc, Fab′, $F(ab')_2$, etc.

Differences in sensitivity to proteolytic digestion have been observed and related to IgG subclasses. Thus, monoclonal IgG2a are the most sensitive to papain, an almost complete cleavage into Fab, and Fc is obtained without addition of a reducing agent in the buffer. In the same conditions IgG1 is uncleaved by papain.[73]

Monoclonal IgG2c are very sensitive to pepsin digestion. A very fast cleavage into $F(ab')_2$ and pFc′ is obtained and $F(ab')_2$ is converted into Fab′ after a prolonged time of digestion (18 h). In the same conditions, monoclonal IgG1 and IgG2a are very resistant, and previous incubation at acid pH (pH 2.8) has been found necessary to obtain an optimal cleavage with pepsin of monoclonal IgG1 and IgG2a.[73]

Monoclonal IgG2b and IgG2c are split with trypsin into Fab(t) and Fc(t).[74] Such a cleavage is not observed for proteins of IgG1 and IgG2a subclasses. These subclasses are apparently uncleaved with trypsin when studied in nondenaturing conditions but in denaturing conditions (such as polyacrylamide gel electrophoresis in SDS) two fragments of 120,000 and 13,000 Da are observed. The fragment of 13,000 Da corresponds to the whole VH domain (Rousseaux et al., in preparation).

Monoclonal IgG2b are cleaved by *Staphylococcus aureus* V8 proteinase into a $F(ab')_2$ and Fc fragments. Such a cleavage is only obtained for proteins of IgG2b subclass, while other subclasses are uncleaved.[75]

2. IgE

Several reports[76-78] have stated an impossibility to produce, by proteolytic cleavage, a Fc from rat monoclonal IgE. The main fragment released by various enzymes (papain, pepsin, trypsin, subtilisin...) is a $F(ab')_2$-epsilon fragment. Recently another fragment identified as a dimer of the C epsilon 4 domains has been described.[79] Such a fragment is obtained by short time digestion with papain or trypsin. The C epsilon 4 domain is unable to bind to mast cell receptors, as well as the $F(ab')_2$-epsilon. This result gives indirect evidence that binding sites to mast cell receptors are located in the C epsilon 3 domain.

H. SENSITIVITY TO DENATURATION OF RAT MONOCLONAL IgG SUBCLASSES AND IgE

Conformational studies of rat IgG subclasses have been performed on monoclonal immunoglobulins either in their native state or after treatment with various denaturing agents.[80] The purpose of these studies was to define whether IgG subclasses could be distinguished by some conformational features.

Thermal denaturation of rat monoclonal IgE has also been investigated in order to establish the relationship between molecular changes and loss of biological activity (affinity for mast cells).

1. IgG Subclasses

Monoclonal IgG of each subclass and some of their fragments (Fab, Fc) have been studied by circular dichroism.[80] In their native state, no clear-cut differences in circular dichroic spectra were observed. However, effects of denaturating treatments were found to be different according to IgG subclass. Thus, only monoclonal IgG1 and IgG2c are affected by thermal denaturation (heating at 56°C); the modification of circular dichroic spectra suggests an aggregation of IgG molecules. Monoclonal IgG1 and IgG2c are also very sensitive to acid pH. Deep changes in circular dichroic spectra are observed at pH 2.8, related to denaturation of secondary and tertiary structures. Treatment with the chaotropic salt NaSCN induces an increase of the beta-sheet structure in all IgG subclasses except IgG2c.

2. IgE

Denaturation of rat IgE at 56°C has been found to depend on the pH, ionic strength, presence of calcium ions, or reducing agents.[81,82] The mechanism of thermal denaturation has been related to the formation of biologically inactive polymers.[83] No polymers are formed upon heating at

56°C either F(ab')$_2$-epsilon or C epsilon 4 domain, thus suggesting an aggregation of IgE molecules via the C epsilon 3 domain.[79] A different interpretation of thermal denaturation has been given by Burt et al.[84] These authors, using antibodies directed against synthetic peptides of the C epsilon 4 domain, have found an increase of antibody binding to heated IgE. These results have been interpreted by Burt et al.[84] as an unfolding of the C epsilon 4 domain, the sequences recognized by antipeptides antibodies being more exposed in heated IgE than in native IgE. Nevertheless, in our opinion another interpretation of these data could also be that antipeptides antibodies bind with more avidity to polymeric IgE molecules produced by heating. Thus, results from Burt et al.[84] could be in agreement with the polymerization mechanism proposed by our laboratory.

I. PRIMARY STRUCTURE OF LOU RAT IMMUNOGLOBULINS

1. Introduction

The availability of monoclonal or homogeneous immunoglobulins is of paramount importance to establish the primary sequence of this type of protein. Human and mouse plasmacytomas were the first material to be sequenced. Mouse and rabbit homogeneous antibodies also provided relevant data.

LOU rat immunocytomas secreting all the H-chains and the two L-chains recognizable to date represent a large reservoir of information in this context.

Immunoglobulin primary sequence determination allowed the obtainment of the molecular proof of the antibody variability as well as the requirement for multiple genes coding for L- and H-chains. Furthermore, when immunoglobulins from different species were compared, a clearer view of the genetic events affecting the evolutionary relationships of the different Ig genes could be proposed.

The data collected on rat immunocytomas are especially relevant to this domain, allowing the comparison of two rodent species, i.e., mouse and rat, separated by 10 million years, as opposed to the sequence alignment reported between more distant species: human, mouse, and rabbit.

In this section, we report the published work available on LOU rat immunoglobulin light and heavy chains completed by some unpublished data, especially about the light chain sequences.

2. Light Chain Sequences

a. Ck-Regions

The complete sequences of two kappa LOU rat Bence Jones proteins, S210 and S211, are available. The S210 sequence[85] (Figure 4) seems to represent the major rat Ck-type sequence when compared to the S211 sequence (see Figure 7 and reference 89) and the LOU nucleotide sequence.[86] S211 protein exhibited two amino acid substitutions in positions 138 (Lys for Asn) and 209 (Asn for Ser). The available Ck-sequences from IR731 IgD light chain,[90] from LOU,[86] and from Lewis rat pool,[46] as well as the amino acid compositions of Ck-peptides from S204 and S208 proteins[33] confirmed the uniqueness of the S211 sequence.

Other differences between LOU rat Bence Jones amino acid sequences and LOU rat nucleotide sequence were observed at positions 110 (Asm to Asp), 156 (undetected Arg to Arg), and 205 (Gln to Glu), which may account for by trivial sequence errors. Another discrepancy is found at position 207, where a Trp residue was observed in four LOU K-chains and was absent from all other rat Ck-regions studied. As explained later, these differences do not seem to be allotype-related (Figure 4) and rather these data suggest the existence of several rat Ck-isotypes and/or an incomplete inbreeding process at the time the first few immunocytomas (S204, S208, S210, and S211) were observed.

Partial amino acid sequences of rat Ck-allotypes detected by Beckers et al.[43] were performed on pooled K-chains from Lewis and DA rat strains.[46] A reappraisal of these data came with the determination of the nucleotide sequences of a 1.2 kb BspRI fragment from LOU, LEW, and DA Ck-genes.[86]

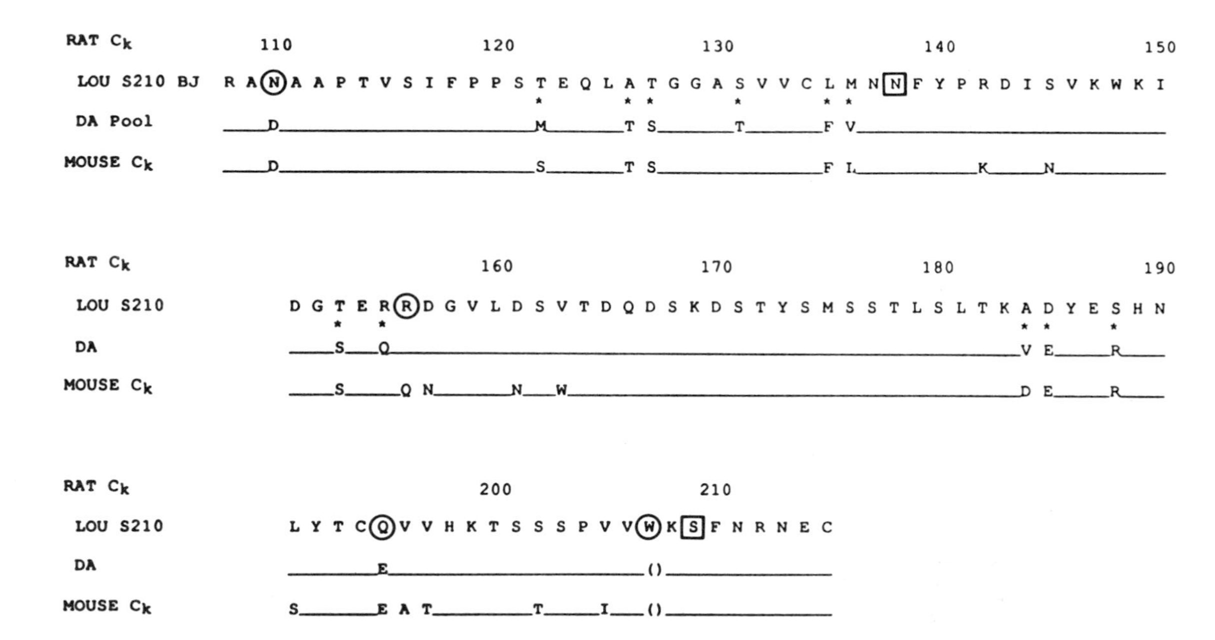

FIGURE 4. Comparison of rat and mouse Ck-region sequences. Squared residues: differences between S210 and S211 sequences. Circled residues: discrepancies between S210 amino acid sequence and LOU Ck-nucleotide sequence.[86] (LOU S210 Bence Jones protein: Starace;[85] DA pool: Gutman et al.;[46] mouse Ck (MOPC21): Svasti and Milstein.[87] The one-letter code for the amino acids is taken from Dayhoff.[88]

Rat Ck-allotypes were defined as complex allotypes involving 11 positions unevenly distributed in three clusters along the Ck-region (Figure 4). Two of them (153 to 155, and 184, 185, 188) are external and some observed specificities may determine the serological specificities. Six out of the eleven substitutions were homologous between DA rat and mouse while two positions were identical between LOU rat and mouse sequences, showing an unequal divergence between both rat allotypes.

When examined at the nucleotide sequence level, the rat and mouse Ck-coding regions exhibited 88.5% homology, as compared to 80% at the protein level (see Figure 5). A surprisingly high degree of homology (between 80.3 and 88.4%) for the non-coding region and a high degree of conservation of silent sites within the coding regions were other salient features of this comparison.[86]

A comparison of Ck-sequences for four species, i.e., rat, mouse, human and rabbit (Figures 5 and 7), showed the closest homology between both Rodent species. This homology lies at a 80% level for protein sequence comparison between rat and mouse and falls at 63 and 48% when rat proteins are compared with human and rabbit Ck-regions. The evolution of the Ck-genes will be re-evaluated in relation with the data collected for the Vk-genes (see Section 2.d).

b. C-Lambda-Regions

Very few lambda light chains were identified among several hundreds of monoclonal LOU rat monoclonal immunoglobulins serologically screened until now, although a kappa/lambda ratio of 95:5 was reported previously by Hood et al.[35] for light chains prepared from rat pooled normal serum.

No lambda Bence Jones proteins have been detected so far and the limited sequence information was collected from the light chains of two IgG1 monoclonal LOU rat immunoglobulins, i.e., IR18 and IR31.

The sequence of the ^{14}C-labeled C-terminal octapeptide was established[92] and showed a single substitution, as compared with the mouse lambda 2 type.[93] The last residue was a Val for the rat instead of a Leu for the mouse. Three and five amino acid differences were observed respectively with the mouse lambda 1 and the human lambda chain C-terminal peptides.[91] These data confirmed at the primary sequence level the existence of rat lambda light chains.

c. Jk-Segments

Information about kappa chain J-segments was obtained from the nucleotide sequences of cloned DNA, either from LOU rat liver[94] or from LEWIS rat embryonic kidney.[95,96]

The rat Jk-cluster is 2.1 kb, as compared to 1.4 kb for the mouse Jk- cluster, and is located at about 2.5 kb from the 5′ end of the Ck-gene. Eight Jk-coding sequences from which six are expressible and identified in the rat species. The distance between these segments is 0.3 kb.

The rat Jk-coding sequences are designated from the 5′ end to the 3′ end J1-J2b-J2a-J2-J3r-J3′-J4-J5r. The corresponding mouse genes are the J1-J2-J3r-J3′-J4-J5r genes.[97,98] Two remnant or pseudogenes (suffix "r") are identified in both species. In human, the order of the Jk-regions is J1-J2-J3-J4-J5.[99]

The rat Jk-cluster evolved from an ancestral cluster of five Jk-genes by two successive unequal crossing-over events, resulting in the duplication of the J2 coding segment. The first crossover involved a 14 bp region spanning the 3′ end of the coding region of J1 and J2 sequences. The second event had a target of 370 bp tight homology region including the two newly duplicated genes. These events occurred approximately 2×10^6 and 1×10^6 years ago, respectively.

The three recognition sequences are highly preserved in the rat as in mouse and human J-genes. These are the RNA splicing signals at the 3′ end of the coding regions, the heptanucleotide located 5′ to each J-gene and the nonanucleotide present 30 bp 5′ to each J-gene and absent from the remnant genes.

```
                 110                 120                 130                 140                 150
RAT       R A N A A P T V S I F P P S T E Q L A T G G A S V V C L M N N F Y P R D I S V K W K I
MOUSE     ____D_______________________S_______T S_______________F L___________K______N__________
HUMAN     __T V_____S_____F___________D_______K S___T_____________L_____________E A K___Q_____V
RABBIT    G P D V_________L_________A A D___V_______T V T I_____V A___K Y F__( )__V T___T___E V

                                     160                 170                 180                 190
RAT             D G T E R R D G V L D S V T D Q D S K D S T Y S M S S T L S L T K A D Y E S H N
MOUSE           ____S_____Q N_____________________________________________________V E___E R____
HUMAN           __N A L Q S G N S Q E _____E___________________L__________T___S_________E K___K
RABBIT          ______T Q T T___I E N___K___P_______A___C_____N L__________T_____S T Q___N_____K

                                     200                 210
RAT             L Y T C Q V V H K T S S S P V V W K S F N R N E C
MOUSE           S_______E A T_________T_____I__( )______________
HUMAN           V___A___E___T___Q G L_________T( )__________G____
RABBIT          E_______K___T( )Q G T T__( )___( )Q_________G D__
```

FIGURE 5. Alignment of Ck-region sequences from four species. (—): sequence homologous to rat S210 BJ[85] and RI1b LOU.[86] Mouse, human, and rabbit Ck-region sequences from Kabat et al.[91]

An unusually strong preservation of flanking sequences relative to coding sequences is observed in the Jk-cluster. The overall homology between corresponding rat and mouse J-gene flanking sequences is 88.8% compared with 92.3% between the coding regions. This observation confirms the functional significance of the noncoding sequences.

The amino acid sequences available in this region, S210[85] and S211,[89] do not correspond to any germline rat Jk-segments. Undetected Jk-genes 5′, the sequenced Jk-cluster, as well as somatic mutation may account for the variants observed in rat Jk-segments.

d. Vk-Regions

The Vk-region amino acid sequences of monoclonal immunoglobulins secreted by LOU rat immunocytomas throw light upon the extent of the V-gene repertoire available in the rat species and its eventual relationships with other species V-gene repertoire, particularly with that of the mouse where the majority of the hybridomas have been obtained until now.

The comparison of the two complete Bence Jones protein sequences, i.e., S210 and S211, allows assessing that these Vk-regions belong to two different Vk-subgroups. The homology between their four framework segments (FR) does not exceed 75%. These differences are scattered through the N-terminal 23 residues and clustered in the other FRs, 42-44, 78-83, and 98-103.

More informative is the alignment of the N-terminal FR (23-24 positions) from 29 Vk-immunoglobulin light chains and Bence Jones proteins (Figure 6) which present data from Alcaraz et al.,[90] Starace and Querinjean,[89] Starace,[85] Wang et al.,[100,101] as well as a collection of unpublished sequences.

The different Vk-FR1 sequences may be subgrouped according to eight prototype sequences differing in at least five positions and including from two (VkV, VkVIII) to seven proteins (VkII).

If these subgroups are easily recognized by two to nine subgroup specific residues, some analogies with mouse and human Vk-subgroups are quite striking. For example, rat and mouse VkI- and VkIII-subgroups differ only by two residues and their VkV subgroups by three residues. A similar difference is observed between rat VkII and human VkI subgroups. The latter differs only at a single position from the mouse VkV subgroup. The human VkI-subgroup seems to have been positively selected throughout the evolution.

In contrast to the preservation of a given Vk-gene, we observe a larger subgroup number in both rodent species as compared to the human species, without taking into account several unique unclassifiable sequences.

The repertoire of Vk-genes seems to be even broader in rat species as the eight proposed subgroups are deduced from 29 Vk-sequences as opposed to seven subgroups established by the comparison of 428 kappa L-chains.[91]

Outside the close homology observed between VkI (91%), VkIII (91%) and VkV (89%) rat and mouse subgroups, the average relationship between all other subgroups amounts to 55% homology with a range from 83% (rat VkIII and mouse VkV) to 22% (mouse VkII and VkVI).

In view of the relationships previously put forward for the Ck-genes (see Section 2.a), it may be emphasized that Vk- and Ck-genes did not evolve at the same rate (Figure 7). Within the Vk-genes the evolution pressure seems to have been applied with predilection on some genes which have evolved with a minimum variation. This seems to be the case of the human VkI-gene exhibiting 87.5% and 96.3% homology, respectively with the rat VkII- gene and the mouse VkV-gene. These figures are to be compared with the homologies between human Ck-genes and rat (62%) and mouse (61%) corresponding genes.

The availability of a larger number of rat light chain sequences throws some light on the variability of evolution rate of Vk-genes between two closely related species.

The Vk-subgroup homology lies within the species around 55% for rat and 56% for mouse and rises up to 58% between both Rodent species, if we exclude the subgroups presenting

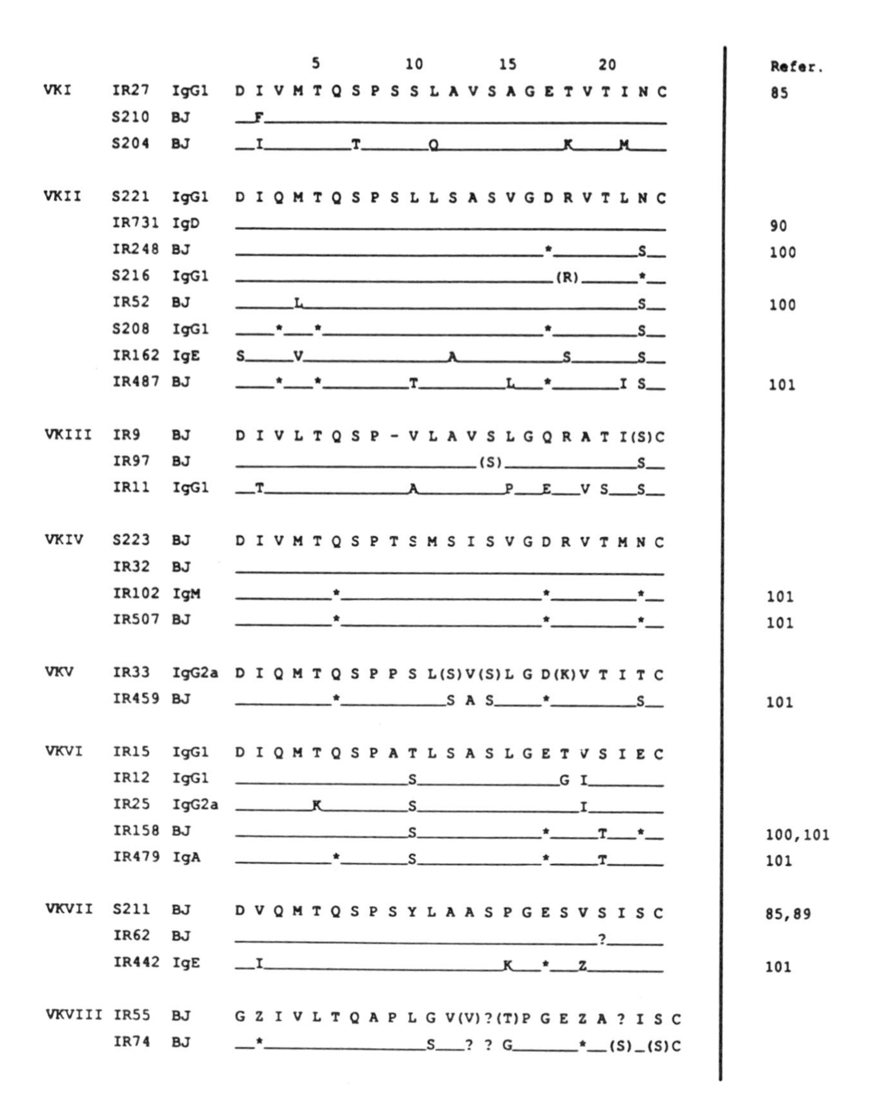

FIGURE 6. The eight rat Vk-subgroups.

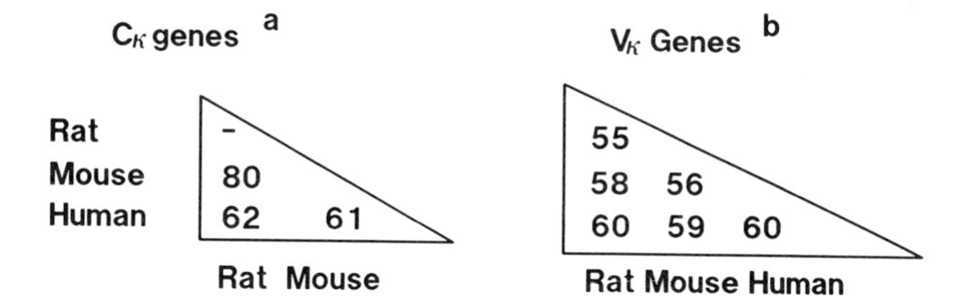

FIGURE 7. Homologies between Ck- and Vk-amino acid sequences from three different species: (a) Ck-gene homology calculated on 108 positions for mouse/rat and human/rat comparisons and on 107 positions for mouse/human comparisons; (b) Vk-gene-FR homology (%) calculated on 80 positions for all sequences compared.

abnormally high homologies from this estimation. In contrast, both Rodent species Vk-genes are as similar to the human Vk-genes (59 or 60%) as these genes are between themselves.

In conclusion, if Vk- and Ck-genes evolve at similar rates within the human species, the rodent Vk-genes present 22% less homology compared with their Ck-genes homology. This divergence is even slightly higher (2-3%) when intraspecies similarities are tabulated. Selective pressure applied significantly differently in the Rodent species on Vk- and Ck-genes.

3. Heavy Chain Sequences

a. CH-Regions

i. Gene Order and C-Gamma-Genes

Most of the data concerning the rat CH-regions came from gene sequence analysis.

The order of rat CH-genes assigned to chromosome 6[102] was established by Brüggemann et al.,[28] by screening a rat cosmid library with mouse probes. The different rat C-gamma-genes were detected by cross-hybridization to various mouse C-gamma-probes and identified in transfection experiment where the CH-genes were linked to a variable region and expressed in light chain-producing myeloma cells.

The gene order of rat Ig H-chain locus may be compared with those assigned for mouse[103] and human.[104-106]

From the comparison presented in Figure 8, a high homology between the rodent H-chain loci is apparent with the clustering of the C-gamma-genes and the absence of pseudogenes as observed in human locus. Furthermore, the distance between the genes is similar and exon sequences are preserved more than intron sequences. Up to 91% homology is observed between rat and mouse introns spanning the Ig H-chain enhancer.

When the subclass partial nucleotide sequences are compared, the rat C-gamma1- and C-gamma2a-genes, and their high homology reflecting a recent gene duplication are very similar to the mouse C-gamma1-genes; rat gamma2b is homologous to the mouse C-gamma2a/C-gamma2b pair, whereas a 87% homology was recently established[32] between rat C-gamma2c and mouse C-gamma3.

ii. C Delta-Genes

The identification of IgD-secreting immunocytomas in LOU rat strain[20] provided a unique material to obtain sequence information on IgD isotype.

Partial Ck and Vk chain (see Sections 2a and 3d above) amino acid sequences were obtained from IgD IR731 rat monoclonal immunoglobulin.[90,107]

The native molecule behaved as a noncovalent dimer (M_r: 300 kDa). The molecular weight discrepancy with other species IgDs as reviewed by Spiegelberg[108] may be accounted for by a deletion of 100 to 120 residues involving most of the C-delta-H2 domain and a few residues at the beginning of the C-delta-H3 domain.

The IR731 heavy chain is blocked. The sequence of the cysteinyl-containing tryptic peptides of this 50 kDa delta-chain confirms the susceptibility to proteolytic cleavage of a hinge region poor in proline and cystein residues, but rich in lysine residues. The hinge region carries the 13 kDa carbohydrate moiety.

Extensive sequence comparison with mouse[109,110] and human delta-chains[111,112] was possible through the sequence determination of a cDNA clone using a delta mRNA prepared from a IR731 tumor cell line.[20,113,114] Two species of IgD mRNAs, approximately 1.8 kb and 2.7 kb long coding for secreted and membrane-bound delta chains, respectively, have been isolated in other species.[115-118] Similar observations were performed by Sire et al.[114] with rat IgD.

Comparison between mouse and rat IgD polypeptide sequences showed a broad range of homology depending on the considered domains. Moderate (50 to 55%) homology was observed between C-delta-H1, C-delta-hinge, and part of C-delta-H3 (residues 176 to 204), but the homology rose up to 85 to 90% between parts of C-delta-H3 (residues 138 to 175 and 205 to 244), C-distally coded exon[114] with an overall homology of 70% for the whole gene.

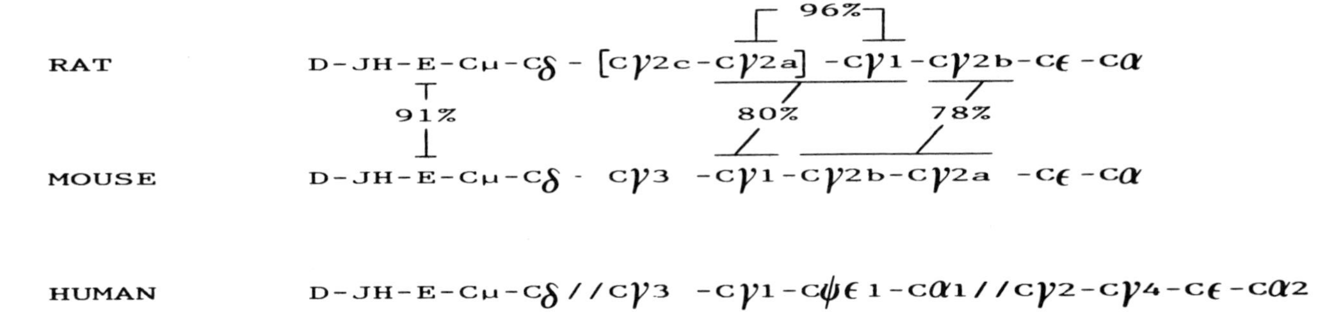

FIGURE 8. The order of the immunoglobulin H-chain genes in three species (rat: Brüggemann et al.;[28,32] mouse: Shimizu et al.;[103] human: Flanagan and Rabbits[104] and Ravetch et al.[105]).

iii. C-Epsilon-Genes

The studies at the molecular level of the immediate hypersensitivity involving the IgE have been hampered for a long time by the lack of a suitable animal model.

The high incidence of IgE-secreting immunocytomas in LOU rat strain[18,53] significantly modified a situation where IgE monoclonal proteins were identified in rare human patients.[119]

Rat IgE sequence was mainly studied at the nucleotide level, with limited confirmation by amino acid sequence analysis.[120-122]

RNA preparations from IR2 and IR162 IgE-secreting solid tumors helped to build a LOU cDNA library.[120,121] Another library obtained from Sprague-Dawley liver DNA was screened with a probe containing the C-epsilon-H2 to C-epsilon-H4 exons.[121,122] A 2100 bp long sequence covering all four CH-domains of rat epsilon chain was sequenced. Two separate mRNA species, one major 2.3 kb and one minor 2.8 kbp long, were identified in rat immunocytomas. They encode the secreted and the membrane bound forms of the epsilon chain. The membrane exons, M1 and M2, are located some 2 kb away from the 3′-end of the C-epsilon H4 exon.

A comparison between rat and mouse[123] epsilon-chain sequences shows a high degree of homology: C-epsilon-H4, 81%; M1, 87%; and M2, 89% at the protein level. A major difference was observed between the C-epsilon-H1 and C-epsilon-H2 exons, where a 366 bp long mouse repetitive sequence was inserted. This insertion contains two inverted repeats, 14 bp long, and has the properties of transposable elements from which approximately 100 copies have been detected in the mouse genome. This peculiarity leaves a homology of 74% for C-epsilon-H1 and 70% for C-epsilon-H2 between rat and mouse sequences.

The homology between rat and human[124] epsilon-chain domains is also higher for the third and fourth constant domains (50 and 54%) as opposed to 43 and 38% for the C-epsilon-H1 and C-epsilon-H2 domains. The conservation of the C-terminal part of the epsilon-chain points to its important role for IgE-binding to receptors at the surface of mast cells and basophils.[125]

Furthermore, the three species differ in the length of the C-terminus with rat and mouse having an extra peptide (10 and 6 residues long, respectively) relative to the human sequence. Furthermore, no pseudogenes could be detected in the genomic organization of the epsilon-chain genes of LOU and Sprague-Dawley rat strains. This pattern is similar to the mouse gene organization,[123] but differs markedly from the human epsilon-chain genes, where a possible C gamma-C-epsilon-C-alpha duplication was proposed, accounting for the two C-epsilon pseudogenes.[124,126-128]

When compared with other immunoglobulin region evolution (see Figure 7 for Vk- and Ck-genes and Figure 8 for C-gamma-genes), the C-epsilon-gene has diverged more rapidly than other constant domain genes to evolve at a rate comparable to the one observed for the Vk-framework residues.

b. VH-Regions

A set of seven VH N-terminal 30 residues from LOU rat monoclonal immunoglobulins of various isotypes[34] are available for comparison with mouse and human subgroup sequences (Figure 9). These sequences may also be compared with VHIII sequences from pooled heavy chains from various species.[129,132]

From 14 myeloma heavy chains sequenced until now, half of them are unblocked and five show a clear relationship with the human and mouse VHIII subgroups (Figure 9). They differ at most in five positions from the VHIII prototype, while IR33 H-chain differing in eight positions cannot be classified in any existing subgroup. In contrast, S216 polypeptide chain is highly homologous, 5 substitutions only, to mouse MOPC 315 alpha-chain[131] belonging to the VHI(A) subgroup.[91] The Sprague-Dawley anti-Streptococcal group A carbohydrate antibody H-chain[130] presents only four differences with the rat VHIII prototype.

Position 16 is strikingly variable and IR64 protein being the only one showing a glycine residue corresponding to this highly conserved position in other species prototypic sequences. This position could be part of the variant positions associated with some genetic marker.

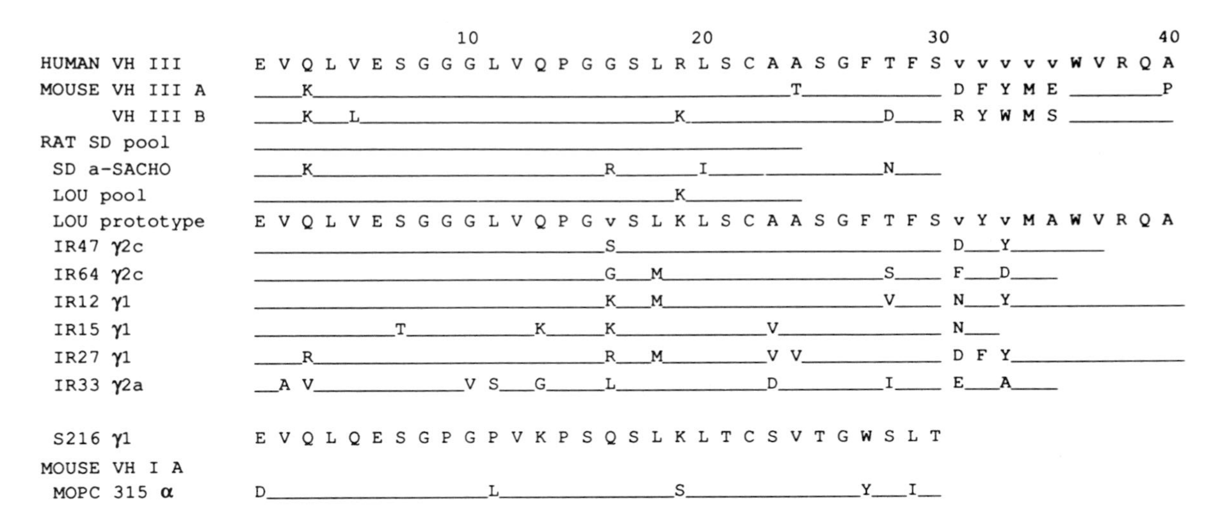

FIGURE 9. Comparison of human, mouse, (Kabat et al., 1987) and rat VHIII N-terminal sequences. The rat sequences are taken from Querinjean et al.,[34] except the Sprague-Dawley (SD) pool from Capra et al.;[129] anti-SACHO: anti-Streptococcal group A carbohydrate antibodies from Sprague-Dawley rats from Klapper et al.;[130] mouse MOPC 315 from Francis et al.[131]

TABLE 7
Biological Properties of Rat Immunoglobulin Classes and Subclasses

	Ref.	IgM	IgD	IgG1	IgG2a	IgG2b	IgG2c	IgE	IgA
Antibody activity	5, 59,133	yes	yes	yes	yes	yes	yes	yes	yes
Valency for antigen	7, 59	5 + 5	—	2	2	2	2	2	(2)n
Half-life (d)	62, 63, 134	2.6	1.6[b]	13[a]	5	15[a]	3[b]	0.5	NT[c]
Complement fixation (human origin)	9	yes	—	yes	yes	yes	no	yes	yes
Cross placenta	135—137	no	—	yes	yes	—	—	no	no
Presence in colostrums or milk	135—138	no	—	yes	yes	—	—	yes	yes
Absorption by the intestine of the suckling rats	135—138	no	—	yes	yes	—	—	yes	yes
Fix to homologous mast cells	18, 25, 53,139	—	—	—	yes	—	—	yes	—
Half-life in homologous skin (d)	134	—	—	—	2.40 ± 0.25	—	—	7.40 ± 0.89	—
Passive cutaneous anaphylaxis in rat	17, 18 , 138, 140	—	—	—	yes	no	no	yes	—

[a] Platteau and Bazin, obtained by injections of unpurified ascitic fluid of rat IgG1 or IgG2b MAbs anti-DNP (manuscript in preparation).
[b] Platteau and Bazin, obtained by injections of unpurified ascitic fluid or rat IgD or IgG2c Igs (manuscript in preparation).
[c] Not tested.

Rat myeloma protein sequences confirmed the sequences obtained from pooled H-chains. LOU and Sprague-Dawley rat strains pools differ in a single position, i.e., Lys to Arg substitution in position 19.[34] Similar proportions of unblocked H-chains belong to the VHIII subgroup: 36% (5/14) among LOU rat myeloma proteins, 35% in LOU pool, and 20% in Sprague-Dawley pool. The variability observed in myeloma proteins contrasts with the homogeneity observed in pooled H-chain sequence. A clear difference is observed at position 16, where a clear glycine is identified in the pool as compared with a highly variable residue in rat myeloma proteins.

The rat VH-sequences confirm the Vk analysis (see Section 2.d above) showing the existence of highly preserved genes, for instance, VHIII- and mouse VHI(A)-genes, as well as the unique characteristic of variability (i.e., position 16).

4. General Conclusion

LOU rat monoclonal immunoglobulins and tumors represent a source of invaluable material to provide sequence information, throwing light on the immunoglobulin repertoire potential to build unique antibodies.

V. BIOLOGICAL PROPERTIES

A. GENERALITIES

Some biological properties of rat immunoglobulin isotypes can be found in Table 7. Some other properties of rat antibodies obtained by the use of rat monoclonal antibodies will be described in Part V of this volume. Among the most interesting physiological properties of the rat immune system, the serum immunoglobulin concentrations and the distribution of rat antibodies among the various isotypes can be considered for antibody purification and immunization before fusion for hybridoma obtainment. These two points will be considered in the following pages.

B. IMMUNOGLOBULIN SERUM CONCENTRATIONS IN NORMAL AND IMMUNIZED RATS

Serum concentrations of immunoglobulins can vary considerably in normal rats and in other species, depending on the bacteriological status of the colony. However, some basic data can be considered in every species. In the rat species, an IgM serum concentration of 1 mg or less per milliliter is normal and will be found even in axenic (germ free) animals. The IgD serum concentration is low in micrograms per milliliter. IgG serum concentrations depend essentially on the bacteriological status of the breeding. IgG1 and IgG2a are generally the most represented IgG subclasses. IgE serum concentration in "clean" animals is extremely low in nanograms per milliliter. This concentration can very rapidly increase after a certain antigenic stimulation such as helminth infestations. IgA serum concentration is low in rats, generally less than 0.5 mg/ml. Table 8 gives some serum Ig levels of normal or immunized rats.

C. HUMORAL IMMUNE RESPONSES IN RATS

In comparison with animal species like rabbit or goat, rats and mice are rather poor producers of precipitating antibodies. However, with adequate immunization protocols, it is quite possible to obtain rather good immune responses.

Evidently, it is impossible to summarize all experimental immunizations of rats described in the scientific literature. This description will be limited to a few examples and unpublished data.

1. Ontogeny of the Rat Humoral Immune Responses

Rats are very poor antibody producers at birth,[143,144] although their immune responses are certainly dependent on many factors. Thus, Pauwels et al.[145] have obtained IgE immune response antiovalbumin by inoculating BN rats at birth with antigen plus *Bordetella pertussis* organisms, and aluminium hydroxide as adjuvants. Quite clearly, the ability to synthesize specific antibodies increases with age. In experiments using sheep erythrocytes, polymerized flagellin, bovine serum albumin, or lipopolysaccharide of *Escherichia coli,* no rat produced humoral antibody response prior to at least 2 weeks of age.[143,144,146] However, in the case of *Brucella abortus,* Halliday[147] or Hervey[148] observed an immune response even when the antigen was injected at birth, but the production was delayed until 7 or 10 d after antigen administration. Using flagella of *Salmonella typhosa,* Winebright and Fitch[149] were able to detect specific antibodies in rats injected at birth and tested as early as 4 d after inoculation.
Between the age of 2 weeks and 3 months, the capacity of antibody synthesis slowly increases to reach its maximum when rats are 1, 2, or sometimes 3 months old, depending on many parameters either exterior to the subject such as antigens, adjuvants, route of administration or inherent to the subject such as its own genome, and its immunological memory.

In conclusion, rats can be generally considered to be mature as soon as they are 2 1/2 months old, and have certainly reached maturity when they are 3 months old. When younger, they are sometimes immunologically mature, but this is not always true.

2. Distribution of the Specific Antibodies Against a Given Antigen, Among the Various Immunoglobulin Classes and Subclasses

Early studies on the rat humoral immune response showed that the first antibodies to appear in the rat species as in others were of the IgM and then of the IgG isotype. However, these results are not always observed. Nossal et al.[150] observed that flagella and polymerized flagellin from *Salmonella adelaide* first caused an IgM immune response, and flagellin was unable to induce such a response.

a. Introduction

The technology of monoclonal antibody production and the development of new tests such

TABLE 8
Serum Immunoglobulin Levels in Rats

Rats	Ig concentration (mg/ml except of IgD and IgE isotypes which are given in μg/ml)								
	IgM	IgD	IgG1	IgG2a	IgG2b	IgG2c	IgE	IgA	Ref.
Normal adult WISTAR	0.56	—	5.85	8.00	—	2.60	20	0.13	7
Normal adult hooded lister	0.85 ± 0.10[a]	—	2.80 ± 0.31	10.39 ± 1.72	—	2.23 ± 0.45	0.81 ± 0.19	0.15 ± 0.06	141
Nippostrongylus infected hooded lister (25 d after infection)	0.90 ± 0.06	—	9.11 ± 1.85	13.72 ± 1.93	—	4.31 ± 0.60	71.60 ± 10.02	0.33 ± 0.11	141
Adult outbred nude (rnu/rnu)	1.9 ± 0.2	1.6 ± 0.6	2.8 ± 0.6	2.5 ± 0.4	2.0 ± 0.4	2.8 ± 0.7	0.2 ± 0.6	0.06 ± 0.01	142
Adult outbred rnu/+	1.8 ± 0.3	1.8 ± 0.4	1.3 ± 0.2	5.1 ± 0.9	1.6 ± 0.3	1.4 ± 0.3	0.3 ± 0.06	0.03 ± 0.01	142

[a] Mean and standard error.

as radio immunoassay or Enzyme Linked Immuno-Sorbent Assay (ELISA) have provided scientists with new tools to study the rat immune responses.

Three-month-old male and female inbred WISTAR/R rats were injected with a T-dependent (DNP2-OVA) or a T-independent (DNP-HES) antigen. Blood samples were taken at various time intervals after immunization and anti-DNP responses were measured using an ELISA.

b. Material and Method

i. Antigens, Adjuvants and Immunizations

T-dependent antigen — Lightly dinitrophenylated chicken ovalbumin (DNP2-OVA) was prepared as described by Ishizaka and Okudaira.[151] One group of rats was immunized with an intraperitoneal injection of a freshly prepared solution of 5 μg DNP2-OVA together with a dose of 10^{10} chemically killed *Bordetella pertussis* organisms and 2.5 ml Al(OH)3 (*B. pertussis* vaccine "Perthydral", Institut Pasteur Production, Paris) as adjuvant. Another group of female rats received an intraperitoneal injection of a freshly prepared solution of 5 μg DNP2-OVA together with as adjuvant 1 mg $Al(OH)_3$ in a volume of 0.5 ml. $Al(OH)_3$ was prepared according to the method of Levine and Vaz, as described in Pauwels et al.[152] A booster injection with 5 μg of DNP2-OVA without adjuvant was given intraperitoneally 90 d after primary immunization.

T-independent antigen — Dinitrophenylated hydroxyethyl starch (DNP-HES) was kindly given by Professor J.H. Humphrey of the Royal Postgraduate Medical School in London and was prepared as described in Humphrey.[153] The rats were injected intravenously, in the lateral tail vein, with 5 μg of DNP-HES in aqueous solution. A booster injection with 5 μg of DNP-HES was given intravenously, 90 d later.

ii. Measurements

Enzyme-linked ImmunoSorbent Assay (ELISA) — Isotype levels of anti-DNP antibodies were measured by an ELISA. Multiwell polystyrene plates (Linbro, Flow Laboratories) were coated overnight at 4°C with 2.5 μg of DNP28-bovine gammaglobulin (DNP28-BGG) in 100 μl of 0.1 *M* sodium carbonate buffer (pH 9.5). After four washings with phosphate buffered saline (PBS), the plates were saturated during 2 h at 37°C with 200 μl of a 10% fetal calf serum-phosphate buffered saline (FCS-PBS) solution.

Plates were washed again four times with PBS and stored at –20°C until further use. When necessary, they were incubated for 1 h at 37°C with 100 μl of serial dilutions of test antiserum or reference antibody preparation. All dilutions were made in 10% FCS-PBS. After another four washings, the plates were incubated for 1 h at 37°C with peroxidase-labeled monoclonal mouse antibodies to rat immunoglobulin isotypes. The plates were washed and the peroxidase was revealed with 100 μl of a freshly prepared substrate solution containing 5 μg O-phenylene diamine (OPD) (Sigma) in 10 ml citrate phosphate buffer 0.1 *M* pH 5.5, and 0.06% ureumper-oxidase. The concentration of the anti-DNP antibodies of the various immunoglobulin isotypes was measured in comparison with purified rat monoclonal anti-DNP antibodies.

iii. Antisera and Reference Antibodies

Rat anti-DNP monoclonal antibodies of different isotypes were generated and characterized in our laboratory (see Chapter 19.VIII). They were purified as described elsewhere by immunoaffinity chromatography using the kappa allotype system as described in Chapter 15. Monoclonal mouse antibodies to rat IgM, IgA, IgE, IgG1, IgG2a, and IgG2b, respectively called MARM-4, MARA-2, MARE-1, MARG1-1, MARG2a-8, and MARG2b-8 were produced, characterized and purified in the laboratory and labeled with peroxidase as described by Nakane and Kawoi.[154]

iv. Results

Isotype levels of anti-DNP antibodies in rats immunized with a T-dependent antigen: DNP2-OVA + *Bordetella pertussis* or DNP2-OVA + $Al(OH)_3$ — As it can be seen in Figures

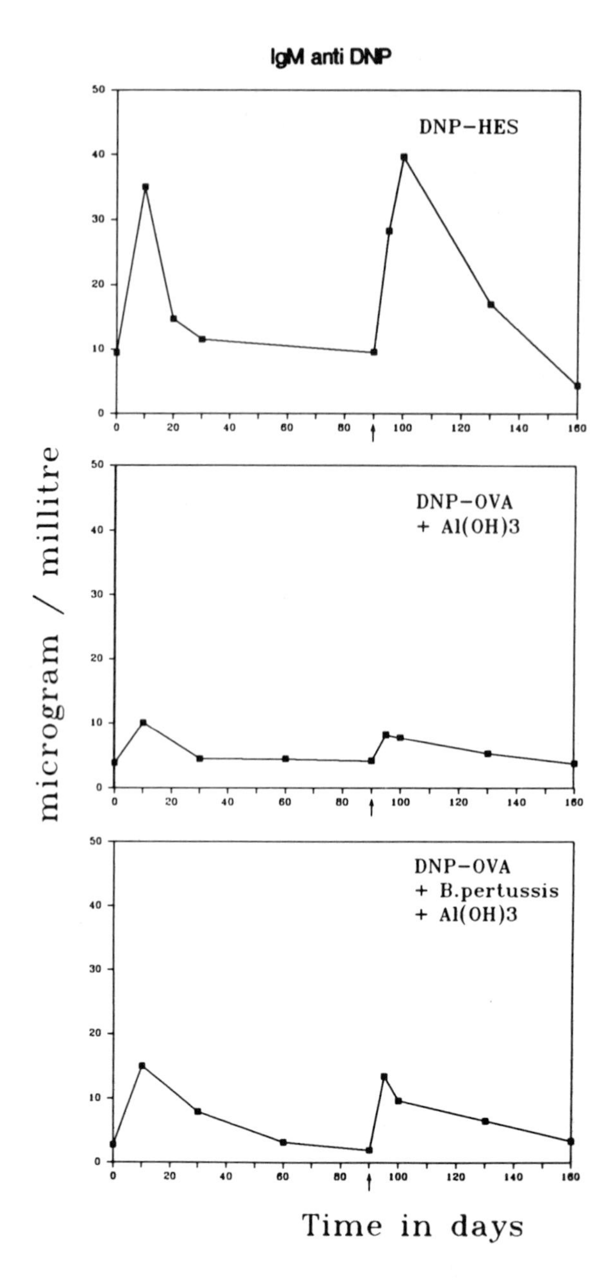

FIGURE 10. Specific IgM rat antibodies against DNP hapten in function of the time in days. Rats were immunized at day O with DNP-hydroxyethyl starch (top), DNP-ovalbumin and aluminium hydroxide (middle), or DNP-ovalbumin plus *Bordetella pertussis* organisms and aluminium hydroxide (bottom). A booster injection was given at day 90 with the same antigen as at day O, but without adjuvant.

10 to 15, the immune responses against DNP2-ovalbumin are more important in the IgM, IgG1, IgG2b, and IgE classes or subclasses when the antigen is given with Bordetella organisms and aluminium hydroxide than when it is given with this last adjuvant only. However, this difference does not exist for the IgG2a or IgA (sub)classes. The highest response is obtained in IgG1 using aluminium hydroxide and *B. pertussis* organisms as adjuvant.

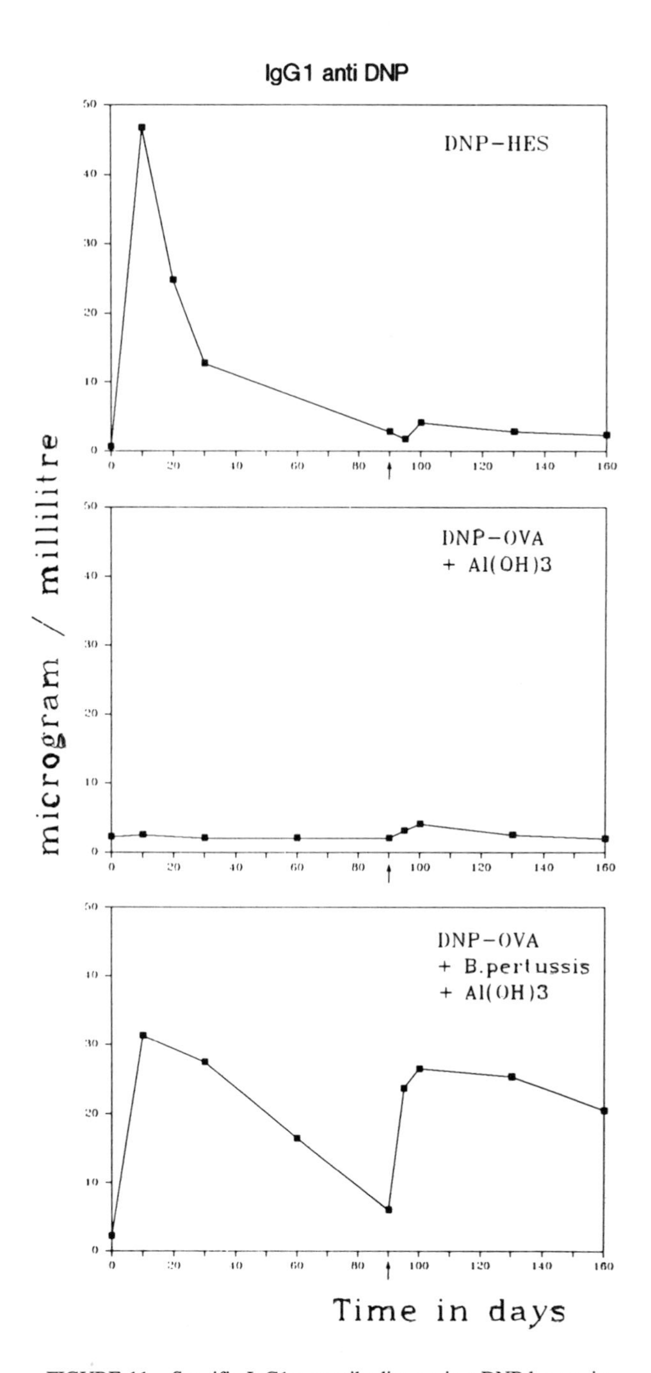

FIGURE 11. Specific IgG1 rat antibodies against DNP hapten in function of the time in days. Rats were immunized at day O with DNP-hydroxyethyl starch (top), DNP-ovalbumin and aluminium hydroxide (middle), or DNP-ovalbumin plus *Bordetella pertussis* organisms and aluminium hydroxide (bottom). A booster injection was given at day 90 with the same antigen as at day O, but without adjuvant.

Isotype levels of anti-DNP antibodies in rats immunized with T-independent antigen (DNP-HES) — Without adjuvant, the immune responses against DNP-hydroxyethyl starch are higher than with a T-dependent carrier in all classes, including IgE which is generally considered

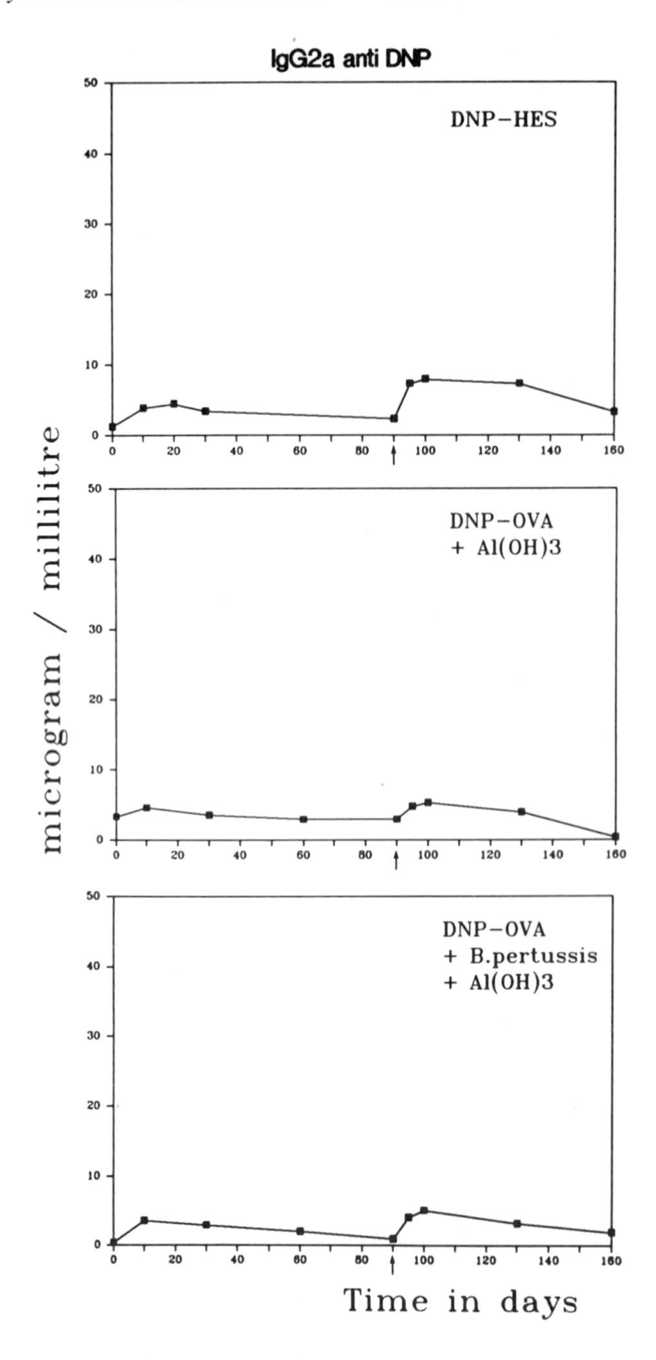

FIGURE 12. Specific IgG2a rat antibodies against DNP hapten in function of the time in days. Rats were immunized at day O with DNP-hydroxyethyl starch (top), ovalbumin and aluminium hydroxide (middle), or DNP-ovalbumin plus *Bordetella pertussis* organisms and aluminium hydroxide (bottom). A booster injection was given at day 90 with the same antigen as at day O, but without adjuvant.

as a typical T-dependent isotype. Curiously, the secondary immune responses are important in the case of IgM, IgE, IgA, but not in IgG subclasses (Figures 10 to 15).

These results show how it is difficult to predict which types of immune responses can be

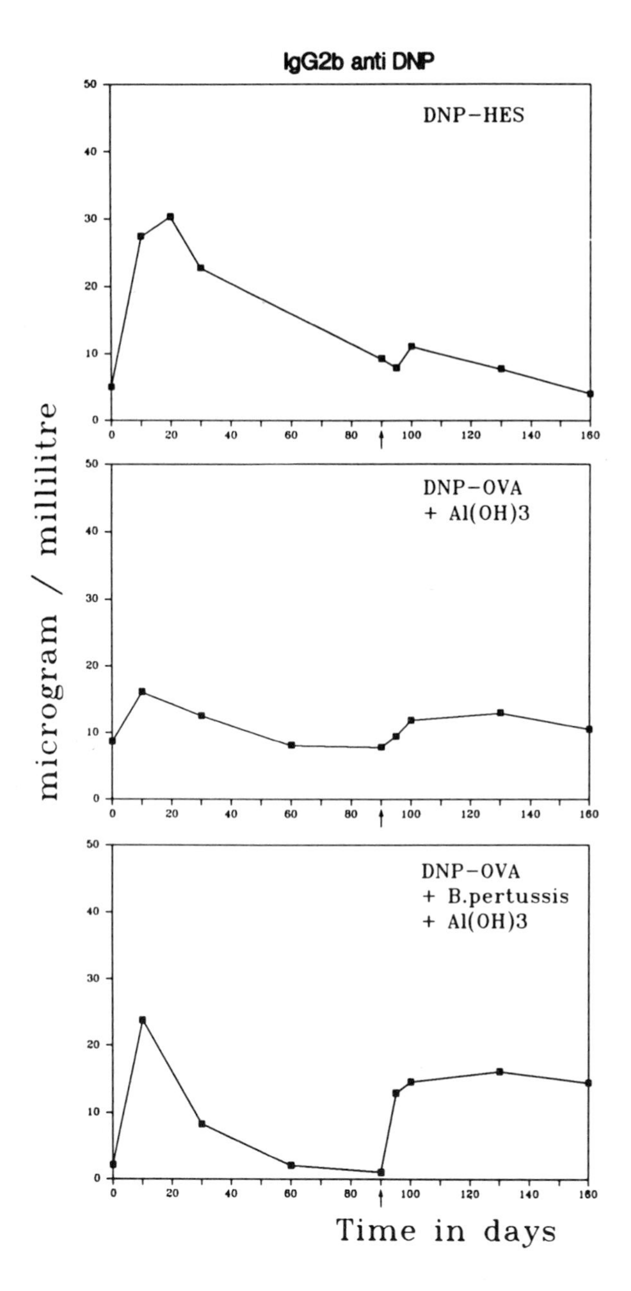

FIGURE 13. Specific IgG2b rat antibodies against DNP hapten in function of the time in days. Rats were immunized at day O with DNP-hydroxyethyl starch (top), DNP-ovalbumin and aluminium hydroxide (middle), or DNP-ovalbumin plus *Bordetella pertussis* organisms and aluminium hydroxide (bottom). A booster injection was given at day 90 with the same antigen as at day O, but without adjuvant.

obtained after an immunization. Depending on the given adjuvant, the dose of antigen, and the used carrier, the immune responses are quite different against the same hapten.

3. Tolerance Induction in Rats

If the concept of tolerance against antigens, i.e., a complete absence of an immune response

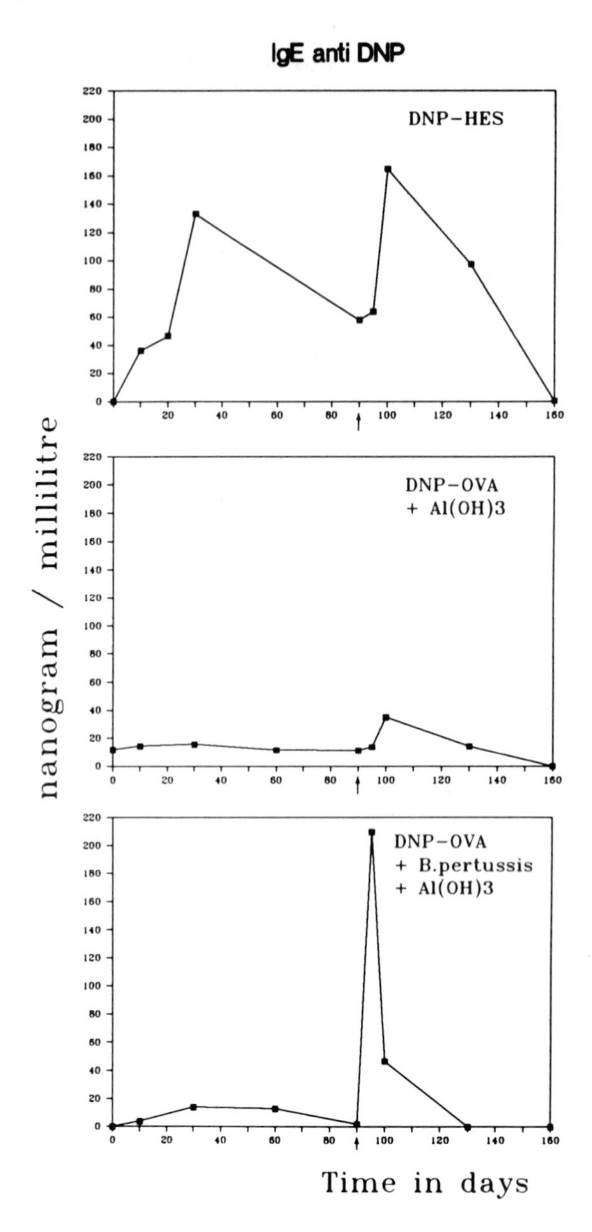

FIGURE 14. Specific IgE rat antibodies against DNP hapten in function of the time in days. Rats were immunized at day O with DNP-hydroxyethyl starch (top), DNP-ovalbumin and aluminium hydroxide (middle), or DNP-ovalbumin plus *Bordetella pertussis* organisms and aluminium hydroxide (bottom). A booster injection was given at day 90 with the same antigen as at day O, but without adjuvant.

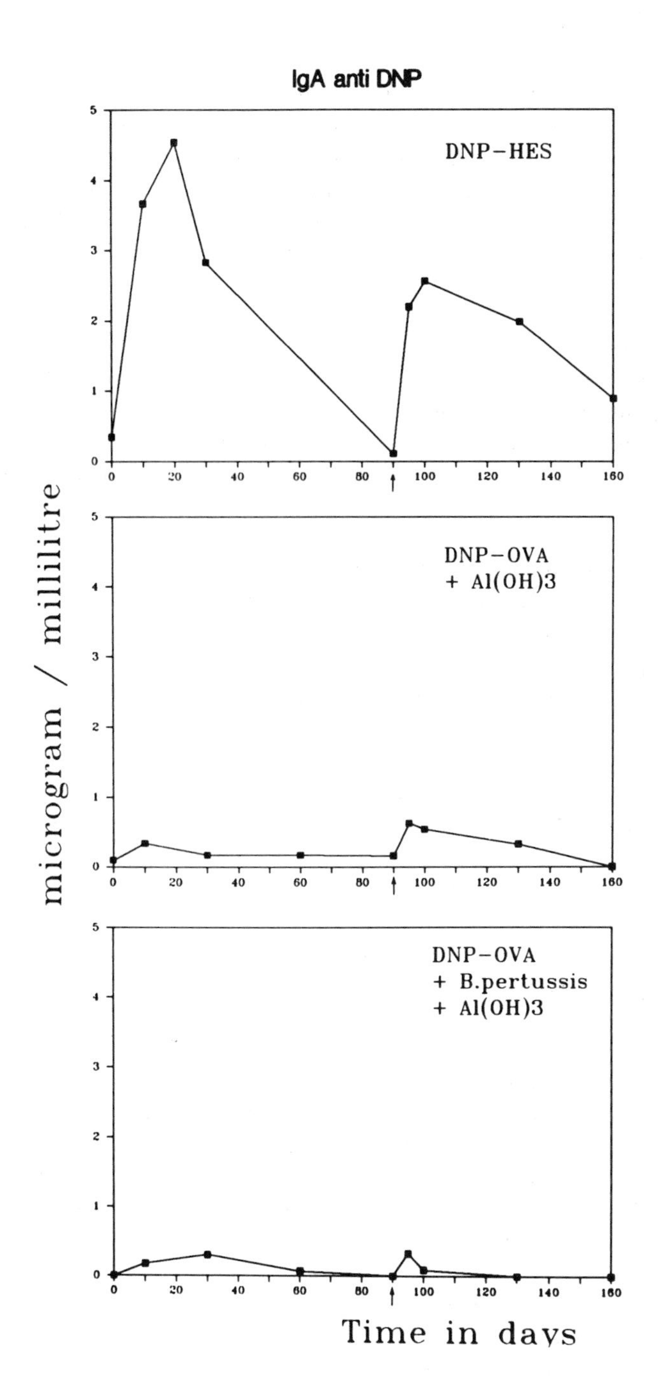

FIGURE 15. Specific IgA rat antibodies against DNP hapten in function of the time in days. Rats were immunized at day O with DNP-hydroxyethyl starch (top), DNP-ovalbumin and aluminium hydroxide (middle), or DNP-ovalbumin plus *Bordetella pertussis* organisms and aluminium hydroxide (bottom). A booster injection was given at day 90 with the same antigen as at day O, but without adjuvant.

to an antigen, is now well accepted, its artificial induction is not always easily obtained. Many reports have been published describing negative results such as those of Bauer et al.,[155] who found no diminished capacity for hemolysin production of rats exposed to sheep erythrocytes before or at the time of birth and challenged with the same antigen 9 to 10 weeks later. Among the various parameters which can be considered for tolerance induction, the nature and the dose of antigen, the repetition of its administration, and the age of the treated individuals are probably the most important.

a. Nature of the Antigen

Unaggregated molecules are more tolerogenic than aggregated ones. Ultracentrifugation of soluble proteins will greatly diminish their immunogeneicity. Reciprocally, slight aggregations such as those obtained by heating at 56°C will increase their immunogeneicity if they are not denaturated by the treatment. Unfortunately, ultracentrifugation cannot be used for particular antigens.

b. Dose of Antigen and Repetition of its Administration

One must know, that tolerance can be obtained with two distinct zones of antigen dosage. Low zone tolerance is induced with doses of antigen which must be repeatedly administered over several weeks. High zone tolerance is achieved with large doses of antigen. Low dose tolerance is often partial and transient. High dose tolerance is quickly obtained and can be long lasting.

The easiest method to induce tolerance is to inject, by the intraperitoneal route, the antigen at birth and to repeat the injections, for example twice a week. The doses must be right. Using monomeric flagellin from Salmonella organisms, Nossal and Ada[156] found with an experimental protocol of daily doses from the birth day for 2 weeks and then injections repeated twice weekly, a tolerance induction in two distinct zones of antigen dosage: minute antigen doses namely 10^{-14} to 10^{-12} g/g body weight/d induce a low zone tolerance and all doses above 10^{-9} g/g body weight/d cause profound high zone tolerance. Doses between these two levels do not induce change in the subsequent immune responsiveness of the treated rats. In order to induce a tolerance, the highest dosage is recommended but evidently it must be available and not toxic for the animals.

c. Age of the Subject When Receiving its First Antigen Administration

To achieve a tolerance, the first administration of antigen must be given as soon as possible after birth. In practice, it is convenient to delay the antigen injection until after the first milk suck of the babies which can easily be observed in the stomach of the young rats. It is better to handle the baby rats with gloves in order to avoid any suspect odor for the mother rat.

d. Results

The immunological status of an animal rendered tolerogenic for an antigen must be considered in statistical terms. Large variations of status from tolerance to immunity must be envisaged. Partial tolerance is a common result. Moreover, it can be transient when the antigen administration is stopped. Thus, tolerance is difficult to obtain.

VI. CONCLUSIONS

Rat immunoglobulins are now well characterized and most of their physicochemical and biological properties, which can be interesting to know when making rat hybridomas or using rat monoclonal antibodies, are readily available.

REFERENCES

1. **Grabar, P., and Courcon, J.,** Etude des sérums de cheval, lapin, rat et souris par l'analyse immunoélectrophorétique, *Bull. Soc. Chim. Biol.*, 40, 1993, 1958.
2. **Escribano, M. J. and Grabar, P.,** L'analyse immunoélectrophorétique du sérum de rat normal, *C.R. Acad. Sci.*, 255, 206, 1962.
3. **Arnason, B. G., de Vaux St. Cyr, C., and Grabar, P.,** Immunoglobulin abnormalities of the thymectomized rats, *Nature (London)*, 199, 1199, 1963.
4. **Arnason, B. G., de Vaux St. Cyr, C., and Relyveld, E. H.,** Role of the thymus in immune reactions in rats.IV.Immunoglobulins and antibody formation, *Int. Arch. Allergy Appl. Immunol.*, 25, 206, 1964.
5. **Binaghi, R. and Sarando de Merlo, E.,** Characterization of rat IgA and its non-identity with the anaphylactic antibody, *Int. Arch. Allergy Appl. Immunol.*, 30, 589, 1966.
6. **Jones, V. E.,** Rat 7 S immunoglobulins: Characterization of gamma2- and gamma1-anti-hapten antibodies, *Immunology*, 16, 589, 1969.
7. **Bazin, H., Beckers, A., and Querinjean, P.,** Three classes and four (sub)classes of rat immunoglobulins: IgM, IgA, IgE and IgG1, IgG2a, IgG2b, IgG2c, *Eur. J. Immunol.*, 4, 44, 1974.
8. **Orlans, E.,** The antigenic interrelations of some mammalian IgG subclasses detected with cross-reacting fowl antisera to human and mouse IgG-Fc, *Immunology*, 28, 761, 1975.
9. **Carter, P. and Bazin, H.,** Immunology, in *The Laboratory Rat*, Vol. 2, Academic Press, New York, 1980, 181.
10. **Nash, D. R., Vaerman, J. P., Bazin, H., and Heremans, J. F.,** Identification of IgA in rat serum and secretions, *J. Immunol.*, 103, 145, 1969.
11. **Orlans, E. and Feinstein, A.,** Detection of alpha, kappa and lambda chains in mammalian immunoglobulins using fowl antisera to human IgA, *Nature (London)*, 233, 45, 1971.
12. **Binaghi, R. A. and Benacerraf, B.,** Production of anaphylactic antibody in the rat, *J. Immunol.*, 92, 920, 1964.
13. **Mota, I.,** Mast cells and anaphylaxis, *Ann. N.Y. Acad. Sci.*, 103, 264, 1963.
14. **Bloch, K. J. and Wilson, R. J.,** Homocytotropic antibody response in the rat infected with the nematode *Nippostrongylus brasiliensis*. III. Characteristics of the antibody, *J. Immunol.*, 100, 629, 1968.
15. **Jones, V. E. and Ogilvie, B. M.,** Reaginic antibodies and immunity to *Nippostrongylus brasiliensis* in the rat. II. Some properties of the antibodies and antigens, *Immunology*, 12, 583, 1967.
16. **Stechschulte, D. J. and Austen, K. F.,** Immunoglobulins of rat colostrum, *J. Immunol.*, 104, 1052, 1970.
17. **Jones, V. E. and Edwards, A. J.,** Preparation of an antiserum specific for rat reagin (rat gamma E?), *Immunology*, 21, 383, 1971.
18. **Bazin, H., Querinjean, P., Beckers, A., Heremans, J. F., and Dessy, F.,** Transplantable immunoglobulin-secreting tumours in rats. IV. Sixty-three IgE-secreting immunocytoma tumours, *Immunology*, 26, 713, 1974.
19. **Ruddick, J. H. and Leslie, G. A.,** Structure and biological functions of human IgD. XI. Identification and ontogeny of a rat lymphocyte immunoglobulin having antigenic cross-reactivity with human IgD, *J. Immunol.*, 118, 1025, 1977.
20. **Bazin, H., Beckers, A., Urbain-Vansanten, G., Pauwels R., Bruyns, C., Tilkin, A. F., Platteau, B., and Urbain, J.,** Transplantable IgD immunoglobulin-secreting tumours in rats, *J. Immunol.*, 121, 2077, 1978.
21. **Banovitz, J. and Ishizaka, K.,** Detection of five components having antibody activity in rat antisera, *Proc. Soc. Exp. Biol. Med.*, 125, 78, 1967.
22. **Austin, C. M.,** Patterns of migration of lymphoid cells, *Aust. J. Exp. Biol. Med. Sci.*, 46, 581, 1968.
23. **Bistany, T. S. and Tomasi, T. B.,** Characterization of rat immunoglobulin A, *Fed. Proc. Fed. Am. Soc. Exp. Biol.*, 28, 280, 1969.
24. **Bistany, T. S. and Tomasi, T. B.,** Serum and secretory immunoglobulins of the rat, *Immunochemistry*, 7, 453, 1970.
25. **Stechschulte, D. J., Orange, R. P., and Austen, K. F.,** Immunochemical and biologic properties of rat IgE.I.Immunochemical identification of rat IgE, *J. Immunol.*, 104, 1082, 1970.
26. **Bazin, H., Beckers, A., Deckers, C., and Moriamé, M.,** Transplantable immunoglobulin-secreting tumours in rats.V.Monoclonal immunoglobulins secreted by 250 ileocecal immunocytomas of the LOU/Wsl rats, *J. Natl. Cancer Inst.*, 51, 1359, 1973.
27. World Health Organization, Nomenclature of human immunoglobulins, *Bull. W.H.O.*, 48, 373, 1973.
28. **Brüggemann, M., Free, J., Diamond, A., Hobard, J., Cobbold, S., and Waldmann, H.,** Immunoglobulin heavy chain locus of the rat: stricking homology to mouse antibody genes, *Proc. Natl. Acad. Sci. U.S.A.*, 83, 6075, 1986.
29. **Ollo, R., Auffray, C., Morchamps, C., and Rougeon, F.,** Comparison of mouse immunoglobulin gamma 2a and gamma 2b chain genes suggests that exons can be exchanges between genes in a multigenic family, *Proc. Natl. Acad. Sci. U.S.A.*, 78, 2442, 1981.
30. **Der Balian, G., Slack, J., Clevinger, B., Bazin, H., and Davie, J. M.,** Antigenic similarities of rat and mouse IgG subclasses associated with anti-carbohydrate specificites, *Immunogenetics*, 152, 209, 1980.

31. **Moon, N., Der Balian, G., Venturini, D., Bazin, H., and Davie, J. M.,** Antigenic-similarities of rat and mouse IgG subclasses associated with anti-carbohydrate specificities, *Immunogenetics,* 11, 199, 1980.
32. **Brüggemann, M., Delmastro-Galgre, P., Waldmann, H., and Calabi, F.,** Sequence of a rat immunoglobulin gamma 2c heavy chain constant region cDNA: extensive homology to mouse gamma 3, *Eur. J. Immunol.,* 18, 317, 1988.
33. **Querinjean, P., Bazin, H., Beckers, A., Deckers, C., Heremans, J. F., and Milstein, C.,** Transplantable immunoglobulin-secreting tumors in rats. Purification of chemical characterization of four kappa chains from LOU/Wsl rats, *Eur. J. Biochem.,* 31, 354, 1972.
34. **Querinjean, P., Bazin, H., Kehoe, J. M., and Capra, J. D.,** Transplantable immunoglobulin-secreting tumours in rats.VI.N-terminal sequence variability in LOU/Wsl rat monoclonal heavy chains, *J. Immunol.,* 114, 1375, 1975.
35. **Hood, L., Gray, W. R., Sanders, B. G., and Dreyer, W. J.,** Light chain evolution, *Cold Spring Harbor Symp. Quant. Biol.,* 22, 133, 1967.
36. **Oudin, J.,** L'allotypie de certains antigènes protéidiques du sérum, *C.R. Acad. Sci.,* 242, 2606, 1956.
37. **Barabas, A. Z. and Kelus, A. S.,** Allotypic specificity of serum protein in inbred strains of rats, *Nature (London),* 215, 155, 1967.
38. **Wistar, R., Jr.,** Immunoglobulin allotype in the rat. Localization of the specificity to the light chain, *Immunology,* 17, 23, 1969.
39. **Armerding, D.,** Two allotypic specificities of rat immunoglobulins, *Eur. J. Immunol.,* 1, 39, 1971.
40. **Gutman, G. A. and Weisman, I. L.,** Inheritance and strain distribution of a rat immunoglobulin allotype, *J. Immunol.,* 107, 1390, 1971.
41. **Humphrey, R. L. and Santos, G. S.,** Serum protein allotype markers in certain inbred rat strains, *Fed. Proc. Fed. Am. Soc. Exp. Biol.,* 30, 248, 1971.
42. **Rokhlin, O. V., Vengerova, T. L., and Nezlin, R. S.,** rl Allotypes of light chains of rat immunoglobulins, *Immunochemistry,* 8, 525, 1971.
43. **Beckers, A., Querinjean, P., and Bazin, H.,** Allotypes of rat immunoglobulins. II. Distribution of the allotypes of kappa and alpha chain loci in different inbred strains of rat, *Immunochemistry,* 11, 605, 1974.
44. **Nezlin, R. S., Vengerova, T. I., Rockhlin, O. V., and Machulla, H. K. G.,** Localization of allotypic markers of kappa light chains of rat immunoglobulins in the constant part of the chain, *Fed. Eur. Biochem. Soc. Meet. (Proc.),* 36, 93, 1974.
45. **Gutman, G. A., Bazin, H., Rockhlin, C. V., and Nezlin, R. S.,** A standard nomenclature for rat immunoglobulin allotypes, *Transplant. Proc.,* 15, 1685, 1983.
46. **Gutman, G. A., Loh, E., and Hood, L.,** Structure and regulation of immunoglobulins: kappa allotypes in the rat have multiple amino acid differences in the constant region, *Proc. Natl. Acad. Sci. U.S.A.,* 72, 5046, 1975.
47. **Bazin, H., Beckers, A., Vaerman, J. P., and Heremans, J. F.,** Allotypes of rat immunoglobulins. I. An allotype at the alpha-chain locus, *J. Immunol.,* 112, 1035, 1974.
48. **Beckers, A. and Bazin, H.,** Allotypes of rat immunoglobulins. III. An allotype of the gamma 2b-chain locus, *Immunochemistry,* 12, 671, 1975.
49. **Leslie, G. A.,** Allotypic determinants (Igh-3) associated with the IgG2c subclass of rat immunoglobulins, *Mol. Immunol.,* 21, 577, 1984.
50. **Cramer, D. V.,** Biochemicaloci of the rat, *Rat Newsl.,* 18, 3, 1987.
51. **Pear, W. S., Wahlström, G., Szpirer, J., Levan, G., Klein, G., and Sumegi, J.,** Localization of the rat immunoglobulin heavy chain locus to chromosome 6, *Immunogenetics,* 23, 393, 1986.
52. **Perlmann, G., Sumegi, J., Szpirer, J., Levan, G., and Klein, G.,** The rat immunoglobulin kappa light chain locus is on chromosome 4, *Immunogenetics,* 22, 97, 1985.
53. **Bazin, H. and Beckers, A.,** IgE myelomas in rats, Nobel Symposium No. 33, Molecular and Biological Aspects of the Acute Allergic Reaction, Stockholm, Johansson, S. G. O., Stranberg, K., and Uvnas B., Eds., Plenum Company, New York, 1976, 125.
54. **Bazin, H.,** The secretory antibody system, in *Immunological Aspects of the Gastro-intestinal Tract and Liver,* Ferguson, A. and McSween, R. N. M., Eds., Med. Tech. Publ., Lancaster, England, 1976, 32.
55. **Hunt, S. V., and Williams, A. F.,** The origin of cell surface immunoglobulin of marrow-derived and thymus-derived lymphocytes of the rat, *J. Exp. Med.,* 139, 479, 1974.
56. **Williams, A. F. and Gowans, J. L.,** The presence of IgA on the surface of rat thoracic duct lymphocytes which contains internal IgA, *J. Exp. Med.,* 142, 335, 1975.
57. **Nezlin, R. and Rokhlin, O.,** Allotypes of light chains of rat immunoglobulins and their application to the study of antibody biosynthesis, *Contemp. Top. Mol. Immunol.,* 5, 161, 1976.
58. **Bazin, H., Cormont, F., and De Clercq, L.,** Rat monoclonal antibodies.II.Rapid and efficient method of purification from ascitic fluid or serum, *J. Immunol. Methods,* 71, 9, 1984.
59. **Oriol, R., Binaghi, R., and Coltorti, E.,** Valence and association constant of rat macroglobulin antibody, *J. Immunol.,* 104, 932, 1971.

60. **Kobayashi, K., Vaerman, J. P., Bazin, H., Lebacq-Verheyden, A. M., and Heremans, J. F.,** Identification of J-chain in polymeric immunoglobulins from a variety of species by cross-reaction with rabbit antisera to human J-chain, *J. Immunol.,* 111, 1590, 1973.
61. **Vaerman, J. P., Heremans, J. F., Bazin, H., and Beckers, A.,** Identification and some properties of rat secretory component, *J. Immunol.,* 114, 265, 1975.
62. **Bazin, H., Beckers, A., Moriamé, M., Platteau, B., Naze-De Mets, J., and Kints, J. P.,** LOU/C/Wsl rat monoclonal immunoglobulins, *FEBS Proc. Meet.,* 86, 117, 1974.
63. **Cremer, N. E., Taylor, D. O. N., Lennette, E. H., and Hagens, S. J.,** IgM production in rats infected with Moloney leukemia virus, *J. Natl. Cancer Inst.,* 51, 905, 1973.
64. **Rousseaux, J. and Bazin, H.,** Rat immunoglobulins, *Vet. Immunol. Immunopathol.,* 1, 61, 1979.
65. **Conrad, D. H., Bazin, H., Sehon, A. H., and Froese, A.,** Binding parameters of the interaction between rat IgE and rat mast cell receptors, *J. Immunol.,* 114, 1688, 1975.
66. **Isersky, C., Kulczycki, Jr., A., and Metzger, H.,** Isolation of IgE from reaginic rat serum, *J. Immunol.,* 112, 1901, 1974.
67. **Bradford, M. M.,** A rapid and sensitive method for the quantitation of microgram quantities of protein utilizing the principle of protein-dye binding, *Anal. Biochem.,* 72, 248, 1976.
68. **Rousseaux, J., Bazin, H., and Biserte, G.,** Differences in sensitivity to reduction of rat immunoglobulin G subclass, *FEBS Lett.,* 98, 359, 1979.
69. **Virella, G. and Parkhouse, R. M. E.,** Sensitivity to reduction of human immunoglobulin G of different heavy chain subclass, *Immunochemistry,* 10, 213, 1973.
70. **Williamson, A. R. and Askonas, B. A.,** Differential reduction of inter-chain disulfide bonds of mouse immunoglobulin G, *Biochem. J.,* 107, 823, 1968.
71. **Rousseaux-Prévost, R., Rousseaux, J., Bazin, H., and Biserte, G.,** Differential reduction of the inter-chain disulfide bonds of rat immunoglobulin E: relation to biological activity, *Mol. Immunol.,* 21, 233, 1984.
72. **Takatsu, R., Ishizaka, T., and Ishizaka, K.,** Biological significance of disulfide bonds in human IgE molecules, *J. Immunol.,* 114, 1838, 1975.
73. **Rousseaux, J., Biserte, G., and Bazin, H.,** The differential enzyme sensitivity of rat immunoglobulin G subclasses to papain and pepsin, *Mol. Immunol.,* 17, 469, 1980.
74. **Rousseaux, J., Rousseaux-Prévost, R., Bazin, H., and Biserte, G.,** Tryptic cleavage of rat IgG: a comparative study between subclasses, *Immunol. Lett.,* 3, 93, 1981.
75. **Rousseaux, J., Rousseaux-Prévost, R., Bazin, H., and Biserte, G.,** Proteolysis of rat IgG subclasses by S. aureus V8 proteinase, *Biochim. Biophys. Acta,* 748, 205, 1983.
76. **Bennich, H. H., Ellerson, J. R., and Karlsson, T.,** Evaluation of basic serum IgE levels and the IgE antibody response in the rat by radioimmunoassays, *Immunol. Rev.,* 41, 261, 1978.
77. **Karlsson, T., Ellerson, J. R., Dahlbom, I., and Bennich, H.,** Analysis of the serum IgE levels in non-immunized rats of various strains by a radioimmunoassay, *Scand. J. Immunol.,* 9, 217, 1979.
78. **Perez-Montfort, R. and Metzger, H.,** Proteolysis of soluble IgE-receptor complexes: localization of sites on IgE which interact with the Fc receptor, *Mol. Immunol.,* 19, 1113, 1982.
79. **Rousseaux-Prévost, R., Rousseaux, J., and Bazin, H.,** Studies of the IgE binding sites to rat mast cell receptor with proteolytic enzymes and with a monoclonal antibody directed against epsilon heavy chain: evidence that the combining sites are located in the C epsilon 3 domain, *Mol. Immunol.,* 24, 187, 1987.
80. **Rousseaux, J., Aubert, J. P., and Loucheux-Lefèbvre, M. H.,** Comparative study of the conformational features of rat immunoglobulin G subclasses by circular dichroism, *Biochim. Biophys. Acta,* 701, 93, 1982.
81. **Prouvost-Danon, M., Abadie, A., and Da Ponte, J. V.,** Heat inactivation of rat and mouse IgE, *Mol. Immunol.,* 17, 247, 1980.
82. **Binaghi, R. A. and Demeleumster, C.,** Influence of the medium on the heat and acid denaturation of IgE, *J. Immunol. Methods,* 65, 225, 1983.
83. **Rousseaux-Prévost, R., Rousseaux, J., Bazin, H., and Biserte, G.,** Formation of biologically inactive polymers is responsible for the thermal inactivation of rat IgE, *Int. Archs. Allergy Appl. Immunol.,* 70, 268, 1983.
84. **Burt, D. S., Hastings, G. Z., and Stanworth, D. R.,** Use of synthetic peptides in the production and characterization of antibodies directed against predetermined specificities in rat immunoglobulin E., *Mol. Immunol.,* 23, 181, 1986.
85. **Starace, V.,** Structure Primaire de Deux Chaînes Légères Monoclonales de Type Kappa de Rat LOU/Wsl et ses Implications Phylogéniques, Ph.D. thesis, University of Louvain, Brussels, Belgium, 1977.
86. **Sheppard, H. W. and Gutman, G. A.,** Allelic forms of rat kappa chain genes: evidence for strong selection at the level of nucleotide sequence, *Proc. Natl. Acad. Sci. U.S.A.,* 78, 7064, 1981.
87. **Svasti, J. and Milstein, C.,** The complete amino acid sequence of a mouse kappa light chain, *Biochem. J.,* 128, 427, 1972.
88. **Dayhoff, M. D.,** *Atlas of Protein Sequence and Structure,* Vol. 5, National Biomedical Research Foundation, Silver Spring, MD, 1976, 180.

89. **Starace, V. and Querinjean, P.,** The primary structure of a rat kappa Bence Jones protein: phylogenetic relationships of V- and C-region genes, *J. Immunol.,* 115, 59, 1975.
90. **Alcaraz, G., Bourgois, A., Moulin, A., Bazin, H., and Fougereau, F.,** Partial structure of a rat IgD molecule with a deletion in the heavy chain, *Ann. Immunol. (Inst. Pasteur),* 131C, 3031, 1980.
91. **Kabat, E. A., Wu, T. T., Reid-Miller, M., Perry, H. M., and Gottesman, K. S.,** Sequences of proteins of immunological interest, U.S. Department of Health and Human Services, NIH, 4th ed., Bethesda, MD, 1987.
92. **Querinjean, P., Bazin, H., Starace, V., Beckers, A., Deckers, C., and Heremans, J. F.,** Lambda light chains in rat immunoglobulin, *Immunochemistry,* 10, 653, 1973.
93. **Schulenburg, E. P., Simms, E. S., Lynch, R. G., Bradshaw, R. A., and Eisen, H. N.,** Amino acid sequence of the light chain from a mouse myeloma protein with anti-hapten activity: evidence for a third type of light chain, *Proc. Natl. Acad. Sci. U.S.A.,* 68, 2623, 1971.
94. **Sheppard, H. W. and Gutman, G. A.,** Rat kappa-chain J-segment genes: two recent gene duplications separate rat and mouse, *Cell,* 29, 121, 1982.
95. **Breiner, A. V., Brandt, C. R., and Schechter, I.,** Somatic DNA rearrangement generates functional rat immunoglobulin kappa chain genes: the J kappa gene cluster is longer in rat than in mouse, *Gene,* 18, 165, 1982.
96. **Burstein, Y., Breiner, A. V., Brandt, C. R., Milcarek, C., Sweet, R. W., Warszawski, D., Ziv, E., and Schechter, J.,** Recent duplication and germ-line diversification of rat immunoglobulin kappa chain gene joining segments, *Proc. Natl. Acad. Sci. U.S.A.,* 79, 5993, 1982.
97. **Max, E. E., Maizel, J. V., and Leder, P.,** The nucleotide sequence of a 5.5 kilobase DNA segment containing the mouse kappa immunoglobulin J and C region genes, *J. Biol. Chem.,* 256, 5116, 1981.
98. **Sakano, H., Huppi, K., Henrich, G., and Tonegawa, S.,** Sequences at the somatic recombination sites of immunoglobulin light chain genes, *Nature (London),* 280, 288, 1979.
99. **Hieter, P. A., Maizel, J. V., and Leder, P.,** Evolution of human immunoglobulin kappa J region genes, *J. Biol. Chem.,* 257, 1516, 1982.
100. **Wang, A. C., Fudenberg, H. H., and Bazin, H.,** The nature of "species-specific" amino acid residues, *Immunochemistry,* 12, 505, 1975.
101. **Wang, A. C., Fudenberg, H. H., and Bazin, H.,** Partial amino acid sequences of kappa chains of rat immunoglobulins: genetic and evolutionary implications, *Biochem. Genet.,* 14, 209, 1976.
102. **Wiener, F., Babovitz, M., Spira, J., Klein, G., and Bazin, H.,** Non-random chromosomal changes involving chromosomes 6 and 7 spontaneous rat immunocytomas, *Int. J. Cancer,* 29, 431, 1982.
103. **Shimizu, A., Takahashi, V., Yaoita, Y., and Honjo, T.,** Organization of the constant region gene family of the mouse immunoglobulin heavy chain, *Cell,* 28, 499, 1982.
104. **Flanagan, J. G. and Rabbits, T. H.,** Arrangement of human immunoglobulin heavy chain constant region genes implies evolutionary duplication of a segment containing gamma, epsilon and alpha genes, *Nature (London),* 300, 709, 1982.
105. **Ravetch, J. V., Siebenust, V., Korsmeyer, S., Waldmann, T., and Leder, P.,** Structure of the human immunoglobulin mu locus: characterization of embryonic and rearranged J and D genes, *Cell,* 27, 583, 1982.
106. **Honjo, T.,** Immunoglobulin genes, *Ann. Rev. Immunol.,* 1, 499, 1983.
107. **Alcaraz, G., Collé, A., Boned, A., Kahn-Perles, B., Sire, J., Bazin, H., and Bourgois, A.,** Tryptic and plasmic cleavage of a rat myeloma IgD, *Mol. Immunol.,* 18, 249, 1981.
108. **Spiegelberg, H. L.,** The structure and biology of human IgD, *Immunol. Rev.,* 37, 1, 1977.
109. **Tucker, P. W., Liu, C. P., Mushinski, J. F., and Blattner, F. R.,** Mouse immunoglobulin D: messenger RNA and genomic DNA sequences, *Science,* 209, 1353, 1980.
110. **Cheng, H. L., Blattner, F. R., Fitzmaurice, L., Mushinski, J. F., and Tucker, P. W.,** Structure of genes for membrane and secreted murine IgD heavy chains, *Nature (London),* 296, 410, 1982.
111. **Lin, L. C. and Putnam, F. W.,** Primary structure of the Fc region of human immunoglobulin D. Implications for evolutionary origin and biological function, *Proc. Natl. Acad. Sci. U.S.A.,* 78, 504, 1981.
112. **Putnam, F. W., Takahashi, N., Tetaert, D., Debuire, B., and Lin, L. C.,** Amino acid sequence of the first constant region domain and the hinge region of the delta heavy chain of human IgD, *Proc. Natl. Acad. Sci. U.S.A.,* 78, 6168, 1981.
113. **Sire, J., Alcaraz, G., Bourgois, A., and Jordan, B. R.,** Rat IgD myeloma protein: cell-free translation of the delta-mRNA and biochemical analysis of intracellular and membrane delta chain, *Eur. J. Immunol.,* 11, 632, 1981.
114. **Sire, J., Auffray, C., and Jordan, B. R.,** Rat immunoglobulin delta heavy chain gene: nucleotide sequence derived from cloned cDNA, *Gene,* 20, 377, 1982.
115. **Mushinski, J. F., Blattner, F. R., Owens, J. D., Finkelman, F. D., Kessler, S. W., Fitzmaurice, L., Potter, M., and Tucker, P. W.,** Mouse immunoglobulin D: construction of a cloned delta chain cDNA and its identification using hybridization selection of mRNA, *Proc. Natl. Acad. Sci. U.S.A.,* 77, 7405, 1980.
116. **Moore, K. W., Rogers, J. Hunkapiller, T., Early, P., Nottenburg, C., Weissman, I., Bazin, H., Wall, R., and Hood, L.,** Expression of IgD may use both DNA rearrangement and RNA splicing mechanisms, *Proc. Natl. Acad. Sci. U.S.A.,* 78, 1800, 1981.

117. **Maki, R., Roeder, W., Traunecker, A., Sidman, C., Wabl, H., Raschke, W., and Tonegawa, S.,** The role of DNA rearrangement and alternative RNA processing in the expression of immunoglobulin delta genes, *Cell,* 24, 353, 1981.
118. **Fitzmaurice, L., Owens, J., Blattner, F. R., Chen, H. L., Tucker, P. W., and Mushinski, J. F.,** Mouse spleen and IgD-secreting plasmacytomas contain multiple IgD delta chain RNAs, *Nature (London),* 296, 459, 1982.
119. **Johansson, S. G. O. and Bennich, H.,** Immunological studies of an atypical (myeloma) immunoglobulin, *Immunology,* 13, 381, 1967.
120. **Hellman, L., Pettersson, U., and Bennich, H.,** Characterization and molecular cloning of the mRNA for the heavy (epsilon) chain of rat immunoglobulin E, *Proc. Natl. Acad. Sci. U.S.A.,* 79, 1264, 1982a.
121. **Hellman, L., Pettersson, U., Engström, A., Karlsson, T., and Bennich, H.,** Structure and evolution of the heavy chain from rat immunoglobulin E., *Nucl. Acids Res.,* 10, 604, 1982.
122. **Steen, M. L., Hellman, L., and Pettersson, U.,** Rat immunoglobulin E heavy chain locus, *J. Mol. Biol.,* 177, 19, 1984.
123. **Ishida, N., Ueda, S., Hayashida, H., Miyata, T., and Honjo, T.,** The nucleotide sequence of the mouse immunoglobulin epsilon gene: comparison with the human epsilon gene sequence, *EMBO J.,* 1, 1117, 1982.
124. **Max, E. E., Battey, J., Ney, R., Kirsch, I. R., and Leder, P.,** Duplication and deletion in the human immunoglobulin E genes, *Cell,* 29, 691, 1982.
125. **Davies, D. R. and Metzger, H.,** Structural basis of antibody function, *Ann. Rev. Immunol.,* 1, 87, 1983.
126. **Flanagan, J. G. and Rabbits, T. H.,** The sequence of a human immunoglobulin epsilon heavy chain constant region gene, and evidence for three non-allelic genes, *EMBO J.,* 1, 655, 1982.
127. **Nishida, Y., Miki, T., Hisajima, H., and Honjo, T.,** Cloning of human immunoglobulin epsilon chain gene: evidence for multiple C-epsilon-genes, *Proc. Natl. Acad. Sci. U.S.A.,* 79, 3833, 1982.
128. **Ueda, S., Nakai, S., Nishida, Y., Hisajima, H., and Honjo, T.,** Long terminal repeat-like elements flank a human immunoglobulin epsilon pseudogene that lacks introns, *EMBO J.,* 1, 1539, 1982.
129. **Capra, J. D., Wasserman, R. L., and Kehoe, J. M.,** Phylogenetically associated residues within the VHIII subgroup of several Mammalian species, *J. Exp. Med.,* 138, 410, 1973.
130. **Klapper, D. G., Stankus, R. P., Leslie, G. A., and Capra, J. D.,** Structural studies on induced antibodies with defined idiotypic specificites. IV. The heavy chains of anti-Streptococcal group A carbohydrate antibodies from Sprague-Dawley rats bearing a cross-reacting idiotype, *Scand. J. Immunol.,* 5, 925, 1976.
131. **Francis, S. H., Leslie, R. G. Q., Hood, L., and Eisen, H. N.,** Amino acid sequence of the variable region of the heavy (alpha) chain of a mouse myeloma protein with anti-hapten activity, *Proc. Natl. Acad. Sci. U.S.A.,* 71, 1123, 1974.
132. **Capra, J. D., Wasserman, R. L., Querinjean, P., and Kehoe, J. M.,** Structure and evolution of immunoglobulin heavy chain variable regions, in *Antibody Structure and Molecular Immunology — Proceedings of the Ninth FEBS Meeting,* Gergely, J. G. and Medgyesi, G. A., Eds., North-Holland/Elsevier, 1975, 3.
133. **Nussenzweig, V. and Binaghi, R. A.,** Heterogeneity of rat immunoglobulins, *Int. Arch. Allergy Appl. Immunol.,* 27, 355, 1965.
134. **Tada, T., Okumura, K., Platteau, B., Beckers, A., and Bazin, H.,** Half-lives of two types of rat homocytotropic antibodies in the circulation and in the skin, *Int. Arch. Allergy Appl. Immunol.,* 48, 116, 1975.
135. **Binaghi, R. A., Oettgen, H. F., and Benacerraf, B.,** Anaphylactic antibody in the young rat, *Int. Arch. Allergy Appl. Immunol.,* 29, 105, 1966.
136. **McGhee, J. R., Michalek, S. M., and Ghanta, V. K.,** Rat immunoglobulins in serum and secretions: purification of rat IgM, IgA, and IgG and their quantitation in serum, colostrum, milk, and saliva, *Immunochemistry,* 12, 817, 1975.
137. **Michalek, S. M., Rahman, A. F. R., and McGhee, J. R.,** Rat immunoglobulins in serum and secretions: comparison of IgM, IgA and IgG in serum colostrum milk and saliva of protein malnourished and normal rats, *Proc. Soc. Exp. Biol. Med.,* 148, 1114, 1975.
138. **Binaghi, R. A. and Boussac-Aron, Y.,** Isolation and properties of a7S rat immunoglobulin different from IgG, *Eur. J. Immunol.,* 5, 194, 1975.
139. **Halpern, J. and Metzger, H.,** The interaction of IgE with rat basophilic leukemia cells. VI. Inhibition by IgGa immune complexes, *Immunochemistry,* 13, 907, 1976.
140. **Moorse, H. C., Bloch, K. J., and Austen, K. F.,** Biologic properties of rat antibodies.II.Time-course of appearance of antibodies involved in antigen-induced release of slow reacting substance of anaphylaxis (SRS-A); association of this activity with rat IgGa, *J. Immunol.,* 101, 658, 1968.
141. **Jarrett, E. E. E. and Bazin, H.,** Serum immunoglobulin levels in Nippostrongylus basiliensis infection, *Clin. Exp. Immunol.,* 30, 330, 1977.
142. **Bazin, H., Platteau, B., Pauwels, R., and Capron, A.,** Immunoglobulin production in nude rats with special attention to the IgE isotype, *Ann. Immunol. (Inst. Pasteur),* 131C, 31, 1980.
143. **Rowley, D. A. and Fitch, F. W.,** The mechanism of tolerance produced in rats to sheep erythrocytes, *J. Exp. Med.,* 121, 671, 1964.
144. **Williams, G. M.,** Ontogeny of the immune response. II. Correlations between the development of the afferent and efferent limbs, *J. Exp. Med.,* 124, 57, 1966.

145. **Pauwels, R., Bazin, H., Platteau, B., and Van der Straeten, M.,** The effect of age on IgE production in rats, *Immunology,* 36, 145, 1979.
146. **Galassi, N. V. and Nota, N. R.,** Ontogenia de la respuesta de anticuerpos contra antigenos no patogenos en la rata, *Catedra de Microbiologia, Parasitologia e Immunologia,* 6, 112, 1987.
147. **Halliday, R.,** The relationship between the occurence of mortality and the development of active immunity in the young rat, *Proc. R. Soc.,* 161, 208, 1964.
148. **Hervey, E. J.,** The immune response to a killed bacterial antigen in foetal and neonatal rats, *Immunology,* 2, 589, 1966.
149. **Winebright, J., and Fitch, F. W.,** Antibody formation in the rat I agglutinin response to particulate flagella from salmonella typhosa, *J. Immunology,* 89, 891, 1962.
150. **Nossal, G. J. V., Ada, G. L., and Austin, C. M.,** Antigens in immunity. II. Immunogenic properties of flagella, polymerized flagellin and flagellin in the primary response, *Aust. J. Exp. Biol. Med. Sci.,* 42, 283, 1964.
151. **Ishizaka, K. and Okudaira, H.,** Reaginic antibody formation in the mouse.II.Enhancement formation by priming with carrier, *J. Immunol.,* 110, 1067, 1973.
152. **Pauwels, R., Bazin, H., Platteau, B., and Van der Straeten, M.,** The influence of different adjuvants on the production of IgD and IgE antibodies, *Ann. Immunol. (Inst. Pasteur),* 130C, 49, 1979.
153. **Humphrey, J. H.,** Tolerogenic or immunogenic activity of hapten conjugated polysaccharides correlated with cellular localization, *Eur. J. Immunol.,* 11, 212, 1981.
154. **Nakane, P. J., and Kawai, T. J.,** Peroxydase-labeled antibody. A new method of conjugation, *J. Histochem. Cytochem.,* 22, 1084, 1974.
155. **Bauer, J. A., Peckhman, P. E., and Osler, A. G.,** Effect of fetal or neonatal immunization on antibody response of the adult rat, *Proc. Soc. Exp. Biol. Med.,* 92, 714, 1956.
156. **Nossal, G. J. V. and Ada, G. L.,** The role of antigens in immunological tolerance, in *Antigens, Lymphoid Cells, and the Immune Response,* Academic Press, New York and London, 1971, 196.

Chapter 2

THE LOUVAIN (LOU) RATS

H. Bazin

TABLE OF CONTENTS

I. ORIGIN

The first LOUVAIN rats were imported into the University of Louvain as a unique couple[1] around 1937[2] or 1941.[3] They were designated to be of Wistar origin but without any precise data, and in fact, without certitude. For a long time they were bred as an outbred stock and were usually named "L".

Radiobiology and cancer studies were carried out on them and for this reason, they were kept for a long time in order to check spontaneous or induced-tumor incidence. From 1955, Maisin et al.[3,4,5] and Maldague et al.[6] identified special tumors appearing in the ileocecal area of these rats. They were of lymphoid origin and were called "leucosarcoma" or "undifferentiated myeloma". A few years later, these authors showed the secretion of "paraproteins" by some of these tumors.[7,8]

In 1970, Bazin and Beckers started an exhaustive study of these lymphoid tumors. They bred rats obtained from various nuclei kept in different laboratories of the Faculty of Medicine of the University of Louvain. They were all derived from the original breeding of the late Professor Joseph Maisin. Bazin and Beckers bred 28 different and distinct lines of animals which were observed for their tumor incidence in the ileocecal area. The model was named LOUVAIN in honor of the university, and abbreviated to LOU to satisfy the international nomenclature rules.[9] Two sublines were chosen from the 28 lines, and the LOU/C presenting the maximum ileocecal lymphoid incidence was chosen as the histocompatibility reference, while the LOU/M was selected for its low tumor incidence and good fertility. To date (1988), they have been mated brother × sister for 33 generations.

Both LOU/C and LOU/M are strictly histocompatible as it can be judged by reciprocal skin graft indefinite acceptance. WSL, which can be added to LOU/C or LOU/M, refers to Bazin's own experimental animal colony.[10] WSL are the initials of Woluwé Saint Lambert, the Brussels suburb where the Faculty of Medicine of the University of LOUVAIN is located.

II. THE LOUVAIN RATS

The LOU/C as well as the LOU/M rats are albino rats (Figure 1, A and B). The distribution pattern of LOUVAIN rats for the 28 established rat polymorphisms is given in Bender et al.[11] Their immunoglobulin allotype markers were published by Bazin et al.,[12] Beckers et al.,[13] Beckers and Bazin,[14] Gutman et al.[15] (Table 1).

III. CONGENIC STRAINS ON THE LOUVAIN BACKGROUNDS

A number of congenic strains have been developed by introducing gene(s) from various rat strains into the LOU/C genetic background by a series of crosses. The most commonly used is the LOU/C.IgK1b(OKA) strain which is a LOU/C having the kappa gene loci from the OKA strain (Table 2) and commonly named LOU/C.IgK-1b.

IV. HANDLING, REPRODUCTIBILITY, AND GROWTH

The LOUVAIN rats are easy to handle as they are fairly placid (Figure 2). However, in the animal house they must be treated as domestic animals. They must be taken directly with the hands without gloves as much as possible, and by the body, from underneath or above, not by the skin or tail. They require a quiet atmosphere with as little stress as possible. They are easy to breed. The litter size of two rat colonies located in the Institut Pasteur de Lille in France, and in our laboratory is given in Table 3. By comparison, the litter sizes are 5.2 and 5.3, respectively, for BALB/c and BALB/cJ16 (Table 4).

Figures 3 and 4 give the growth curves obtained for male and female rats of the LOU/C and LOU/C.IgK-1b strains.

A

B

FIGURE 1. (A) Three-month-old LOU/C rats. (B) Three-month-old mother LOU/C rat with one of her baby rats.

TABLE 1
Allotype Markers of LOU/C, LOU/M, and OKA Rat Strains

Immunoglobulin chains	Kappa	Alpha	Gamma2b
Loci	IgK-1	IgH-1	IgH-2
Allele a	LOU/C, LOU/M	LOU/C, LOU/M	LOU/C,LOU/M
Allele b	DA, OKA	OKA	OKA

All operative procedures such as injections or bleedings are made under ether anesthesia. A convenient device is a glass container with a pledget of cotton soaked in ether. The method is good, but can be dangerous as the percentage of ether can increase in the bottle and can become so high that the rats can be killed by asphyxia. A much better system is to use an apparatus similar

TABLE 2
LOU/C Rats and Their Congenic for the Kappa Allotype: LOU/C.IgK-1b(OKA)

	Genetic background	Kappa allotype	Histocompatibility
LOU/C	LOU/C	IgK-1a	LOU/C
LOU/C.IgK-1b(OKA)	LOU/C	IgK-1b	LOU/C
OKA	OKA	IgK-1b	OKA

FIGURE 2. This female LOU/C rat has just been taken at random from animal breeding. It is not frightened and it seems interested in its new environment.

to that shown in Figure 5, where compressed air passes through ether in the first flask and then goes across the dessicator to be released outside. The system is convenient and allows keeping rats in the dessicator for a long time without risk of death. The compressor can be a small pump or an air pump of the type used for a domestic aquarium (F. Billaut, Institut Pasteur de Lille).

V. AVAILABILITY OF THE LOUVAIN RATS

The LOUVAIN rats have always been freely distributed from our colony. LOU/C and LOU/M rats can be obtained:

1. Federal Republic of Germany — Zentralinstitut für Verzuchstierzucht, Herman Ehlers Alle 57, 3000 Hannover 91

TABLE 3
Reproduction of LOU/C, LOU/M, and Congenic Rats to Them

Strains	Location of breeding	No. of litter	No. of babies at birth	No. of young at weaning	No. of young per litter at weaning
LOU/C	Experimental Immunology Unit University of Louvain	100	728	720	7.20
LOU/C	Institut Pasteur de Lille[a]	183	1404	1228	6.71
LOU/M	Institut Pasteur de Lille[a]	225	1848	1504	6.68
LOU/C.IgK-1b	Experimental Immunology Unit University of Louvain	100	694	674	6.74

[a] Data given by Billaut, F., personal communication.

TABLE 4
Reproduction Characteristics of BALB/c Mice and LOUVAIN Rats

	Inbred strains	Gestation time	First mating	Litter at weaning
BALB/c	yes	21 d	6—8 weeks	5
LOU/C or LOU/M	yes	21—22 d	8—10 weeks	6-7

2. France — Centre National de la Recherche Scientifique, Centre de Sélection et d'Elevages des Animaux de Laboratoire, 45045 Orléans-la-Source
3. Japan — Central Institute for Experimental Animals, 1430 Nogawa, Kawasaki, Kanagawa 213
4. People's Republic of China — Medical Laboratory Animal Center, Chinese Academy of Medical Sciences, Dong Dan San Tiao 9, Beijing
5. U.S. — National Institute of Health, Small Animal Section - Veterinary Resources Branch, Division of Research Services, Building 14A, Room 103, Bethesda, MD 20205 (Dr. K. HANSEN)

They may also be obtained from many animal houses of industries or universities and from commercial breeders (for example, Harlan OLAC Ltd, Bicester, U.K. or Charles River, U.S.).

Evidently, nucleus breeding of LOU/C, LOU/M, and LOU/C.IgK-1b(OKA) can be obtained upon request at the Experimental Immunology Unit, Faculty of Medicine, University of Louvain, 30/56 Clos Chapelle aux Champs, Brussels 1200, Belguim.

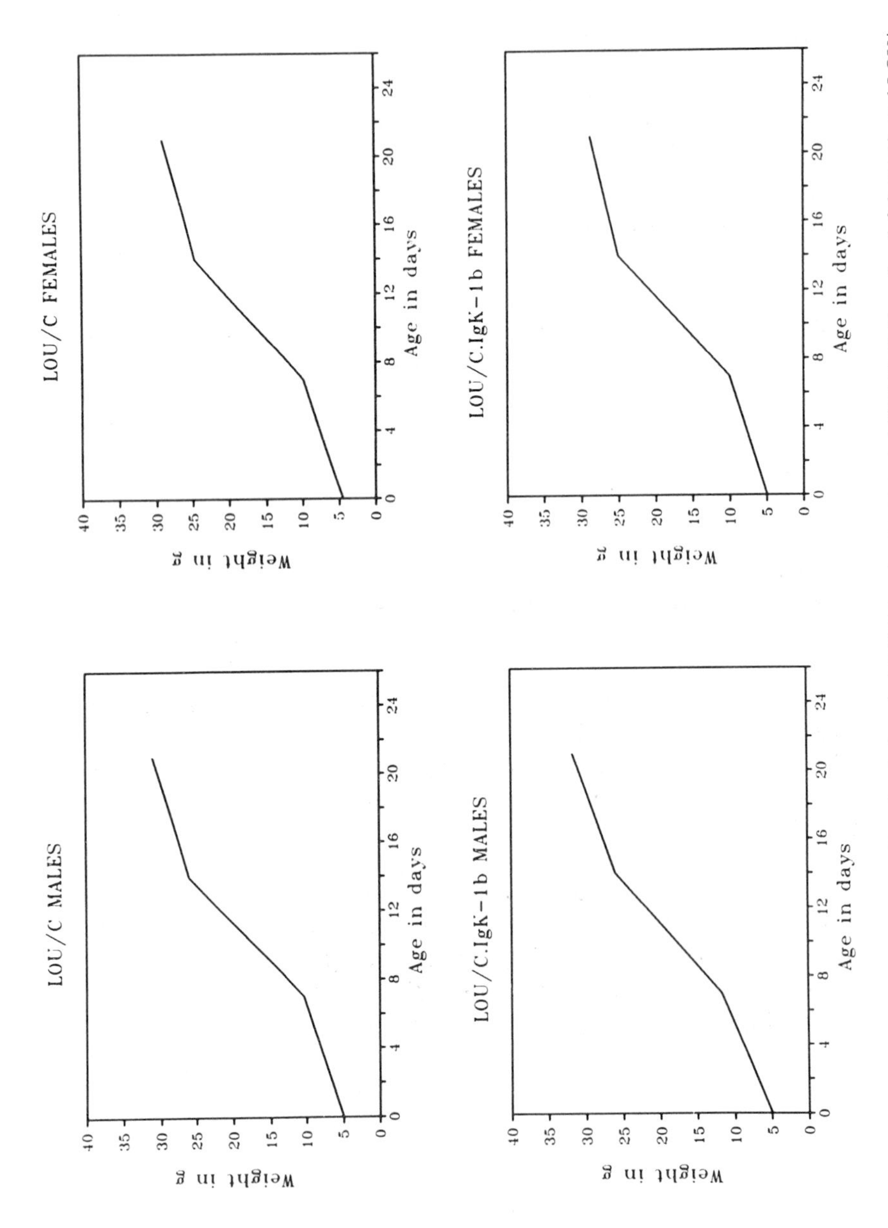

FIGURE 3. Weight in grams of LOU/C males (left top), LOU/C females (right top), LOU/C.IgK-1b males (left bottom), and LOU/C.IgK-1b females in function of their ages in days from 0 to 21 d.

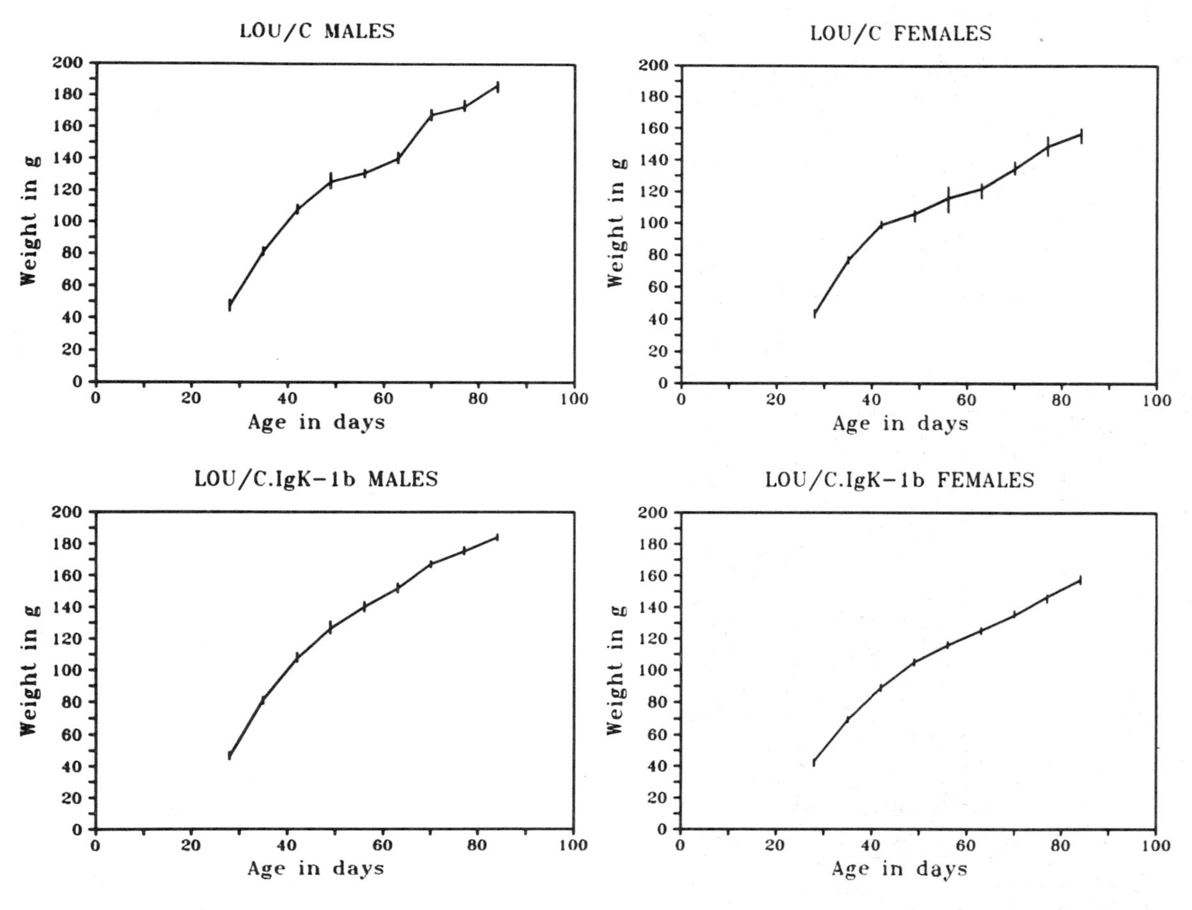

FIGURE 4. Weight in grams of LOU/C males (left top), LOU/C females (right top), LOU/C.IgK-1b males (left bottom), and LOU/C.IgK-1b females in function of their ages in days from 28 to 84 d.

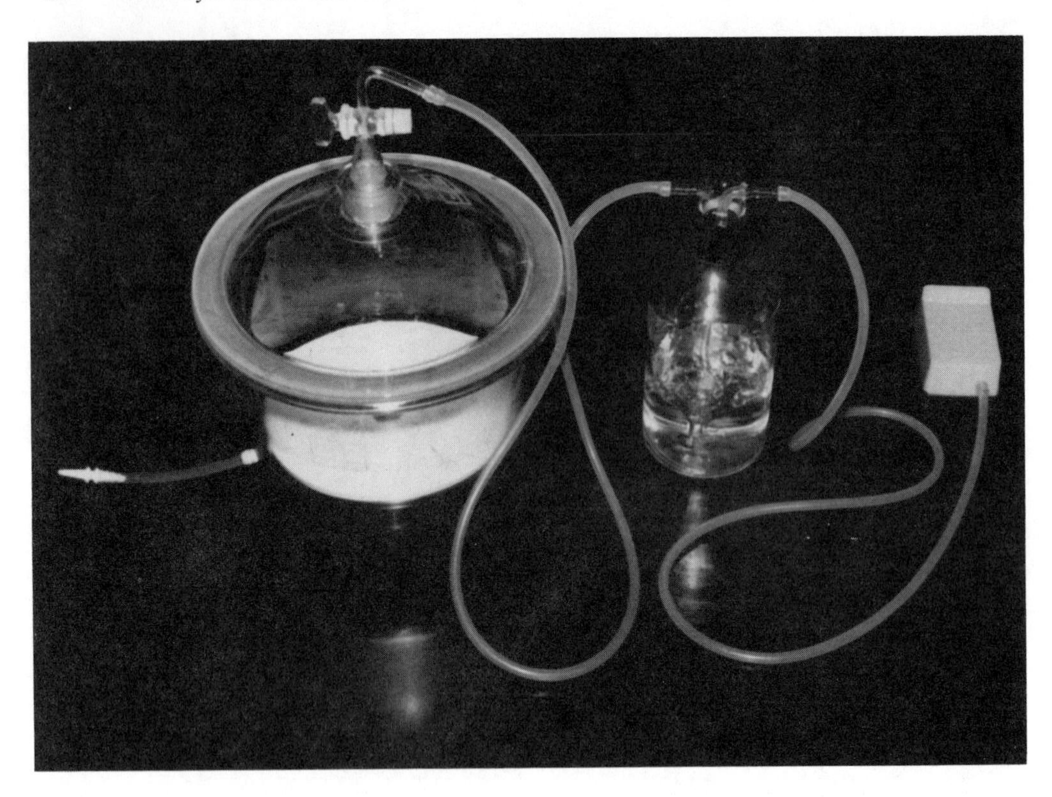

FIGURE 5. The device includes a small air pump for a domestic aquarium which pushes air across a bubble bottle filled with anesthesic ether, and then mixed to ether vapor in a dessicator where the rats can be introduced. Mixture of air and ether leaves the dessicator by its bottom and can be evacuated to the exterior with the help of a plastic tube.

REFERENCES

1. **Maisin, H.,** *Contribution à l'Etude du Syndrome Médullaire après Irradiation,* Arscia Editions, Brussels, 1959.
2. **Maldague, P.,** Radiocancérisation expérimentale du rein par rayon × chez le rat. I. Radiocancers du rein chez le rat, *Pathol. Eur.,* 1, 321, 1966.
3. **Maisin, J., Maisin, H., Dunjic, A., and Maldague, P.,** La radiologie comme méthode de travail en physiopathologie et en cancérologie expérimentale, *Bull. Acad. Suisse Sci. Med.,* 11, 247, 1955.
4. **Maisin, J., Maldague, P., Maisin, H., and Dunjic, A.,** Au sujet de l'influence cancérigène de l'irradiation in toto à dose unique, *C.R. Séances Soc. Biol. Paris,* 149, 1052, 1955.
5. **Maisin, J., Maldague, P., Dunjic, A., and Maisin, H.,** Syndromes mortels et effets tardifs des irradiations totales et subtotales chez le rat, *J. Belge Radiol.,* 40, 346, 1957.
6. **Maldague, P., Maisin, J., Dunjic, A., Hong-Que, P., and Maisin, H.,** Le leucosarcome chez le rat. L'incidence de ce néoplasme après irradiation totale, subtotale et locale, *Sang,* 24, 571, 1958.
7. **Maisin, J., Maldague, P., Deckers, C., and Hong-Que, P.,** Le leucosarcome du rat, *Symp. Lymph. Tumours Afr.,* 341, 1964.
8. **Deckers, C.,** Etude électrophorétique et immunoélectrophorétique des protéines du rat atteint de leucosarcome, in *Prot. Biol. Fluids,* Vol. 11, Peeters E. Ed., Pergamon Press, Oxford, 1963, 105.
9. **Altman, P. L. and Katz, D. D.,** Inbred genetically defined strains of laboratory animals. I. Mouse and rats, in *Biological Books III, Fed. Amer. Soc. Exp. Biol.,* Bethesda, Maryland, 1979, 233.
10. **Festing, M. and Staats, J.,** Standardized nomenclature for inbred strains of rats, *Transplantation,* 16, 221, 1973.

11. **Bender, K., Günther, E., Adams, A., Baverstock, P. R., Den Bieman, M., Bissbort, S., Brdicka, R., Butcher, G. W., Cramer, D. V., Von Deimling, O., Festing, M. F. W., Gutmann, R. D., Hedrich, H. J., Kendall, P. B., Kleige, R., Moutier, R., Womack, J. E., Yamada, J., and Van Zutphen, B.,** Biochemical markers in inbred strains of the rat *(Rattus norvegicus), J. Immunogenetics,* 19, 257, 1984.
12. **Bazin, H., Beckers, A., Vaerman, J. P., and Heremans, J. F.,** Allotypes of rat immunoglobulins. I. An allotype at the alpha-chain locus, *J. Immunol.,* 112, 1035, 1974.
13. **Beckers, A., Querinjean, P., and Bazin, H.,** Allotypes of rat immunoglobulins. II. Distribution of the allotypes of kappa and alpha chain loci in different inbred strains of rat, *Immunochemistry,* 11, 605, 1974.
14. **Beckers, A. and Bazin, H.,** Allotypes of rat immunoglobulins. III. An allotype of the gamma 2b chain locus, *Immunochemistry,* 12, 671, 1975.
15. **Gutman, G. A., Bazin, H., Rockhlin, C. V., and Nezlin, R. S.,** A standard nomenclature for rat immunoglobulin allotype, *Transplant. Proc.,* 15, 1685, 1983.
16. **Crispens, C. G.,** Reproduction and growth characteristics: mouse, in *Inbred and Genetically Defined Strains of Laboratory Animals — Part I Mouse and Rat,* Altman, P. and Katz, D. D., Eds., Federation of American Societies for Experimental Biology, Bethesda, Maryland, 1979, 45.

Chapter 3

RAT IMMUNOCYTOMAS (IR)

H. Bazin, W. S. Pear, G. Klein, and J. Sümegi

TABLE OF CONTENTS

I. INTRODUCTION

Monoclonal immunoglobulins or myeloma proteins or paraproteins synthesized by tumors have been detected in many animal species: men,[1] cows,[2] horses,[3-5] pigs,[6] dogs,[7-11] cats,[12,13] rabbits,[14] ferrets,[15] hamsters,[16,17] and mice.[18] These spontaneous tumors are incidental growths which derive from B lymphocytes and are called myeloma tumors, plasmacytomas, or immunocytomas depending on their characteristics and origins. Their monoclonal immunoglobulins have been used for various studies, but being rare and not transplantable, these tumors have always been considered of low interest.

In rodents, where inbred strains have been available since the beginning of the twentieth century, monoclonal immunoglobulin-secreting tumors are rare but not uncommon. Reticulum cell sarcomas of the SJL/J mice[19,20] and plasma cell leukemias of (CBA × DBA)F1 or BALB/c mice[21] essentially synthesize IgG monoclonal immunoglobulins. However, the production of these tumors was limited and unstable. Moreover, the transplantation of these tumors and particularly of the SJL/J type was difficult, even in thoroughly histocompatible animals. On the contrary, the plasma cell neoplasms described by Dunn[22] were curiously similar to the LOUVAIN IR tumors. They originate at the ileocecal junction of old C3H mice with an incidence inferior to 1% and show a lymphoid nature ranging from undifferentiated appearance to features characteristic of plasma cells. Some of them synthesize monoclonal immunoglobulins.[23] At least, such a C3H plasmacytoma has been widely used the 5563, an IgG2a secreting tumor.

These growths were not specific to the C3H breeding of Dunn, because at least one immunoglobulin-secreting tumor was identified in old C3H bred at the Institut du Radium of Paris (Bazin and Maldague, unpublished results).

The first model of immunoglobulin-secreting tumors really available for scientists was developed by Potter et al.[23] using the incidental observation of Merwin and Algire.[24] They used mineral oil to create a local irritation of the peritoneal cavity and developed hundreds of very valuable tumors used all over the world, but the model of Potter cannot be transferred from species to species. The same type of mineral oil injection does not seem to give rise to immunoglobulin-secreting tumors in other species, and even in mice only a few strains seem susceptible to this tumor induction.[25]

LOU/C ileocecal rat immunocytomas seem to be the only other model of immunoglobulin-secreting tumors occurring with a high incidence. However, they greatly differ from the BALB/c model. They are spontaneous and usually appear in their ileocecal lymph node.

II. LOU/C ILEOCECAL IMMUNOCYTOMAS

A. INCIDENCE OF THE ILEOCECAL LYMPHOID TUMORS IN RATS

1. In the Rat Species

In scientific literature, description of lymphoid tumors arising in the ileocecal area of rats has been observed since the 1930s. Over the years, these tumors have been reported as lymphosarcomas or reticulum cell sarcomas and have been particularly well studied in a series of articles[26,27] in the U.S. and in France.[28,29] Certainly the majority of these tumors were immunocytomas, although none of them were investigated for their immunoglobulin-secreting properties. Most important, the mean incidence of these ileocecal lymphoid tumors was always estimated lower than 3%.

2. In the LOUVAIN Rats

The special tumors of the LOUVAIN rats have been called immunocytomas, which can be abbreviated to IR for "Immunocytome de Rat" or to RIC for "Rat Immunocytoma".

The IR incidence of the LOU/C substrain was found to be ranging from 28.7%[30] to 34.4%[31]

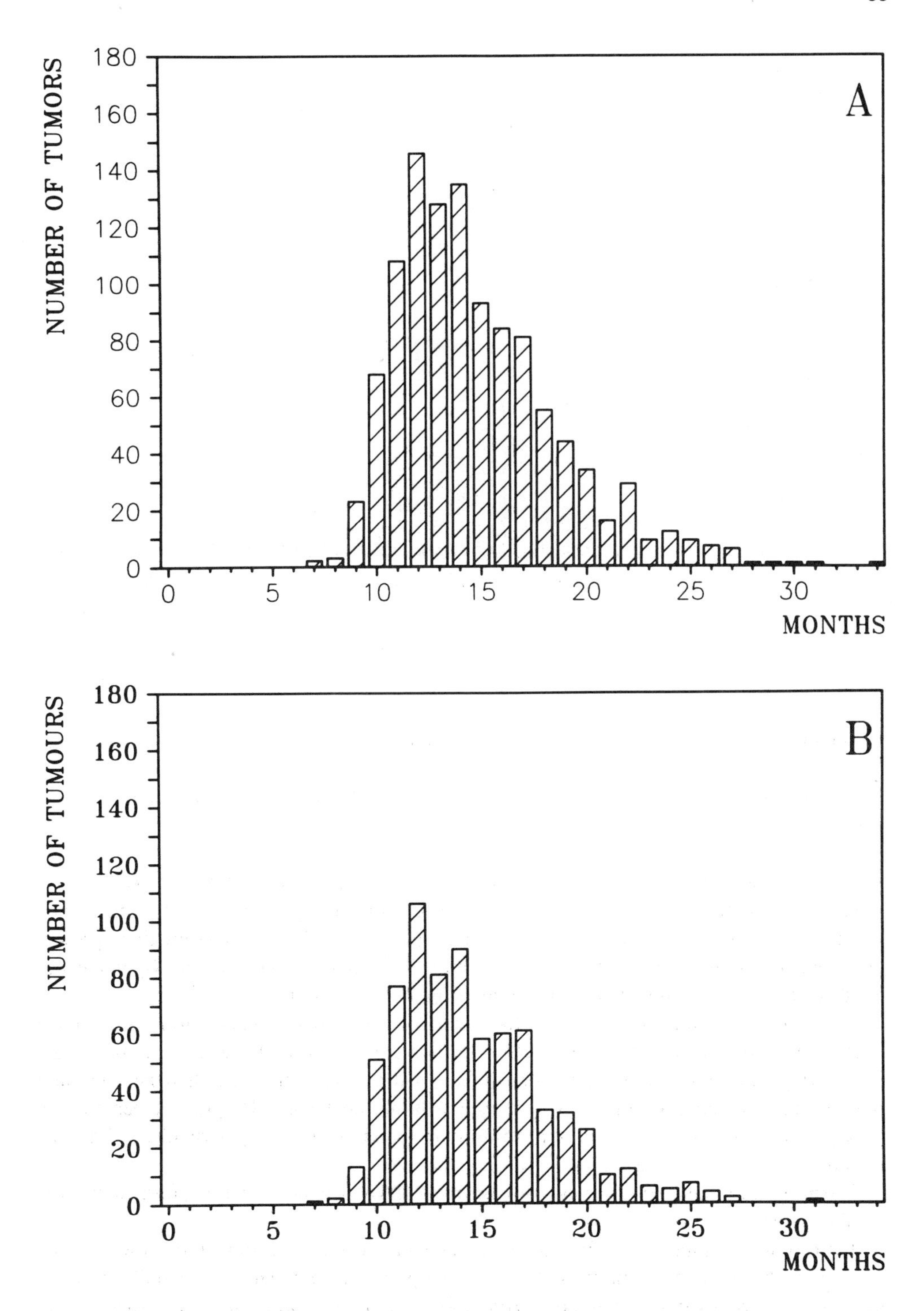

FIGURE 1. (A,B,C) The age distribution of male and female LOU/C immunocytomas.

in male and 14.3%[32] to 17%[31] in female rats. However, these results do not show the real incidence in the colony as the tumor appearance was never homogenous. The maximum incidence found was 57% in male and 29% in female but in some other cases it was very low or nil.[33] The age distribution of LOU/C immunocytomas is illustrated in Figure 1 A, B, and C.

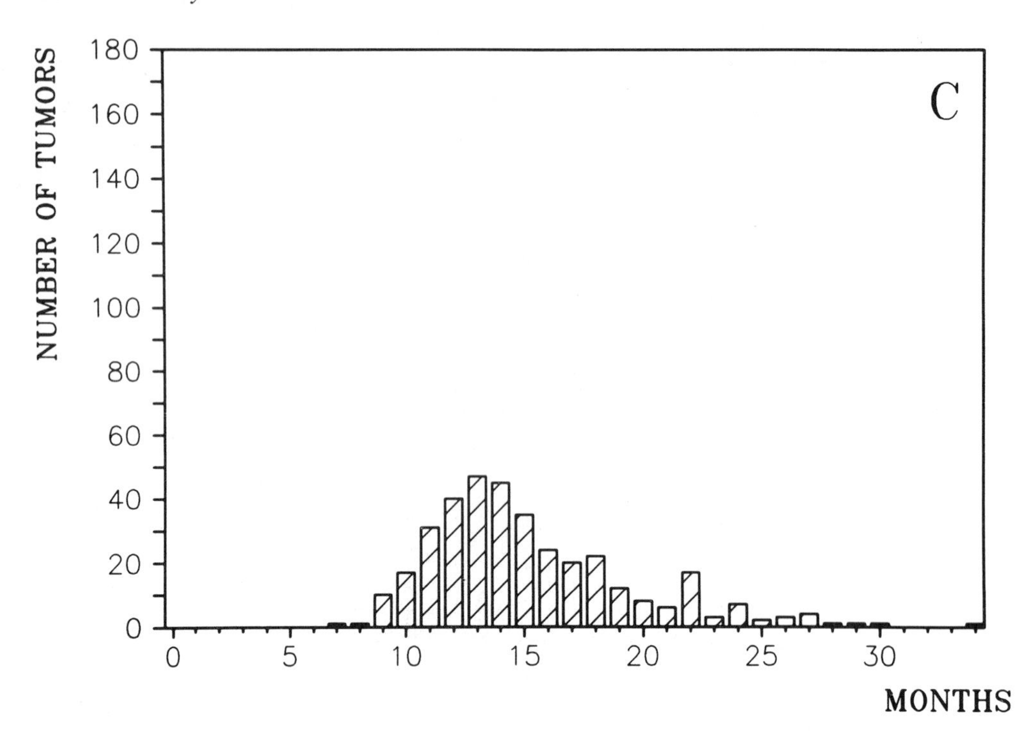

FIGURE 1 (continued)

The incidence appears in 8 month-old rats, its maximum being at 12 to 15 months and decreasing thereafter.

In LOU/M rats, the IR incidence has always been low, between 1 and 5% in male and less in female.

B. MACROSCOPIC AND MICROSCOPIC DESCRIPTION OF IR TUMORS

Generally, tumors appear in the peritoneal cavity as solid masses which can be mobilized by the fingers. They can be detected by palpation of the abdomen. Ascitic fluid is sometimes present in variable quantities which can go up to 100 ml. These tumors grow very quickly in days or weeks and kill their bearers within the month of the first detection.[34]

In most cases, at autopsy the tumor is located in the ileocecal area, in the form of one to two or even three solid nodules (Figure 2). This location corresponds to the ileocecal lymph node, which is a group of three to five nodes; one of these nodes is generally larger than the others and is found near the cecum, and the others being along the terminal ileum. They do not seem to have a special anatomy and be any different from other lymph nodes. They particularly drain the cecum, which is well developed in the rat species.

The primitive tumor seems to appear in the ileocecal lymph node and rapidly form metastasis in other places.

The histological aspect of the IR masses is characteristic of a lymphoid tumor. However, the cells cannot be described as plasma cells, but rather poorly differentiated lymphoid cells in the majority of cases. The immunoglobulin-secreting cells exhibit a granular nucleus with relatively small nuclei, a ring of deeply basophilic and pyroninophilic cytoplasm.[32,35,36] In immunofluorescence, with fluorescein conjugated rabbit antisera anti-rat immunoglobulin, many cells appear to have large quantities of cytoplasmic immunoglobulins. In the electron microscope, immunoglobulin-secreting cells display a well-developed endoplasmic reticulum. By contrast, cells of non-secreting tumors are polymorphic. Their nuclei are granular with very large nucleoli. Their cytoplasm is poorly basophilic with an underdeveloped endoplasmic reticulum, although ribosomes may occur in large numbers.[31,37]

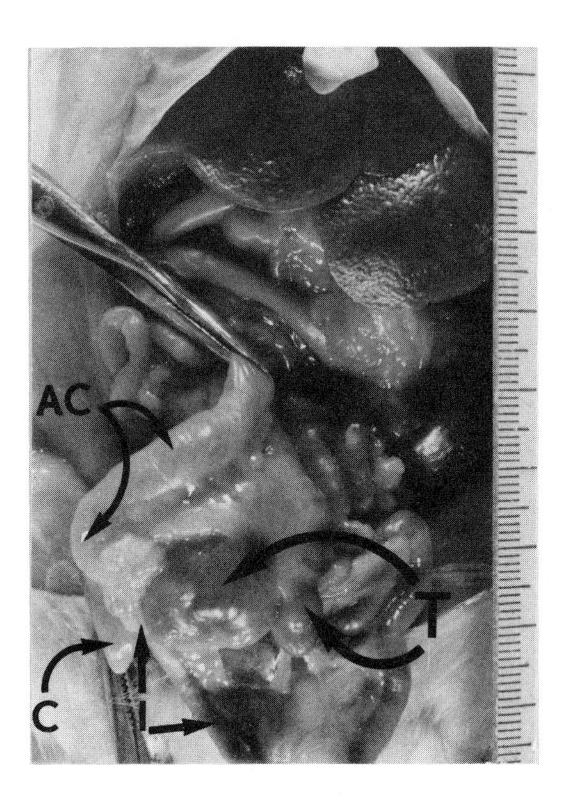

FIGURE 2. Primitive IR19 rat immunocytoma. The lateral scale is in centimeters. AC: ascending colon; T: tumor; C: cecum; and I: small intestine.

TABLE 1
Immunoglobulin-Secreting Properties of 1163 Louvain Immunocytomas

	Immunocytomas	Nondetermined	Secreting tumor	Nonsecreting tumor
	1163	20	764	379
%	100	1.7	65.7	32.6

C. SECRETION OF IMMUNOGLOBULIN BY IR TUMORS

Tumors with immunoglobulin-secreting properties represent about 66% of the total (Table 1). Among them, Bazin and Beckers[31] found 86% of immunoglobulin-secreting tumors, and 14% were only Bence Jones protein secretors. The isotype distribution of complete monoclonal immunoglobulins synthesized by 758 immunocytomas appearing in LOUVAIN and LOUVAIN F1 hybrids is given in Table 2. The highest incidence was found in the IgE isotype followed by those of the IgG1 class.

Light chain of the kappa type in complete and in Bence Jones molecules were detected,[38-40] but only two myelomas IgG1 of the lambda type were identified,[41] although not all the monoclonal immunoglobulins synthesized by IR tumors were typed for their light chain.

This LOUVAIN immunocytomas isotype distribution differs significantly from those secreted by the human myelomas or by the BALB/c mouse plasmacytomas.[42]

The serum of monoclonal immunoglobulins was found to be quite variable in rats bearing primitive IR tumors, depending on the particular secreting properties of each individual tumor and on the metabolism of the considered monoclonal immunoglobulin. Values obtained for

TABLE 2
Isotype Distribution of the Monoclonal Immunoglobulins Synthesized by 758 Immunocytomas

	Secreting tumors	IgM	IgD	IgG1	IgG2a	IgG2b	IgG2c	IgE	IgA
No. of immunocytomas	758	24	6	261	50	4	37	342	34
%	100	3.2	0.8	34.4	6.6	0.5	4.9	45.1	4.5

various tumors were given in Bazin and Beckers[31] or in Beckers and Bazin.[33] Monoclonal immunoglobulin concentrations of 10 mg/ml or more were often detected (Figure 3 A and B).

The secreting properties of IR tumors transplanted *in vitro* or *in vivo* over time are quite variable. Most of the IgM and IgG secreting tumors retain these properties for years and some of them for more than 15 years without any decrease. IgE and IgD secreting IR tumors were less stable and more prone to lose their secreting properties, although good IgE as well as IgD producers were selected.

Antigen-binding properties of monoclonal immunoglobulins secreted by IR tumors were found: two myeloma proteins, respectively of the IgM (IR473) and of the IgA (IR22) isotypes, were identified to have properties of rhumatoid factors for rat IgG1 molecules. Curiously no other binding capacities were detected, particularly for the common antigens found to be able to react with BALB/c monoclonal immunoglobulins (Potter, M., National Institute of Health, U.S.).

D. TRANSPLANTATION, IN VITRO CULTURE AND STORAGE OF IR TUMOR CELLS

1. *In Vivo* Transplantation

The first method used to keep IR tumors was their transplantation into histocompatible hosts. LOU/C and LOU/M are highly inbred strains and in most cases, tumors appearing in them could be transplanted in animals of the same strains (LOU/C and/or LOU/M) without difficulty. Gut of more than 1000 transplantations, 90 to 95% of the primitive IR tumors can be adapted. The latent periods of growth are generally shortened after two or three passages from 14—21 to 7—10 d. No F1 hybrid effects have been observed when small numbers of IR tumor cells were inoculated into hybrids of the first generation between LOU/C or LOU/M rats and rats of various other inbred strains.[42] The technique developed for IR tumor transplantation is now used for rat hybridoma transplantations.

2. *In Vitro* Cell Culture

During the 1960s, techniques of long-term *in vitro* culture of hematopoietic human cells were established by Moore et al.[43] Using the Moore procedure, Burtonboy et al.[44] first obtained continuous cell lines from rat immunocytomas. The ascitic form of the IR tumor was found to be particularly suitable for the adaptation of the IR cells to *in vitro* culture. These cells retained the morphological and immunoglobulin-secreting properties of original tumor cells and were capable of inducing malignant tumors when injected into histocompatible hosts. Most if not all, IR tumors could be adapted to *in vitro* cell culture, generally after one attempt, but sometimes after more.

At least 2 or 3 dozen IR tumors were adapted to continuous cell lines (Bazin, unpublished results).[44,45] Two IR tumors were particularly cultivated *in vitro*. The first is the S210, a kappa Bence Jones producer,[35] adapted to cell culture[44] and given to C. Milstein in 1972,[46] was used for cell fusion experiments with a BALB/c plasmacytoma[47] with the new appellation of 210 RCY3 Ag1, and then named Y3.Ag1,2,3 as a fusion line. However, this rat fusion line is still a kappa Bence Jones producer with a high capacity for synthesizing a kappa light chain in large

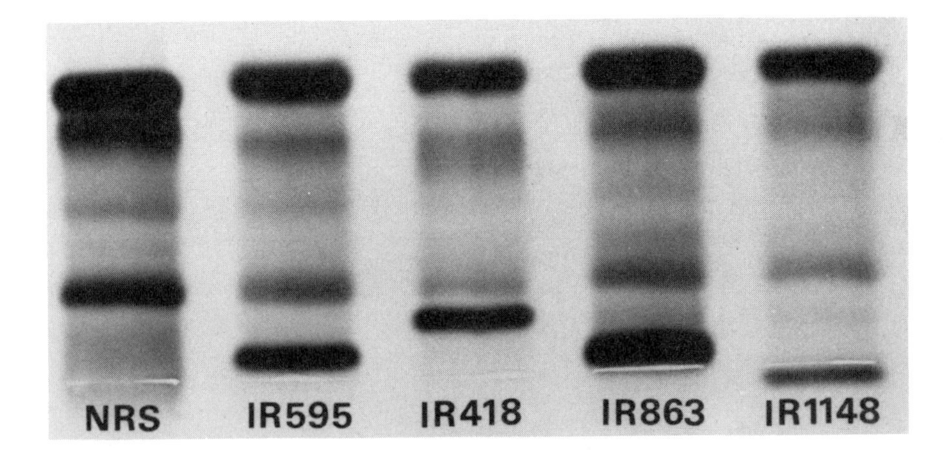

A

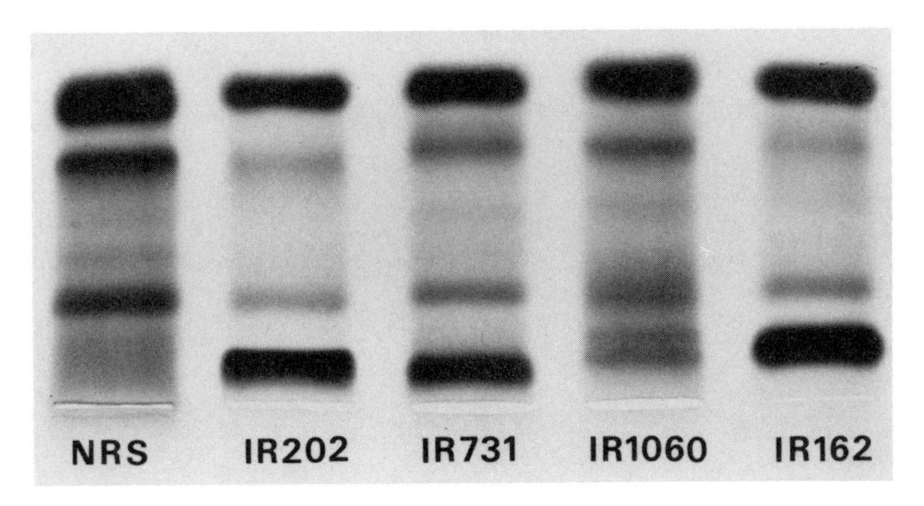

B

FIGURE 3. (A,B) Agarose electrophoresis of sera from normal rat serum (NRS), and LOU/C rats carrying the IR595, IgG1 monoclonal immunoglobulin (MIg), IR418 (IgG2a MIg), IR863 (IgG2b MIg), IR1148 (IgG2c MIg), IR202 (IgM MIg), IR731 (IgD MIg), IR1060 (IgA MIg), and IR162 (IgE MIg).

quantities, as was the original S210 tumor which was selected for that property when chosen to be given to C. Milstein.

The second cell line was derived from the IR983 tumor, a nonsecreting IR tumor from which the IR983F nonsecreting rat fusion cell line was selected.[48,49]

3. Storage

LOUVAIN IR tumor cells can be stored in liquid nitrogen for a long period of time. Last year, primitive tumor cells directly stored in liquid nitrogen and kept for 14 years, were thawed and transplanted in histocompatible animals in which three out of four were recovered. The technique employed to freeze IR cells directly obtained from a solid tumor, ascitic fluid, or *in vitro* cell culture has been used with success for rat hybridoma cells and is described hereafter.

E. ETIOLOGY OF IR TUMORS

This etiology is poorly understood although many interesting results are now comprehended.[42]

1. Natural Occurrence of the Ileocecal Lymphoid Tumors in Rodents

In some circumstances and in a limited number of inbred strains, monoclonal immunoglobulin-secreting tumors were found in rodents. Dunn[22] described plasma-cell neoplasms originating in the ileocecal junction in old C3H mice which secreted monoclonal immunoglobulins.[23] Likewise, in the rat species, lymphoid tumors originating in the ileocecal area have been observed for a long time. Moreover, LOU/C rats seem to be especially susceptible to this kind of cancer incidence.

2. Genetic Control of IR Tumor Incidence

The studies on crosses and backcrosses of Fisher and August rats first conducted by Curtis et al.,[50] showed an inherited susceptibility to "mesenteric lymph node sarcoma". Later, Dunning and Curtis[27] and Curtis and Dunning[26] realized that "some influences other than those postulated by Orthodox genetic hypothesis play a determining role in the incidence of these neoplasms,...".

With the help of the LOUVAIN model and various inbred strains of rats in the same conditions, it was shown that LOU/C rats had a high incidence of IR tumors, LOU/M, AUG, and AxC a low or very low incidence and the OKA, a null incidence.[33] Crosses between LOU/C and AUG, AxC, or LOU/M rats gave hybrids with an intermediate incidence between that of LOU/C and that of AUG, AxC, or LOU/M. By contrast (LOU/C × OKA)F1 had a very low incidence, 1.08%.[42] In order to analyze the resistance of the OKA to IR tumor incidence, congenic strains between this strain and LOU/C rats were selected for major histocompatibility complex (MHC) and light kappa and heavy immunoglobulin chains. The LOU/C.Rt-1^k(OKA) having OKA MHC of the OKA strain in LOU/C background showed no IR tumor incidence, indicating that the OKA MHC or a gene linked to it determines resistance to IR tumor development. The prevalence of this resistance was demonstrated to be predominant over the LOU/C susceptibility, by the very low IR incidence in (LOU/CxOKA)F1 hybrids. The congenic rat with respectively the kappa or the heavy chains of immunoglobulins of OKA strain into the LOU/C genetic background has a tumor incidence identical to the LOU/C strain.[51] Thus, the resistance of the OKA strain to the appearance of IR tumor is not correlated with the special properties of the immunoglobulin gene.

3. Chromosomal Translocation in the Spontaneous Rat Immunocytoma (RIC) Leads to the Juxtaposition of c-myc and Immunoglobulin Heavy Chain (IgH) Sequences

Chromosome translocation is consistently associated with a variety of hemopoetic malignancies.[52] They include Burkitt's lymphoma, chronic lymphocytic leukemia, diffuse B-cell lymphoma, follicular lymphoma and chronic myelogenous leukemia in human,[53] and pristane induced mouse plasmacytoma.[54] In Burkitt's lymphoma (BL) the telomeric end of the long arm of chromosome 8 is translocated to chromosome 14, 22, or 2.[55] In chronic myelogenous leukemia the distal end of the long arm of chromosome 9 is transposed to chromosome 22 that has suffered a deletion of its distal segment. The resulting 22q-marker is designated as the Philadelphia chromosome.[56] In mouse plasmacytoma (MPC) the distal end of the chromosome 15 recombines with the telomeric end of chromosome 12 or with the short arm of chromosome 6.[54]

Molecular studies on BL and MPC associated translocations have shown that c-myc is involved in a reciprocal exchange with one of the three immunoglobulin loci in 100% of the cases.

Wiener et al.[57] have found a consistent translocation in seven early passage rat immunocytoma arising from a reciprocal exchange between the distal part of the q-arm of chromosome 7 and the telomeric region of chromosome 6 with breakpoints at 7q3.3 and 6q3.2, respectively.[57] Pear et al.[58] found the same chromosomal translocation in 12 additional RICs.

The banding pattern of the immunocytoma associated (6;7) translocation resembled the typical (12;15) translocation of the MPCs, suggesting the involvement of syntenic chromosome regions in the two species. No variant translocations have been found in the RIC system so far.

We have localized c-myc to chromosome 7[59] and the IgH locus to chromosome 6.[60]

a. The Chromosomal Breakpoints Cluster at the 5′ End of the C-myc Gene

The 17 kb long EcoRI fragment of the rat genomic DNA appears to accomodate the entire c-myc transcriptional unit.[59,61]

The chromosomal breakpoint was localized within this 17 kb long EcoRI fragment in 12 of the 14 RICs investigated so far.[58] The protein coding exons were intact in all RICs as in BLs and MPCs.[54] As shown in Figure 4, the translocation breakpoints were clustered in 10 of the 12 tumors within a 1.5 kb long XbaI-HindIII fragment proximal to the first exon.[58] In one tumor (IR49) the breakpoint disrupted the first exon. In the IR221 tumor 5′ sequences were intact and the chromosomal breakpoint was distal to the third exon. The c-myc promoters remained intact. The 1.5 kb XbaI-HindIII fragment and sequences upstreams are expected to contain c-myc regulatory elements, as suggested on the basis of their sequence homology to the relevant human and mouse c-myc region.[62-64] The same region is also a preferred proviral insertion site in murine T-cell lymphomas[65-68] and in Moloney-virus-induced rat thymomas.[61] Only in 1 of the 13 RICs was the chromosomal breakpoint distal to the c-myc promoters (Figure 4). More than 90% of the MPCs and the majority of sporadic BLs[54] had their chromosomal breakpoints in the first exon or in the first intron. The 17 kb long EcoRI fragment remained intact in two tumors IR9 and IR304. Restriction enzyme mapping showed that the chromosomal breakpoint must be more than 20 kb 5′ or 23 kb 3′ of the c-myc coding sequences in these tumors.[58]

b. c-myc Expression

The RIC-associated IgH/myc translocation was accompanied by a high constitutive c-myc mRNA level.[58] The c-myc mRNA was of the same size 2.4 kb in normal cells and in the majority of RICs.[58] In constrast IR49, in which the c-myc transcription unit is broken like many BLs and MPCs, yield new mRNA species of 2.1 kb. Removal of the normal c-myc promoters must have triggered one or more cryptic promoters within the intron one in this tumor. In IR304 the breakpoint was at least 20 kb from c-myc, but the mRNA level was equally high as in the other tumors.[58] A previously unidentified form of dysregulation is foreseen in IR221 where the unique breakpoint 3′ of the third exon is associated with a low constitutive level of c-myc expression.

c. Translocation Targets in the IgH Locus

In BLs and MPCs the switch regions, JH region and infrequently the IgH enhancer region can serve as myc rearrangement targets.[54] The switch C alpha served as the target in the majority of MPCs. This is a quasi functional relationship with the Ig product. Most MPCs produce IgA, but the second unbroken chromosome is responsible for the synthesis of the alpha chain. A similar relationship is found in the human system.[54] Most BLs produce IgM, and Smu is the most frequent rearrangement site. The majority of the RICs produce IgE. It is therefore of great interest that six c-myc rearranged to the epsilon locus of the seven IgE producers.[58] Analysis of the IR50 tumor showed that the recombination placed the two genes in opposite transcriptional orientation as in BLs and MPCs.[58]

Cloning of the rearranged c-myc gene in tumors IR75, IR223, and IR209 has shown that the switch region is not involved in these translocations. In the IR75 the translocation occurred between c-myc and DNA sequences 5′ of the switch gamma region. The two sequences were in identical transcriptional orientation.[69] In IR223 the breakpoint was localized to the intron one of the C epsilon gene.[70] The involvement of this region suggests that the switch recombination

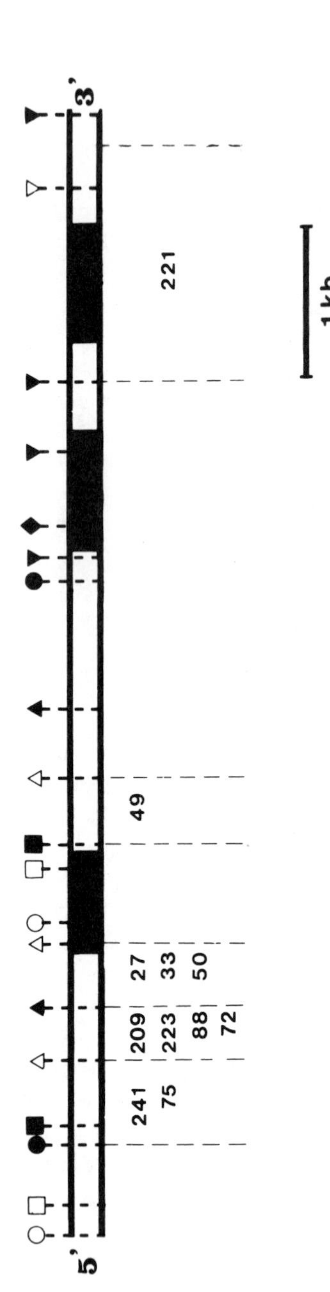

FIGURE 4. Detailed restriction map of the region surrounding the c-myc gene containing the chromosome 7 breakpoint in 12 of 14 RICs. Restriction sites: O HindIII, ❑ SacI, ● XbaI, ■ BglII, Δ BamHI, ▲ PvuII, ♦ EcoRV, ▼ PstI, ∇ XhoI.

mechanism may confer a recombination potential on sequences outside the switch regions themselves. This is also suggested by the case of IR209, a gamma2a-producer that expresses c-myc mRNA at similar level as most other RICs. Cloning and sequencing of the rearranged c-myc of this tumor showed that a recombination occurred between c-myc and a repetitive element corresponding to the LINE family[71] of moderately repetitive DNA sequences.

d. Conclusions

Our cytogenetic and molecular studies of the RIC tumors have showed that the same genetic loci (c-myc and IgH) were juxtaposed by chromosomal translocation as in BL and MPC. These tumors occur in three different species, represent at least two different stages of B-cell maturation and their natural histories and modes of induction are different. Our findings are consistent with the view that chromosomal translocation alters the regulation of c-myc. The consistency of the translocation suggests that it represents the rate-limiting step in the malignization process. The c-myc gene product is probably involved in the control of cell proliferation and/or DNA replication.[72] Its elevated constitutive expression may therefore promote continous proliferation. The variation in the DNA sequences involved in the translocation (switch, DNA sequences without apparent function, intron and repetitive DNA) suggests several potential modes for c-myc dysregulation by chromosomal translocation. Deletion of c-myc 5′ regulatory sequences is one of the potential mechanisms. Most of the c-myc breakpoints have led to the removal of the 5′ flanking sequences postulated to contain c-myc regulatory elements in the RICs that have been studied.

Enhancing effects of the juxtaposed sequences represents a second potential mechanism of activation. We did not find a functional enhancer in the transposed sequences of IR75 and IR209, however.

A third possible mechanism of c-myc activation is related to the insertion of the gene into an active IgH chromatin region. The question as to whether there are specific sequences within the IgH constant region, in addition to the IgH enhancer that are capable of this activation, or whether the overall chromation configuration of the region is responsible for c-myc activation remains to be investigated.

4. Attempts to Modify the IR Tumor Incidence in LOUVAIN Rats

Several attempts were made to modify the incidence or the latent period of the IR tumor in LOU/C rats. By removing the ileocecal lymph nodes of 5- to 21-day-old rats, Moriamé et al.[73] observed a considerable reduction of IR tumor incidence in LOU/C rats when compared with their controls. On the contrary, intraperitoneal injections of Pristane or Complete Freund's adjuvant, single or fractionated doses of irradiation never increased IR tumor incidence, and even decreased their mean latent period.[31]

III. GENERAL CONCLUSIONS ON THE LOUVAIN IR TUMOR MODEL

The LOUVAIN rat model of immunoglobulin-secreting tumors has provided some new opportunities for research and application:

1. The availability of monoclonal immunoglobulins of eight different (sub)classes in the rat species and especially the first IgE and IgD monoclonal immunoglobulins identified in an animal species
2. The development of a rat-rat hybridoma technology
3. The cytogenetic and molecular studies of a spontaneous immunoglobulin-secreting tumor, giving with the Burkitt lymphoma and the BALB/c plasmacytoma, a third model of c-myc-immunoglobulin juxtaposition and suggesting a central role for this chromosomal translocation in the genesis of these B-cell tumors.

REFERENCES

1. **MacIntyre, W.,** Case of mollities and fragilitas ossuim accompanied with urine strongly charged with animal matter, *Med. Chir. Soc. Trans.*, 33, 211, 1850.
2. **Pedini, B. and Romanelli, V.,** I. Plasmacitoma negli animali domestici. Osservazioni e considerazioni su di un caso riscontrato nel vitello, *Arch. Vet. Ital.*, 6, 193, 1955.
3. **Cornelius, C. E., Goodbury, R. F., and Kennedy, P. C.,** Plasma cell myelomatosis in a horse, *Cornell Vet.*, 49, 478, 1959.
4. **Dorrington, K. J. and Rockey, J. H.,** Studies on the conformation of purified human and canine gamma A-globulins and equine gamma T-globulin by optical rotatory dispersion, *J. Biol. Chem.*, 243, 6511, 1968.
5. **Dorrington, K. J. and Tanford, C.,** The optical rotary dispersion of human gamma-M-immunoglobulins and their subunits, *J. Biol. Chem.*, 243, 4745, 1968.
6. **Englert, H. K.,** Die leukose des Schweines, *Zentralbl. Veterinaermed.*, 2, 607, 1955.
7. **Bloom, F.,** Intramedullary plasma cell myeloma occuring spontaneous in a dog, *Cancer Res.*, 6, 718, 1946.
8. **Groulade, J., Morel, P., Creyssel, R., and Groulade, P.,** Un cas de paraglobulinémie chez le chien, *Bull. Acad. Vet. Fr.*, 6, 354, 1959.
9. **Jennings, A. R.,** Plasma-cell myelomatosis in the dog, *J. Comp. Pathol. Ther.*, 59, 113, 1949.
10. **Medway, W., Weber, W. T., O'Brien, J. A., and Krawitz, L.,** Multiple myeloma in a dog, *J. Am. Vet. Med. Assoc.*, 150, 386, 1967.
11. **Osborne, C. A., Perman, V., Sautter, J. H., Stevens, J. B., and Hanlon, G. F.,** Multiple myeloma in the dog, *J. Am. Vet. Med. Assoc.*, 153, 1300, 1968.
12. **Farrow, B. R. H. and Penny, R.,** Multiple myeloma in a cat, *J. Am. Vet. Med. Assoc.*, 158, 606, 1971.
13. **Kehoe, J. M., Hurvitz, A. I., and Capra, J. D.,** Characterization of three homogeneous feline immunoglobulins, *J. Immunol.*, 109, 511, 1972.
14. **Pascal, R. R.,** Plasma cell myeloma in the brain of a rabbit, *Cornell Vet.*, 51, 528, 1961.
15. **Methiyapun, S., Pohlenz, J. F., and Bertschinger H. V.,** Ultrastructure of the intestinal mucosa in pigs experimentally inoculated with an edena disease-producing strain of Escherichia coli (O139:K12:H1), *Vet. Pathol.*, 22, 517, 1985.
16. **Garcia, H., Baroni, C., and Rappaport, H.,** Transplantable tumors of the Syrian golden hamster (Meso-cricetus auratus), *J. Natl. Cancer Inst.*, 27, 1323, 1961.
17. **Mohr, U. and Dontenwill, W.,** Organ-und Blutveranderungen beim KG-13-Plasmocytom des Goldhamsters, *Z. Krebsforsch.*, 66, 29, 1964.
18. **Dunn, T. B.,** Plasma-cell neoplasms beginning in the ileocaecal area in strain C3H mice, *J. Natl. Cancer Inst.*, 19, 371, 1975.
19. **McIntire, R. R. and Law, L. W.,** Abnormal serum immunoglobulins occuring with reticular neoplasms in inbred strain of mouse, *J. Natl. Cancer Inst.*, 39, 1197, 1967.
20. **Wanebo, H. J., Gallmeier, W. M., Boyse, E. A., and Old, L.,** Paraproteinemia and reticulum cell sarcoma in an inbred mouse strain, *Science*, 154, 901, 1966.
21. **Rask-Nielsen, R., McIntire, R. R., and Ebbesen, P.,** Plasma cell leukemia in BALB/c mice inoculated with subcellular material. II. Serological changes, *J. Natl. Cancer Inst.*, 41, 495, 1968.
22. **Dunn, T. B.,** Plasma-cell neoplasms beginning in the ileocaecal area in strain C3H mice, *J. Natl. Cancer Inst.*, 19, 371, 1957.
23. **Potter, M., Fahey, J. L., and Pilgrim, H. J.,** Abnormal serum protein and bone destruction in transmissible mouse plasma-cell neoplasma (multiple myeloma), *Proc. Soc. Exp. Biol. Med.*, 94, 327, 1957.
24. **Merwin, R. M. and Algire, G. H.,** Induction of plasma-cell neoplasms and fibrosarcomas in BALB/c mice carrying diffusion chambers, Proc. Soc. Exp. Biol. Med., 101, 437, 1959.
25. **Potter, M.,** Immunoglobulin producing tumors and myeloma proteins of mice, *Physiol. Rev.*, 52, 631, 1972.
26. **Curtis, M. R. and Dunning, W. F.,** Transplantable lymphosarcomata of the mesenteric lymph nodes of rats, *Am. J. Cancer*, 40, 299, 1940.
27. **Dunning, W. F. and Curtis, M. R.,** The respective roles of longevity and genetic specificity in the occurrence of spontaneous tumors in the hybrids between two inbred lines of rats, *Cancer*, 6, 61, 1946.
28. **Roussy, G. and Guérin, P.,** Les lymphosarcomes du rat, *Bull. Assoc. Fr. Etude Cancer*, 30, 17, 1942.
29. **Guérin, M.,** *Tumeurs Spontanées des Animaux de Laboratoire*, A. Legrand et Cie, Paris, 1954.
30. **Bazin, H., Rousseaux, J., Kints, J. P., and Herno, J.,** Studies on the incidence of rat ileocecal malignant immunocytoma: vertical transmission of the high tumour incidence, *Cancer Lett.*, 8, 353, 1980.
31. **Bazin, H. and Beckers, A.,** *Nobel Symposium 33, Molecular and Biological Aspects of the Acute Allergic Reaction*, Stockholm, Johanson S.G.O., Stranberg, K., and Uvnas, B., Eds., Plenum Company, New York, 1976, 125.
32. **Bazin, H., Beckers, A., Deckers, C., and Moriamé, M.,** Transplantable immunoglobulin-secreting tumours in rats. V. Monoclonal immunoglobulins secreted by 250 ileocecal immunocytomas in the LOU/Wsl rats, *J. Natl. Cancer Inst.*, 51, 1359, 1973.

33. **Beckers, A. and Bazin, H.,** Incidence of spontaneous immunocytomas in hybrids of LOU/C rats and rat strains with spontaneous tumor incidence, *J. Natl. Cancer Inst.*, 60, 1505, 1978.
34. **Bazin, H.,** *Rat Immunocytomas in Mechanisms of B Cell Neoplasia,* Melchers, F. and Potter, M., Eds., Roche, Bâle, Switzerland, 1985, 208.
35. **Bazin, H., Deckers, C., Beckers, A., and Heremans, J. F.,** Transplantable immunoglobulin-secreting tumours in rats. I. General features in LOU/Wsl strain rat immunocytomas and their monoclonal proteins, *Int. J. Cancer,* 10, 568, 1972.
36. **Maisin, J., Maldague, P., Dunjic, A., Hong-Que, P., and Maisin, H.,** Le leucosarcome chez le rat. L'incidence de ce néoplasme après irradiation totale, subtotale et locale, *Sang,* 24, 751, 1958.
37. **Burtonboy, G., Beckers, A., Rodhain, J., Bazin, H., and Lamy, M. E.,** Rat ileocecal immunocytomas. An ultrastructural study with special attention to the presence of viral particles, *J. Natl. Cancer Inst.*, 61, 477, 1978.
38. **Querinjean, P., Bazin, H., Beckers, A., Deckers, C., Heremans, J. F., and Milstein, C.,** Transplantable immunoglobulin-secreting tumours in rats. Purification and chemical characterization of four kappa chains from LOU/Wsl rats, *Eur. J. Biochem.*, 31, 354, 1972.
39. **Wang, A. C., Fudenberg, H. H., and Bazin, H.,** The nature of "species-specific" amino acid residues, *Immunochemistry,* 12, 505, 1975.
40. **Wang, A. C., Fudenberg, H. H., and Bazin, H.,** Partial aminoacid sequences of kappa-chains of rat immunoglobulin: genetic and evolutionary implications, *Biochem. Genet.*, 14, 209, 1976.
41. **Querinjean, P., Bazin, H., Starace, V., Beckers, A., Deckers, C., Heremans, J. F.,** Lambda light chains in rat immunoglobulins, *Immunochemistry,* 10, 653, 1973.
42. **Bazin, H., Pear, W. S., and Sumegi, J.,** LOUVAIN rat immunocytomas, *Adv. Cancer Res.*, 50, 279, 1987.
43. **Moore, G. E., Garner, R. E., and Franklin, H. A.,** Culture of normal human leucocytes, *J. Am. Med. Assoc.*, 199, 87, 1967.
44. **Burtonboy, G., Bazin, H., Deckers, C., Lamy, M., and Heremans, J. F.,** Transplantable immunoglobulin-secreting tumours in rats. III. Establishment of immunoglobulin-secreting cell lines from LOU/Wsl strain rats, *Eur. J. Cancer,* 9, 259, 1973.
45. **Bennich, H., Karlsson, T., Bazin, H., and Zeuthen, J.,** *In vitro* culture of IgE secreting cells from two rat myelomas, *Protides Biol. Fluids,* 25, 559, 1978.
46. **Bazin, H.,** More on rat monoclonal antibodies (Letter to the Editor), *Immunol. Today,* 4, 274, 1983.
47. **Cotton, R. G. H. and Milstein, C.,** Fusion of two immunoglobulin-producing myeloma cells, *Nature,* 244, 42, 1979.
48. **Bazin, H.,** Les anticorps monoclonaux, *Louvain Med.*, 100, 3, 1981.
49. **Bazin, H.,** Production of rat monoclonal antibodies with the LOU rat non-secreting IR983F myeloma cell line, in *Protides of the Biological Fluids,* 29th Colloquium 1981, Peeters, E., Ed., Pergamon Press, Oxford, 1982, 615.
50. **Curtis, M. R., Bullock, F. D., and Dunning, W. F.,** A statistical study of the occurrence of spontaneous tumors in a large colony of rats., *Am. J. Cancer,* 15, 67, 1931.
51. **Bazin, H., Kints, J. P., and Rousseaux, J.,** Genetic control of the resistance to spontaneous immunocytoma (plasmacytoma IR tumour) development in LOU/C rats, *Anticancer Res.*, 6, 45, 1986.
52. **Rowley, J. D.,** A new conistent chromosomal abnormality in chronic myelogenous leukemia identified by quinacrine fluorescence and Giemsa staining, *Nature,* 243, 290, 1973.
53. **Croce, C.,** Role of chromosome translocations in human neoplasia, *Cell,* 49, 155, 1987.
54. **Cory, S.,** Activation of cellular oncogenes in hemopoietic cells by chromsome translocation, *Adv. Cancer Res.*, 47, 189, 1986.
55. **Klein, G.,** Specific chromosomal translocation and the genesis of B-cell derived tumors in men and mice, *Cell,* 32, 311, 1983.
56. **Groffen, J., Heisterkamp, N., and Stam, K.,** Oncogene activation by chromosomal translocation in chronic myelocytic leukemia, *Cold Spring Harbor Symp. Quant. Biol.*, 51, 911, 1986.
57. **Wiener, F., Babonits, M., Spira, J., Klein, G., and Bazin, H.,** Non-random chromosomal changes involving chromosomes 6 and 7 in spontaneous rat immunocytomas, *Int. J. Cancer,* 29, 431, 1982.
58. **Pear, W. S., Wahlstrom, G., Nelson, S. F., Axelsson, H., Szeles, A., Wiener, F., Bazin, H., Klein, G., and Sumegi, J.,** The 6:7 chromosomal translocation in spontaneously arising rat immunocytomas: evidence for c-myc breakpoint clustering and correlation between isotypic expression and c-myc target, *Mol. Cell. Biol.*, 8, 441, 1988.
59. **Sumegi, J., Spira, J., Bazin, H., Szpirer, J., Levan, G., and Klein, G.,** Rat c-myc oncogene is located on chromosome 7 and rearranges in immunocytomas with a t(6:) translocation, *Nature,* 306, 494, 1983.
60. **Pear, W. S., Wahlstrom, G., Szpirer, J., Levan, G., Klein, G., and Sumegi, J.,** Localization of the rat immunoglobulin heavy chain locus to chromosome 6, *Immunogenetics,* 23, 393, 1986.
61. **Steffen, D.,** Proviruses are adjuvacent to c-myc in some murine leukemia virus-induced lymphomas, *Proc. Natl. Acad. Sci. U.S.A.*, 81, 2097, 1984.

62. **Yang, J. Q., Bauer, S. R., Muschinski, J. F., and Marcu, K. B.,** Chromosome translocations clustered 5 of the murine c-myc gene qualitatively affect promoter usage: implications for the site of normal c-myc regulation, *EMBO J.*, 4, 1441, 1985.
63. **Remmers, E. F., Yang, J. O., and Marcu, K. B.,** A negative transcriptional control element located upstream of the murine c-myc gene, *EMBO J.*, 5, 895, 1986.
64. **Bentley, D. L. and Groudine, M.,** Novel promoter upstream of the human c-myc gene and regulation of c-myc expression in B-cell lymphomas, *Mol. Cell. Biol.*, 6, 3481, 1986.
65. **Corcoran, L. M., Adams, J. M., Dunn, A. R., and Cory, S.,** Murine T-lymphomas in which the cellular myc oncogene has been activated by retroviral insertion, *Cell*, 37, 113, 1984.
66. **Li, Y., Holland, C. A., Hartley, J. W., and Hopkins, N.,** Viral integration near c-myc in 10—20% of MCF 247 induced AKR lymphomas, *Proc. Natl. Acad. Sci. U.S.A.*, 81, 6808, 1984.
67. **Selten, G., Cuypers, H. T., Zijlstra, M., Melief, C., and Berns, A.,** Involvement of c-myc in MulV-induced T-cell lymphomas in mice. Frequency and mechanism of activation, *EMBO J.*, 3, 3215, 1984.
68. **Wirschubsky, Z., Tsichlis, P., Klein, G., and Umegi, J.,** Rearrangement of c-myc, pim-1 and Mlvi-I and trisomy of chromosome 15 in MCF- and Moloney MulV-induced murine T-cell leukemias, *Int. J. Cancer*, 38, 739, 1986.
69. **Pear, W. S., Ingvarsson, S., Steffen, D., Munke, M., Francke, U., Bazin, H., Klein, G., and Sumegi, J.,** Multiple chromosomal rearrangements in a spontaneously arising t(6:7) rat immunocytoma juxtapose c-myc and immunoglobulin heavy chain sequences, *Proc. Natl. Acad. Sci. U.S.A.*, 83, 7376, 1986.
70. **Pear, W. S., Nelson, S. F., Axelsson, H., Wahlstrom, G., Bazin, H., Klein, G., and Sumegi, J.,** Aberrant class switching juxtaposes c-myc with a middle repetitive element (LINE) and a IgH intron in two spontaeously arising rat immunocytomas, *Oncogene*, 2, 499, 1988.
71. **D'Ambrosio, E., Waitzkin, S. D., Witney, F. R., Salemme, A., and Furano, A. V.,** Structure of the highly repeated, long interspersed DNA family (LINE or L1Rn) of the rat, *Mol. Cell. Biol.*, 6, 411, 1986.
72. **Classon, M., Henriksson, M., Sumegi, J., Klein, G., and Hammarskjold, M. L.,** Elevated c-myc expression facilitates the replication of SV40 DNA in human lymphoma cells, *Nature*, 330, 272, 1987.
73. **Moriamé, M., Beckers, A., and Bazin, H.,** Decrease of the incidence of malignant ileocaecal immunocytoma in LOU/C rats after surgical removal of the ileocaecal lymph nodes, *Cancer Lett.*, 3, 139, 1977.

Part II
Production of Rat-Rat Hybridomas

Chapter 4

FUSION CELL LINES

C. Digneffe, F. Cormont, B. Platteau, and H. Bazin

TABLE OF CONTENTS

I. INTRODUCTION

There are many fusion cell lines currently utilized for monoclonal antibody-secreting hybridoma production. They are all derived from mouse, rat, or human cell lines. Human hybridomas are still at an early stage of development. On the contrary, both mouse and rat systems are widely used. The mouse system is historically the first to have been developed by Köhler and Milstein[1] as a method for the production of monoclonal antibodies of predefined specificity, and several mouse fusion lines are available (Table 1).

However, the rat system has been reported to have definite advantages at least in some circumstances (see Introduction). A clone of rat immunocytoma, suitable for the derivation of rat-rat hybrid cells, has been developed from a LOU tumor called S210, which was determined by Bazin et al.[10] to be a kappa Bence Jones producer. The tumor was adapted to *in vitro* cell culture by Burtonboy et al.[11] and was given to Milstein in 1972.[12] The name of these tumoral cells has been changed from S210 into Y3-Ag1,2,3. This tumor still secretes its kappa monoclonal immunoglobulin and the hybridoma cell lines derived from it express not only the specific heavy and light chains of immunoglobulin produced by the lymphoid cell fusion partner, but also the myeloma kappa chain.[6] It is clearly preferable in a fusion experiment to use a myeloma partner which by itself does not produce myeloma immunoglobulin chains. This is what everyone does when using the mouse hybridoma technology. Firstly, the avidities of the MAb will always be lower than with a nonsecreting fusion cell line. Secondly, in case of human applications, it is obviously preferable not to inject unrelated myeloma proteins at the same time as useful MAb. No national or international authorities will ever recommend their human uses. Thus, the adhoc working party on Biotechnology/Pharmacy of the Committee for Proprietary Medicinal Products of the Commission of the European Communities in a note (III/859/86-EN, June 1987) to applicants for marketing authorizations on "the production and quality control of MAbs of murine origin intended for use in man" has clearly recommended not to use fusion lines such as the fusion Y3 line, which still secrete myeloma protein. Indeed, its recommendation reads: "the myeloma cell line used for fusion should be preferably selected as one which does not synthetize any immunoglobulin chains".

Another fusion cell line has been derived by Bazin.[7,8] It is a nonsecreting LOU/C immunocytoma that appeared in a 72 week-old male LOU/C rat. This tumor was first adapted to ascitic growth in LOU/C rats and then to *in vitro* cell culture. This rat fusion cell line is called IR983F (abbreviated to 983).

A nonproducing rat fusion cell line called YB2/0 has been derived from a hybridoma obtained by fusion of LOU/C Y3 cells with spleen cells from an AO rat.[9,13] Hybridomas obtained by fusion with this cell line can only be transplanted *in vivo* in (LOU/C × AO)F1 rats, because of the origin of the fusion YB2/O cell line. Using this cell line as a rat fusion line has the disadvantage that, in order to produce rat MAbs in vivo, it is necessary to breed the two parental strains and to cross them in order to obtain F1 hybrids.

Interspecific fusions using mouse fusion cells and spleen cells from a rat often immunized with murine antigens have also been performed.[14-17] In our laboratory, fusions have been made between the mouse myeloma (PAIO) and spleen cells from a LOU rat immunized with peroxidase in order to study their properties (Chapter 19.VI). Obviously, there are several reasons for the interest in preparation of intraspecific hybridomas (rat-rat), which are as easy to obtain as the interspecific one (mouse-rat). The most common motivation is to obtain stable rat-rat hybrids without chromosome loss. A second important point is the possibility to produce large amounts of MAb by growing the hybridoma cells as tumors *in vivo*. A third potential reason is the possibility to save clones from bacterial contamination by *in vivo* transplantation in a histocompatible and immunocompetent system.

There are now quite a range of rodent cell lines available as fusion partners as shown in Table 1.

TABLE 1
Suitable Rodent Cell Lines for Fusion

Species	Cell line: Name	Cell line: Short name	Immunoglobulin expression	Ref.
Mouse	P3-X63/Ag8	X63	IgG1 (kappa)	1
	P3-NS1/1.Ag4.1	NS1	kappa (non-) secreted)	2
	P3-X63/Ag8.653	NS0/1	none	3
	SP2/O-Ag14	SP2/O	none	4
	FO	—	none	5
	PAIO	—	none	CIBA-GEIGY, 1982
Rat	210-RCY3-Ag1,2,3	Y3	kappa	6
	IR983F	983	none	7,8
	YB2-O	YB2	none	9

II. SELECTION PROCEDURE

If it is desired to produce a long term hybrid cell line from two cell types and with one or both of them being from a continuous cell line, a selection procedure is required. The description of the (HAT)-selective medium and the use of mutant cell lines to isolate hyrids by Littlefield[18] solves this problem. The principle behind HAT selection is as follows. Cells can synthesize deoxyribonucleic acid (DNA) either by using a salvage pathway or by *de novo* synthesis. The drug aminopterin blocks *de novo* synthesis, thus requiring cells to use a salvage pathway which depends on the presence of DNA enzymes, hypoxanthine-guanine phosphoribosyltransferase (HGPRT) and thymidine kinase (TK). In the presence of aminopterin and excess of hypoxanthine and thymidine (HAT medium) cells will proliferate and survive as long as they have the enzymes TK and HGPRT. However, if a cell mutant is deficient in either TK or HGPRT, it will not survive in HAT medium (Figure 1). These enzyme deficiencies are not detrimental to the cell under the standard culture conditions since cells can still make DNA using folic acid via *de novo* synthesis. It is only when the *de novo* pathway is blocked with aminopterin that cells with these enzyme deficiencies cannot proliferate. Hybrids between these cells and spleen cells, which contain the wild type salvage pathway enzymes, can be selected from the parental cells as the only cells that actively multiply in HAT medium. In general, it is easier to select mutants for HGPRT-negative cells rather than for TK-negative ones. In most mammals, the enzyme HGPRT is coded by the X chromosome and there is normally only one active copy of the X chromosome per cell.[19] Consequently, only one mutation is required in theory for an HGPRT-negative mutant to be generated.

The mutants are usually selected among those able to grow in the presence of 8-azaguanine (Sigma grade II No. A1007) at a concentration of 20 μg/ml.

The selection in azaguanine can lead to the production of cell lines with normal levels of enzyme which will grow in HAT medium.[21,22] In that case, the cells are selected in presence of 6-thioguanine (Sigma No. A-4882) at gradually increasing concentrations starting around 4 μg/ml and leading up to levels from 10 to 20 μg/ml after 4 to 6 weeks.

If there is extensive cell death, the surviving cells can be purified on Ficoll. The resistant cells are cloned several times by limiting dilution, and clones are chosen for high growth rate and for HAT sensitivity by placing them in the selection medium. Y3, YB2, and 983 cell lines are all HGPRT-negative cell lines.

TK-negative cell lines are selected in a similar way with bromodeoxyuridine as the selective agent.

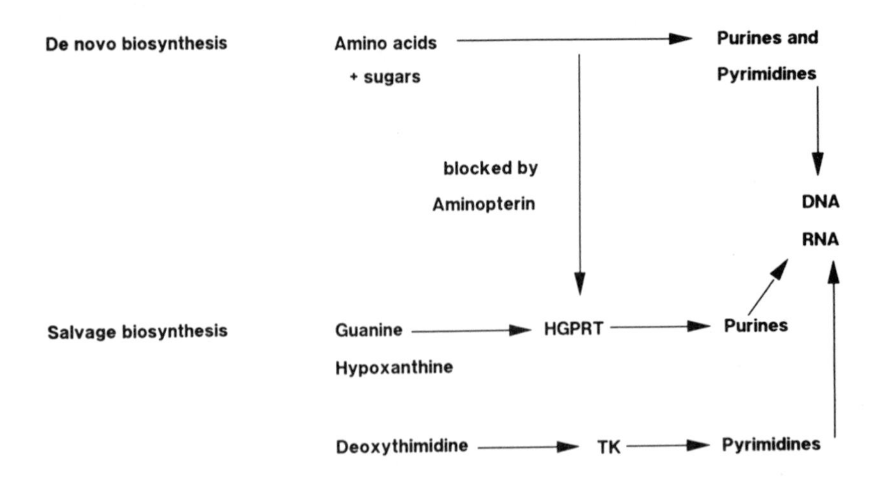

FIGURE 1. Nucleic Acid Biosynthesis — The pathways available to the cell include the *de novo* synthesis and salvage routes. HGPRT = hypoxanthine guanine phosphoribosyl transferase and TK = thymidine kinase.

III. MAINTENANCE OF MYELOMA FUSION CELL LINE

A. MEDIUM

The 983 cells grow well in supplemented Dulbecco's Modified Eagle Medium (DMEM) which contains 4.5 g/l of glucose and 2 m*M*/l of glutamine, without Hepes (Gibco Cat. No. 041-1965). This medium is supplemented with:

1. 5% heat inactivated horse serum (30 min at 56°C)
2. 5% heat inactivated fetal calf serum
3. 1% non essential aminoacids (NEAA) (Gibco No. 043-1140)
4. 1% sodium pyruvate 100 m*M* (Gibco No. 043-1360)
5. 0.1% gentamycin sulfate 50 mg/ml (Gibco No. 043-5750)
6. 1% glutamine (Gibco No. 043-5030)

All serum batches used for culture purposes must be tested for their ability to give the 983 cell line a optimal doubling time of around 14 h (range 12 to 16 h), a cloning efficiency of 60 to 70% and a fusion efficiency of 5 clones or more per 10^6 lymphocyte cells or more. The HAT medium is prepared by adding 2% of HT supplement (Gibco No. 043-01065H) and 1% of aminopterin (Gibco No. 043-01362D).

B. EXPONENTIAL GROWTH

The successful generation of hybrids is strongly depending on the high viability and even exponential growth of the parental myeloma culture. Exponential growth implies that the average doubling time for the cell culture is constant. The different ways to obtain this are:

1. Adjust the cell concentration to 5×10^5 cells/ml every day for 6 days[23]
2. Adjust the cell concentration to 2.5×10^5 cells/ml in 10 ml every 2 d, 6 d before the fusion
3. Put the cell concentration back to 10^5 cells/ml in 10 ml of complete medium on days 0, 3, and 5. On day 7, cells may be collected for hybridization.

The doubling time of mammalian cells in culture is virtually never less than 12 h. For fusion, the 983 cell line must have a doubling time of around 14 h. The doubling time (dt) may be calculated from the formula:

$$dt = 0.693 \cdot t/\log_e N/No$$

where t = elapsed time, No= starting number of cells, N = final number of cells, and e = 2.7183

Inappropriate maintenance of parental cells is a frequent cause of the failure of hybrids to grow following HAT selection. In this context it is important to be familiar with the myeloma cell line being used in regards to normal cell contour and morphology, size, cluster formation, and rate of growth.

C. CONTAMINATION CONTROL

There is no doubt that the failure of many hybridoma experiments is due to contamination and most of this is probably mycoplasma. Fungal and bacterial contaminations may also occur, but are more easily avoided.

Mycoplasmas are small prokaryotes without cell walls which grow in and around the cytoplasm of the mammalian cells which they have infected. Infection with mycoplasma causes culture to die out gradually, but this is controllable by injecting cells into histocompatible and immunocompetent rats and recovering the mycoplasma-free cells when the ascite is taken off. Several passages are sometimes necessary to "clean" a cell line from mycoplasma.

There are several tests for mycoplasma but in our laboratory we used a fast cytological detection of mycoplasma.[24] Method:

1. Cell suspension is prepared at a concentration of 2.5×10^6 cells/ml.
2. Supernatant is removed.
3. The pelleted cells are resuspended in 10 ml of warm (37°C) hypotonic medium prepared just before use (sodium citrate solution 1% is adjusted to pH 7.4 with NaOH 3 *M*).
4. After 10 min, cells are centrifuged and the supernatant is removed except for 1 ml.
5. The pelleted cells are mixed very slowly with a Pasteur pipette.
6. Fixing Carnoy medium (ethanol acetic acid 3/1) is added dropwise to the cells (10 ml).
7. The steps 5 and 6 are repeated three times.
8. The cells are kept with 1 ml of Carnoy fixing for half an hour.
9. One drop of the suspension is put on a frozen slide and dried at room temperature.
10. For the staining, the slides are put for 10 min in the dye solution (Giemsa 3 ml/100 ml PBS).
11. The slides are rinsed with tap water and dried at room temperature before being observed under microscope at visible light.
12. Microscopic examination is performed using phase optics.

The infected cultures clearly show an abundance of darkly skinned granular material located in the peripheral parts of the cytoplasm and in the intercellular spaces.

IV. STABILITY OF THE 983 CELL LINE

The possibility of reversion to an aminopterin-resistant line is always present and it is preferable to confirm the sensitivity of the 983 cells to HAT before fusion. Normally after 2 d, the cells die rapidly in HAT medium while they grow at an exponential rate in HT medium.

We have never observed this phenomenon for the 10 years that we have been using the 983 cell line, but it is probably possible for a proportion of the 983 cells to revert, and once a year myeloma cells are submitted to selection in 8-Azaguanine (20 μg/ml) to kill revertants and the cells are recloned.

V. AVAILABILITY

The IR983 (abbreviated to 983) cell line can be obtained from:

1. Europe — Prof. Hervé Bazin, Experimental Immunology Unit, Faculty of Medicine University of Louvain, Clos Chapelle aux Champs 30.56, Brussels 1200, Belgium, Telelephone 32.2.764.34.30, Telex UCL 23722, Fax 32.2.764.39.46
2. U.S. —Zymed Laboratories, Inc., 52 South Linden Avenue, Suite 5, South San Francisco, CA 94080, U.S., Tel. (415)871-4494, Telex 4993181

REFERENCES

1. **Köhler, G. and Milstein, C.,** Continuous cultures of fused cells secreting antibody of predefined specificity, *Nature,* 256, 495, 1975.
2. **Köhler, G. and Milstein, C.,** Derivation of specific antibody-producing tissue culture and tumor lines by cell fusion, *Eur. J. Immunol.,* 6, 511, 1976.
3. **Kearney, J. F., Radbruch, A., Liesegang, B., and Rajewsky, K.,** A new mouse myeloma cell line that has lost immunoglobulin expression but permits the construction of antibody-secreting hybrid cell lines, *J. Immunol.,* 123, 1548, 1979.
4. **Shulman, M., Wilde, C. D., and Köhler, G.,** A better cell line for making hybridomas secreting specific antibodies, *Nature,* 276, 269, 1978.
5. **Fazekas de St Growth, S. J. and Scheidegger, D.,** Production of monoclonal antibodies: strategies and tactics, *J. Immunol. Methods,* 35, 1, 1980.
6. **Galfré, G., Milstein, C., and Wright, B.,** Rat × rat hybrid myeloma and a monoclonal anti-Fd portion of mouse IgG, *Nature,* 277, 131, 1979.
7. **Bazin, H., Crzych, J. M., Verwaerde, C., and Capron, A.,** A LOU rat non-secreting myeloma cell line suitable for the production of rat-rat hybridomas, *Ann. Immunol. (Inst. Pasteur),* 131D, 359, 1980.
8. **Bazin, H.,** Production of rat monoclonal antibodies with the LOU rat non-secreting IR983F myeloma cell line, in *Protides of the Biological Fluids,* Peeters, H., Ed., Pergamon, Oxford, 1982.
9. **Kilmartin, J. V., Wright, B., and Milstein, C.,** Rat monoclonal antibodies derived by using a new non-secreting rat cell line, *J. Cell Biol.,* 93, 576, 1982.
10. **Bazin, H., Deckers, C., Beckers, A., and Heremans, J. F.,** Transplantable immunoglobulin-secreting tumours in rats. I. General features in LOU/Wsl strain rat immunocytomas and their monoclonal proteins, *Int. J. Cancer,* 10, 568, 1972.
11. **Burtonboy, G., Bazin, H., Deckers, C., Beckers, A., Lamy, M., and Heremans, J. F.,** Transplantable immunoglobulin-secreting tumours in rats. III. Establishment of immunoglobulin-secreting cell lines from LOU/Wsl strain rats, *Eur. J. Cancer,* 9, 259, 1973.
12. **Bazin, H.,** More on rat monoclonal antibodies (Letter to the editor), *Immunol. Today,* 4, 274, 1983.
13. **Galfré, G. and Milstein, C.,** Preparation of monoclonal antibodies: strategies and procedures, in *Methods in Enzymology,* Vol. 73, Immunochemical Techniques, (Part B), Langone, J. J. and Vunakis, H. V., Eds., Academic Press, New York, 1981.
14. **Galfré, G., Howe, S. C., Milstein, C., Butcher, G. W., and Howard, J. C.,** Antibodies to major histocompatibility antigens produced by hybrid cell lines, *Nature,* 266, 550, 1977.
15. **Springer, T., Galfré, G., Secher, S., and Milstein, C.,** Monoclonal xenogeneic antibodies to murine cell surface antigens: identification of novel leukocyte differentiation antigens, *Eur. J. Immunol.,* 8, 539, 1978.
16. **Schroder, J., Autio, K., Jarvis, J. M., and Milstein, C.,** Chromosome segregation and expression of rat immunoglobulins in rat/mouse hybrid myelomas, *Immunogenetics,* 10, 125, 1980.
17. **Kincade, P. W., Lee, G., Sun, L., and Watanabe, T.,** Monoclonal rat antibodies to murine IgM determinants, *J. Immunol. Methods,* 42, 17, 1981.
18. **Littlefield, J. W.,** Selection of hybrids from matings of fibroblasts *in vitro* and their presumed recombinants, *Science,* 145, 709, 1964.
19. **Lyon, M. F.,** Gene order in the X chromosome of the mouse, *Nature,* 190, 372, 1961.
21. **Littlefield, J. W.,** The inosinic acid pyrophosphoryase activity of mouse fibroblasts partially resistant to 8-azaguanine, *Proc. Natl. Acad. Sci. U.S.,* 50, 568, 1963.
22. **Cox, R. and Masson, W.,** Do radiation-induced thioguanine-resistant mutants of cultured mammalian cells arise by HGPRT gene mutation or X-chromosome rearrangement?, *Nature,* 276, 629, 1978.
23. **Lebacq-Verheyden, A. M., Neirinck, A., Ravoet, A. M., and Bazin, H.,** Rat hybridoma technology: culturing of rat myeloma cell line IR983F prior to cell fusion, *Hybridoma,* 2, 355, 1983.
24. **Fogh, J. and Fogh, H.,** A method for direct demonstration of pleuropneumonia-like organisms in cultured cells, *Proc. Soc. Exp. Biol. Med.,* 117, 899, 1968.

Chapter 5

IMMUNIZATION OF RATS

F. Ackermans, F. Nisol, and H. Bazin

TABLE OF CONTENTS

I. INTRODUCTION

There is no universal method of immunization for rat-rat hybridoma production. Many protocols have been published and their diversities reflect the difficulty in finding a good one. In many cases, the experiments have never been repeated and the so-called "new and efficient method of immunization" is often no more than a lucky result which cannot be reobtained because of a number of uncontrolled parameters. However, a number of basic points must be taken into account when a protocol of immunization has to be chosen.

In Chapter 1, we have summarized data on rat immune responses. Many methods can be used to obtain high humoral immune responses in immunized rats, but they must be adapted for hybridoma production because the best time for cell fusion is not when the highest concentration of circulating antibodies or even of antibody forming cells is produced. As it will be noted in Chapter 6, at the time of fusion the best criterium of immunization is the absolute amount of lymphoblasts, although it is in fact much more a stimulation than an immunization index, at least when the antigenic preparation is not pure.

II. ANIMALS

A. RAT STRAINS

Rat immunocytoma fusion cell lines are all derived from LOU origin. Both IR983F and Y3-Ag1,2,3 are derived from LOU tumors which can grow in LOU (LOU/C, LOU/M) or histocompatible rats with LOU/C.IgK-1b. The Y2B cell line results from a fusion between the Y3-Ag1,2,3 cell line and B lymphocytes from an AO rat. Its use is impaired by this origin as the only possible histocompatible rats susceptible to grow Y2B made hybridomas are (LOU × AO)F1 hybrids, and only if the immunized B lymphocytes are from the LOU or the AO inbred strains. When another rat is used, no histocompatible rat exists. Table 1 gives the histocompatible rats to hybridomas made by fusing rat cells from the 983, Y3, or Y2B cell lines and B lymphocytes from various origins.

For unclear reasons, it is sometimes easier to adapt hybridomas to *in vivo* culture in a purely histocompatible system (LOU hybridomas in LOU rats) than in F1 hybrids, for example LOU hybridomas in (LOU × DA)F1 or in (LOU × BN)F1.

Problems of histocompatibility can be overcome by a sublethal irradiation (4 to 6 Gy, depending on the animals and ionizing radiations). However, in such immunodeficient animals, no salvage of contaminated hybridoma cell lines can be obtained. But, in many circumstances, such irradiated rats could be very useful for the production of MAbs in nondirectly histocompatible systems.

Considering these various points, LOU rats are the best choice to make rat hybridomas. However, a better immune response could be obtained in other rat strains. The availability of F1 hybrids must be considered for a possible salvation of interesting clones and *in vivo* production before choosing the rat strain to immunize.

B. AGE OF THE RATS

Rats between 8 and 12 weeks old are generally used for immunization. Up to 6 months, they can be used without great problems. Younger rats, even at weaning age (4 weeks old), can sometimes give an immune response similar to that of adult rats. Up to 1 year old, rats have an adult immune potential and can be used for immunization. However, it is better not to use rats older than 1 1/2 years old at the time of the fusion. Concretely, 12-week-old rats are the most valuable for immunization; from 3 to 6 months old, they have the best immune responses; and from 6 to 12 months old, they can still be used. It is generally better to avoid the use of older rats.

C. SEX OF THE RATS

Female rats are commonly preferred to male, as sometimes female rodents produce more

TABLE 1
Histocompatible Rats with Hybridomas Made by Fusing Cells from the 983, Y3 and Y2B Fusion Lines and Immunized B Lymphocytes from Various Origins

Immunized B lymphoid from	Hybridoma done with: Fusion lines 983	Y3	Y2B	Histocompatible rat with the hybridoma	Remark
LOU[a]	×			LOU or LOU F1 hybrid[b]	—
LOU[a]		×		LOU or LOU F1 hybrid[b]	kappa secretor
LOU[a]			×	(LOU × AO)F1 hybrid	
Other inbred rat (not from the LOU inbred strain)	×			F1 hybrid of the same strain[c]	—
Other inbred rat (not from the LOU inbred strain)		×		F1 hybrid of the same strain[c]	kappa secretor
AO rat			×	(LOU × AO)F1 hybrid	—
Other inbred rat (not from the AO inbred strain)			×	None	—

[a] LOU means LOU/C, LOU/M, and congenic line for immunoglobulin loci on the LOU/C genetic background.
[b] F1 hybrid made by a cross between LOU rat and any other rat (even outbred).
[c] F1 hybrid made by a cross between LOU rat and any other rat from the inbred strain which has been used as source of immunized spleen cells.

antibodies than male. However, this fact is rarely observed. LOU male rats housed together do not fight and can be kept for a long time without any problems. To our knowledge, the great majority, if not all rat inbred strains, have similar behavior.

III. ANTIGENS

There are still many doubts about the best techniques to present an antigen to a rodent in order to obtain the best immune response. Many experimental data have been published on the rat immune responses (see Chapter 1). Moreover, the majority of facts known about the mouse can be considered, at first sight, as also applicable to the rat species.

A. HAPTEN-CARRIER RELATED PROBLEMS

Immunization schedules cannot be considered separately from the nature of antigens. Indeed, insoluble molecules seem more immunogenetic than soluble ones. Soluble antigens, proteins, lipids, or polysaccharides require different adjuvants and immunization protocols.

To obtain rat MAbs, highly-purified antigen is not required for immunization. But, if a solution containing different components is used for immunization, the major antigens are frequently known to dominate. In order to obtain monoclonal antibodies against minor antigens of a solution, it is often necessary to purify them.

Rats must be inoculated with soluble antigens and potent adjuvants must be used: complete Freund's adjuvant, alumine hydroxide, and *Bordetella pertussis* organisms (see adjuvants). Likewise, good immunizations can be achieved if aggregated forms are used. Heating soluble antigens to 56°C for 30 min will give aggregates, but will sometimes modify their antigenic properties.

Coupling soluble antigens or small molecules to a carrier will also greatly increase the immunogenicity of the component. Convenient carriers for the rat species are ovalbumin or the keyhole limpet hemocyanin.

Ovalbumin can be coupled to a molecule or a hapten and will increase the immune response of the animal : 300 mg (or more) of antigens in 3 ml Na_2CO_3 0.2 *M* pH 10.7 and 150 mg ovalbumin grade V (Sigma) in 15 ml Na_2CO_3 0.2 *M* pH 10.7 was mixed 4 h at room temperature (in the dark). Filtration was performed on a Sephadex G25 chromatography with PBS (0.05 *M*, pH 7.5).

Nippostrongylus brasiliensis (Nippo) or Ascaris extract can also be used as a carrier, often to obtain IgE isotype of monoclonal antibodies (see Chapter 19.XI).

Molecules poorly immunogenetic can be directly coupled to cellulose[1] or charcoal (De Clercq and Bazin, unpublished results). Soluble proteins (8 mg in aqueous solution) were mixed with 400 mg of activated coal (Aktiv Kohle, Merck) and incubated for 2 h at room temperature. The solution was filtered and washed with physiologic water. Activated coal conjugated with protein was collected and resuspended in 4 ml physiologic water. The amount of 0.5 ml of this solution was used per injection. In some cases, the results were interesting, but this technique cannot be considered general.

B. DOSES OF ANTIGENS.

At the level of an antigen-presenting cell, a T-helper lymphocyte or a B lymphocyte, the quantity of antigen susceptible to induce an immune response is probably identical, whatever the species from which the cells are derived. That means the dose of antigen to be administered to a rat is identical to that administered to a mouse if the antigen can easily reach a lymphoid organ (footpad route, for example); the dose will be slightly larger when a dilution appears because of the size of the rat in comparison with that of the mouse (intraperitoneal route, for example). The doses also depend on the antigen. Doses of 50 to 500 μg in 0.1 to 1 ml of PBS are generally used. Doses that are too high can induce tolerance and must be avoided. However, in cases of oral route administration, the dose can be very large. Doses of less than 1 μg can be used for IgE induction, but they need special adjuvant.

C. ADJUVANTS

1. Complete Freund's Adjuvant

Many adjuvants can be used for rat immunizations. Until now, the most efficient has been the Freund Complete Adjuvant (FCA) which is widely used. For normal immunization, i.e., to obtain MAbs from the IgM and/or IgG subclasses, FCA is very convenient. It is still the most powerful adjuvant and is used as a reference for all new adjuvants. Preparation of FCA and antigen is important and must be made correctly. Reagents:

1. Complete Freund's adjuvant (Difco, USA)
2. Solution of antigen (1 mg/ml), if possible in PBS or saline
3. Doses of antigens used with this adjuvant vary between 50 and 500 μg per injection.

It is traditional to use equal volumes of antigen and adjuvant emulsion. A typical volume of emulsion is 400 μl to 1 ml per rat, per injection.

To obtain the adequate preparation which should be thick and creamy, different methods could be used. It is important to know that is always the aqueous phase (antigen in PBS or saline) that is injected into the oil and not the oil into the aqueous phase. As described by Goding,[2] a double hubbed needle joining two glass syringes could be used with success. A Vortex system, which consists in injecting small aliquots of antigen solution in the complete Freund's adjuvant with repeated vortexing, is also frequently utilized.

The emulsion can be tested by gently letting small drops of the preparation fall onto the surface of warm water (30 to 40°C). The majority of them must remain intact at the surface of the liquid.

Remark — Complete Freund's adjuvant can be dangerous, if it is accidentally injected in the fingers or spread in the eyes. Care must be taken to avoid these dangers.

2. Alumine Hydroxide

This is an old adjuvant which is commonly employed in vaccine preparation. It can be prepared as described by Pauwels et al.[3] Alumine hydroxide is obtained by mixing equal volumes of 2 *N* $AL_2 (SO_4)_3$ and 2 *M* NaOH. After washing, the preparation is homogenized with a mixer and NaCl and Tris are added (final concentration 0.15 *M* NaCl and 0.01 *M* Tris). NaOH 1 *M* is used to adjust the final pH of the solution to 7.5.

Dry weight is determined with the final $Al(OH)_3$ solution. Pharmaceutical preparations (or commercial preparations, Amphogel-Wyeth) are also available. Doses of 5 to 6 mg per injection mixed with the antigen solution can be used. Doses of 5 to 100 µg of antigen are commonly injected with these adjuvants.

3. *Bordetella pertussis* Organisms

These organisms are very good adjuvants which are often employed for IgE immune response. Unfortunately, it is more and more difficult to find these organisms on the market as the majority of firms sell associated vaccines including *Bordetella pertussis* organisms, amongst others. However, it is possible to order large quantities from commercial firms, which sometimes will agree to prepare a special batch for this use. Another possibility is to ask a microbiologist who can prepare the organisms for you.

One dose of 10^{10} *Bordetella pertussis* organisms is enough, injected separately or together with the first injection of antigen by the intraperitoneal route.

IV. ROUTE OF IMMUNIZATION

Many possibilities exist concerning the route of immunization, the easiest being the intraperitoneal. But various other routes could be used when special types of immunization are requested. All rats are immunized under light ether anesthesia as described in Chapter 2.

A. THE INTRAPERITONEAL ROUTE

On the whole, this is similar to the intravenous route for soluble antigens of relatively small molecular weight. In the case of the administration of antigen mixed with FCA, it will stay in the peritoneal cavity and create many granulosae in it. This route is suitable for insoluble and soluble antigens. It can be used with any adjuvants. It will lead to a systemic immunization. The final boost before fusion is generally given intravenously or intraperitoneally and in that case the best organ in which to find immunized cells is the spleen. All isotypes can be found after such an immunization, but there is a special emphasis on IgM-IgG isotypes (Figure 1).

B. THE INTRAVENOUS OR INTRACARDIAC ROUTE

The majority of the antigen will reach the spleen. This route is rarely used for primary immunization; it is often used for boosting before fusion (Figure 2).

C. THE SUBCUTANEOUS ROUTE

It can be used instead of the intraperitoneal route in order to avoid too many injections by this route, sometimes giving adherences between omentum, peritoneum, and the various organs of the peritoneal cavity (Figure 3).

D. THE FOOTPAD ROUTE

The footpad route is very efficient when the available quantity of antigen is limited. Too large a quantity of adjuvant, such as FCA, must be avoided as the local reaction is acute and can be unnecessarily painful for the animal. Only one injection is made with FCA. For this route, it is better to use *Bordetella pertussis* (10^{10}) as adjuvant. The immune response is pratically localized in the popliteal lymph nodes. It leads to IgM-IgG type immune responses (Figure 4).

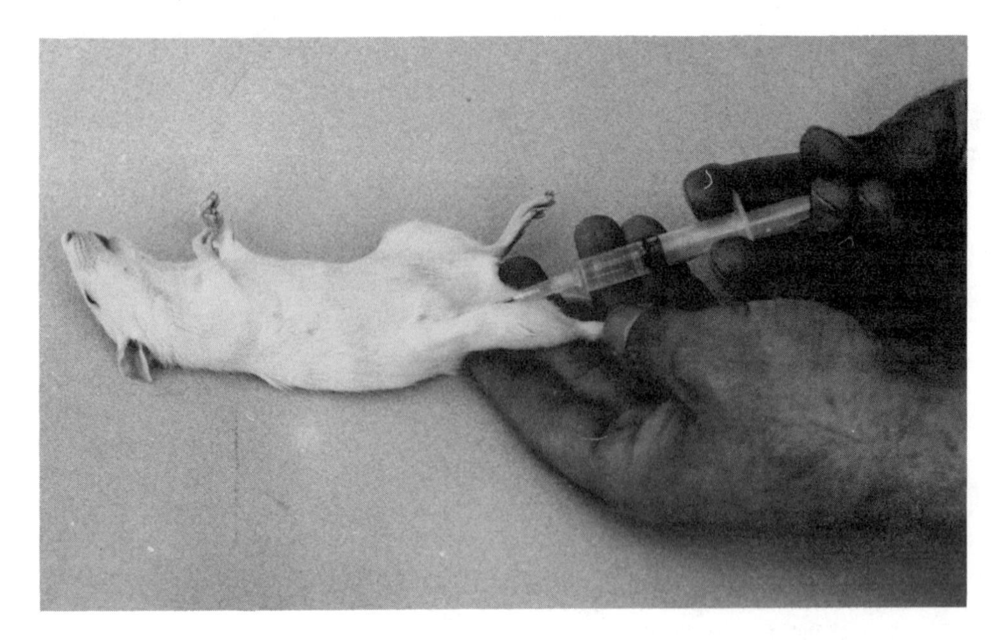

FIGURE 1. The intraperitoneal route.

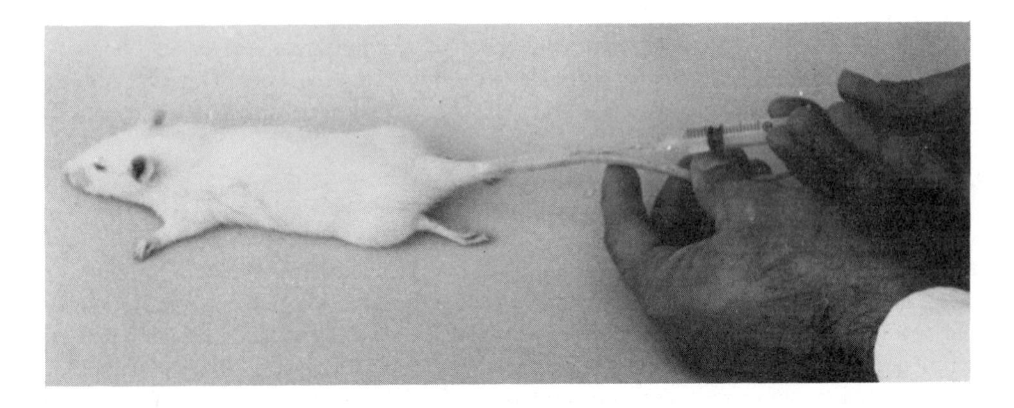

FIGURE 2. The intravenous route.

E. THE INTRASPLENIC ROUTE

Efficient in the rat as well as in the mouse species,[4,5] it gives a good primary immune response with pratically only IgM-secreting hybridomas.

F. THE ORAL ROUTE

The oral route obtained by feeding the rat with the antigen could give interesting results, and particularly because it produces IgA-secreting hybridomas. The dose and the duration of antigenic administration must be prolonged for a long time. However, in the mouse species, Dean et al.[6] have proposed to immunize in the Peyer's patches and to use the thoracic duct lymph as the source of immunized cells. It is much easier to take the mesenteric lymph node, where after oral immunization there are many IgA-secreting cells[7,8] (Figure 5).

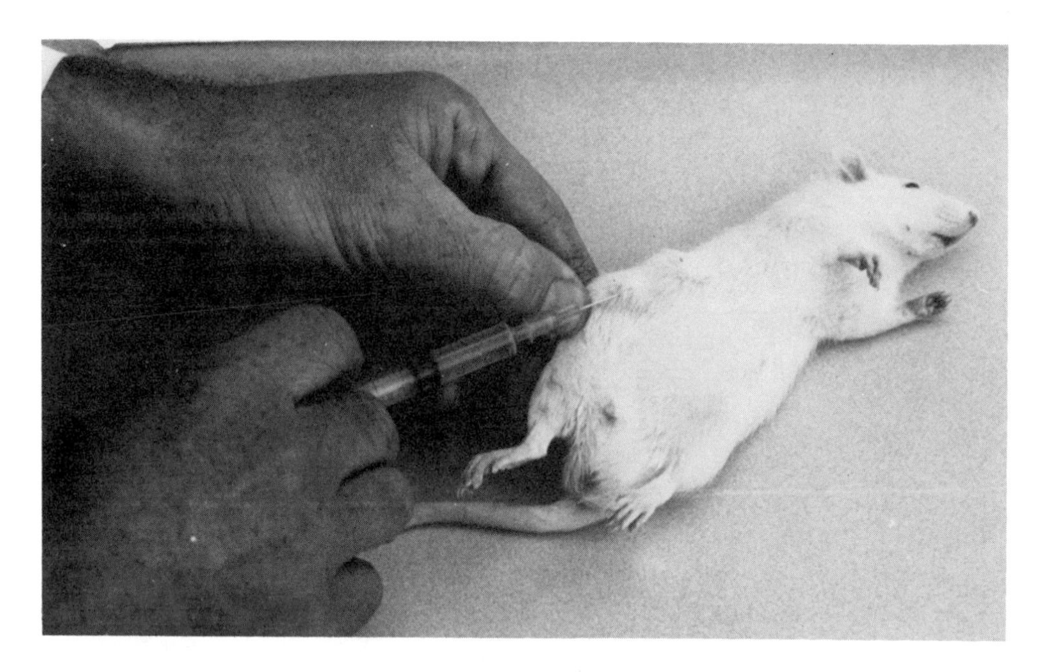

FIGURE 3. The subcutaneous route.

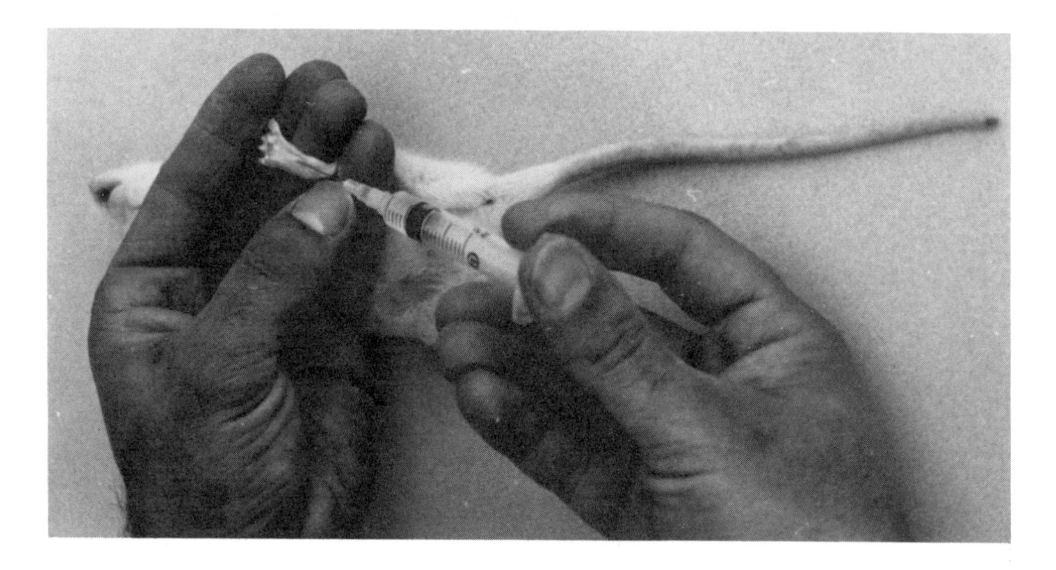

FIGURE 4. The footpad route.

V. TOLERANCE

In many circumstances, it could be beneficial to induce a tolerance against contaminating or dominating antigens which are difficult or impossible to eliminate from the antigenic preparations.

As described in Chapter 1, tolerance is difficult to induce, especially when the substance is highly immunogenic. Various methods have been described, but a high and repeated administration of antigen as soon as the birth seems to be the most efficient.

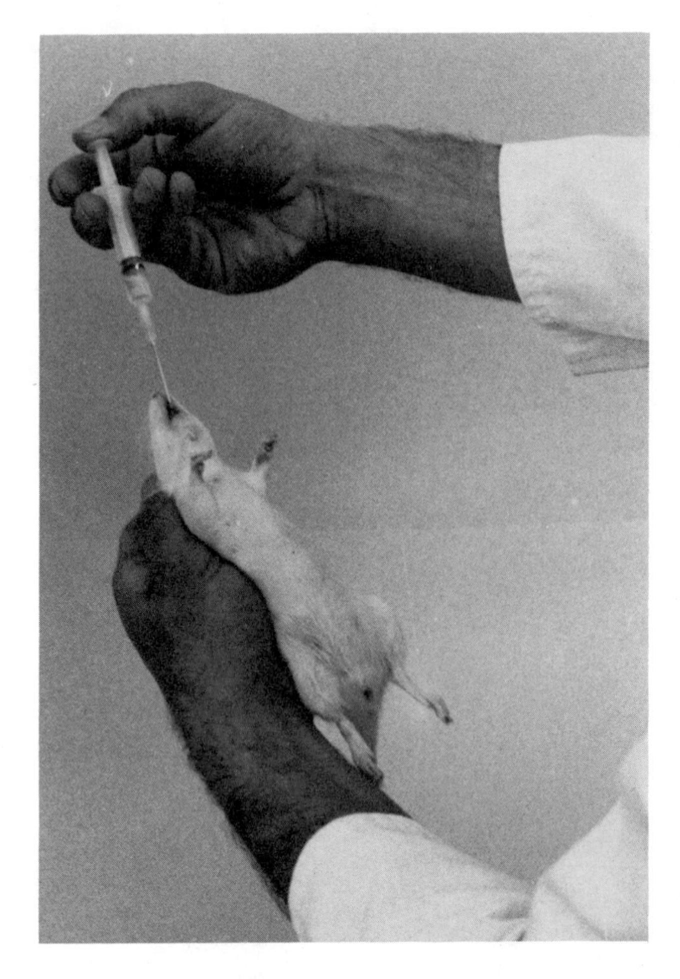

FIGURE 5. The oral route.

VI. TRICKS FOR DEPLETING IMMUNOSUPPRESIVE T LYMPHOCYTES

Various techniques have been described. They can be efficient in immunizing methods which are highly susceptible to T-suppressor cells, as well as the IgE immune responses: injection of antithymocyte serum,[9] which can be replaced by MRC-OX8 monoclonal antibody injection[10] or a low dose of irradiation.[9] However, enhancement of the IgE immune response by irradiation can be obtained in low or medium responder, but not in high responder animals.[11] With such techniques, we have obtained very interesting individual results, but we have never been able to elaborate a reproducible method allowing us to repeat the experiments with different antigens and animals.

VII. THE CHOICE OF A METHOD OF IMMUNIZATION

The choice of a method of immunization depends on various parameters: the nature and availability of the antigen and the isotypes and avidity of monoclonal antibodies desired. A few examples will be given.

A. GENERAL PROTOCOL FOR SOLUBLE ANTIGEN

1. Short Term Immunization (Footpad or Intrasplenic Injection)

Footpad immunization — Female or male rats between 3 and 6 months old are used for this type of immunization. Soluble antigens on their own will rarely give interesting hybridomas. It is necessary to insolubilize the antigen before using this technique.

At day 0:150 μl of antigen (1 mg/ml) + 150 μl of *Bordetella pertussis* (10^{10}); day 3: 150 μl of antigen (1 mg/ml) + 150 μl of alumine hydroxide; day 6: idem; day 9: idem; day 12: fusion with popliteal lymph nodes.

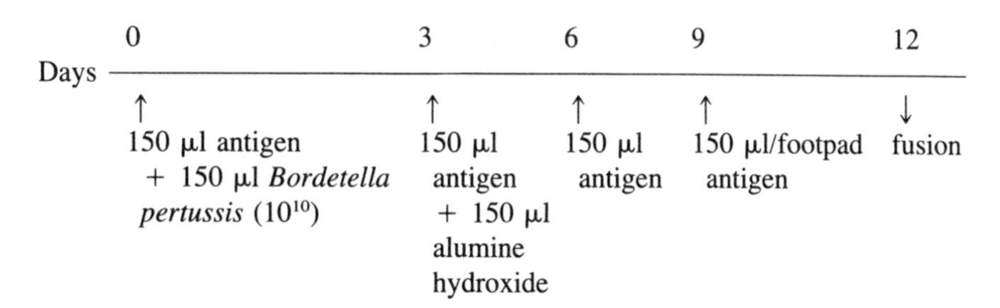

No test of immunization is possible before the fusion. The number of lymphoblasts can give an indication of the stimulation, as well as of previous similar immunization.

Intrasplenic route — This technique is widely described by Spitz et al.[4] and Spitz and Spitz.[5] An amount of 20 μg of antigens in 50 or 100 μl PBS is used for an intrasplenic injection. In short, a rat (between 3 and 6 months) is anesthesized with ether and the animal is placed on its right side. The spleen is exposed and the needle is inserted deeply into this organ; the antigen is moved as the needle is pulled out in order to distribute the antigen through a large part of the spleen. The peritoneum and the muscle layer are sutured together and then the skin is closed. Fusion with spleen cells was made 4 d after the single injection.

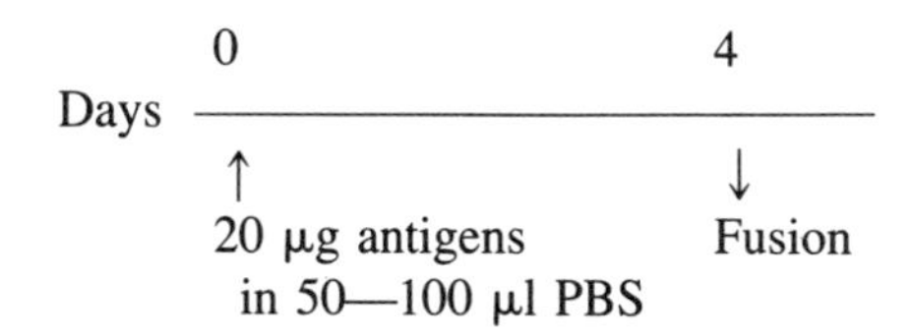

2. Long Term Immunization

Three month-old female rats are immunized with two or more injections of antigen plus adjuvants, at 10 to 15 d intervals. Two weeks after the second or the third injection, serum samples are taken from the animals and tested for their immune response. If necessary, rats are given one or several more injections and they are again tested for their immune response. The rats giving the highest antibody response are selected and the others are discarded. The rest period is generally 4 to 6 months. The spleen cells are collected for fusion 4 d after an intravenous or intraperitoneal boost.

In practice, three or more 10 to 12 week-old LOU/C or LOU/M female rats are given twice 50 to 500 (generally 100) μg of antigen in 0.2 ml of buffer mixed with 0.2 ml of FCA by the intraperitoneal route at an interval of 2 weeks. If the antigen is not well characterized or if its immunogenecity is unknown, it is more convenient to test three doses of antigen as 50, 250, and 500 μg per injection.

The rats are bled by the retro-orbital sinus 10 d later and their sera are tested by Ouchterlony, immunodot, ELISA, or RIA and are compared with a normal serum.

If the immune responses are unsatisfactory, the rats are given a new series of two or three injections by the subcutaneous or intramuscular routes with a dose of antigen which gave the best result after the first series of injections. For the second series of injections, alumine hydroxide is generally chosen as adjuvant. The interval between injections is still 10 days to 2 weeks. The rats are tested 10 d after the last injections for their immune responses and the rats which have the highest titer in antibodies are kept for the fusion. As far as mouse fusion[12] is concerned, we notice that overimmunization does not help to produce rat hybridoma clones.

The antibody test can include a selection not only for specific antibodies against the antigen, but also for their isotypes (see Chapters 9 and 10).

The rest period is generally from 4 to 6 months. Rats are preferably never used for fusion when they are 1 year old or more.

The boost is administered by the intravenous, intracardiac, or even sometimes intraperitoneal route and the fusion is made 3 or 4 d later. The dose of antigen must be large, and when it is possible from 100 to 500 μg in PBS or saline.

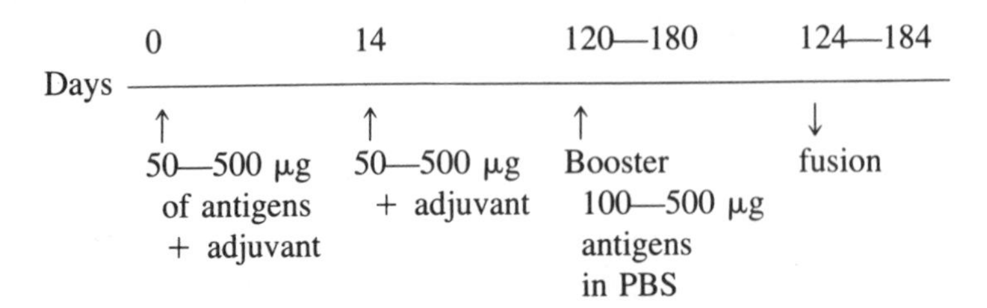

B. GENERAL PROTOCOL FOR INSOLUBLE ANTIGENS

1.Viral Antigens

See Chapter 22. I.

2. Bacterial Antigens

A simple, effective, and short protocol of immunization consists of injecting bacteria into the two footpads of a LOU/C rat as it has been described for soluble antigens. For the first immunization 150 μl of fresh bacteria (10^7 to 10^9) in PBS was mixed with 100 μl of *Bordetella pertussis* (10^{10}) (Institut Pasteur Production, France) and the solution was injected into the footpad (250 μl per footpad). At days 3, 6, and 9, only 150 μl of bacteria (10^7 to 10^9) was used for the immunization. At day 12 popliteal lymph node cells were taken and fused with the myeloma cells.

Days	0	3	6	9	12
	↑	↑	↑	↑	↓
	250 μl bacteria + *Bordetella pertussis*	150 μl bacteria	150 μl bacteria	150 μl/footpad bacteria	Fusion

3. Cell Antigens

Frozen cells containing dimethylsulfoxide (DMSO) and 5% fetal calf serum were used for LOU/C rat immunization. A short protocol of immunization gave a great number of hybridoma clones. At day 0, 2×10^7 cells were injected intraperitoneally into the animal, and at day 21, the operation was repeated with 5×10^7 cells. After 4 d the spleen cells were collected for fusion.

Days	0	21	26
	↑	↑	↓
	2×10^7 cells with DMSO	5×10^7 cells with DMSO	Fusion

REFERENCES

1. **Gurvich, A. E. and Korukova, A.,** Induction of abundant antibody formation with a protein-cellulose complex in mice, *J. Immunol. Methods,* 87, 161, 1986.
2. **Goding, J. W.,** *Monoclonal Antibodies: Principles and Practice,* Academic Press, London, 1983.
3. **Pauwels, R., Bazin, H., Platteau, B., and Van der Straeten, M.,** The influence of different adjuvants on the production of IgD and IgE antibodies, *Ann. Immunol. (Inst. Pasteur),* 130C, 49, 1979.
4. **Spitz, M., Spitz, L., Thorpe, R., and Eugin, E.,** Intrasplenic primary immunization for the production of monoclonal antibodies, *J. Immunol. Methods,* 70, 39, 1984.
5. **Spitz, M. and Spitz, L.,** Intrasplenic immunization for the production of monoclonal antibodies, in *Methods of Hybridoma Formation,* Bartal, A. H. and Hirshaut, Y., Eds., Humana Press, Clifton, New Jersey, U.S., 1987, 249.
6. **Dean, C. J., Cyure, L. A., Styles, J. M., Hobbs, S. M., North, S. M., and Hall, J. G.,** Production of IgA secreting hybridomas: a monoclonal rat antibody of the IgA class with specificity for RT1c, J. Immunol. Methods, 53, 307, 1982.
7. **Rits, M., Cormont, F., Bazin, H., Meykens, R., and Vaerman, J. P.,** Rat monoclonal antibodies.VI.Production of IgA secreting hybridomas with specificity for the 2,4 dinitrophenyl (DNP) hapten, *J. Immunol. Methods,* 89, 81, 1986.
8. **Bazin, H., Levi, G., and Doria, G.,** Predominant contribution of IgA antibody-forming cells to an immune response detected in extra-intestinal lymphoid tissues of germ-free mice exposed to an antigen by the oral route, *J. Immunol.,* 105, 1049, 1970.
9. **Tada, T.,** Regulation of reaginic formation in animals, *Prog. Allergy,* 19, 122, 1975.
10. **Brideau, R. J., Carter, P. B., McMaster, W. R., Mason, D. W., and Williams, A. F.,** Two subsets of rat T lymphocytes defined with monoclonal antibodies, *Eur. J. Immunol.,* 10, 609, 1980.
11. **Bazin, H., Platteau, B., and Pauwels, R.,** Genetic control of the IgE reaginic immune response and its radiation-induced enhancement, *Transplant. Proc.,* 13, 1369, 1981.
12. **Oi, V. T., Jones, P. P., Goding, J. W., and Herzenberg, L. A.,** Properties of monoclonal antibodies to mouse Ig allotypes, H-2 and Ia antigens, in *Current Topics in Microbiology and Immunology,* Vol. 81, Melchers, F., Potter, M., and Warner, N. L., Eds., Springer-Verlag, New York, 1978, 115.

Chapter 6.I

FUSION PROCEDURE

A. M. Ravoet and H. Bazin

TABLE OF CONTENTS

I. FUSION PROCEDURE*

A. INTRODUCTION

The fusion of antigen-stimulated B lymphoblasts and myeloma cells gives rise to hybrids combining both the properties of continuous growth of the malignant partner and those of secretion of antibodies directed against a predefined antigen.[1] The method for production of rat × rat hybrids, using the 983 rat immunocytoma line,[2-4] is derived from that described by Galfré et al.[5] The absolute number of hybrid clones depends mainly on the growth rate of the immunocytoma cells; the percentage of specific antibody-secreting hybrids depends on the fraction of freshly-stimulated large lymphoblasts[6] and hence on the immunization conditions.

B. GROWTH RATE OF THE 983 RAT IMMUNOCYTOMA FUSION CELL LINE

Exponential growth of a cell line means that the total amount of cells (N) increases according to the formula N = No × 2t/T with T being the mean doubling time. Hence, when plotting log N vs. t, a linear relationship is found with slope = (log2)/T.

Only in optimal conditions, i.e., in the presence of saturating amounts of nutrients and growth-promoting factors and in the absence of toxic agents, is a linear relationship found (see Chapter 4). Therefore, kinetics of growth of the immunocytoma line using different sera, different cell concentrations (daily adjustment), etc., should be observed and mean doubling time determined. In our laboratory, the mean doubling time for the 983 cells is around 14 h (range of 12 to 16 h) (Lebacq et al.[7] and recent results).

Sensitivity of the immunocytoma line to HAT medium should be checked before fusion and subcloning in 8-Azaguanine performed once a year.

At least 5, but preferentially 10 d before fusion, immunocytoma cells are thawed and grown at their optimal cell concentration. On the fusion day 2×10^7 to 1×10^8 cells must be available.

C. ANTIGEN-STIMULATED B LYMPHOBLASTS

Different immunization schedules have been discussed extensively in Chapter 5. The boost injection can be given either in a single massive dose 4 d before fusion, or in fractionated doses during the 4 d preceding fusion.[6] While the measure of the antibody concentration in the serum is a good indicator of the immune response of the animal after the first injections, the amount of serum antibody at the time of fusion shows no correlation with the amount of antibody-producing hybrids obtained. The percentage and absolute amount of lymphoblasts in the spleen or other lymphoid organs used for fusion, seem the most reliable parameters for estimation of the immunization efficiency.[6] After boosting with 5×10^7 cells (see Chapter 5), we obtained spleens weighing from 500 to 800 mg (unstimulated spleens for animals of the same age are around 350 mg), and we recovered 3×10^8 splenocytes among which 10 to 25% were blasts. Usually, around 30% of the clones secreted anti-target antibodies.[8]

According to the work of Andersson and Melchers,[9] hybrids derive totally from the large cells (sedimenting at 1 g), either because fusion with blast cells is more efficient or because these cells enter more readily in the S phase when, after fusion with a myeloma cell, the immunocytoma nucleus of the heterokaryon does. It is, however, not necessary and not advisable to remove erythrocytes, small lymphocytes and T lymphocytes from the splenocyte suspension before fusion, since these additional manipulations lower the total amount of hybrids produced.

D. HYBRIDIZATION (TABLE 1)

Fusion involves aggregation of cells, fusion of cell membranes, and fusion of cell nuclei.

* All references made in this chapter can be found at the end of Chapter 6.II.

TABLE 1
Scheme for Fusion

6×10^7 exponentially growing 983 cells, washed twice in PBS.
$\pm 3 \times 10^8$ splenocytes of one immunized LOU rat, washed once in PBS.
Mix, remove 1/50 of the mixture for control wells, and centrifuge the mixture.
To the dry pellet, at 37°C, while shaking, add dropwise 2 ml of 43% PEG-10% DMSO in PBS pH 7.5 (over 90 s) and mix during 30 s, add 2 ml of PBS, over 90 s, then add 40 ml of PBS, over 5 min.
Centrifuge.
Resuspend at 1—1.5 × 10^6 cells/ml in DMEMc (without Hepes), 15% horse serum or 15% fetal calf serum, 20% conditioned medium, HAT 2 ×. Keep at 37°C, 5% CO_2, for 1—4 h. Seed 0.1 ml of fused suspension into wells, containing 0.1 ml of WISTAR rat peritoneal cells (20,000 cells per well).

1. Effect of Polyethyleneglycol

Aggregation of splenocytes and immunocytoma cells, followed by fusion of the cell membranes, is obtained by centrifugation of the mixture and addition of polyethyleneglycol (PEG) to the dry pellet.[10] PEG is supposed to bind the H_2O molecules adjacent to the membrane, thereby destabilizing the lipid bilayer and inducing the fusion of the spleen membranes of aggregated cells.[11] These authors have shown that a PEG concentration of 50% (w/w) is required to bring about efficient fusion of membranes in 1 min. We observed that fusion of 3 × 10^8 splenocytes with 1 × 10^8 983 cells using 2 ml of PEG is much more efficient than using 1 ml of PEG, as shown in an example given in Figure 1 (Xia Hanzhang, unpublished results).

Besides and inherent in its high efficiency of induction of membrane fusion, PEG is extremely toxic. Gefter et al.[12] observed that mouse plasmacytoma cells in the presence of 50% PEG first clumped and fused to form asymetric shapes and after 1 min lysed. The optimal contact time can be extended by diluting the PEG with a serum-free medium.[11] With this dilution step being very critical, the medium should be added quite slowly to avoid osmotic shock. The presence of dimethylsulfoxide (DMSO) reduces this problem[13] and enhances the fusion efficiency according to Norwwod et al.[14]

We found that PEG dissolved in PBS and adjusted to a pH of 7.5 to 7.8 induces more hybrids than PEG dissolved in Eagle's Minimum Essential Medium (EMEM), buffered with 10 m*M* HEPES at a pH around 7.8 (absence of CO_2). This could be due to the presence of HEPES, which is toxic when allowed to enter the cells.[15] Klebe and Mancuso,[16] however, found HEPES to be the best buffer among those tested. Some authors[17] use HCO_3^-/CO_2 to buffer the PEG solution. Use of PBS or glucose-containing PBS[18] seems more reliable since it is easily reproducible.

2. Splenocyte: Immunocytoma Ratio

The ratio of splenocytes vs. immunocytoma cells seems to slightly influence the total number of hybrids obtained and more importantly the percentage of specific antibody-secreting hybrids.

Within one experiment, splenocytes were fused with 983 cells at a 2:1 and 4:1 ratio. Seeding was performed, keeping the immunocytoma cell input constant. Results are shown in Table 2; hybrids grew faster and in slightly higher amounts after a 4:1 fusion-ratio, possibly because of a feeder-layer effect of the splenocytes.

The second observation is that the percentage of antitarget MAb-secreting clones is much higher when a 4:1 fusion ratio is used rather than a 2:1 fusion-ratio. This could reflect the fact that at a 4:1 fusion-ratio the stimulated lymphoblasts fuse more readily than the unstimulated lymphocytes, and while at a 2:1 ratio, the selective fusion with lymphoblasts is lowered, giving rise to a higher percentage of irrelevant hybrids. Hence, a ratio of 4: to 5:1 was adopted.

3. Effect of Temperature

Since the work of Fazekas de Saint Groth and Scheidegger,[13] it is known that fusion is more efficient when performed at room temperature or at 37°C, than at 0°C.

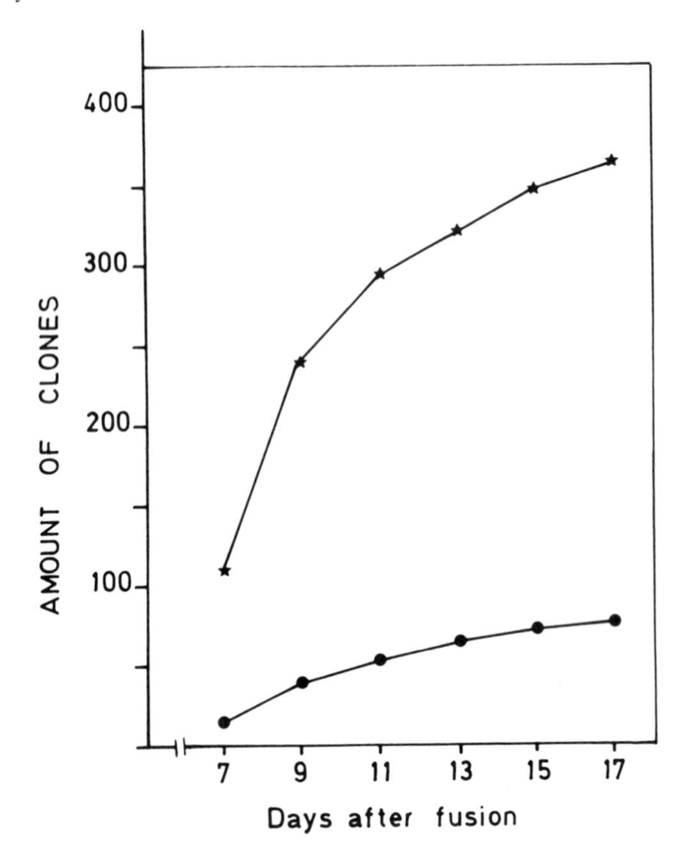

FIGURE 1. Yield of hybrid clones after fusion using different amounts of 43% PEG. Splenocytes from two immunized rats (±3.3 × 10^8 cells) and 1.1 × 10^8 983 cells were mixed and divided into two samples. One mixture was fused using 1 ml of PEG-DMSO in PBS (●–●), the other using 2 ml of PEG-DMSO in PBS (★ –★). Total numbers of hybrids obtained from 7 to 17 d after fusion are given (unpublished results of Xia Hanzhang).

4. Timing

For fusion 0.6 × 10^8 rat 983 cells with 3 × 10^8 rat spleen cells, for example, we have adopted the following timing: to the dry pellet we add drop by drop and while gently mixing 2 ml of 43% PEG — 10% DMSO in PBS over 90 s, and leave the cell suspension at 37°C for 30 s; then we dilute the PEG-DMSO by dropwise addition of 2 ml PBS over 90 s, and add 40 ml PBS over 5 min. After this dilution step such a fused mixture can be kept, if necessary, for 5 to 10 min at 37°C without harmful or beneficial effect. Different quantities of cells can be fused with the same timing. Using this procedure, growth of hybrids is observed in 30 to 80% wells (10^5 cells/well).

While fusion of membranes is achieved within a few minutes, the fusion of nuclei only occurs when there is a synchronized mitosis, i.e., in the first hours or days after fusion. We found that after centrifugation of the fused mixture and resuspension in culture medium containing serum, HAT, and conditioned medium (see below), and before seeding, the suspension is best left in bottles for 1 to 4 h at 37°C with 5% CO_2.

E. TECHNICAL ASPECTS

In this section, we describe the technical details used in our laboratory. Each scientist should adapt conditions to the laboratory facilities available to him or her.

TABLE 2
Effect of the Splenocyte: Immunocytoma Ratio on Hybrid Growth and Monoclonal Antibody Production

	Splenocyte: 983 ratio (nb of plates)	
	2:1 (10)	4:1 (9)
Nb of wells with clonal growth per plate ± S.D.	51 ± 10	61 ± 8
Estimated number of clones per plate	74	98
Nb of 0.2 well contents transferred to large wells	511	537
Nb of large wells with clonal growth (% of transferred wells)	284 (56%)	281 (52%)
Nb of wells containing antitarget MoAb (% of growing hybrids)	37 (13%)	90 (32%)

Note: A splenocyte suspension prepared from LOU rat immunized with human tonsil cells was divided into two samples and mixed with 983 cells as to obtain a splenocyte: immunocytoma ratio of 2:1 (1×10^8 splenocytes + 5×10^7 983 cells) or a 4:1 ratio (16.4×10^7 splenocytes + 4.1×10^7 983 cells). Both mixtures were fused using 1 ml of PBS-10% DMSO-43% PEG pH 7.55. After washing out the PEG, cells were resuspended in HAT medium and plated at concentrations of 4×10^4 983 cells per well. The number of clones was estimated from the number of wells with clonal growth according to Poisson's law. All clones were transferred from the 0.2 ml wells into 2-ml wells 7 to 21 d after fusion and the supernatant was screened for antitarget cell reactivity as soon as clones had grown large enough (at least 1/8 of the well).

1. Material

a. Equipment

The references in the parentheses are given as an example.

1. Laminar flow hood (E.S.I., Cachan, France).
2. Incubator(s) (FORMA, U.S. or Heraus B5060 EK/CO_2, Germany) adjusted at 37°C, humidified with 5% CO_2 in air atmosphere. The incubators should be cleaned regularly and CO_2 level checked with a Bacharach Fyrite test kit, Ambac Ind., Pittsburgh, for example.
3. Water bath at 37°C; for the fusion-procedure itself, we found it convenient to use a bottle-warmer.
4. Balance.
5. pH meter.
6. Vacuum pump with connected vacuum flasks and tubing (for aspiration of culture supernatants).
7. Centrifuge (bench-top centrifuge).
8. Inverted phase-contrast microscope (Leitz, FRG) with oculars of 10 × and objectives of 2.5 ×, allowing view of the whole surface of a small well (96-well plate) and objectives of 10 and 32 × for detailed analysis.
9. Laboratory microscope.

b. Minor Equipment

1. Scissors, forceps
2. Glass bottles with screw caps

c. Disposable Material (All Sterile)

1. Culture plates: 96-well plate (flat-bottom) (Costar, U.S.; Nunc, Denmark; Falcon, U.S.).
2. 1 ml-, 2 ml-, 5 ml-, and 10 ml-pipettes (Falcon)
3. 50 ml conical tubes (Falcon)
4. Glass tube for autoclaving of PEG
5. Petri dishes of 50-mm diameter
6. Pasteur pipettes

d. Animals

1. LOU/C or LOU/M rats for immunization (see Chapter 5)
2. Rats for feeder-layer preparation (for example from the outbred Wistar strain).

e. Reagents and Media

1. PBS isotonic (prepared from Dulbecco's PBS; 10 ×; Gibco No. 042-04200 M)
2. PEG: polyethylene glycol 4000 (Merck No. 9727)
3. DMSO: dimethylsulfoxide (Merck No. 2950 or Merck No. 802912)
4. Alcohol for sterilization (70%)
5. DMEM or RPMI 1640 (Gibco No. 041-1965 or No. 041-1875)
6. Horse serum or fetal calf serum (selected batch) (Gibco No. 034–060050H); all sera should be heat-inactivated
7. Glutamine 200 m*M* (Gibco No. 043-5030 H)
8. Gentamycin 50 µg/ml (Gibco No. 043-5750H) or penicillin-Streptomycin 500 ×: 1 g streptomycin and 1×10^6 units of Penicillin are dissolved in 20 ml 0.9% NaCl. The solution is filtered, aliquoted, and stored –20°C.
9. Nonessential amino acids: 100 × (Gibco No. 043-1140)
10. Sodium pyruvate 100 m*M* (Gibco No. 043-01360H)
11. HT 50 × (Gibco No. 043-01065H) or hypoxanthine (Sigma No. H-9377) and thymidine (Sigma No. T-9250)
12. Aminopterin 100 × (Gibco No. 043.013621) or aminopterin (Sigma No. A-1784)
13. Trypan blue solution (Gibco No. 043-05250-H)
14. Türk's solution (Merck No. 9277)
15. Fusion medium: in a clean glass beaker, weigh 5 g polyethylene glycol 4000, add 7 ml PBS, add 1 ml DMSO; heat at 50°C in a water-bath with occasional stirring until complete melting; sterilize by filtration on 0.22 µm-filters (Millipore) or distribute into glass tubes with screw caps and autoclave for 15 min at 120°C; store at 37°C for up to 2 months
16. Alternative fusion medium in EMEM: same as described above but with PBS replaced by MEM Rega3 10 m*M* HEPES (Gibco No. 045-09993M); use within 2 weeks
17. Aminopterin (1000 ×): make up a stock solution of aminopterin by dissolving 17.6 mg aminopterin in 10 ml 0.1 *N* NaOH; add around 80 ml of water and keep the mixture in a water bath at 45°C until completely dissolved; adjust pH to 7 with 0.1 *N* HCl and volume at 100 ml; sample and store at –20°C.
18. Aminopterin solution can be bought ready-made (Gibco No. 043-1362D)
19. HAT medium (50 ×): dissolve 136 mg of hypoxanthine in 100 ml of H_2O by heating at

70°C; after cooling, add 39 mg of thymidine and 10 ml of aminopterin ×1000; adjust at 200 ml with DMEM; filter, sample, and store at –20°C

20. HAT solution can be bought ready-made (Gibco 043-1060H)
21. HT medium (50 ×): same as described for HAT but without the aminopterin.
22. DMEMc (complete DMEM): DMEMc refers to DMEM supplemented with 2 m*M* glutamine, 100 µg streptomycin/ml, 100 U penicillin/ml and non-essential amino acids 1 ×

2. Manipulations to be done before Fusion

a. Sterilization of Tools

The tools include: scissors, forceps, funnels, gauze, and Pasteur pipettes. The tools can be wrapped in aluminium paper or in sterilization paper bags.

b. Immunization of Rat

See Chapter 5.

c. Maintenance of Rat Immunocytoma Line 983 in Exponential Growth Conditions

Depending on conditions of culture (sera,...), the optimal cell concentration varies between 1×10^5 and 5×10^5. See Chapter 4.

d. Preparation of Feeder-Layers (Sterile)

Prepare 100 ml of DMEMc without HEPES and containing 15% horse serum or fetal calf serum or a mixture v/v of both, then fill a 20 ml syringe with this medium. Anesthetize a rat in ether and sprinkle the abdominal skin with alcohol. Inject the 20 ml of medium into the peritoneal cavity and massage the abdomen of the rat (rat still alive). Wait 2 to 5 min and kill the rat by a massive ether dose (by the respiratory tract). Wet liberally the abdominal skin with alcohol and open largely, and then wet the abdominal wall with alcohol. With sterile scissors and forceps, cut a small hole in the abdominal wall and while holding it with the forceps, aspirate the injected medium with a sterile syringe. From 12 to 16 ml should be recovered at a concentration of 1 to 1.5×10^6 nucleated cells/ml. Count the nucleated cells (dilution in Türk solution). Dilute the cell suspension to 200.000 cells/ml and seed into 96-well culture plates at 0.1 ml/well. Keep the plates in the incubator (5% CO_2, humidified atmosphere, 37°C).

e. Notes

Feeder-layers are usually prepared on the day before fusion and for developing and cloning, feeder-layers can be used up to 4 d after preparation. When 24-well culture plates are used, 0.5 to 1 ml is added per well.

3. Manipulations on the Day of the Fusion

The key ideas are gentleness, rapidity, and 37°C.

a. Preparation of Media

1. PBS
2. PBS-5% horse serum or 5% fetal calf serum
3. DMEMc-15% horse serum (without Hepes): the procedure below uses horse serum (heat-inactivated) throughout, but fetal calf serum or a mixture of horse serum and fetal calf serum can be used as well
4. Conditioned medium: freshly harvested supernatant of 983 myeloma cell line in exponential growth
5. Hybrid culture medium can be prepared during washing procedures, with freshly made up

DMEMc and freshly thawed horse serum. Depending on the acidity (color) of the conditioned medium, add from 20 to 50% filtered conditioned medium to DMEMc-15% horse serum. HAT is added at double concentration. Example for 250 ml final medium: 160 ml DMEMc without Hepes + 32 ml horse serum + 10 ml HAT 50 × + 50 ml filtered conditioned medium.

b. Removal of Spleen

In the animal house or an appropriate room, anesthetize the immunized LOU rat. Lay him down on his right side and sprinkle liberally with alcohol. Resect the skin largely as to see the spleen through the abdominal wall and wet with alcohol. With sterile scissors and forceps, cut a slit over the spleen, pull the spleen out, and transfer it to the sterile Petri dish containing 5 ml of PBS — 5% horse serum.

c. Teasing of Spleen

Inject 10 ml of medium into the spleen, causing inflation of the spleen. Then tease apart using forceps and scissors. Alternatively cut the spleen into small pieces and squeeze gently with the piston of syringe. The spleen-cell suspension is passed on gauze, adapted in a funnel on top of a 50-ml conical tube. The squeezing technique gives a high cell recovery ($>3 \times 10^8$ cells per spleen) with acceptable cell viability (80 to 90%), while the perfusion technique gives a somewhat lower recovery (2×10^8) with a higher cell viability (>90%).

d. Count of the Splenocytes

After dilution in Türk's solution, the total amount of nucleated cells is counted and after dilution in Trypan blue, the total amount of dead cells is counted, and then deduce total amount of viable cells. Make a cell smear. A posteriori and after staining with May-Grünwald-Giemsa, determine percentage of lymphoblasts. Centrifuge the splenocytes for 7 min at 500 g and discard supernatant, then resuspend into PBS and repeat centrifugation.

e. Immunocytoma Fusion Cells

Count the 983 cells of the different culture flasks, choose the best ones (>90% viability, aspect under microscope, and color of supernatant) and transfer around 1×10^8 cells into 50-ml tubes. Refeed the 983 cells left over with DMEMc, 15% horse serum. These steps are best done before the manipulations of removal and teasing of the spleen. 983 cells are kept at 37°C, in a 5% CO_2 atmosphere, while the operator is in the animal house; they are centrifuged while the spleen is teased. Centrifuge the 983 suspension at 250 *g* for 5 min. Keep the supernatants (= conditioned medium). Wash the pellet of 983 cells twice with PBS and eventually resuspend the cells in PBS.

f. Mixing Splenocytes and Immunocytoma Cells; Control Wells

The total amount of splenocytes and 983 cells is not significantly lowered by the washing steps. Hence the expected ratio of splenocytes to immunocytoma cells can be calculated during the centrifugations. Mix the proper amount of 983 cells and the splenocytes so as to have a ratio of around 5:1. Remove 1/50 of this mixture for control wells and keep in DMEMc-HS at 37°C, 5% CO_2. The exact ratio of immunocytoma to splenocytes can be obtained by counting the large and small cells after mixing.

g. Fusion Step

Centrifuge the 983: splenocyte mixture for 7 min at 500 *g*, then discard the supernatant completely. Further manipulations are carried out under continuous shaking at 37°C (water heater). Dropwise with a 2-ml pipette and over a 90 s-duration, add 2 ml of the fusion medium (PEG-DMSO-PBS). Wait for 30 s at 37°C and change the pipette, then dropwise with a 2-ml pipette and over a 90 s duration, add 2 ml of PBS. Add 4 × 10 ml PBS over a 5 min duration, first slowly then faster and centrifuge for 5 min at 250 *g*. Discard the supernatant totally. Resuspend the fused mixture in 20 ml of hybrid-culture medium, then count in Trypan blue and Türk solution (usually viability is around 85% and recovery, in absolute number of cells, around 80%). Dilute the cells in the hybrid-culture medium so as to obtain around 2×10^5 myeloma cell/ml. Leave the cells for 1 to 4 h in the incubator (37°C, 5% CO_2).

h. Plating of the Fused Cell Suspension and Controls

Distribute the fused cell suspension over the feeder-layers: two drops per well, except in the first two wells of each plate. Resuspend the control mixture in the hybrid-culture medium at exactly the same concentration as the fused cell suspension. Add two drops to the first two wells of each plate.

Chapter 6.II

RAT HYBRIDOMA PRODUCTION: SCANNING ELECTRON MICROSCOPY OF A FUSION EXPERIMENT

E. Mrena, J. Ch. Fang, G. Burtonboy, M. Bodeus, N. Delferrière, and G. M. Marchal

In the course of various fusion experiments made to obtain rat hybridomas secreting monoclonal antibodies against hepatite B surface antigen (HBs),[19] cell samples were taken and examined by scanning electron microscopy.[20,21]

The myeloma cell line used in the work is IR983F, an azaguanine-resistant clone derived from a rat immunocytoma and supplied to us by Prof. H. Bazin.[22]

The spleen cells were obtained by teasing the spleen of LOU rat immunized against the viral antigen by repeated intraperitoneal injections, the animal being sacrificed 3 d after the last boost. The fusion was performed by adding PEG as described earlier in Chapter 6.I. Briefly, spleen cells and myeloma cells were thoroughly washed in warm (37°C) EMEM medium, mixed, then centrifuged at a low speed (1000 rpm) and slowly resuspended in 50% polyethyleneglycol (4000 Merck) containing 5% DMSO. This cell population is distributed in microplates and incubated in a selective medium.

For the examination by the scanning electron microscope,[23] cell suspension was filtered by free flow through a FLOTRONIC silver filter (pore size 0.8 μm) cut to a diameter of 10 mm to fit the size of the microscope preparation stub. As soon as the filter became clogged the membrane was transferred into a fixative solution containing 1.5% glutaraldehyde in Milloning's phosphate buffer.[24]

Fixation proceeded for at least 2 h in the cold, then the preparations were rinsed in Milloning's buffer solution and postfixed in 1% osmium tetroxide dissolved in the same buffer. The samples were then dehydrated in graded acetone, dried by the critical point method[25] (CO, 74 atm) and finally coated by sputtering with gold.[26,27] The specimens were examined and photgraphed in an EM501 Philips scanning electron microscope operating at 30 kV and beam size of 20 nm.[20,28]

The myeloma cells 983 (Figure 1) were taken during exponential growth. The aspect of the cells is rather heterogenous, although these cells are in *in vitro* culture since about 10 years ago and have been cloned many times. Their surface is either covered with short projections or almost bare with all the intermediates. The cell diameter varies from 8 to 16 μm.[29,30]

The suspension of spleen cells observed after washing with the culture medium contained a mixture of various types of cells.[21] Numerous erythrocytes and lymphoid cells of different sizes were observed (Figure 2).[31]

The mixed population of splenic and myeloma cells was examined at the end of the fusion experiment. The sample taken was similar to what had been distributed in microplates.

The cells were resuspended in HAT medium (see Chapter 7). A few heterokaryons were detected by examination of a drop in a phase-contrast microscope.[32]

In the electron microscope such heterokaryons, although not frequent, were easily discerned (around 100 in 1 preparation). Their aspect was quite typical (Figures 3). They clearly appear to be made of a small and a large cell fused together. The small one covered with projections is similar to the lymphoid spleen cell. The big one has the size of the myeloma cell, but its surface seems different with no typical surface projections but instead has a ridge-like structure. The rest of the preparation showed a mixture of various cells including lymphocytes and a few erythrocytes.

Finally hybridoma clones were examined (Figure 4). The cells have a mean diameter of 21 μm. As far as their surface is concerned they look very much like the myeloma cells.

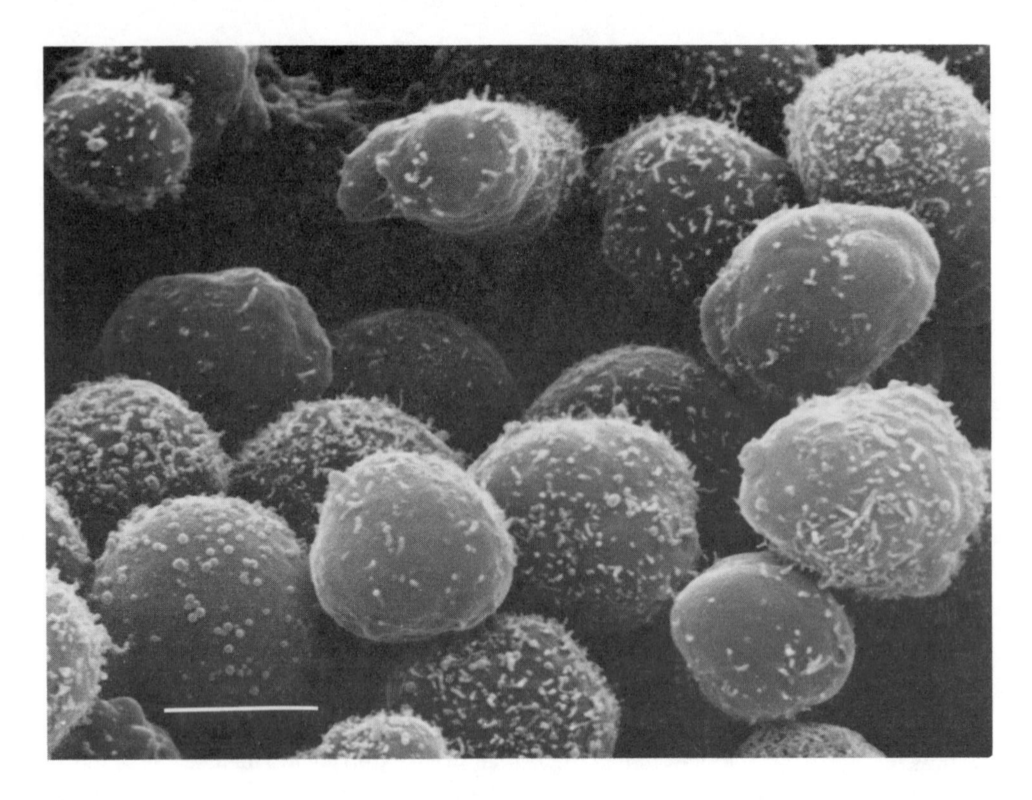

FIGURE 1. Image of myeloma cells IR983F. (The scale bar represents 10 μm.)

The present study using a scanning electron microscope shows some interesting details impossible to obtain by light microscopy or even by a transmission electron microscopy. In the latter case the number of cells to examine if one wants to find a heterokaryon is quite large, so that the study of these kinds of structures is difficult to consider.

On the other hand the resolving power of a scanning electron microscope is superior to that of the light microscope, thus permitting to reveal more details of the cell surface,[33,34] allowing to distinguish types of cells and to study the surface aspect of two cells of which a heterokaryon is made.

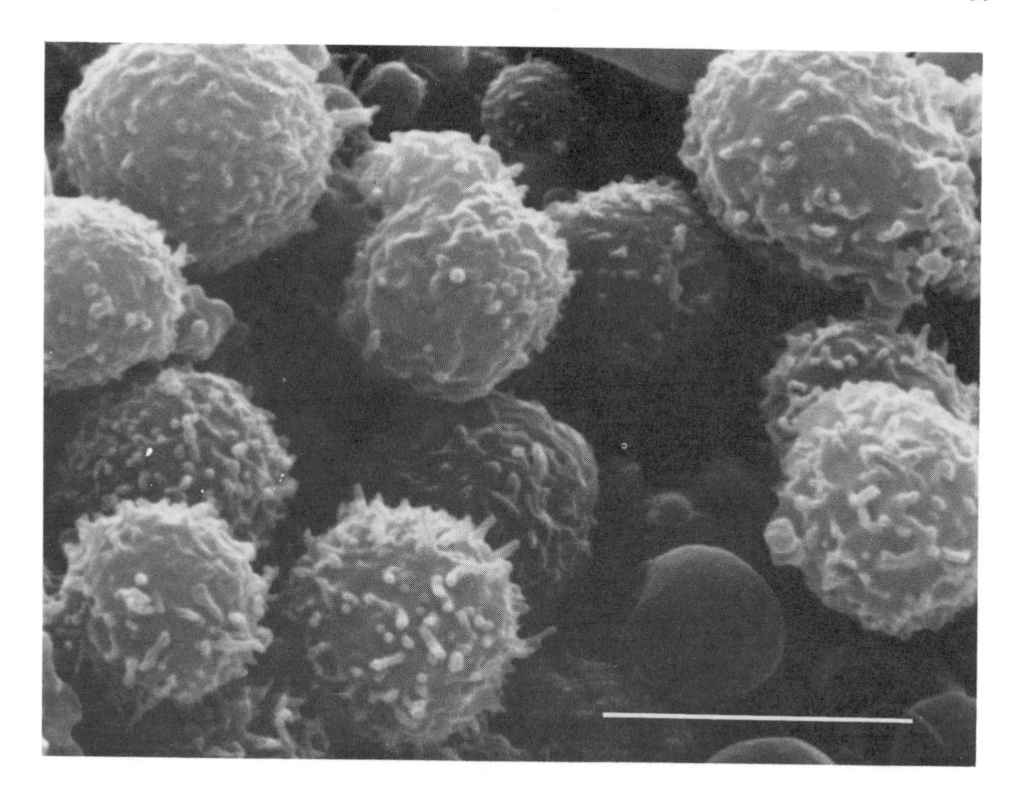

FIGURE 2. Aspect of the splenic cells used in the fusion. (The scale bar represents 10 μm.)

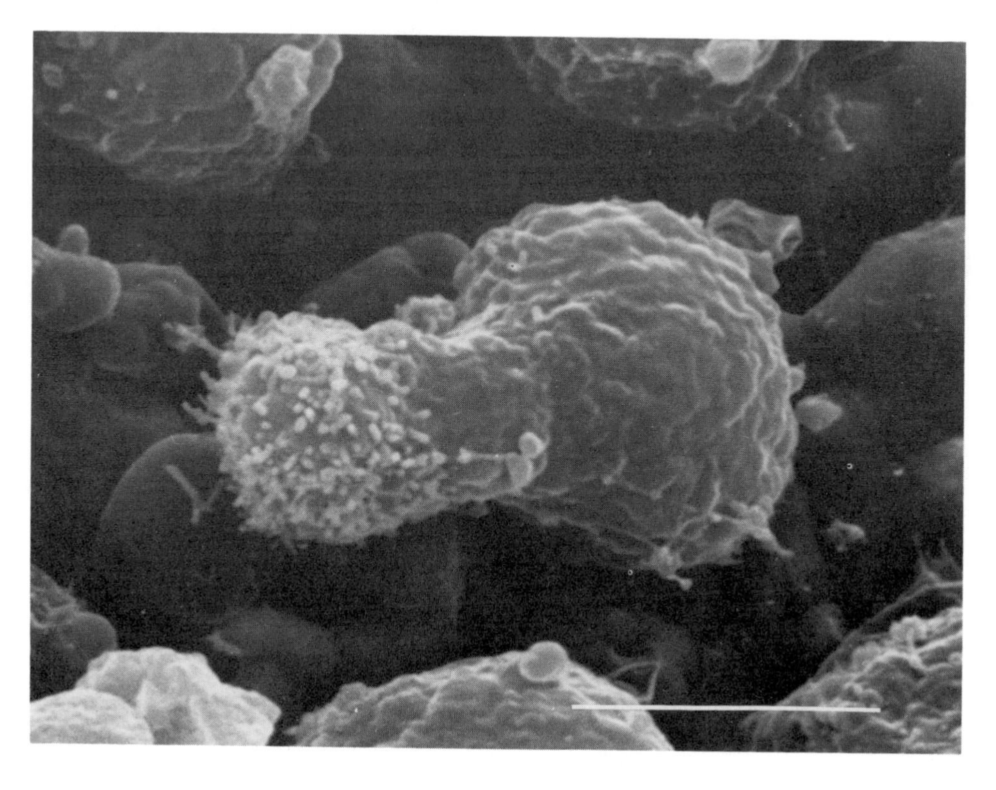

A

FIGURES 3.(A and B) Heterokaryons observed after polyethylene glycol fusion of the splenic cells with 983 cells. (The scale bar represents 10 μm.)

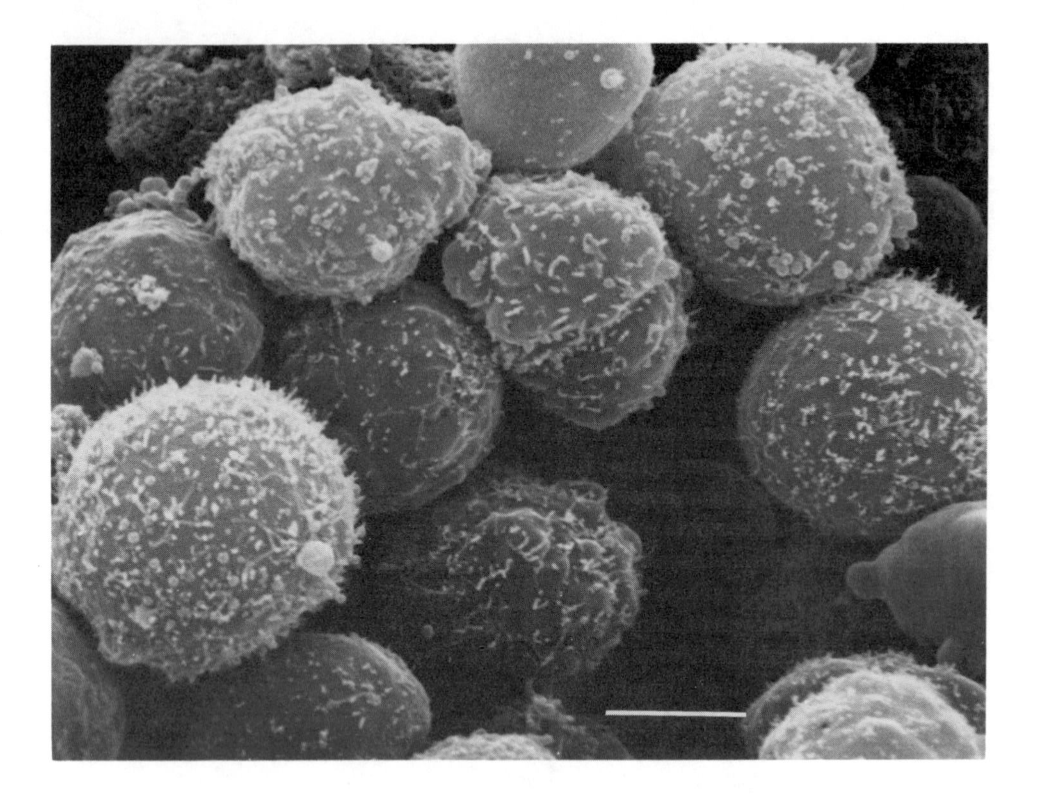

FIGURE 3B

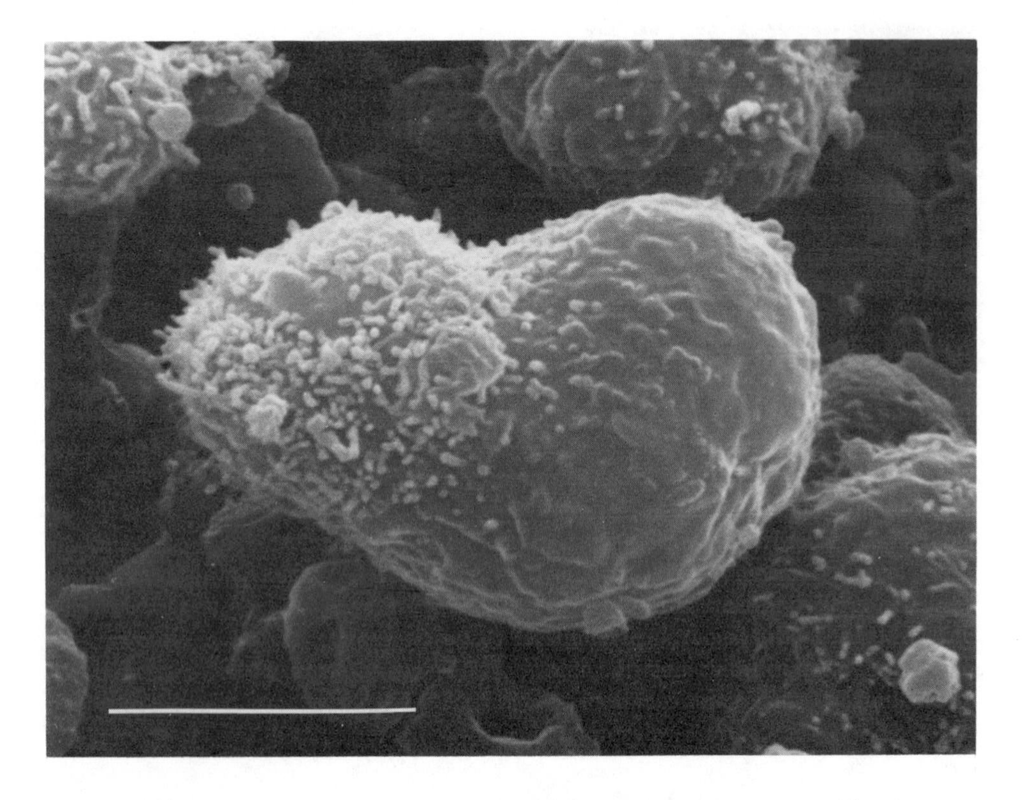

FIGURE 4. Hybridoma cells secreting monoclonal antibody. (The scale bar represents 10 μm.)

REFERENCES

1. **Köhler, G. and Milstein, C.,** Continuous cultures of fused cells secreting antibody of predefined specificity, *Nature,* 256, 495, 1975.
2. **Bazin, H.,** Production of rat monoclonal antibodies with the LOU rat non secreting IR983F myeloma cell line, *Protides of the Biological Fluids Proc.,* 29th Colloqium, 1981, Vol. 29, Peeters, E. Ed., Pergamon Press, Oxford, 1982, 615.
3. **De Clercq, L., Cormont, F., and Bazin, H.,** Generation of rat-rat hybridomas with the use of the LOU IR983F non secreting fusion cell line, *Methods Enzymol.,* 121, 234, 1986.
4. **Bazin, H.,** Rat-rat hybridoma formation and rat monoclonal antibodies, in *Methods of Hybridoma Formation,* Bartal, A. H. and Hirshaut, Y., Eds., Humana Press, Clifton, NJ, 1987, 337.
5. **Galfré, G., Milstein, C., and Wright, B.,** Rat × rat hybrid myelomas and a monoclonal anti-Fd portion of mouse IgG, *Nature,* 277, 131, 1979.
6. **Stähli, C., Staehelin, T., Miggiano, V., Schmidt, J., and Häring, P.,** High frequencies of antigen-specific hybridomas: dependence on immunization parameters and prediction by spleen cell analysis, *J. Immunol. Methods,* 32, 297, 1980.
7. **Lebacq-Verheyden, A. M., Neirynck, A., Ravoet, A. M., and Bazin, H.,** Rat hybridoma technology: culturing of rat myeloma cell line IR983F prior to cell fusion, *Hybridoma,* 2, 355, 1983.
8. **Lebacq, A. M. and Bazin, H.,** Obtention of rat monoclonal antibodies reactive with human leukaemic lymphoblasts, *Bull. Cancer,* 70, 93, 1983.
9. **Andersson, J. and Melchers, F.,** The antibody repertoire of hybrid cell lines obtained by fusion of X63-AG8 myeloma cells with mitogen-activated B-cell blasts, *Curr. Top. Microbiol. Immunol.,* 81, 130, 1978.
10. **Galfré, G., Hawe, S. C., Milstein, C., Butcher, G. W., and Howard, J. C.,** Antibodies to major histocompatibility antigens produced by hybrid cell lines, *Nature,* 266, 550, 1977.
11. **Blow, A. M. J., Botham, G. M., Fisher, D., Goodall, A. H., Tilcook, C. P. S., and Lucy, J. A.,** Water and calcium ions in cell fusion induced by polyethylene glycol, *FEBS Lett.,* 94, 305, 1978.
12. **Gefter, M. L., Margulies, D. H., and Scharff, M. D.,** A simple method for polyethylene glycol-promoted hybridization of mouse myeloma cells, *Somatic Cell Genet.,* 3, 231, 1977.
13. **Fazekas de St Groth, S. and Scheidegger, D.,** Production of monoclonal antibodies: strategy and tactics, *J. Immunol. Methods,* 35, 1, 1980.
14. **Norwood, T. H., Zeigler, C. J., and Martin, G. M.,** Dimethyl sulfoxide enhances polyethylene glycol-mediated somatic cell fusion, *Somatic Cell Genet.,* 2, 263, 1976.
15. **Westerwoudt, R. J.,** Improved fusion methods. IV. Technical aspects, *J. Immunol. Methods,* 77, 181, 1985.
16. **Klebe, R. J. and Mancuso, M. G.,** Chemicals which promote cell hybridization, *Somatic Cell Genet.,* 7, 473, 1981.
17. **Clark, M. and Waldmann, H.,** Production of murine monoclonal antibodies, in *Monoclonal Antibodies,* Vol. 13, Beverley, P. C. L., Ed., Churchill Livingstone, London, 1986, 1.
18. **Lovborg, U.,** Monoclonal antibodies: production and maintenance, W. Heinemann Medical Books, London, 1982.
19. **Tiollais, P., Pourcel, C., and Dejean, A.,** The hepatitis B virus, *Nature,* 317, 489, 1985.
20. **Pease, R. F. W.,** Fundamentals of scanning electron microscopy, in *Scanning Electron Microscopy,* Johari, O. and Corvin, I., Eds., ITT Research Institute, Chicago, 1971, 9.
21. **Burkhardt, E.,** Scanning electron microscopy of peripheral blood leukocytes of the chicken, *Cell Tissue Res.,* 147, 1979.
22. **Bazin, H.,** Production of rat monoclonal antibodies with the LOU rat non secreting IR983F myeloma cell line, in *Protides of the Biological Fluids,* Peeters, H., Ed., Pergamon, Oxford, 1982, 615.
23. **Brunk, U., Collins, V. P., and Arro, E.,** The fixation, dehydration, drying and coating of cultured cells for SEM, *J. Microsc.,* 132, 121, 1980.
24. **Millonig, G. and Marinozzi, V.,** Fixation and embedding in electron microscopy, in *Advances in Optical and Electron Microscopy,* Baker, R. and Cosslett, V. E., Eds., A.P.2, 251, 1968.
25. **Cohen, A. L.,** Critical point drying, principles and procedures, Scanning el. micr., SEM Inc., AMF O'Hare (Chicago), II, 303, 1979.
26. **Echlin, P.,** Coating techniques for SEM and X-Ray microanalysis, Scanning Electron Microscopy, SEM Inc., AMF O'Hare (Chicago), II, 109, 1978.
27. **Boyde, A.,** Pros and cons of critical point drying and freeze drying for SEM, Scanning Electron Microscopy, SEM Inc., AMF O'Hare (Chicago), II, 303, 1978.
28. **Wetzel, B., Canon, G. B., Alexander, E. L., Erickson, B. W., and Westbrook, E. W.,** A critical approach to the scanning electron microscopy of cells in suspension, in *Scanning Electron Microscopy,* Johari, O., Corvin, I., Eds., ITT Research Institute, Chicago, 1974, 581.
29. **Gersteberger, R. and Paweletz, N.,** Scanning electron microscopic observation of cells grown *in vitro.* VI. Aggregate formation in confrontation cultures of human diploid and tumor cells, *Eur. J. Cell Biol.,* 26, 136, 1981.

30. **Yoshikawa, T., Oyamada, T., Yoshikawa, H., Koyama, H., and Komoriya, Y.,** Scanning electron microscopic study of bovine leukemic cells, *Am. J. Vet. Res.*, 44, 1358, 1983.
31. **Michaelis, T. W., Lanimer, N. R., Metz, E. N., and Balcerzak, S. P.,** Surface Morphology of human leucocytes, *Blood,* 37, 23, 1971.
32. **Mc Donald, L. W. and Hayes, T. L.,** Correlation of scanning electron microscope and light microscope images of individual cells, in human blood and blood clots, *Exp. Mol. Pathol.*, 10, 186, 1969.
33. **Hayat, M. A.,** Principles and techniques of scanning electron microscopy, Biological applications, Van Nostrand Reinhold Company, 1, 273, 1974.
34. **De Harven, E., Lampen, N., and Sato, T.,** Scanning electron microscopy of cells infected with a murine leukemia virus, *Virology,* 51, 240, 1973.

Chapter 7

SELECTIVE GROWTH OF HYBRIDS

A.M. Ravoet and H. Bazin

TABLE OF CONTENTS

I. INTRODUCTION

Methods for the growth of mouse × mouse hybrids have been described in detail by Fazekas de St Groth and Scheidgger,[1] Goding,[2] Westerwoudt,[3] Zola and Brooks,[4] and Bartal and Hirschaut.[5]

Although most of these methods can be used for the growth of rat × rat hybrids using the 983 cell line,[6] rat hybrids seem more demanding as to the culture medium.[7]

II. HYPOXANTHINE-AMINOPTERIN-THYMIDINE (HAT) MEDIUM

Since the probability of hybrid formation induced by PEG is around 10^{-5}, the use of a selective medium which is toxic for immunocytoma cells and which can promote the growth of hybrid cells is essential. In HAT medium,[8] only those cells containing the chromosomes encoding for synthesis of nucleic acids by the salvage pathway will survive. Most hybridoma fusion lines, including the 983 rat immunocytoma line, have been selected for the absence of hypoxanthine guanine phosphoribosyl transferase (HGPRT), and hence die in the presence of aminopterine within a few hours.

Immunocytoma × splenocyte hybrids have the 2 chromosomal sets enabling them to survive indefinitely and to give rise to hybridoma clones. During the first few weeks however, some hybrids will disappear due to chromosomal instability.

HAT is added at once at the normal concentration, to prevent the growth of immunocytoma cells and the exhaustion of the medium.

III. MEDIUM, BUFFER

Besides HAT, the culture medium must contain all the nutrients necessary for normal cell growth, plus some growth stimulators. DMEM supplemented with glutamine sodium pyruvate and nonessential amino acids provides the nutrients. For convenience, penicillin and streptomycin or gentamycin are added to all culture media. HEPES should not be added to the medium, as its toxic effect lasts even after washing the cells free from PEG and DMSO: the hybrid yield was 2.5 times as low in DMEM containing 12 m*M* HEPES as in HEPES-free DMEM (Table 1). However for subsequent development (8 d after fusion) and cloning of hybrids, HEPES has a beneficial effect on growth. Since HCO_3-/CO_2 is the main buffer in the medium, the CO_2 pressure in the incubator must be adjusted accurately at 5 to 6%, and it is recommended not to take the culture plates out of the incubator too often or for too long.

IV. SERA

The addition of sera to the DMEM or RPMI medium is absolutely necessary and is supposed to provide hormones or growth factors. Sera from different animal sources or different batches of the same animal origin differ from one another in their capacity to promote hybrid growth. As an example (Table 2), a batch of fetal calf serum found appropriate for growth of human cell lines and culture of bone marrow progenitor cells, was found four times less efficient than a batch of horse serum for rat hybridoma growth. A mixture of fetal calf serum and horse serum had an intermediate efficiency. However, in other experiments a mixture of horse serum and fetal calf serum gave rise to a higher amount of hybrids than fetal calf or horse sera used separately. So, it is difficult to propose a simple rule to choose a good serum and experimental tests of each batch are needed. It must be done before fusion by comparing clonal growth of 983 cells or an established hybridoma line after plating at limiting dilutions. However, this is not an absolute test, and the only genuine criterion is the growth of newly fused hybrids. We are testing all our

TABLE 1
Effect of Hepes on Newly Produced Hybrids

	Number of wells with clonal growth per plate	
	CDMEM	**DMEM + HEPES**
PC1	13, 20, 23,33, 30	8, 4, 9, 8
PC2	35, 17, 21, 38	22, 17
X ± S.D.	26 ± 9	11 ± 6

Note: Splenocytes were fused with 983 cells and resuspended into DMEMc, 15% horse serum, HAT 2X and 60% conditioned medium. One half of the fused suspension was diluted twice into DMEMc, 15% horse serum, HAT 2X - 25 m*M* HEPES; the other half was diluted in the same medium without HEPES. Both suspensions were seeded onto feeder-layers (PC1 or PC2) prepared with 12.5 m*M* HEPES or without HEPES. Clonal growth was checked after 13 d.

TABLE 2
Effect of Different Sera on the Growth of Newly Produced Hybrids

	Number of wells with clonal growth per plate	
HS1	**HS1 + FCS2**	**FCS2**
33, 22	10, 18	3, 3
38, 29	14, 8	5, 13
29, 25	13, 10	9, 1
41, 35	16, 16	2, 10
	12	11, 11
32 ± 7	13 ± 3	7 ± 4

Note: After fusion, the cell suspension was split into 3 equal parts. Each was resuspended in DMEM containing HAT, 20% conditioned medium and 16% serum from horse (HS1), from fetal calf (FCS2) or a mixture of 8% FCS1 and 8% HS1. 0.1 ml of the fused cell suspension was added to wells containing 0.1 ml of peritoneal cells (pool of peritoneal cells of 5 outbred WISTAR rats). Clonal growth was checked after 7 d.

serum batches by the growth of the 983 cells, limiting dilution of hybridomas, or then by the growth of newly fused hybrids.

Sera can be kept at –20°C for months without loss of growth-promoting characteristics.

V. CONDITIONED MEDIUM

Conditioned medium is an ill-defined parameter: it should be understood as a filtered supernatant of the exponentially growing immunocytoma cell line. Its quality therefore depends on the growth conditions of the immunocytoma. We observed that the addition of the supernatant of 983 cells, reset 18 to 24 hours before harvest at 3×10^5 cells/ml, is largely beneficial to the growth of newly formed hybrids and makes feeder-layers redundant. However, as far as possible both feeder-layers and conditioned medium are added to the culture medium after fusion.

TABLE 3
Comparison of the Hybrid Growth on Feeder-Layers Derived from Different Rat Strains

		Nb of wells with clonal growth per plate after 13 d	Average ±S.D.
Wistar	CP4	13, 20	24 ± 7
	CP5	23, 33, 30	
LOU	CP2	13, 15	14 ± 5
	CP3	6, 18, 20	
L + W	CP1+CP2	35, 17	28 ± 9
	CP3+CP4	21, 38	

Note: All peritoneal cells were seeded at 20,000 cells per well. The mixture L + W contained 10,000 cells of each rat per well.

VI. FEEDER-LAYER

An efficient way to provide the scarce hybrid cells with growth-promoting factors is with the use of feeder-layers.[1,9]

Rat peritoneal cells are easily collected and give satisfactory results. They contain macrophages and fibroblasts. Besides secreting essential growth-factors, macrophages act as scavengers by endocytosing the spleen erythrocytes. They practically do not multiply *in vitro,* but survive for about a week.

We found outbred WISTAR or WISTAR/R inbred rats to provide peritoneal cells better than those of LOU rats (using LOU splenocytes and 983 fusion cells). As an example, in one experiment, 24 ± 7 hybrids per 96 well-plate (5 96-well plates; 2 donor rats) were obtained on WISTAR peritoneal cells, against 14 ± 5 hybrids 1 plate (5 plates; 2 rats) on LOU rats; mixture of LOU and WISTAR peritoneal cells gave 28 ± 9 hybrids 1 plate (4 plates; 2 + 2 rats) (Table 3). A high variability, however, exists between feeder-layers from different animals from the same outbred or inbred strain. Hence, to be significant, more experiments involving a larger number of animals of both strains, would be needed.

As for the concentration of peritoneal cells, around 10,000 to 20,000 nucleated cells in 0.1 ml/well gave the best results. Outbred WISTAR rats are generally used as giving results comparable or even better than those of the other strains, and moreover being the least expensive.

VII. NUMBER OF CELLS PER WELL

In the first rat × rat hybridization experiments using the LOU-derived Y3-Ag1,2,3. immunocytoma line10, 2×10^6 splenocytes (2×10^5 immunocytoma cells) were seeded into 2-ml wells and only a few wells showed hybrid growth. As techniques improved this method gave rise to multiple clones in all wells. The use of feeder-layers made it possible to drastically lower the cellular input per well (and to use 0.2-ml wells). By aiming at a growth in 50% of the wells, the probability of achieving monoclonality from the seeding time on, is high (70% of the wells with clonal growth will contain only one clone). In our experiments, after distributing 1×10^5 cells per well, 30 to 80% of the wells exhibited clonal growth. Clonal growth of 80% is acceptable only when the percentage of hybrids secreting a specific antibody is very low. However, if around 30% of the clones secrete antitarget antibodies (for example, using cellular antigens) and if one aims at the production of MAb with restricted reactivity and a rapid selection of these interesting hybrids, no more than 50% of the wells should contain hybrids.

VIII. NEGATIVE CONTROLS

Two types of negative controls are necessary: (1) one has to be sure that the HAT medium kills the fusion immunocytoma cells; and (2) unfused antigen-stimulated splenocytes continue to secrete their antibodies for 8 to 15 d, giving rise to a (sometimes high) background reactivity on the target. Therefore, we put aside a small part (1/50) of the mixture of immunocytoma and spleen cells before the addition of PEG. This mixture is resuspended in the HAT medium at the same cellular concentration as the fused mixture and distributed into 2 microwells on each plate. No growth should be observed in these wells and supernatant provides an adequate negative control for screening.

IX. FEEDING

When HAT is present from the seeding time on, the main reason for feeding is to remove polyclonal antibodies secreted by unfused splenocytes. In fusion experiments directed against soluble antigens replacement of hybrid culture supernatants 4,6,8, and 10 d after fusion bring the polyclonal immune response back to acceptable levels in most cases. Sometimes up to eight feedings are necessary to reach low background values. When using cellular targets however, we found it more convenient to follow an alternative procedure consisting in scheduling one feeding 5 d after fusion and transferring of all clones into 2-ml wells as soon as they have grown large enough (7 to 17 d after fusion). We obtain about 50 to 80% clonal growth after transfer. This procedure has three advantages:

1. The reactivity due to the polyclonal immune response is brought back to an acceptable background level.
2. The arrest of growth or of secretion of antibody by the hybrid clone, which is presumably due to loss of chromosomes, happens mostly before the first screening; hence the amount of screening work is reduced.
3. The availability of 2 ml of hybrid culture supernatant allows a short preliminary analysis of reactivity on different targets to be done, thereby reducing the number of selected hybrids and hence the work of cloning and developing hybrids.

X. TECHNICAL ASPECTS

A. FUSION AND SEEDING OF THE FUSED SUSPENSION

Detailed procedures are described in Chapter 6.

B. FEEDING OF HYBRIDS AND TRANSFER TO LARGE WELLS

Culture medium for feeding consists of DMEM-15% horse or fetal calf serum or a mixture of both sera, HAT (1×), and 20% conditioned medium (freshly harvested from exponentially growing 983 cells and filtered). Aspirate half the content of the wells and add new culture medium (2 to 3 drops) 4 to 6 d after fusion and about 8 to 10 d after fusion, the hybrid clones should be large enough to allow their transfer into large wells. Prepare feeder-layers and distribute them into large wells (100,000 cells per well). The next day all hybrid clones covering at least 1/5 of the surface of the small well are transferred into large wells. Culture medium containing HT (instead of HAT) and conditioned medium are added. Efficiency of the transfer can be checked by means of the inverted microscope.

REFERENCES

1. **Fazekas de St Groth, S. and Scheidegger, D.,** Production of monoclonal antibodies: strategy and tactics, *J. Immunol. Methods,* 35, 1, 1980.
2. **Goding, J. W.,** Antibody production by hybridomas, *J. Immunol. Methods,* 39, 285, 1980.
3. **Westerwoudt, R. J.,** Improved fusion methods. IV. Technical apsects, *J. Immunol. Methods,* 77, 181, 1985.
4. **Zola, H. and Brooks, D.,** Techniques for the production and characterization of monoclonal hybridoma antibodies: techniques and Applications, in *Monoclonal Hybridoma Antibodies,* Hurrell, J. G. R., Ed., CRC Press, Boca Raton, FL, 1982, 1.
5. **Bartal, A. H. and Hirschaut, Y.,** *Methods of Hybridoma Formation,* Humana Press Clifton, NJ, 1987.
6. **Bazin, H.,** Production of rat monoclonal antibodies with the LOU rat non secreting IR983F myeloma cell line, in *Protides of the Biological Fluids, Proceedings of the 29th Colloquium,* 1981, Peeters, H., Ed., Pergamon Press, Oxford, 1982.
7. **De Clercq, L., Cormont, F., and Bazin, H.,** Generation of rat-rat hybridomas with the use of LOU IR983F nonsecreting fusion cell line, *Methods Enzymol.,* 121, 234, 1986.
8. **Littlefield, J. W.,** Selection of hybrids from matings of fibroblasts *in vitro* and their presumed recombinants, *Science,* 145, 709, 1964.
9. **Andersson, J. and Melchers, F.,** The antibody repertoire of hybrid cell lines obtained by fusion of X63-AG8 myeloma cells with mitogen-activated B-cell blasts, *Curr. Top. Microbiol. Immunol.,* 81, 130, 1978.
10. **Galfré, G., Milstein, C., and Wright, B.,** Rat × rat hybrid myelomas and a monoclonal anti-Fd portion of mouse IgG, *Nature,* 277, 131, 1979.

Chapter 8

CLONING

F. Cormont, C. Digneffe, B. Platteau, and H. Bazin

TABLE OF CONTENTS

I. INTRODUCTION

Before considering antibodies as being monoclonal, one has to be sure of the monoclonality of the hybrid cell clone. The two main reasons for polyclonality are (1) two or more hybrid cells have been dropped in the same well after fusion; or (2) in a single growing clone, a daughter cell has lost the chromosomes encoding for antibody secretion or has mutated for its antibody production.

Between 10 to 20 d after fusion, hybridomas of interest are already identified (Figure 1) and then cloned. Cloning means the isolation of one hybridoma cell, and hence the subsequent growth of its daughter cells in one colony only. Cloning is carried out by limiting dilution technique, in which the cells are diluted in medium to such an extent that if growth is observed in culture wells, statistically it is virtually certain to be due to only one cell. Cloning can also be done in soft agar to produce individual clones. In both methods, feeder-layer cells may be used to provide hybridoma cell growth with a favorable environment.

II. LIMITING DILUTION

Cloning hybridoma cells either from rat or from mouse species is rather similar. Several techniques have been described in detail by Goding.[1]

The theoretical approach to cloning by limiting dilution is based on the Poisson's distribution: a = e – b, where "a" is the fraction of wells with no growth, "b" is the average number of clones per well and "e" has the customary mathematical notation.

If 37% of the wells show no growth, the average amount of clones per well is b = – ln 0.37 = 1 clone per well. Hence, in order to obtain a reasonable probability to show that it is a real clone hybridoma, at least 37% of wells should show no growth.

Cloning efficiency is the ratio of experimentally observed growth to theoretical growth, expressed in percentage. The recommended cloning procedure is to seed the hybridomas at 1 and 0.5 cells per well. Indeed, a batch of 10 cells per well is also used to test the system.

A. MATERIALS

1. Minor equipment: Bürker cell
2. Disposal material: sterile tubes of 50 or 12 ml; sterile 96-well tissue culture with flat-bottomed wells (COSTAR); 1-, 2-, 5-, and 10-ml sterile pipettes

B. REAGENTS AND MEDIA

1. Solution of trypan blue at 0.1% (Gibco No. 043-05250-H)
2. Sterile HT medium
3. Complete medium DMEM (see chapter 4)
4. WISTAR rats for feeder layer preparation
5. Suspension of hybridoma cells

C. PROTOCOL

Seed the feeder cells (rat peritoneal cells, See Chapter 6) in 96-well plates. Feeders may be prepared a few days before cloning. Count the cells from the hybridoma to clone and plate the cells at 10, 1, and 0.5 cells per 200 µl into 96-well plates of feeders, and change the medium after 6 d. Macroscopic colonies, which can be detected by looking at the undersurface of the plate, should become visible 1 or 2 weeks after the beginning of the culture. The dilution at which growth is observed in about 50% of the wells may be assumed to contain single colonies. Generally, clones should be assayed after 2 weeks, but they will survive and grow for several

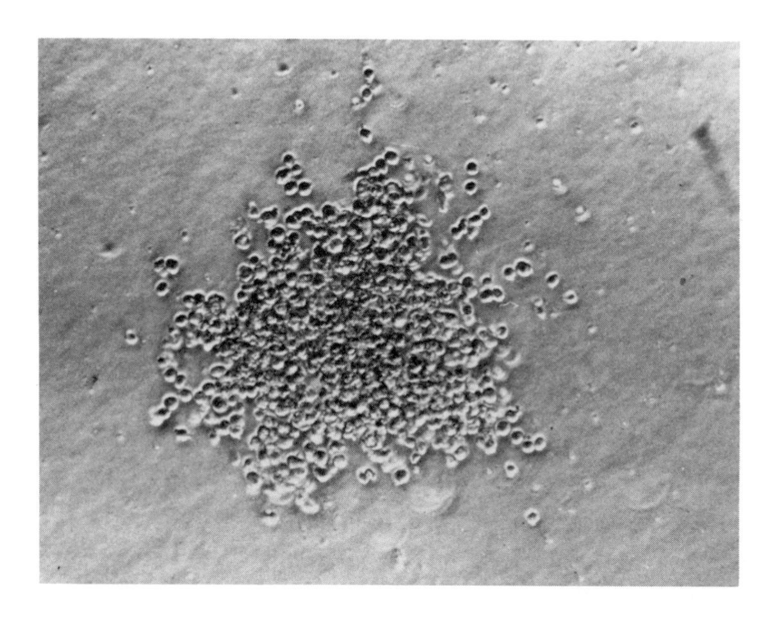

FIGURE 1. A hybridoma made by fusing spleen cells from an immunized LOU/C rat with 983 fusion cells.

TABLE 1
Limiting Dilution: Number of Wells with Hybridomas Growth/96 Well Plate

	Cell suspension at 5 cells/ml (0.5 cell/well)	Cell suspension at 10 cells/ml (1 cell/well)
Theorical number	48	96
Theorical number (Poisson distribution)	38.2	60.7
F 28.1	5	13
F 28.2	30	56
F 28.3	40	84
F 28.4	25	29
F 28.5	31	57
F 28.6	30	55
F 28.7	29	59

days afterwards. Recloning should always be carried out because of only one cloning by limiting dilution does not absolutely guarantee monoclonality.

D. RESULTS

Results in Table 1 are taken from an experiment in which all the interesting hybrids were cloned as soon as possible and were transferred into 24-well plates for further expansion. The cloning efficiency under limiting dilution conditions is close to 50% and varies for different hybridomas.

Nearly all screened clones were positive. We have never observed the interesting hybrid being overgrown by a nonsecreting hybrid. However in some cases, when two or three clones are present in the fusion wells, a negative reactivity can be due to the existence of a mixture of secreting and nonsecreting clones, but the interesting hybrids can be cloned successfully afterwards.

III. SOFT AGAR CLONING

This technique relies on diluting the hybrid cells to a level where colonies grow at distinct sites. Soft agar cloning is less frequently used than limiting dilution. Soft agar presents technical difficulties in that it is necessary to have molten agar for the suspension of the cells and the cells are obviously vulnerable to clonage at a temperature above 37°C. In addition, batches of agar are highly variable and some may be toxic.

We actually use an agarose gelling at low temperature derived from agar, which circumvents both of these problems and may prove more suitable than the conventional agar cloning. After the colonies have grown to a size which is easily visible by the naked eye, the clones can be picked from the agar and require additional reculturing in liquid medium before antibody production can be tested.

A. MATERIALS

1. Minor equipment: Bürker cell and a container with warm water (37°C)
2. Disposal material: sterile Petri dishes of a 90-mm diameter, sterile pipettes, and sterile tubes of 50, 12, and 5 ml

B. REAGENTS AND MEDIA

1. Solution of trypan blue at 0.1%
2. Complete medium DMEM
3. Bidistilled water
4. Sea plaque agarose (Seaplaque FMC corporation, Bioproducts, U.S.)

C. PROTOCOL

Dissolve 10 ml of 0.8% agarose in boiling bidistilled water, autoclave at 120°C for 15 min and put into 37°C water bath. Warm up 90 ml of DMEM at 37°C. Mix agarose and DMEM (DMEM-agarose 0.8%), and put 12 ml into a number of Petri dishes and then put the DMEM-agar back at 37°C. Allow the agarose to solidify in dishes for approximately 30 min at room temperature and then resuspend the cells at different concentrations of cells per milliliter in DMEM. Mix 0.1 ml of this suspension with 3 ml of DMEM agar and pour mixture onto the lower agarose layer of the Petri dishes to cover their whole area. Allow growth for about 15 d in humid CO_2 incubator, at 37°C. Pick colonies with a Pasteur pipette out of dishes where only a few colonies grew and then transfer into 0.2 ml of DMEM in a 96-well culture plate.

D. RESULTS

Table 2 shows the frequency of clonal growth in soft agar when seeding was done at various concentrations of cells. The clone numbers were checked after 2 or 3 weeks and were visible by the naked eye. Then they were transferred into a liquid medium; supernatants were tested by ELISA and found positive for all the clones.

Cloning efficiencies are much lower in agar than in liquid media, requiring the use of much higher cell concentrations (10^3 to 10^4 cells per milliliter instead of 5 or 10 in the limiting dilution method).

IV. SWITCH

Detection and isolation of hybridoma cells which have a class or a subclass switch are becoming a useful adjunct to the hybridoma technology. It is often not possible to obtain antibodies of the desired class or subclass. It is therefore very useful if one can take a hybridoma

TABLE 2
Cloning of Hybridoma Cells in Soft Agar: Number of Clones per Petri Dish

	Number of cells (Petri dish)					
Hybridomas	**125**	**250**	**500**	**1000**	**10,000**	**100,000**
i1C11	1	5	8	10	+[a]	++[a]
i 7G9	1	0	7	9	ND[b]	ND
i 8G11	0	0	0	1	ND	ND
I 9E8	3	4	0	5	ND	ND

[a] +, number of clones superior to 200; ++, very high number of clones.
[b] ND = not done

cell line producing a MAb of a certain class or subclass and detect a switch variant from it in culture in order to get the production of other classes or subclasses. In the mouse system, class switch variants occur spontaneously (with a frequency between 10^{-7} and 10^{-5} cells per generation) in myeloma and hybridoma cells *in vitro*.[2] Class switch variants have been isolated by the agar overlay technique[3] and by means of a fluorescence-activated cell sorter (FACS).[2] The former method is limited by the small number of cells that can be analyzed, and the second method requires the availability of a FACS. Müllter and Rajewsky[4] have described a simple method which allows the isolation of class-switch variants from large numbers of cells without the need for sophisticated equipment. The isolation of variants by sequential sublining is based on the following principle: cells of the wild-type population are divided into several aliquots. After growth of the cells, an aliquot which contains a class-switch variant is identified by detection of the secreted variant antibody. Hybrid clones of a positive well are expanded and the process is repeated at progressively lower cell inputs down to 1 to 5 cells per well. Finally the variant line is cloned by limiting dilution.

This method allowed isolating switch variants from mouse-rat hybridoma lines secreting rat MAbs. Switch variants from IgM to IgG2a, from IgG2a or IgG2b to IgE, and from IgE to IgA were obtained.[5] Class-switch variants have also been isolated from rat-rat hybridomas, using red cell-linked antibodies to identify the desired secreted Ig by reverse passive hemagglutination. This technique is sufficiently sensitive to detect one switch variant in the presence of up to 10^4 parental cells in the microcultures. The frequency of spontaneous switch variants in the rat hybridomas is approximately ten times lower than in mouse hybridomas and the rate of switching back to the parental isotype is substantial.[6]

In our laboratory, we tried to isolate switch variants from IgG1 (rat IgG1 anti-DNP) to IgG2a or IgG2b and from the cell line (LC-66), which produces an IgM antibody specific for CD24 (human granulocytes, lymphocytes B), but our attempts have been unsuccessful, with the lines being extremely stable.

REFERENCES

1. **Goding, J. W.,** *Monoclonal Antibodies: Principles and Practice,* Academic Press, New York, 1983.
2. **Radbruch, A., Liesegang, B., and Rajewsky, K.,** Isolation of variants of mouse myeloma X63 that express changed immunoglobulin class, *Proc. Natl. Acad. Sci. U.S.A.,* 77, 2909, 1980.
3. **Koskimies, S. and Birshtein, B. K.,** Primary and secondary variants in immunoglobulin heavy chain production, *Nature,* 264, 480, 1976.

4. **Müllter, C. E. and Rajewsky, K.,** Isolation of immunoglobulin class switch variants from hybridoma lines secreting anti-idiotype antibodies by sequential sublining, *J. Immunol.,* 131, 877, 1983.
5. **Pluschke, G. and Bordmann, G.,** Isolation of rat immunoglobulin class switch variants of rat-mouse hybridomas by enzyme-linked immunosorbent assay and sequential sublining, *Eur. J. Immunol.,* 17, 413, 1987.
6. **Hale, G., Cobbold, S. P., Waldmann, H., Easter, G., Matejtschuk, P., and Coombs, R. R. A.,** Isolation of low-frequency class-switch variants from rat hybrid myelomas, *J. Immunol. Methods,* 103, 59, 1987.

Chapter 9.I

SELECTION OF ANTI-TARGET ANTIBODY-PRODUCING HYBRIDS

M. V. Chavez, T. Delaunay, and H. Bazin

TABLE OF CONTENTS

I. SCREENING OF MONOCLONAL ANTIBODIES DIRECTED AGAINST SOLUBLE ANTIGENS*

A. INTRODUCTION

A first screening of culture supernatants should be done as soon as clones appear and cell damage should be avoided well before the medium turns to bright yellow. However, controls must be carried out in order to detect polyclonal-specific antibodies secreted by unfused splenocytes. A second screening after two or three changes of culture media is often necessary. The chosen method must allow a rapid selection of all interesting hybrids, and hence has to be simple and sensitive. Methods used in our laboratory for the detection in the culture supernatants of monoclonal antibodies (MAbs) against soluble antigens are discussed later in the chapter.

B. ENZYME IMMUNOASSAY

The enzyme immunoassay methods are based on immunologic and enzymatic principles.[1-3] The characteristic properties of MAbs (specificity and affinity) and of enzymes (catalytic growth and detectability) allow fast and reproducible results. The enzyme immunoassay has proven to be a suitable alternative to radioimmunoassay discussed below.

In our laboratory (see Chapters 10 and 24. V.), two enzyme immunoassay methods were performed to detect antibodies from culture supernatants against soluble antigens: ELISA (Enzyme Linked Immuno Sorbent Assay) and immunodot. Both methods need enzyme-conjugated antibodies. The conjugate should be active and stable. Peroxidase and beta galactosidase enzymes are usually employed.

Three main steps can be distinguished in a single antibody assay:

1. Noncovalent adsorption of the soluble antigens on a solid phase: the surface of a plastic microwell (ELISA) or a nitrocellulose paper (immunodot).
2. Incubation with the culture supernatants to be tested.
3. Revealing step which can involve one or several layers, the last one having to be conjugated to the enzyme.

1. ELISA

a. Advantages

Specificity — Enzyme-conjugated polyclonal sheep or rabbit antibodies directed against rat immunoglobulins and without cross-reactivities with non-rat Ag are now available: PAKO, Mercia-Brocades, U.K.; and MILES Stoke Poje, U.K. The quality (specificity and titration point) of each batch has to be tested before use. Using enzyme-linked anti-rat MAbs, (Seralab, U.K.; Serotec, U.K.; Biosys, France; Zymed, U.S.) highly specific and reproductive results are obtained. Among the mouse MAbs which were produced in our laboratory and which recognized the constant region of heavy or light chains of rat immunoglobulins, we chose to use a pool of mouse MAbs against rat kappa and lambda light chains (MARK-1, MARK-3, and MARL-4). This pool of MAbs detects all rat immunoglobulins in the same way as they are detected by polyclonal rabbit or goat antiserum, but it does not cross-react with non-mouse antigens such as horse, calf, or human Ig.

Sensitivity — The sensitivity depends on the number of second layer antibodies that get attached to the first layer MAbs, and on the ratio of enzyme molecules per second layer antibody molecules. Therefore, when MAbs are used in the second layer and in order to achieve good amplification, the use of a pool of MAbs directed against different epitopes of the first layer MAb is advisable. When a blend of MARK-1, MARK-3, and MARL-4 is used, as little as 10 ng of antibody per milliliter of culture supernatant is detected.

* The references mentioned in this chapter are found at the end of Chapter 9.II.

Rapidity — The results can be easily obtained in 4 h. The 96 wells are read approximately in 3 min, depending on the apparatus utilized.

Quantification — The sensitivity of the ELISA is high enough to detect small concentrations of MAbs. For this purpose standards of known concentrations are needed.

Accuracy — Accurate results may be achieved with the use of standards and with a constant examination of the reagents and instruments.

b. Disadvantages

Chromogens—The response is measured by using an ELISA reader. For this purpose, a chromogen, OPD (orthophenylenediamine dihydrochloride) and ABTS (2,2′-azinobis(3-ethylbenzothiazoline-6-sulfonic acid) diamonium salt), is commonly used. Unfortunately, it is carcinogenic. Gloves must be used to avoid direct contact with hand skin and safety bulbs must be utilized to prevent an accident during the work.

Impurities — All glassware must be cleaned and rinsed thoroughly; any trace of household detergent in contact with peroxidase will result in false negative tests. To detect such problems, each series of experiments must include positive controls.

When OPD is used, it must be prepared extemporaneously. Any impurity in this solution reacts with hydrogen peroxide. This would result in the oxidation of OPD, and thus very high background coloration.[4]

Azide — Any trace of azide will inhibit the peroxidase, resulting in false negative results.

c. Materials

Antigens — At the screening step, they must be highly purified to allow detection of the specific MAbs. ELISA does not require a highly purified antigen solution; however, if it contains impurities a lot of work will be needed to eliminate MAbs against them.

Buffers — Depending on the antigen, different coating buffers are chosen. The origin of the saturating proteins is given as an example:

1. Carbonate-bicarbonate buffer: 0.05 *M*, pH 9.6 or 0.15 *M* phosphate buffer saline (PBS) pH 7.2.
2. Washing buffer: 0.15 *M* PBS pH 7.2, 0.1% Tween-20 (Sigma No. 1379)
3. Saturating buffer: PBS containing different proteins are used, e.g., PBS + 10% skimmed milk (Gloria, France), or PBS + 0.5% bovine serum albumine (BSA) (Biochemical Corporation, U.S.), PBS + 0.5% gelatine (Merck No. 4078), or PBS + 5% serum (not from mouse or rat).

All are good saturating buffers. Old sterile culture medium containing 10% serum can also be employed.

d. Reagents

1. 30% H_2O_2 (Merck No. 7208)
2. OPD (O-phenylenediamine, Sigma p1526) is dissolved extemporaneously into citrate-phosphate buffer 0.15 *M* pH 5 at a concentration of 0.4 mg/ml (4 mg/10 ml) and H_2O_2 at a final concentration of 0.003% (1 μl of 30% H_2O_2 into 10 ml) is added
3. Stopping solution: 0.56 M H_2SO_4
0.5% Na_2SO_3
4. Peroxidase-labeled MAbs (MARK-1, MARK-3 and MARL-4)

e. Procedure

Coat the plate with 100 μl of antigen at a concentration of 5 to 10 μg/ml diluted in a coating buffer and incubate 2 h at room temperature. Wash three times in PBS 0.1% Tween-20. Saturate the plates with 200 µl of saturating buffer between 30 min and 1 h at 37°C depending on coating concentration. Again wash three times in PBS 0.1% Tween-20. Add 100 µl of supernatant per well to test and incubate 2 h at room temperature or 1 h at 37°C. Wash three times in PBS 0.1% Tween-20. Add 100 µl of anti-rat MAbs (MARK-1 + MARK-3 + MARL-4) labeled with peroxidase at 0.5 µg/ml in PBS Tween-20, then incubate 2 h at room temperature or 1 h at 37°C. Wash six times with PBS 0,1% Tween-20 and add 100 µl of the chromogen solution. As soon as the color of the positive controls turns to dark yellow, stop the reaction with 100 µl of stopping solution and read at 495 nm.

2. Immunodot

The different steps of this method are the same as in the ELISA technique. The main difference is the nature of the solid phase: a nitrocellulose membrane which strongly binds proteins (see Chapter 10).

a. Advantages

The immunodot assay requires a much lower amount of antigen than the ELISA. A drop of 1 µl or less, of a 20 µg/ml solution of soluble antigen is enough.

Easy reading—Positive reaction is indicated by a black precipitate; in negative controls and negative samples, no color is detected.

Storage—The antigen-dotted filter may be generally stored dry for several months without any loss of activity.

Sensitivity—The dot assay offers sensitivity at least comparable to ELISA.

b. Disadvantages

The strips, especially when in large numbers, are not as easily handled as ELISA plates.

Quantification—chloronaphtol precipitation prevents quantification of MAbs in supernatants. For the normal screening of culture supernatants this is not a handicap, but when information of the amount of MAbs secreted by the hybrid is required, an alternative immunodot assay involving a radio-labeled second layer antibodies or protein A when it can bind to the MAb (not recommended for rat MAbs) should be used.

Impurities—As in ELISA test, one should avoid impurities and buffers that contain azides (for more details, see ELISA test).

c. Reagents

1. Nitrocellulose membrane filter (Schleicher and Schuell) 0.22 µm porosity 300 mm × 3000 mm (cat No. 401196)
2. NaCl
3. Fetal calf serum or horse serum (Gibco)
4. 4 chloro-1-naphtol (Sigma No. 8890)
5. 30% H_20_2 (Merck No. 7209)
6. Ethanol (Merck No. 971)
7. Tris hydroxymethyl-aminomethane (Janssen Chimica No. 16.762.78)
8. Tween-20 (Sigma No. 1379)
9. Coating buffer: 0.02 *M* Tris, 0.15 *M* NaCl pH 8
10. Incubating buffer: O.O2 *M* Tris; 0.15 *M* NaCl; 5% serum and O.1% Tween-20 pH 8
11. Saturating buffer: 0.02 *M* Tris; 0.15 *M* NaCl; 20% serum and 0.1% Tween-20
12. Substrate solution: dissolve chloronaphtol at 2 mg/ml in ethanol; add 4 volumes of PBS and 1 µl H_2O_2 per ml of final volume

d. Procedure

Dot in sequence 1 μl of antigen on a 5 mm wide strip of nitrocellulose filter, the strips are dried at room temperature for at least 1 h to stabilize the binding. Wash the strips for 30 min at room temperature with constant agitation in 0.02 *M* Tris buffer pH 8 containing 20% serum, 0.1% Tween-20 and NaCl 0.15 *M*, then put 50 μl of antigen of supernatant in 2 ml of 0.02 *M* Tris buffer pH 8, 5% serum, 0.1% Tween-20, NaCl 0.15 *M* and add the strip. Incubate 1 h at 37°C. Wash the strip of nitrocellulose twice with 0.02 *M* Tris buffer pH 8, 5% serum, 0.1% Tween-20 and NaCl 0.15 *M* and add about 10 ml of peroxidase-conjugated MARK-1 + MARK-3 + MARL-4 (diluted 1/2000) to Petri dishes and add the strips (about 8 per box). Incubate 1 h at room temperature. Wash three times in 0.02 *M* Tris buffer pH 8, 5% serum, 0.1% Tween-20 and NaCl 0.15 *M*, and then transfer to another vessel which contains a freshly prepared solution of 0.4 mg/ml 4-chloro-1-naphtol in PBS and 0.03% H_2O_2. Coloration appears quickly. Stop by washing the strip with distilled water.

C. RADIOIMMUNOASSAY (RIA)

The principle of the radioimmunoassay and the ELISA is the same. In the RIA test, antibodies are labeled with a radioactive isotope. ^{125}I is most commonly used.

1. Advantages

Quantification test — The same precision can be obtained in RIA and ELISA, but the sensitivity is often better in RIA than in ELISA.

^{125}I-labeling of second layer antibodies — ^{125}I-iodine is an easy conjugate to immunoglobulins. High specific radioactivities are obtained without alteration of the specificity of the antibodies (see Chapter 18).

2. Disadvantages

Storage — Owing to radioactive decay, the ^{125}I-labeled antibody cannot be stored for too long, about 60 d.

Radiation hazards — Iodine is very volatile and has a high affinity for the thyroid gland. Sampling must be carried out in a hood; a gas mask and gloves must also be worn.

Tritium can also be used if a liquid scintillation counter is available, but the specific activity is very low compared to ^{125}I. However, it allows a more heavy labeling and the labeled antibodies may be stored for years at –20°C. Moreover, tritium labeling does not alter the antibody conformation very much and should be preferred when ^{125}I-labeling inactivates the antibodies.

3. Reagents and Materials

1. Washing buffer: 0.15 *M* PBS pH 7.2 with O.1% bovine serum albumin (BSA) and 0.01% NaN_3; 0.15 *M* PBS pH 7.2 with 0.01% NaN_3
2. Antigen diluted in PBS with 0.01% NaN_3
3. Anti-rat antibodies (MARK-1 + MARK-3 + MARL-4) labeled with ^{125}I
4. 96-well plates DYNATECH (Cat. No. 1-220-24)
5. Disposable material should be preferred

4. Procedure

Coating of plates with antigen — Coat well with 50 μl of antigen diluted at 50 μg/ml in PBS pH 7.2 containg 0.01% NaN_3 and incubate 2 h at room temperature. Wash twice with PBS containing 0.1% BSA and 0.01% NaN_3 and wash one time with PBS with 0.01% NaN_3 and dry. The plate can be stored at –20°C for 2 months maximum.

Incubation with culture supernatants — Add 50 μl of supernatant per well in duplicate for

1 h at room temperature followed by a night at 4°C and wash three times with PBS containing 0.1% BSA and 0.01% NaN_3.

Revelation with antibody labeled with ^{125}I — Add 50 µl per well of (MARK-1 + MARK-3 + MARL-4) labeled with ^{125}I with 50,000 to 100,000 cpm per 50 µl and incubate 1 h at room temperature. Wash four times with PBS containing 0.1% BSA and 0.01% NaN_3, then dry the plates and count in a gamma counter.

D. HEMAGGLUTINATION

This technique is based on the cross-linking of antigen-coated red blood cells by specific antibodies. When hemagglutination is reached, the red blood cells make a continous carpet at the well bottom. If no specific antibodies or too low a concentration of antibodies are present, the red blood cells will pellet fully and form a button at the well bottom. False positive results occur if complement inactivated serum is not used.

1. Advantages

Sensibility — Hemagglutination assay is less sensitive than ELISA, RIA, or Immunodot assay. However, this fact could be an advantage rather than a handicap since only high-affinity antibodies or good producer hybridomas will be detected. Therefore poor quality hybridomas will be excluded.

Rapidity — Red blood cells can be coated 24 h before their use. The test lasts approximately 2 h.

No sophisticated equipment — positive and negative assays are easily distinguished by the naked eye. No photometer or gamma counter is required.

2. Disadvantage

The quality of red blood cells varies from one donor to another. This problem prevents comparison between tests; hence reproducibility is not always good. It is therefore advisable to identify the sheep donors and select the best ones when possible. Erythrocytes require pretreatment with a chemical reagent in order to facilitate the adsorption or conjugation. Soluble antigens are easily attached to red blood cells after $CrCl_3$ pretreatment.

3. Materials and Reagents

1. Fresh sheep red blood cells
2. Antigen dialyzed 48 h against 0.9% NaCl
3. 0.9% NaCl
4. Old $CrCl_3$ solution (> 6 months age) (Merck No. 818090)
5. Culture medium
6. Sterile tube of 50 ml
7. Sterile tube of 12 ml
8. Hemagglutination plate (Gibco cat. No. 2-621-70)

4. Procedure

a. Coating of Sheep Red Blood Cells

Wash red blood cells twice with 0.9% NaCl by centrifugation 15 min at 1000 *g;* wash red blood cells once with 0.9% NaCl by centrifugation 6 min at 150 *g*. In a 15 ml tube, mix 100 µl packed sheep red blood cells, 100 µl of 1 to 2 mg/ml sterile antigen previously dialyzed vs. 0.9% NaCl, and 100 µl of $CrCl_3$ solution by rotation of the hand during 10 min at room temperature. Wash red blood cells three times with 10 ml of sterile 0.9% NaCl for 6 min at 150 *g*. Suspend red blood cells in 4 ml of sterile 0.9% NaCl and store for 3 d at maximum at 4°C.

TABLE 1

	ELISA	Immunodot	RIA	Hemagglutination
Sensitivity	+++	+++	++++	++
Reproducibility	++	++	+++	+
Quantitation	++	—	+++	—
Conservation of labeled antibodies	Up to years	Up to years	60 d max	No labeled antibody
Cost	++	++	+++	+
Dangers	Most of the chromogens are cancerogenous	as ELISA	Radioactivity	—
Time	Hours	Hours	Necessity of several hours to read the tests	Hours
Lecture	Automatic	Naked eye	Automatic	Naked eye
Storage of coating antigen	Months	Months	Months	3 d

b. Hemagglutination

To each well of the hemagglutination plate, add 25 μl of complete medium, 25 μl of hybridoma supernatant to test, and 25 μl of antigen-coated red blood cells. Incubate 1 h at 37°C.

E. CONCLUSIONS

Table 1 gives a general view of the advantages and disavantages of the various techniques which must be selected in the function of every given experiment.

Chapter 9.II

SCREENING OF HYBRID SUPERNATANTS FOR MONOCLONAL ANTIBODIES AGAINST CELLULAR ANTIGENS

A.M. Ravoet

TABLE OF CONTENTS

I. INTRODUCTION

When intact cells are used for immunization, resulting MAbs will be directed against almost any surface antigen. Depending on the field of interest of the scientist, only some MAbs and hence some hybridomas will be selected for further study. Therefore, a good strategy comprises at least two steps of selection. First, all supernatants of culture wells where clonal growth is observed are screened for reactivity with the target cell. In a second step, the wells containing a MAb directed against a target cell antigen are screened for reactivity with a frequent and undesirable cross-reacting cell type. Exclusion of hybrids whose MAbs exhibit undesirable cross-reactions is especially important when the target cell bears strong immunodominant antigens, as for example X-hapten (CD15) or CD24 antigens on lung cancer cells, HLA-A, -B on lymphocytes and platelets, and HLA-DR on B cells.

The risk of such a severe selection is that when two hybridomas, one secreting an interesting MAb and the other secreting a cross-reactive MAb, coexist in one well, they are thrown away. The probability of such an event can be kept low by adapting the amount of cells seeded per well after fusion. In our experiments clonal growth was observed on an average in 60% of the wells. Among these, around 30% (i.e., 18% of the total amount of wells) secreted MAb directed against a target cell antigen. According to Poisson's distribution, this means that on the average 39% of the wells with clonal growth contain two or more hybrids, but only 1 out of 16 specific answers comes from multiple antitarget MAbs. Experimentally we observed similar frequencies. Among 36 ascites from uncloned hybridomas, two were shown to contain two MAbs: one was a mixture of two anti-B cell MAbs which would have been selected anyway, and the other mixture contained a MAb of unknown specificity and an anti-HLA-A, -B MAb.

II. CHOICE OF THE METHOD

The requirements for a good screening method are sensitivity, the method must detect a positive supernatant, and the possibility to perform a large number of tests in one day; and thus speed, minimum effort per well and low cost per well.

Almost all screening procedures involve indirect detection of the MAb by a second layer. After a first incubation of the cells with hybrid-culture supernatant (one drop will saturate 5×10^5 cells), unbound antibodies are removed by washing. A second incubation of the MAb-coated cells involves second layer antibodies conjugated to peroxidase, alkaline phosphatase (ELISA), fluoresceine (IIF), or labeled with ^{125}I or ^{3}H(RBA). After washing, cell-bound fluoresceine, radioactivity, peroxidase, or phosphatase are evaluated using an epifluorescence microscope or a cytofluorograph (IIF), and a gamma-counter (^{125}I) or a liquid scintillation counter (^{3}H). The amount of bound peroxidase/phosphatase is measured by means of a photometer, after incubation with a substrate that changes color after oxidation/hydrolysis.

The result of ELISA or RBA is proportional to the total amount of antigen in the cell suspension, if MAb and second layer are present in saturating amounts. Hence, for homogenous cell suspensions, the density of the antigen on the cell can be calculated. The relative densities of the antigen on a few cell lines are informative of the antigen recognized.[5] However, when heterogenous cell suspensions are analyzed, one cannot distinguish between weak labeling of all cells and strong labeling of a subpopulation of cells. Moreover, some cells contain endogenous peroxidase or phosphatase and the endogenous response will contribute to enhancing the background staining. Therefore, we mostly restrict use of ELISA and RBA to screening for undesirable cross-reactivities of MAbs with other cell types. In this case, the lack of discrimination between a real negative and a MAb staining a small subpopulation of cells results in keeping an uninteresting hybrid and not in missing a possibly interesting one.

Indirect immunofluorescence allows a cell-by-cell analysis. The percentage of labeled cells, the intensity of labeling per cell, and some information about the kind of labeled cells are obtained. Analysis by microscopy is rapid and cheap, but if available a cytofluorograph with data processor provides a maximum of information for a minimum of effort (percentage, shape of histogram, fluorescence on different cell types present in the sample). Since different kinds of cells are analyzed, attention should be paid to background labeling. This background labeling is mostly due to the aspecific binding of the first and second layer antibodies to Fc receptors on monocytes, B cells, etc. Use of (Fab′)2 fragments of second layer antibodies minimizes this problem. However, background should always be estimated by running a negative control parallel. In the first screenings after fusion, the supernatant of wells containing the unfused mixture of splenocytes and immunocytoma (see Chapter 7) provides the best negative control. Later on however, when the concentration of polyclonal immunoglobulin in theses wells becomes negligible, normal rat serum or ascites of a nonsecreting myeloma per hybridoma should be used.

When a large number of supernatants have to be screened, ELISA, RBA, as well as IIF are performed in microtitration plates of 96 wells. If cells are to be washed by centrifugation and aspiration of the supernatants, round-bottomed wells are used. If cells are fixed to the well-bottom and washing is by flicking of the plate, flat-bottomed wells are chosen.

Fixing of the cells (by glutaraldehyde) should be avoided as far as possible, because it destroys most proteic antigens. Moreover, an antigen with a restricted distribution on cell surfaces can have a much broader distribution when cytoplasmic + membrane antigens are detected.

III. INDIRECT IMMUNOFLUORESCENCE

A. MATERIAL

1. Epifluorescence microscope (Leitz) or cytofluorograph (Coulter, FACS, Ortho)
2. Centrifuge for microtitration plates
3. 96 round-bottomed well microculture plates (Costar, Sterilin)
4. Pipettes

B. REAGENTS

1. Supernatants of hybrid clones; supernatant of a control well is used as a negative control
2. PBS (Gibco No. 042-04200 M)
3. Washing buffer: PBS-2% FCS-0.2% NaN_3
4. FITC-labeled goat or rabbit anti-rat Ig antibodies (RARA-FITC): whole Ig or (Fab′)2 fragments (Miles, Sigma, Bio Yeda, Tago)
5. Formaldehyde 1% in PBS, freshly prepared (if fixing is required)
6. Propidium iodide: stock solution: 10 mg propidium iodide is dissolved in 10 ml 0.9% NaCl-5% EDTA. Before use, dilute 300 × in PBS. Propidium iodide will stain dead cells (also formaldehyde-fixed cells).

C. PROCEDURE

With a sterile Pasteur pipette, transfer one drop of each hybrid-culture supernatant to a well of a microtitration plate and add 5×10^5 target cells in 25 μl. Incubate for 45 min at 4°C or 30 min at room temperature. Add 100 μl of washing buffer, centrifuge for 5 min at 250 *g*, and aspirate the supernatants thoroughly, then resuspend by passing the plate on a vortex mixer. Repeat the washing three times with 0.2 ml washing buffer. Add a saturating amount of FITC-labeled anti-rat Ig antibodies and incubate for 45 min at 4°C (or 30 min at 20°C). Wash three times as above. Either resuspend in 15 μl PBS containing propridium iodide and transfer the suspension to a microscope slide and analyze immediately on an epifluorescence microscope, or resuspend in 400 μl PBS (or PBS-1% formaldehyde) and analyze on a cytofluorograph within a few hours (or within a few days). Never combine propidium iodide and PBS-formaldehyde.

IV. RADIOBINDING ASSAY

A. MATERIALS

1. Bench-top centrifuge for microtitration plates
2. Flexible vinyl microtitration plate (round-bottomed) (Dynatech M24)
3. Pipettes
4. Counter: a gamma-counter if ^{125}I-labeled second layers are used (Berthold) and a liquid scintillation counter if ^{3}H-labeled second layers are used (Beckman)

B. REAGENTS

1. Supernatants of hybrid clones; supernatant of a control well as negative control
2. Second layer antibodies (goat or rabbit anti-rat Ig antibodies); labeling with ^{125}I is done by the chloramine-T method,[6] labeling with ^{3}H by the formaldehyde/boro-^{3}H-hydride method[7]
3. PBS (Gibco No. 042-04200 M)
4. Washing buffer: PBS-2% FCS-0.2% NaN_3

C. PROCEDURE

First incubation of cells with hybrid-culture supernatant and three washings carried out as described above (see indirect immunofluorescence procedure). Add ^{125}I-labeled or ^{3}H-labeled rabbit anti-rat Ig antibodies and incubate for 45 min at 4°C (30 min at 20°C). Wash three times as above. Cut the wells apart and place the wells in counting tubes (^{125}I) and count in a gamma-counter or transfer the wells to vials (^{3}H), add scintillation mixture and count in a liquid scintillation counter.

REFERENCES

1. **Avrameas, S. and Uriel, J.,** Méthode de coloration des acides aminés à l'aide de la L-aminoacide-oxhydrase, *C.R. Acad. Sci.,* 262, 2543, 1966.
2. **Engvall, E. and Perlmann, P.,** Enzyme-linked immunosorbent assay (ELISA). Quantitative assay of immunoglobulins G, *Immunochemistry,* 8, 871, 1971.
3. **Van Weemen, B. K. and Schuurs, A. H. W. M.,** Immunoassay using antigen-enzyme conjugates, *FEBS Lett.,* 15, 232, 1971.
4. **Tijssen, P.,** Laboratory techniques in biochemistry and molecular biology, in *Practice and Theory of Enzyme Immunoassay,* Vol. 15, Burdon, R. H. and Van Knippenberg, P. H., Eds., Elsevier, Amsterdam, 1985.
5. **Ravoet, A. M. and Lebacq-Verheyden, A. M.,** Clustering of anti-leukemia and anti-B cell monoclonal antibodies, in *Leukocyte Typing,* Vol. 2, Reinherz, E.L., Haynes, B. H., Nadler, L. M., and Bernstein, I. D., Eds., Springer Verlag, New York, 1986, 213.
6. **Hunter, W. M.,** Radioimmunoassay, in *Handbook of Experimental Immunology,* Vol. 1, Wei, D. M., Ed., Blackwell Scientific Publications, Oxford, 1973, 17, 1.
7. **Means, G. E. and Feeney, R. E.,** Reductive alkylation of amino groups in proteins, *Biochemistry,* 7, 2192, 1968.

Chapter 10

ISOTYPING OF RAT MONOCLONAL ANTIBODIES

P. Manouvriez, F. Nisol, T. Delaunay, and H. Bazin

TABLE OF CONTENTS

I. INTRODUCTION

Each rat immunoglobulin isotype has its specific physicochemical and biological properties. They have been described in Chapter 1. The treatments and use of monoclonal antibodies (MAbs) will be principally ruled by those properties. Knowing the immunoglobulin (Ig) isotype secreted by a hybridoma is therefore a necessity. Moreover, identifying them in culture supernatants is also a way to check the clonality of a culture and to select the desired clones. If more than one isotype is present in one supernatant, the culture generally has to be cloned again. However, some clones have been described to secrete two Ig isotypes, but this is extremely rare (Bazin and Libert, unpublished results).

Different techniques are available to screen the rat MAb isotype, each with its advantages and disadvantages. Among these techniques double radial immunodiffusion is the simplest, ELISA and immunodot are the most sensitive.

Both heterologous polyclonal antibodies and monoclonal anti-rat isotype antibodies can be used. Their qualities and properties govern the isotyping technique which is to be used.

This chapter deals with the obtainment and use of polyclonal anti-rat immunoglobulin isotype antibodies and also with the use of MAb in the different isotyping techniques.

II. POLYCLONAL AND MONOCLONAL ANTI-RAT Ig ISOTYPE ANTIBODIES

A. OBTAINMENT OF POLYCLONAL ANTIBODIES

High quantities of rat immunoglobulins of a given isotype can easily be obtained from culture supernatants or ascites produced by rat hybridomas or immunocytomas.[1,2] They can be purified to more than 99%, as described in Chapters 3 and 15. All rat immunoglobulin isotypes are immunogenic in rabbit and goat.

1. Animals and Material

1. Rabbits
2. Purified rat immunoglobulin at 1 to 5 mg/ml in phosphate buffered saline (PBS)
3. Complete Freund's adjuvant (Difco, U.S.)
4. Syringe and 18 G1 1/2 needle

2. Procedure

Antigen and complete Freund's adjuvant are emulsified v/v (see Chapter 5). To obtain a good immunogen, the emulsion must be stable. If aqueous and oil phases separate after a few minutes, the mixture has to be mixed further (see also Chapter 15). Rabbits are generally used; goats or sheep can also be employed with the same protocols. Two injections of 0.5 to 1 ml of extemporeaneously prepared antigen emulsion are made subcutaneously or intramuscularly at a 2-week interval; the following immunizations are made according to the antibody titer of the collected serum and when it is too low immunizations are repeated.

To avoid the appearance of an anti-idiotype immune response, at least two different antigenic preparations must be used alternatively if monoclonal Igs have been taken as immunogen. Antibody activity is tested for its specificity by immunoelectrophoresis analysis against whole rat serum and different myeloma proteins according to their isotypes.

3. Improvement of the Antibody Specificity

Most antisera contain anti-light chain antibody activities and often cross-react with all isotypes. The undesired activities must be eliminated by extensive immunoadsorption on insolubilized, highly purified rat Igs. Several chromatographic media were tested and the one we found the easiest and cheapest to utilize was the CNBr-activated Sepharose 4B (Pharmacia, Sweden) to which all our rat Ig preparations can be conjugated in accordance with the manufacturers' prescriptions. The monospecificity of the antisera was tested by immunoelectrophoresis analysis and double radial immunodiffusion against normal rat serum and purified monoclonal Igs. If the user wishes to use its polyclonal preparation for immunoenzymatic tests, it will be further tested by ELISA and if necessary, will be adsorbed again to eliminate remaining cross-reactions.

Heterosera against IgM, IgA, IgE, and IgD are relatively easy to render monospecific, even for use in RIA or ELISA tests, but antisera directed against all IgG subclasses or a given IgG subclass are difficult to produce.

To obtain an antiserum which can recognize all IgG subclasses equally well, it is often necessary to pool sera from several immunized animals.

To obtain a good antiserum specific for one IgG subclass, at least two or three animals must be immunized, because sera collected from some animals are impossible to render perfectly monospecific, at least at the sensitivity level of ELISA or RIA. Therefore, if only heterologous anti-isotype antibodies are available, the use of these very sensitive techniques will often be excluded.

B. USE OF MOUSE ANTI-RAT IG ISOTYPE MABS

Mouse monoclonal anti-rat immunoglobulin antibodies against all isotypes have been produced either in our laboratory or elsewhere. Unlike polyclonal antisera, they have no batch to batch specificity variations and they are easy to prepare once the hybridoma is available. Their main disadvantage is often linked to their poor precipitating properties in gel. Thus, they are preferentially used in ELISA, RIA, or immunodot tests.

III. ISOTYPE SCREENING TECHNIQUES

A. DOUBLE RADIAL IMMUNO-DIFFUSION

The Ig content of culture supernatants is generally on the lowest limit of the sensitivity of the double radial immunodiffusion or Ouchterlony test; this is the major limitation of this method. The gel is punched with very large wells in which 80 µl supernatant can be put. The hetero-antiserum used must have very good precipitating properties. As the volumes involved are considerable, the diffusion time is rather long. Although most of the precipitation lines generally appear after one night of diffusion at room temperature, sometimes 4 d are needed for completion. The accuracy of the isotype determination depends on the quality of the antisera used. But in any case the waiting period is rather long. If the result is urgently needed to decide which culture to develop, this is a drawback.

1. Gel Preparation

a. Reagents and Material

1. Barbital buffer, pH 8.6 made with Na-barbital 45 g, 32.5 ml of 1 *M* HCl, NaN_3 2 g and adjusted to 5 l with distilled water
2. Agarose such Indubiose (Agarose EEQ = –0.017, IBF, France)
3. Polyethyleneglycol 6000 (Merck No. 807491)
4. 2-l conical flasks
5. Magnetic stirrer heater
6. A Bunsen
7. 56°C water bath
8. 15-ml tubes

b. Procedure

In a 2-l conical flask, 500 ml of distilled water is mixed with 500 ml of barbital buffer and brought to a boil. The buffer is maintained very hot (but not boiling) on the magnetic stirrer heater, while stirring, about 15 g Indubiose are very progressively added. To favor the immunoprecipitation of the antigen-antibody complexes, 20 g/l polyethyleneglycol 6000 are also added. When the powders are entirely dissolved, the conical flask is put for 1 h at 56°C. One gram azide (NaN_3) is then dissolved in the gel which is aliquoted in 15-ml tubes and stored at 4°C.

The quantity of agarose to mix can be slightly increased or decreased depending on the batch used and the consistency of the gel used.

2. Gel Slide Preparation and Immunodiffusion

a. Reagents and Material

1. 1.5% Agarose gel with 2% PEG 6000 in barbital buffer
2. 0.9% NaCl
3. Staining solution made as follows : glacial acetic acid, 60 ml; Na acetate, 8.2 g; Amido-Schwartz (e.g., from Merck No. 1167), 2 g; adjusted to 2 l with distilled water
4. Destaining solution prepared with 100 ml glacial acetic acid, 10 ml glycerin, and 1900 ml distilled water
5. Clean glass slides (they must be well degreased)
6. 5-ml pipettes
7. Gel punch of 7.7 mm (0.33″) for the large wells and of 2.5 mm (0.1″) for the small well; two patterns can be used and are shown in Figure 1
8. Pasteur pipettes and a water vacuum pump
9. Humid chamber
10. Boiling water bath
11. Filter paper
12. 37°C-incubator
13. Monospecific heterologous antisera anti-rat immunoglobulin isotypes can be obtained from several suppliers (Miles, U.S., Biosys, France, Zymed, U.S., Serotec, U.K.). Their monospecificity must be checked against normal rat serum before use.

b. Protocol

The agar gel is melted in a boiling water bath. Agar is put on the slides (160 µl/cm^2). Use a leveled surface, and punch a pattern as shown in Figure 1 a and b. Suck out agar plugs with a Pasteur pipette connected to a water vacuum pump, then fill the large well(s) with (approximately 80 µl) culture supernatant and the small well(s) with (approximately 7 µl) anti-rat monospecific isotype antisera until the meniscus just disappears. The slide is left overnight in a humid chamber at room temperature.

When the pattern (Figure 1 b) is used to test culture supernatants, the diffusion may require up to 4 d at room temperature, and with the other pattern (Figure 1 a) most of the precipitating lines can be seen after an overnight diffusion. To remove the unprecipitated proteins, the slides are washed for 48 h in 0.9% NaCl, changed once, and thereafter washed at least 4 h in distilled water. All these washes are performed at 37°C. After the last wash the slides are covered with moistened filter paper and left to dry for one night The filter papers are discarded and the slides quickly rinsed with distilled water. Staining is performed for 20 min in staining solution at room temperature, and the excess stain is removed in destaining bath for 10 min., then the slides are left to dry.

If the slides are entirely destained, they can be restained. Often, Ouchterlony tests are read without staining if there is no need to keep the slides.

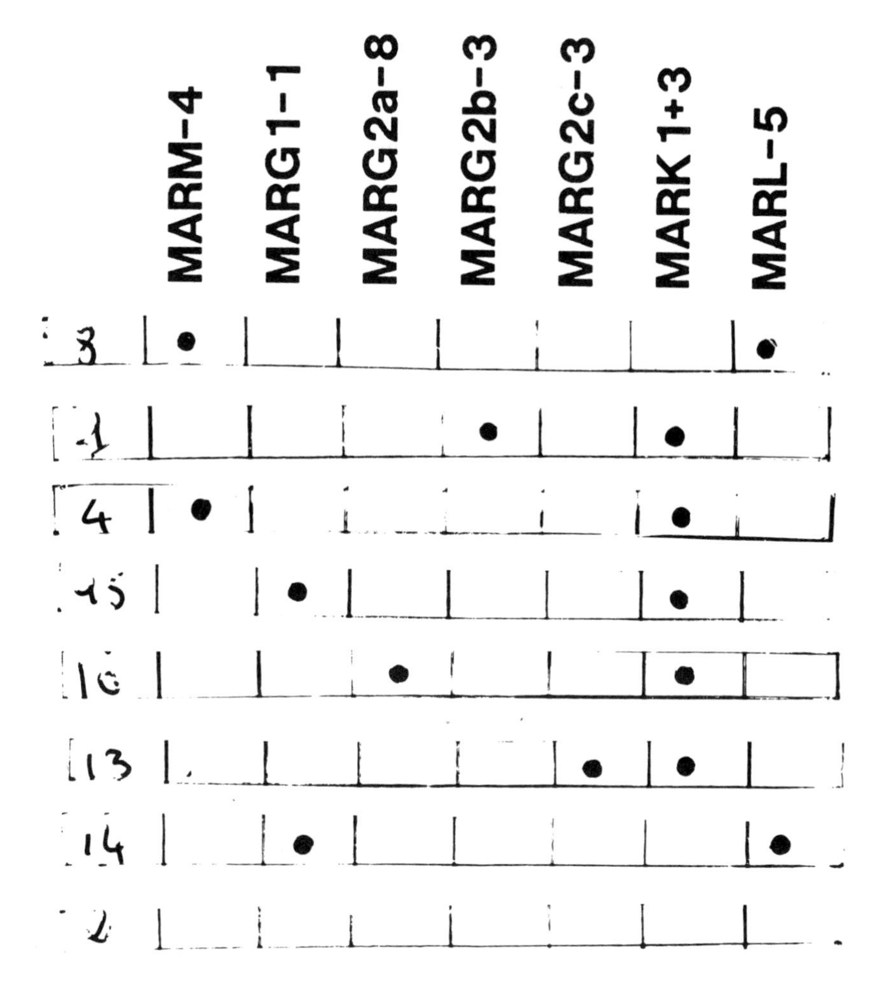

FIGURE 1. Immunoglobulin isotype determination by double radial immunodiffusion. a. The central well was filled with 80 µl of LO-DNP-11 (rat IgG2a anti-DNP MAb) the peripheral wells were filled with 7 µl of (1) rabbit anti-rat IgM, (2) rabbit anti-rat IgG1, (3) rabbit anti-rat IgG2a, (4) rabbit anti-rat IgG2b, (5) rabbit anti-rat IgG2c, and (6) rabbit anti-rat IgA. b. The central well was filled with 7 µl rabbit anti-rat IgG1 the peripheral wells were filled with 80 µl culture supernatant of (1) LO-DNP-39 rat IgM MAb anti-DNP, (2) LO-DNP-1 rat IgG1 MAb anti-DNP, (3) LO-DNP-7 rat IgG2a MAb anti-DNP, (4) LO-DNP-11 rat IgG2b MAb anti-DNP, (5) LO-DNP-64 rat IgA MAb anti-DNP, and (6) LO-DNP-30 rat IgE MAb anti-DNP.

B. ELISA

1. Introduction

The development of strictly specific mouse anti-rat isotype MAbs has made possible the very sensitive immunoenzymatic tests for the isotype determination. The result of the ELISA test is semiquantitative. However, the assay can be done quantitatively by the insertion of standards in the tests. Owing to its pentameric nature, the detection of IgM is more sensitive than that of the other isotypes. When performed without standards, the overestimation of IgM compared to the other isotypes must thus be taken into account.

Rat MAbs of the IgD class have never been observed in the laboratory. IgE and IgA rat MAbs are rarely obtained except after specific stimulations.

The screening of the isotypes is then usually limited to eight different tests performed with the following mouse MAbs at 10 or 20 µg/ml. We always use LOU/C rats for immunization which produce hybridomas secreting IgK-1a MAbs. Ig kappa allotype of commonly used rat strains are given in Table 1.

TABLE 1
Commonly Used Rat Strains of the Two Kappa Allotypes Reviewed in Bazin et al.[3] and Hunt and Fowler[4]

IgK-1a allotype	IgK-1b allotype
AUG (August)	BD
BN	Black and White Hooded lister
Gowans (albino)	COP (Copenhague)
LEW (Lewis)	DA
LIS (hooded lister)	LOU/C.IgK-1b(OKA)[a]
LOU/C[a]	OFA
PB	OKA (OKAMOTO)
PVG/c	PVG/Ciu[b]
PVG-*RI-a*[a](DA)[b]	SD (Sprague Dawley)
	YOS (Yoshida)

[a] Histocompatible[5]
[b] Histocompatible[4]

1. MARM-4: mouse IgG1-kappa anti-rat μ heavy chain
2. MARG1-1: mouse IgG1-kappa anti-rat gamma 1 heavy chain or MARG1-2
3. MARG2a-8: mouse IgG1-kappa anti-rat gamma 2a heavy chain or MARG2a-1 or MARG2a-7
4. MARG2b-3: mouse IgG1-kappa anti-rat gamma 2b heavy chain
5. MARG2c-3: mouse IgG2a-kappa anti-rat gamma 2c heavy chain
6. MARL-5: mouse IgM-kappa anti-rat lambda light chain
7. A cocktail of MARK-1: mouse IgG1 anti-rat kappa light chain and MARK-3: mouse IgG1 anti-rat kappa-1a light chain of the IgK-1a allotype

If necessary, the following other MAbs are also used:

1. MARA-2: mouse IgG1-kappa anti-rat alpha heavy chain
2. MARE-1: mouse IgG1-kappa anti-rat epsilon heavy chain
3. MARD-3: mouse IgG1-kappa anti-rat delta heavy chain

2. Reagents and Material

a. Buffers

Phosphate buffered saline (PBS) — pH 7.2, NaCl 8 g/l, KCl 0.2 g/l, Na_2HPO_4 (0.008 *M*) 1.15 g/l, KH_2PO_4 0.2 g/l, and 40 mg phenol red (Janssen Chemical) dissolved in 1 l distilled water. It is convenient to make a ×10 stock solution for storage and dilute as required.

Saturating solution — This will be chosen among the following buffers; PBS + 5 to 10% skimmed milk, PBS + 0.5% bovine serum albumin, or PBS + 5% serum (not from mice or rats).

Washing solution — PBS + 0.1% Tween 20 (Sigma)

Coating buffer — 0.1 *M* borate buffer pH 9.5

40 mg phenol red per liter can be added in washing and coating buffers.Phenol red is added to the buffer to make the visualization of a correct well-filling easier.

b. Substrate

1. 0.1 *M* citrate buffer, pH 5.5, Tri-sodium citrate 29 g/l, Citric acid 4.1 g/l.

2. 4 mg o-phenylbenediamine dihydrochloride (OPD Sigma Cat No. p1526 or Merck No. 7243) are dissolved extemporaneously in 10 ml citrate buffer and 1 µl H_2O_2 30% is added.
   Caution—This solution must be prepared in plastic ware or in very clean glassware finely washed with sulfochromic acid and thoroughly rinsed with distilled water. The solution must be entirely colorless. Otherwise, it must be discarded and a new solution must be prepared with other material (pay particular attention to the cleanness of the spatule which is used to weigh).
3. Stop solution 0.1 *M* H_2SO_4 + 0.5% Na_2SO_3.

c. Antibodies

The following mouse anti-rat isotype MAbs are all in solution in the coating buffer:

1. MARM-4, at 0.5 µg/ml
2. MARA-1, at 10 µg/ml
3. MARG1-1, at 10 µg/ml or MARG1-2 at the same concentration
4. MARG2a-8, at 10 µg/ml or MARG2a-1 or MARG2a-7 at the same concentration
5. MARG2b-3, at 10 µg/ml
6. MARG2c-2, at 10 µg/ml
7. MARL-5, at 10 µg/ml
8. A v/v mixture of 20 µg/ml of MARK-1 and MARK-3

A mixture of equal quantities of peroxidase-labeled antibodies in PBS MARK-1, mouse anti-rat kappa light chain; MARK-3, mouse anti-rat kappa-1a allotype; and MARL-4, mouse anti-rat lambda light chain.

The optimal use concentration of each of these antibodies in the absence of denaturation, is of the order of 0.5 µg/ml, (i.e., a 2000-fold dilution in PBS, protein, 0.1 *M*% Tween of 50% glycerol stock solution, cf. PO-labeling of immunoglobulin).

A specific peroxidase-labeled polyclonal anti-rat immunoglobulin antiserum may also be used. It must be thoroughly absorbed on mouse immunoglobulins before use.

Supernatants or media to be tested : 800 µl, or 50 µl diluted 16-fold in saturating medium.

d. Material

1. Plastic tubes or grease-free and detergent-free glass tubes
2. 96 well ELISA plates (e.g. Nunc, Flow, Linbro, Falcon, Pro-Bind)
3. A 50-µl pipette
4. Plate washer
5. 490 nm ELISA reader

3. Procedure

a. Coating of ELISA Plates

The wells of each row of a 96-well ELISA plate are coated with 50 µl of each of the following mouse MAb anti-rat immunoglobulin solutions:

1. MARM-4
2. MARA-1
3. MARG1-1 or MARG1-2
4. MARG2a-8 or MARG2a-1 or MARG2a-7
5. MARG2b-3
6. MARG2c-3

7. MARL-5
8. MARK-1 + MARK-3

Attention must be paid to the presence of NaN_3 in the MAb solutions. It blocks the activity of peroxidase in an irreversible way.

Coating is performed for at least 1 h at 37°C or 2 h at room temperature or still better, for one night at 4°C. The plate is washed once with PBS, and then each well is incubated for 1 to 2 h at 37°C with 250 μl saturating solution to saturate the remaining binding sites. Two washes are made with washing solution (PBS 0.1% Tween 20) and one with PBS. At this stage the plates may be dried and stored until further use at –20°C. If the plates are not stored, all the washes are made with washing solution.

b. Test

For security reasons, each test has to be done in duplicate. If necessary, dilutions may be done with PBS-protein. 50 μl of the media to analyze are put in 2 wells coated with each of the eight mouse MAb solutions and incubated 1 h at 37°C; the plates are then washed three times with PBS-Tween 0.1%. Each well is filled with 50 μl PO-labeled antibody solution and left 1 h at 37°C. Each plate is washed four times with PBS-Tween 0.1% and 50 μl OPD solution is added per well. The reaction is stopped after 20 to 30 min with 50 μl stop solution. The OD is measured at 490 nm.

Caution — For any series of assays, have a negative control (e.g., the same culture medium prepared with the same batch of serum as the culture supernatants to be tested). A positive control also has to be performed to assess the activity of the different mouse anti-rat MAbs.

c. Conclusions

The sensitivity of the test is of 10 ng/ml for IgA and IgG subclasses and of 2 ng/ml for IgM. It has been possible to dilute the reference supernatants more than 100-fold. The screening tests have always been performed with 16- to 20-fold diluted supernatants. The optic density of the well coated with the anti-isotype directed against the major clone is generally at saturation (O.D. equal or superior to 2) (Figure 2). Often minor clones of other isotypes are also detected. This would be hindered at higher dilutions. Although the presence of more than one isotype in a culture well generally implies the existence of multiple clones, the presence of one isotype does not mean monoclonality. Two different clones could indeed be of the same isotype.

C. IMMUNODOT

The immunodot assay is slightly more sensitive than the ELISA test. The immunodot also has the advantage of consuming less mouse MAbs than the classic ELISA. However, the nitrocellulose strips are generally manipulated less easily than ELISA plates, but they can be kept for long periods once revealed.

1. Reagents

1. 0.22 μm porosity nitrocellulose membrane (Schleicher and Schuell, FRG or Biorad, U.S.)
2. Saturating buffer: 0.02 *M* Tris buffer pH 8, 10% serum, 0.1% Tween 20, 0.15 *M* NaCl or the same buffers than for ELISA
3. 0.15 *M* PBS, pH 7.2
4. Washing buffer: 0.02 *M* Tris buffer pH 8, 5% goat or fetal calf serum or 5% skimmed milk, 0.1% Tween 20, 0.15 *M* NaCl
5. Mouse MAbs anti-rat Ig isotype in 0.02 M Tris buffer pH 8: 10 μg/ml MARG1-2, 10 μg/ml MARG2a-8, 20 μg/ml MARG2b-3, 10 μg/ml MARG2c-3, 5 μg/ml MARM-4, 50 μg/ml MARL-4, and 50 μg/ml MARK-1

MARM-7
MARG1-1
MARG2a-8
MARG2b-3
MARG2c-3
MARK-1+3
MARL-5
normal mouse Ig

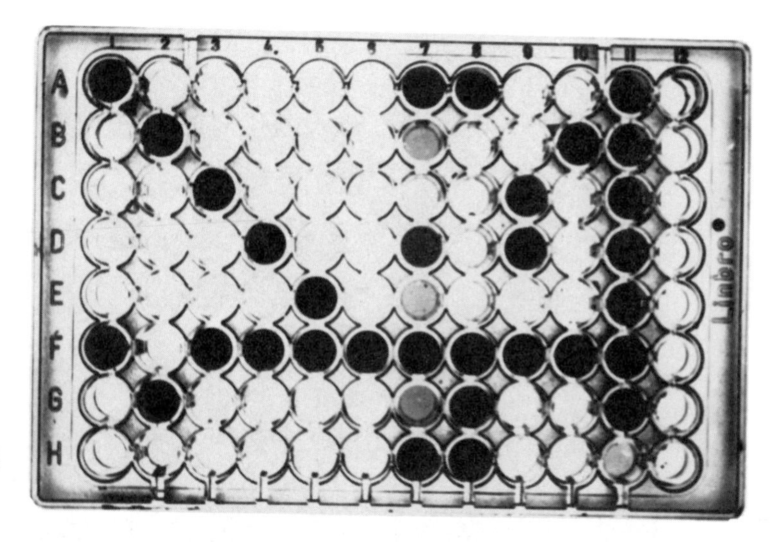

FIGURE 2. ELISA determination of rat immunoglobulin isotype. Each row was coated with mouse MAbs anti-rat immunoglobulin isotype: anti-μ, MARM-7 (A); anti-gamma1, MARG1-1 (B); anti-gamma2a, MARG2a-8 (C); anti-gamma2b, MARG2b-3 (D); anti-gamma2c, MARG2c-3 (E); anti-kappa, a mixture of MARK-1 and MARK-3 (F); anti-lambda, MARL-5 (G) and normal mouse Ig (H) to detect the presence of rheumatoid factors which can give false positive tests. Specifically adsorbed rat immunoglobulins were revealed by their light chain with a mixture of peroxidase labeled MARK-1 + MARK-3 + MARL-4. Each column of 8 wells was incubated with 16-fold-diluted culture supernatant or 40-fold diluted serum: (1) LO-LCA1-1 is an IgM-kappa, (2) IR31, IgG1-lambda, (3) LO-LCA3-3, IgG2a-kappa, (4) LO-Tact-1, IgG2b-kappa, (5) LO-CD24-1/AL1a, IgG2c-kappa, (6) LO-DNP-30, IgE-kappa, (7) IR202, IgM-kappa rheumatoid factor, (8) IR1060, IgA-kappa rheumatoid factor, (9) Uncloned culture supernatant containing IgG2a and IgG2b with Ig kappa light chain, (10) Uncloned culture supernatant containing IgG1 with Ig kappa light chain, (11) Normal rat serum, and (12) Normal mouse serum.

6. Peroxidase-labeled MARK-1 + MARK-3 + MARL-5 at 0.5 μg/ml or peroxidase-labeled polyclonal rabbit antibodies directed against rat isotypes in washing buffer
7. 4-chloro-1-naphtol (Sigma cat No. 8890) at 2 mg/ml in methanol can be stored in dark at –20°C as stock solution
8. 4 ml of the chloronaphtol solution is diluted in 16 ml PBS and 10 μl of 30% H_2O_2 are added.

2. Procedure

Strips of 5 × 80 mm are drawn on the nitrocellulose filter membrane. Each strip is first marked to allow its recognition and orientation. It is dotted with 1 μl deposits of each of the mouse anti-rat isotype MAbs. When the dots are dry (at least 1 h at room temperature), the strips are saturated by a 30 min incubation in saturating buffer and they are washed twice with washing buffer and dried. They can be stored for several weeks at –20°C after drying.

To screen the Ig isotype content of culture supernatants, the strips are put individually in elongated trays containing 2 ml washing buffer. To each tray, 50 to 100 μl of supernatant to test are added. Incubation is performed with agitation for 1 h at room temperature. The strips are then transferred in a screw-top container (e.g. 50 ml Falcon tube) for two washes with washing buffer and transferred to a Petri dish containing the same peroxidase-labeled cocktail of MAbs as for the ELISA test (MARK-1 + MARK-3 + MARL-4) or rabbit peroxidase-labeled polyclonal antibodies. The incubation is performed for the same time, shielded from light.

The staining is different from that used in ELISA. Indeed, the chromogeneous derivative of the oxidized OPD is soluble. 4-chloro-1-naphtol used for the immunodot technique precipitates at the point of oxidization where it remains attached and stains the dots where the peroxidase-labeled antibodies have been adsorbed by rat Igs. Coloration is completed after 15 min. The strips are stabilized by washing in distilled water. The intensity of the staining depends on the

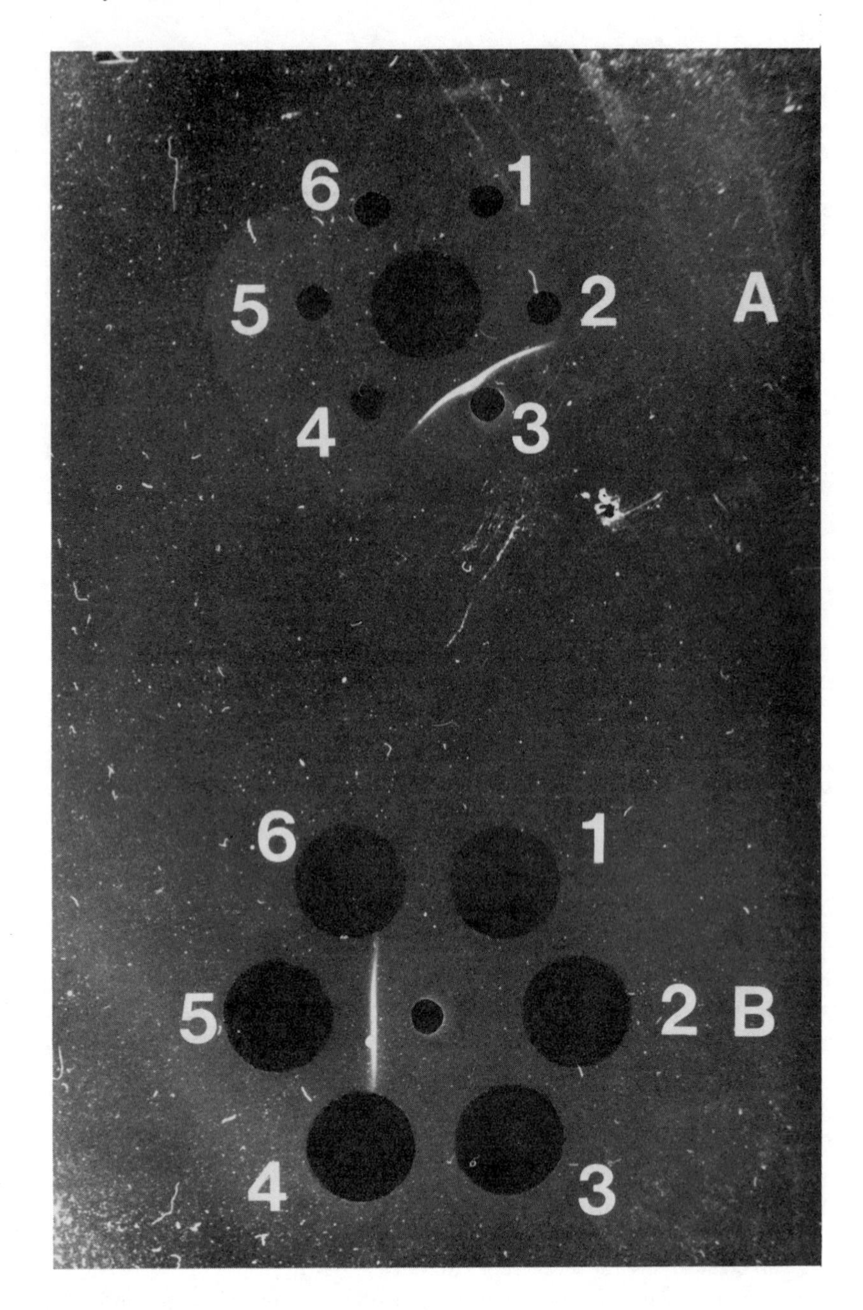

FIGURE 3. Immunodot determination of rat immunoglobulin isotype in culture supernatant. The same mouse MAb anti-rat immunoglobulin isotypes were used for the ELISA test (Figure 2). Each nitrocellulose strip was incubated with 40-fold diluted culture supernatant. Rat immunoglobulins were revealed by their light chain with a mixture of peroxidase labeled MARK-1 + MARK-3 + MARL-5. From the left to the right, the strips revealed the following isotypes in the culture supernatants; 2. nothing, IR983F culture supernatant as negative control; 14. IgG1-lambda; 13. IgG2c-kappa; 10. IgG2a-kappa; 15. IgG1-kappa; 4. IgM-kappa; 1. IgG2b-kappa; and 8. IgM-lambda.

antigen concentration. However, it remains a purely qualitative appreciation. Qualitative examples are shown in Figure 3.

Positive and negative controls must be added to the tests. It is possible to incubate the labeled blend of MAbs in the same trays than the supernatants to test, so there is no loss of time.

IV. CONCLUSIONS

The double immunodiffusion test is the simplest and the least expensive technique for the determination of the Ig isotype content of culture supernatants. Although it can take 4 d before immunoprecipitation is complete, but most generally only 1 d is enough to get the result. Moreover, the Ouchterlony test cannot be performed with MAbs because of their generally poor precipitating properties. It is possible that in the future, pool of MAbs recognizing different antigenic epitopes of the same isotype will allow the Ouchterlony test to be performed with MAb. The heterologous polyclonal antibodies as required for the Ouchterlony test are very difficult to render specific enough to permit accurate IgG subclass determination by ELISA or immunodot, which need the specificity of Mab. Right now, a mixture of MARK-1 and MARK-3 is used to precipitate IgK-1a immunoglobulins in double radial immunodiffusion tests. The most sensitive assays are ELISA and immunodot. The last one can be performed easily at the single test level, but strips are also fragile, especially when dry. They do not need very expensive material. However, the laboratories which are equipped with automated ELISA devices will find it time-saving compared with immunodot. The two techniques being of equal value, the choice will depend on personal preferences and available equipment.

REFERENCES

1. **Bazin, H., Beckers, A., and Querinjean, P.,** Three classes and four subclasses of rat immunoglobulins: IgM, IgA, IgE and IgG1, IgG2a, IgG2b, IgG2c, *Eur. J. Immunol.*, 4, 44, 1974.
2. **Bazin, H., Beckers, A., Urbain-Vansanten, G., Pauwels, R., Bruyns, C., Tilkin, A. F., Platteau, B., and Urbain, J.,** Transplantable IgD immunoglobulin-secreting tumours in rats, *J. Immunol.*, 121, 2077, 1978.
3. **Bazin, H., Nezlin, R., and Brdicka, R.,** Immunoglobulin polymorphisms: rat. II. Allotype distribution, in *Inbred and Genetically Defined Strains of Laboratory Animals, Part I, Mouse and Rat*, Altman, P. L. and Katz, D. D., Eds., Federation of American Societies for Experimental Biology, Bethesda, Maryland, 1979, 306.
4. **Hunt, S. V. and Fowler, M. M.,** A repopulation assay for B and T lymphocyte stem cells employing radiation chimaeras, *Cell Tissue Kinet.*, 14, 445, 1981.
5. **Beckers, A. and Bazin, H.,** Allotypes of rat immunoglobulins. III. An allotype of the gamma2b chain locus, *Immunochemistry*, 12, 671, 1975

Chapter 11

ADAPTATION TO *IN VIVO* CULTURE

P. Manouvriez, J. P. Kints, and H. Bazin

TABLE OF CONTENTS

I. INTRODUCTION

The first screening of the clones obtained after fusion is always done with culture supernatant obtained from the 96 well-plates. The selected clones are generally developed in 24-well plates and eventually in culture flasks for further characterization of the secreted MAbs. When multiple antibodies of the same specificity and the same quality are obtained from a fusion, it is generally the most vigorous hybridoma(s) observed in culture which is selected for *in vivo* production. However, if great vitality is observed *in vitro,* life conditions are not the same as in animals. Great care must be taken with the first *in vivo* production of a hybridoma. Some hybridomas will not develop or will not secrete their MAb *in vivo* without special precautions when the recipient rats are perfectly histocompatible.[1,2]

II. ROUTINE FIRST PASSAGE

If syngeneic hybridomas have to be implanted, i.e., hybridomas obtained by fusion of 983 cells or one of the Y3 fusion cells with lymphocytes of LOU/C rats, animals from this strain will be used. If the lymphocytes are from another strain, F1 hybrids from this strain with the LOU/C (see Chapter 5) will be used. If those hybrids are not available, *in vivo* production can be tried in LOU/C rats but success remains hazardous and the rats generally need to be irradiated.

The first passage of *in vitro*-cultured hybridoma cells is always done by the subcutaneous route. This route is more successful and needs a number of cells lower than that required for the intraperitoneal route (Kints, unpublished results).

A. REAGENTS AND MATERIAL

1. 5.10^6 hybridoma cells per rat to be injected (Cells from culture: after centrifugation, the cells are resuspended in 0.5 to 1 ml physiologic water just before injection. Cryopreserved cells: the cryotube is immersed in 37°C water for quick thawing and the whole content of the tube is immediately injected)
2. A 6- to 8-week-old histocompatible rat
3. A 2-ml syringe with a 23-gauge needle
4. Bell jar and ether (see Chapter 2)

For tumor collection:

1. Sterile surgical instruments, two scissors, two toothed forceps
2. Potter homogenizer
3. 70% alcohol
4. 0.15 *M* phosphate buffer saline pH 7.2

B. PROCEDURE

1. Preliminary Remarks

The hybridoma cells are always implanted subcutaneously or intraperitoneally. When using this last route, it sometimes happens that a few cells develop a subcutaneous solid tumor. To avoid any confusion, subcutaneous injections are always made on the right flank and intraperitoneal injections on the left flank. Therefore, if a solid tumor begins to develop on the left flank, it is known that the rat will also produce an ascitic tumor.

2. Procedure

The rat is anesthetized and the hybridoma cells are always injected subcutaneously in the right flank for convenience. The rat is observed three times a week for tumor growth. It generally appears after 2 or 3 weeks, but up to 3 months can elapse before the hybridoma cells begin to proliferate in the recipient animal.

After adequate tumor growth, the rat is anesthetized and bled to check the MAb secretion. Thereafter it is killed in the bell jar. The fur is swabbed with 70% ethanol; a piece of tumor excised and disrupted in PBS. A Potter homogenizer is used to that end.

Part of the cells is taken for the ascites production which is discussed in Chapter 14, and part is used for liquid nitrogen storage of *in vivo* adapted hybridoma cells.

III. WHAT TO DO IF THE FIRST ATTEMPT OF *IN VIVO* CULTURE FAILS

In histocompatible rats, 5 to 10% of the hybridomas never grow by simple subcutaneous injection (unpublished results). Others begin to grow, but are later rejected by the host and the solid tumor degenerates into an abscess. We used several techniques to adapt these clones to *in vivo* culture. They will be described in this section in the order we tested them.

A. USE OF IMMUNE-DEFICIENT RATS

When the routine procedure (see Section II) is followed, a solid tumor appears in the 5 weeks following the subcutaneous injection of hybridoma cells.

If no tumor grows after that time, a new attempt is made with immunosuppressed syngeneic rats. The rats are always used in the 24 h following a 5 Gy (500 rad) irradiation used as immunosuppression technique. X and gamma rays have been shown equally valuable in this respect (unpublished results). If the irradiation is lower, the animals are not entirely immunosuppressed and if the irradiation is higher, some rats die from the intestinal syndrome. In each case, precious hybridomas could be lost.

When the hybridoma does not grow on irradiated rats, *in vivo* transfer is tried subcutaneously in athymic rnu/rnu rats. These rats lack mature T cells and are thus unable to reject the injected cells.

The monitoring of tumor growth and the hybridoma cell collection are performed as described above for the routine procedure. After this first passage into irradiated rats or nude rats, the *in vivo*-adapted cells generally grow easily as a solid tumor or ascites in normal histocompatible rats.

B. THE INTRAVENOUS AND THE INTRASPLENIC INOCULATION OF HYBRIDOMA CELLS

Witte and Ber[3] and Ber and Lanir[4] showed that intrasplenic inoculation of hybridoma or plasmacytoma cells is an excellent method to promote *in vivo* tumor growth in mice. They needed less than 10,000 hybridoma cells and thus collected ascites containing *in vivo*-adapted hybridoma cells and high titers of MAb, though the mice used for inoculation were not primed with pristane or Freund adjuvant.[5,6] The authors of the intrasplenic rescue of hybridomas promote this method for the recovery of contamined hybridomas or of frozen stocks that could not be re-established in culture by conventional techniques.

Hirsh et al.[2] used the intravenous route for the *in vivo* adaptation of rat hybridomas. In this way, they rescued eight out of the nine hybridomas which had failed to grow in syngeneic or nude rats. Clones grew in liver, bone at the level of the lumbar vertebrae, and in femoral bone marrow. We now use this technique for valuable clones which fails to grow in immunodeficient rats, as described above in Section III.A.

1. Preparation of Injectable Cell Suspensions

The hybridoma cells in saline are set in a volume of 0.05 to 0.1 ml for the intrasplenic route and a volume of 1 ml for the intravenous route.

a. Materials

1. Hybridoma cells from one well or a tube of cryo-preserved cells
2. Culture medium: Dubbelcco's modified Eagle's medium with 10% fetal calf serum or horse serum, 1 mmol sodium pyruvate
3. Sterile 0.9% NaCl
4. Sterile tubes
5 Sterile pipettes
6. Centrifuge

b. Procedure

Cells coming from culture are centrifuged and resuspended in the required volume of 0.9% NaCl before injection.

Cells preserved in liquid nitrogen are quickly thawed in water at 37°C. The thawed-cell suspension is immediately centrifuged for 3 min., the supernatant is discarded, and the cell pellet is washed with 10 ml culture medium. After centrifugation the cells are resuspended in the required volume of physiologic water.

2. Intrasplenic Injection.

Modified from Ber and Witte.[7]

a. Materials

1. 6- to 8-week-old syngeneic rat
2. Hybridoma cells in 0.05 to0.1 ml physiologic water or 0.15 *M* PBS pH 7.2
3. Dissecting board
4. Sterile surgical instruments: toothed forceps, two pairs of scissors, and blunted curved forceps
5. 1 ml syringe with 30 1/2 gauge tuberculin needle
6. 70% ethanol
7. Surgical clips
8. A weight/weight mixture of penicilline G and streptomycine
9. Bell jar and ether, or 36 mg/ml chloral hydrate, or 10 mg/ml ketamine hydrochloride, or 6 mg/ml nembutal

b. Procedure

The rat is thoroughly anesthetized in the bell jar or by intraperitoneal injection of 1 ml/100g body weight of any one of the other anesthetics.

The rat is put on the right flank or on its back, the left foreleg over the right foreleg, and immobilized on the dissecting board. The left side is swabbed with 70% alcohol and a button hole of 1 to 1.5 cm is made with the points of the scissors. The lower pole of the spleen is brought outside the peritoneal cavity. The cell suspension is gently injected (a blister appears) and the needle is carefully withdrawn to avoid any bleeding or leakage. The spleen is pushed back in the peritoneal cavity. Some penicilline-streptomycine is applied on the cut and one or two clips are applied to close it, care being taken that peritoneum, muscles, and skin are correctly positioned. When awake, the inoculated rats are kept in ordinary conditions. They are palpated three times a week to follow tumor or ascites growth.

3. Intravenous Injection

a. Materials

1. 6- to 8-week-old syngeneic rats

2. Bell jar and ether (see Chapter 2)
3. 2-ml syringes with a 23 G1 1/4 needle
4. 38 to 40°C hot water
5. 1 ml hybridoma cell suspension in physiologic water
6. 0.15 *M* pH 7.2 PBS
7. Potter homogenizer.
8. Mechanical constrain device

b. Procedure

The rat is carefully anesthetized in the bell jar. Special care is to be taken if laboratory temperature is higher than 30°C. At this temperature volatility of ether is very high and some rats could die if they are left for too a long period in the bell jar (see Chapter 2). The rat is eventually immobilized in a mechanical constrain device. The tail of the rat is immersed in 38 to 40°C hot water to enhance the tail vein vasodilatation.

The needle bevel upside is inserted in the vein. When the needle is in the vein, it slides easily and injection is done without resistance. The needle is carefully removed to avoid bleeding. After awakening, the rats are maintained in conventional animal house facilities with water and food *ad libidum.*

The animals are observed and palpated three times a week for tumor growth. Tumors appear in liver, bones, and also at the level of the lumbar vertebrae which can result in hind legs paralysis.

The rats are sacrificed when the liver is found to be enlarged under clinical investigations. Single cell suspensions in PBS are obtained from liver with a Potter homogenizer and from femoral bone marrow by successive suctions through a fine needle.

A blood sample is taken to assess the MAb production. Part of the collected cells is injected intraperitoneally in syngeneic rats for ascites production as it has been described in Chapter 14, and part is frozen as it is described below in the following subheading.

IV. NITROGEN PRESERVATION OF *IN VIVO* CULTURE-ADAPTED HYBRIDOMAS

Some hybridomas are not always easily adapted to *in vivo* growth. The injection of hybridomas is sometimes used to rescue a contaminated clone. The first grown solid tumor is therefore very precious and it is very important to preserve the cells in liquid nitrogen. After thawing, the cells can be cultured *in vitro* and *in vivo.*

A. BIOLOGICAL MATERIALS AND REAGENTS

1. Ascites fluid or single cell suspension of solid tumor
2. Ringer: NaCl 9 g, KCl 0.42 g, $CaCl_2$ 0.24 g, $NaHCO_3$ 0.2 g, and glucose 5 g are dissolved in distilled water and volume is adjusted to 1 l
3. 40% dimethylsulfoxide (DMSO) in 0.9% NaCl refrigerated in melting ice before use
4. 2-ml cryotubes
5. Programmable freezing device which will permit a continuous cooling from 0 to –70°C (Air Liquide, Belgium)
6. Liquid nitrogen tanks (Air Liquide, Belgium)
7. Melting ice

B. PROCEDURE

Three milliliters of solid tumor cell suspension (at 10^7 cell/ml) or 3 ml ascites are taken and mixed with 1 ml 40% DMSO in 0.9% NaCl. This cell suspension is aliquoted in 2-ml cryotubes and kept in melting ice until freezing. This time must be as short as possible because DMSO is harmful for the cells. In any case it must be shorter than 20 min.

The cells are frozen at –70°C with a cooling gradient of 2°C/min from 0 to –40°C, then with an increasing speed from –40 to –50°C, and 4°C/min from –50 to –70°C, and then a security plateau at this temperature. The vials at –70°C are immediately stored in liquid nitrogen. If a programmable freezing device is not available, cryotubes can be put in a –70°C freezer in a polystyrene box for 6 h. The cells are immediately thereafter put in liquid nitrogen.

REFERENCES

1. **Noeman, S. A., Misra, D. N., Yankes, R. J., Kunz, H. W., and Gill, T. J., III,** Growth of rat-mouse hybridomas in Nude mice and Nude rats, *J. Immunol. Methods,* 55, 319, 1982.
2. **Hirsch, F., Vendeville, B., De Clercq, L., Bazin, H., and Druet, P.,** Rat monoclonal antibodies. III. A simple method for facilitation of hybridoma cell growth *in vivo, J. Immunol. Methods,* 78, 103, 1985.
3. **Witte, P. L. and Ber, R.,** Improved efficiency of hybridoma ascites production by intrasplenic inoculation in mice, *J. Natl. Cancer Inst.,* 70, 575, 1983.
4. **Ber, R. and Lanir, M.,** Fusion of plasmacytoma and host cells *in vivo:* selection of proliferating and nonproliferating cultures, *J. Natl. Cancer Inst.,* 72, 403, 1984.
5. **Cancro, M. and Potter, M.,** The requirement of an adherent cell substratum for the growth of developing plasmacytoma cells *in vivo, J. Exp. Med.,* 144, 1554, 1976.
6. **Mueller, V. W., Hawes, C. S., and Jones, W. R.,** Monoclonal antibody production by hybridoma growth in Freund's adjuvant primed mice, *J. Immunol. Methods,* 87, 193, 1986.
7. **Ber, R. and Witte, P. L.,** Intrasplenic inoculation. A method for rescuing hybridoma clones, in *Methods of Hybridoma Formation,* Bartal, A. H. and Hirshaut, Y., Eds., Humana Press, Clifton, N.J., 1987, 413.

Chapter 12

FREEZING AND THAWING

A. M. Ravoet, D. Wauters, T. Delaunay, and H. Bazin

TABLE OF CONTENTS

I. INTRODUCTION

Cryopreservation of the hybridomas allows MAbs to be produced for decades without alteration in their specificity. Rat hybridomas kept for over 6 years in liquid nitrogen have been thawed and grown in culture without loss of activity. Some of our *in vivo* grown rat myelomas and hybridomas have been kept frozen for 17 and 8 years respectively, without loss of tumorigenic potential or secretion of MAbs. To date no hybridoma cell line has been lost owing to freezing or thawing.

II. FREEZING OF HYBRIDOMA CELLS

A. RATIONALE

Freezing down *in vitro* developed hybrids requires the same precautions as freezing continuous growing cell lines or lymphocytes: sterile handling, addition of DMSO at 0 to 4°C and progressive lowering of the temperature. Usually, 1×10^5 to 1×10^6 cells are frozen. Alternatively, whole culture plates containing newly produced hybrids are frozen.

Hybridomas, produced as solid tumors or ascites in animals, are available at very high cellular concentrations (10^8 to 10^9/ml). Even the loss of 50 to 90% of the cellular viability does not jeopardize the tumorigenic potential of the frozen cells. No sterile handling is necessary, since the rat immune system readily eliminates all contaminants. However, it is advisable to add DMSO at a low temperature (0 to 4°C).

B. TECHNICAL ASPECTS

1. Freezing of *In Vitro* Cultured Hybrids

a. Material

Sterile ampoules with screw cap (Sterilin, GB) and an automatic biological freezing chamber (Air Liquide, France), and a tank for liquid nitrogen (Air Liquide). Two programs have been used successfully in our laboratories. Program 1 includes the following seven steps:

1. 10 min at 0°C
2. –1°C/min down to –6°C
3. Maximum admission of liquid nitrogen down to –40°C
4 2 min at –40°C
5. –1°C/min down to –40°C
6. –5°C/min down to –60°C
7. –10°C/min down to –100°C

Program 2 includes the following four steps:

1. 5 min on melting ice
2. –1°C/min from 0°C down to –45°C
3. –2.5°C/min from –45°C down to –70°C
4. Transfer to liquid nitrogen

The lack of an automatic freezing chamber can be compensated by insulating the vials with styrofoam and placing them in a –70°C freezer, and 24 h later transferring them to liquid nitrogen.

b. Reagents

1. Ice
2. DMEM-HS: DMEM containing 15% horse serum (or 10 to 15% fetal calf serum)

3. DMEM-HS-DMSO: DMEM containing 15% horse serum (or 10 to 15% fetal calf serum) and 15-20% DMSO.

c. Procedure

Hybridoma cells from 1- or 2-ml culture wells or from 4 to 10 0.2-ml culture wells are collected and centrifuged. The supernatant is discarded. Cells are resuspended into 1 ml of cold DMEM-HS and icecold DMEM-HS-DMSO is added drop by drop while shaking. The suspension is poured into a sterile ampoule and progressive freezing is started at once in the freezing chamber. Frozen cells are kept in liquid nitrogen.

2. Freezing of *In Vivo* Grown Hybridomas

a. Material

1. 2-ml ampoules with screw caps
2. A programmed freezing chamber (see above)
3. Tank for liquid nitrogen

Two procedures have been used successfully in our laboratory.

b. Procedure 1

i. Reagents

Reagents include ice PBS-FCS: PBS containing 10% fetal calf serum, and PBS-FCS-DMSO: PBS containing 10% FCS and 15% DMSO.

ii. Procedure

Only the tumor outer part containing no necrosis is collected. The tumor is minced with scissors in PBS-FCS, and passed through gauze. The cell concentration is adjusted at 5×10^6 to 5×10^7/ml. An equal volume of PBS-FCS-DMSO is added progressively and the suspension is distributed into freezing vials. Progressive freezing is started immediately. Frozen hybridomas are kept in liquid nitrogen.

c. Procedure 2

i. Reagents

Reagents include melting ice, Ringer solution: containing in 1 l: 9 g NaCl, 0.42 g KCl, 0.24 g $CaCl_2$, 0.2 g $NaHCO_3$ and 5 g glucose,.and DMSO: 40% in Ringer solution.

ii. Procedure

The tumor outer parts which contain no necrosis are cut out and ground in NaCl 0.9% (6 ml per tumor) by means of a Potter homogenizer. To 6-ml homogenized cell suspension, 2 ml of DMSO-40% in Ringer is added. The suspension is distributed into freezing vials, left on ice for 5 min and progressive freezing (–1°C/min) is started. Tumor homogenates are kept frozen in liquid nitrogen.

III. THAWING OF FROZEN HYBRIDOMA CELLS

A. RATIONALE

The overall rule is to make frozen *in vitro* produced hybrids grow in culture before transferring them to animals and to use frozen samples of *in vivo*-produced tumor in order to resume *in vivo* production. For the *in vitro* growth of thawed hybrids, a culture medium containing no HEPES should be used (HEPES is toxic to cells if it enters the cells), and addition of growth factors (for example, feeder-layers or/and filtered supernatant of continuous growing rat myeloma cells) is useful but not essential. Since DMSO is not toxic to rats at the concentration used to freeze cells, the samples can be injected s.c. or i.p. into the rats immediately after thawing and without removal of DMSO.

B. TECHNICAL ASPECTS

1. Thawing of Hybrids for in Culture Growth

a. Material

Culture plates, sterile pipettes, etc.

b. Reagents

1. PBS-HS: PBS containing 15% horse serum or 15% fetal calf serum
2. Culture medium: DMEM without Hepes, supplemented with 15% horse serum or 15% fetal calf serum, 20% conditioned medium (i.e., filtered supernatant of exponentially growing 983 cells), 2 m*M* glutamine, 100 U penicillin/ml, 100 μg streptomycin/ml and nonessential amino acids
3. Culture plate containing 20,000 feeder-layer cells per well (see Chapter 6)

c. Procedure

The frozen hybrids are quickly thawed by placing the freezing vial into warm water (37°C). The content of the vial is poured into a 50-ml tube and 10 volumes of PBS-HS are added drop by drop while shaking. After centrifugation for 5 min at 250 *g*, another wash with 10 volumes of PBS-HS is performed. Eventually the cells are resuspended in culture medium and seeded into 96-well culture plates containing feeder-layers.

2. Thawing of Frozen Tumors/Ascites

a. Material

LOU rats and a syringe are used.

b. Reagents

There are no reagents.

c. Procedure

The frozen tumor cells are quickly thawed by placing the freezing vial into warm water (37°C). The content of the tube is immediately injected s.c. or i.p. into a LOU rat.

IV. CRYOPRESERVATION OF NEWLY PRODUCED HYBRIDS IN THE TISSUE CULTURE PLATE

A. RATIONALE

This technique, advisable when many positive clones per tissue plate are obtained after fusion, allows storage of numerous non cloned positive hybridoma cells. These cells can be reused at least 2 months after freezing down if a problem occurs during the further development and cloning of the first selected clones.

B. TECHNICAL ASPECTS

1. Freezing

a. Material

No material is used.

b. Reagents

The reagent used is DMEM-HS-DMSO: DMEM containing 15% horse serum (or 10 to 15% fetal calf serum) and 7.5% DMSO and kept at 4°C.

c. Procedure

Culture supernatants from each well of the whole tissue plate are discarded and 100 μl of DMEM-HS-DMSO is added to each well. Then, the tissue plates are set at –20°C. When the medium begins to freeze 30 min later, the plates are stored at –70°C.

2. Thawing

a. Material

No material is used.

b. Reagents

The reagents used are culture medium (DMEM containing serum) and freshly prepared rat peritoneal feeder cells.

c. Procedure

The frozen hybrids are quickly thawed by filling each well with 100 μl of culture medium and by placing the tissue culture plate at 37°C. As soon as the DMEM-HS-DMSO becomes liquid, the wells are emptied and washed twice with 200 μl of culture medium. Finally, each well is filled with 200 μl of freshly prepared rat peritoneal feeder cells and the hybridoma growth is managed like any culture after primary fusion. The first results show that approximately 90% of the thawed hybridoma cells kept their growth ability, but only 10% remain positive presumably because of overgrowth by nonproducing cells.

Part III
Production, Purificaton, Fragmentation, and Labeling of Rat Monoclonal Antibodies

Chapter 13

IN VITRO PRODUCTION OF RAT MONOCLONAL ANTIBODIES

M. Bodeus, J. Ch. Fang, G. Burtonboy, and H. Bazin

TABLE OF CONTENTS

I. INTRODUCTION

Initially mouse monoclonal antibodies (MAbs) have been described by Köhler and Milstein[1] and a few years later Galfré et al.[2] extended this technology to the rat system. The field of applications of MAbs spreads out from day to day and it would be illusive to draw up an inventory of all the services they have rendered in fundamental research as in the fields of diagnosis and therapy.[3,4]

The principal advantages of MAbs on conventional antisera are their high specificity and their capability to be produced indefinitely with invariable characteristics. Therefore, the production of large amounts of a MAb is an important step in the technology of hybridoma either in the mouse or in the rat system. When a fusion experiment has brought into existence hybrid cells producing MAb directed against a given antigen, besides the fact that these hybrids have to be selected and cloned, it is essential to consider the different possibilities to produce the immunoglobulin (Ig) of interest. Two main approaches are indeed available: *in vivo* production, which will be discussed more extensively in Chapters 11 and 14, and *in vitro* production. Each presents advantages and disadvantages.

II. *IN VIVO* PRODUCTION

LOU rat represents a very practical model for *in vivo* production.[5] The use of histocompatible rats allows to grow hybridomas, either intraperitoneally or subcutaneously, without preliminary irradiation of the animal. By intraperitoneal inoculation it is possible to obtain per rat, up to 50 ml ascitic fluid containing a very high concentration of MAb, from 1 to 15 mg/ml. On the other side, after subcutaneous inoculation and growth of a solid tumor, the animal serum also contains large amounts of MAb but the volume never reachs more than 10 ml. Ascitic fluids and sera can be easily purified by affinity chromatography using mouse MAbs anti-rat Igs[6] (Chapter 15), and especially, the very artful technique based on allotypic differences. But in each case, it requires extensive animal facilities which are not always available. Moreover, although techniques have been described to improve *in vivo* growth,[7,8] it is not always possible to adapt hybridoma to *in vivo* production.

III. *IN VITRO* PRODUCTION

In vitro production may be an interesting, and sometimes the only possibility to obtain the needful quantity of MAb. In the research laboratory, hybridomas are generally cultured in 25 cm^2 flask (Falcon) in which the cells are kept in exponential growth by adjusting every other day at a density of $2. \times 10^5$ cells/ml in 10 ml fresh medium. Dulbecco's modified Eagle's medium with 4.5 g/l glucose, supplemented with 1 m*M* sodium pyruvate, 4 m*M* glutamine, 1 % of a mixture of non-essential amino acids and 10% fetal calf serum, is generally used and gives reproducible results. The flasks can be kept with closed caps in a 37°C incubator without CO^2. But, it is a time consuming procedure, about 1 h weekly by flask if one takes into account preparation of the medium, centrifugation of the culture, counting and seeding the cells (Table 1). Besides, in such *re-fed bottles* the concentration of monoclonal Ig in the supernatant is low, ranging from 1 to 50 μg/ml, with an average of 15 μg/ml. The production is thus limited to a few hundred micrograms for a week.

Various techniques have already been proposed in order to increase the yield of the monoclonal Ig; they are mostly based on a large scale production. Stirring devices, fermenters, dialysis culture, and the recent Opticell system[9] give interesting results, but most of them necessitate trained staff and sophisticated equipment[10,11] and do not give the practical possibility of growing a large number of different antibodies at the same time. Satisfactory at the industrial level, most of the times they are not workable in every laboratory. An easy method for MAbs

TABLE 1
Production of RH16 in a 2-Week Experiment Comparison of Re-fed and Roller Bottle Systems

	Production system	
	Re-fed bottle	**Roller bottle**
Volume per flask (ml)	10	300
Equivalent number	5 × 6	1
Manipulation time (h/2 weeks)	10	0.5
Volume collected (ml)	300	300
MAb mean concentration (μg/ml)	9.3	37.0
Final yield (mg/ml)	2.8	11.1

in vitro production which did not ask for investment neither in work nor in material has been devised in our laboratory.[12]

Preliminary observations have been made with rat hybridoma cells secreting MAbs directed against the canine parvovirus (CPV)[13] (Chapter 22). Some of these cells were accidentally forgotten in the incubator for 2 weeks without renewing the culture medium and it was found that although almost no living cells were left in the flask, the concentration of the monoclonal Ig as well as the antibody activity of the medium were very high.

This observation has been confirmed by a series of experiments. Several rat hybridomas referred to here as "RH" followed by a serial number have been cultured during weeks and kinetic studies have been performed. For this purpose, 25-cm^2 flasks were seeded as described above with 2×10^5 cells/ml in 10 ml fresh medium and incubated without any replacement of the medium; every 2 or 3 d a sample was taken and analyzed. The number of cells and the percentage of dead cells were counted in a hemocytometer using a trypan blue dye exclusion method.[14] The Ig concentration was measured in a radial immunodiffusion assay according to Mancini et al.[15] Since the MAbs were directed against the CPV, the antibody activity was titrated in a hemagglutination inhibition test as described previously;[13] the titer being the reverse of the highest dilution which produced inhibition.

Figure 1 shows the evolution of such a hybridoma culture during a 1 month period. The number of RH 24 cells increases up to 13 millions, but after a maximum at day 5, if the culture medium is not replaced the cell growth stops and the number of living cells decreases quite rapidly. However, at the same time, the concentration of the monoclonal Ig keeps increasing to reach a plateau at day 15.

Another example is presented in Figure 2, where the evolution of RH 35 culture shows that the increase of the Ig concentration corresponds to a rise in the antibody activity which does not seem to be affected by proteases possibly present in the culture supernatant.

In further experiments hybridoma cells taken during exponential growth were resuspended at a density of 2×10^5 cells/ml in larger flasks (Falcon 75 cm^2) in 100 ml fresh medium; these cultures referred to as stationary bottle, were incubated for 2 weeks without changing the medium. The results were quite similar; the amount of rat Ig was found to increase to a level higher than what was obtained in the re-fed bottle. However, the handling of a large number of such units is rather cumbersome and it would be very difficult to obtain enough MAb by this procedure.

A simple technique to overcome this difficulty was to use the so-called *roller bottle* method. The apparatus consists of parallel rollers driven by a chain and on top of which large glass bottles are placed. The Bellco cell production roller apparatus used in this study is shown in Figure 5. The Bellco bottles designated for this purpose are 30 cm high with a diameter of around 10 cm; they are feeded with 300 ml of fresh medium containing 2×10^5 cells/ml and then rotated at 0.5 rpm for 2 weeks.

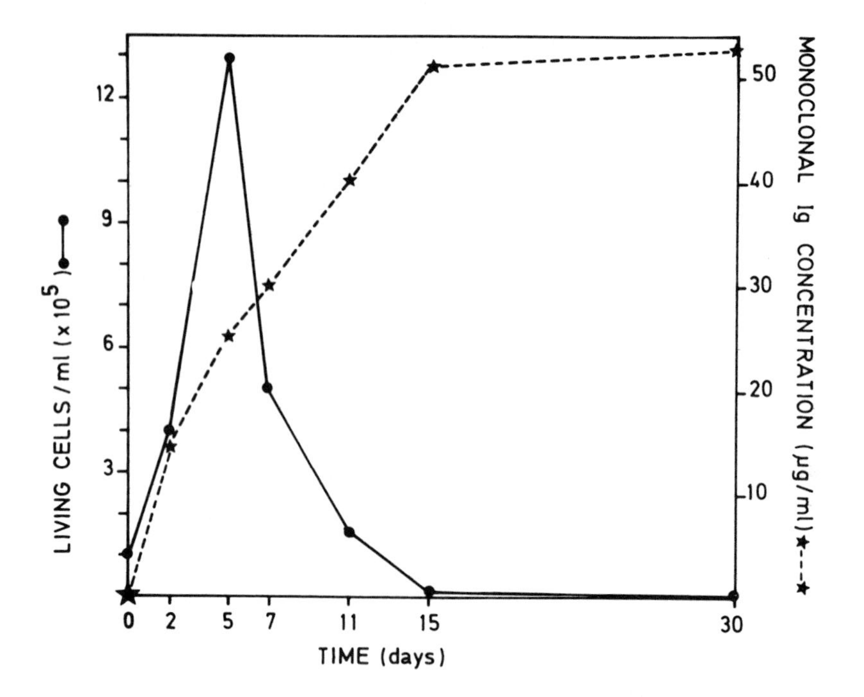

FIGURE 1. When hybridoma cells (RH24) are grown without renewing the culture medium, the number of living cells after a maximum at day 5 diminishes progressively; monoclonal Ig concentration continues to increase and reaches a plateau at day 15.

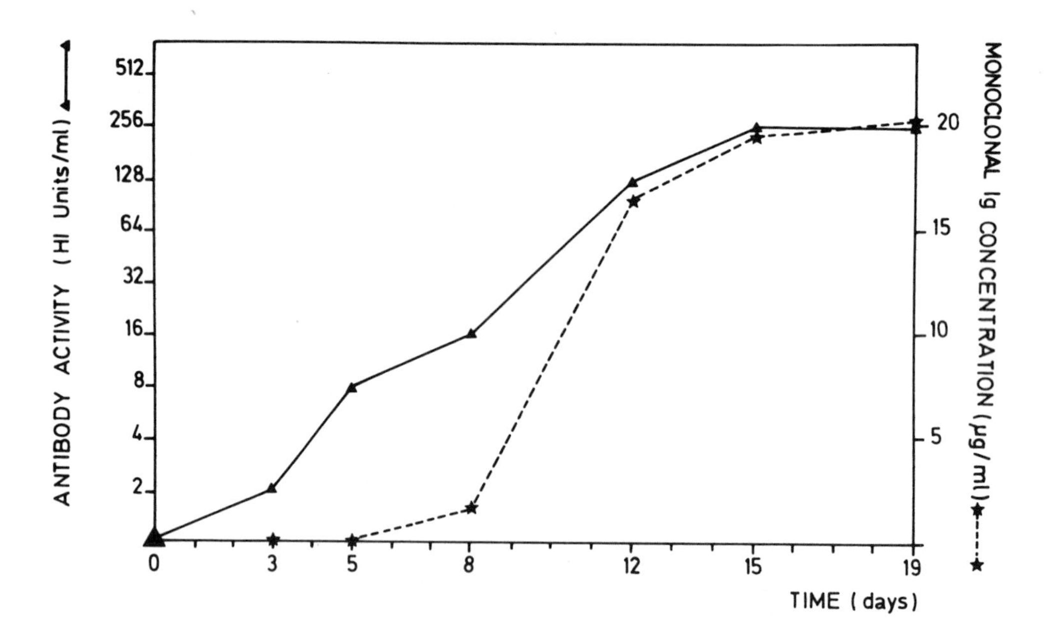

FIGURE 2. When hybridoma cells (RH35) are grown without renewing the culture medium, the Ig concentration and the MAb activity increases in parallel to reach a plateau at day 15.

Comparison of hybridoma cultures in roller bottle vs. stationary bottle system, revealed that the yield was always better in the rotating one; the concentration in monoclonal Ig was up to three times higher (Figure 3).

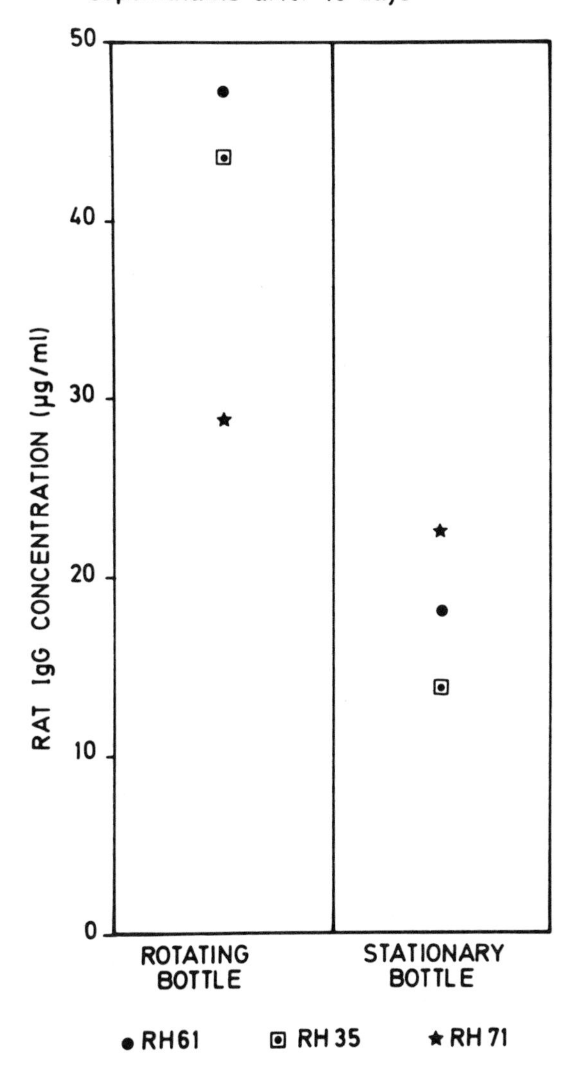

FIGURE 3. Three different hybridoma cells (RH) have been cultured either in roller bottle or in stationary bottle. In the first system the yield is up to three times higher.

Therefore, this type of production in roller bottle has been used on a larger scale for various hybridomas directed against antigens of CPV (Chapter 22), human immunodeficiency virus (Chapter 22), and CEM cells; the concentration of the Ig was in each case higher than that value found in the usual re-fed cultures when new medium was added every 2 or 3 d. Figure 4 summarizes the comparison for eleven anti-CPV hybridomas. In the re-fed bottles the Ig concentration in the supernatant reaches a mean value of 9.3 µg/ml (ranging from 3.7 to 20 µg/ml); in the roller bottles with a mean value of 39.0 µg/ml (ranging from 7.1 to 61.2 µg/ml), the yield is two to eight times higher and in each case the antibodiy activity follows the same increase. Beside the fact that the amount of MAb obtained is more important, it is interesting to dwell on the advantages of the procedure which are obvious if one considers the time and work consuming aspect. Table 1 is a comparison of the conventional and the roller method, stressing the point that for a better yield the rotating bottle was only feeded to start with and does not need

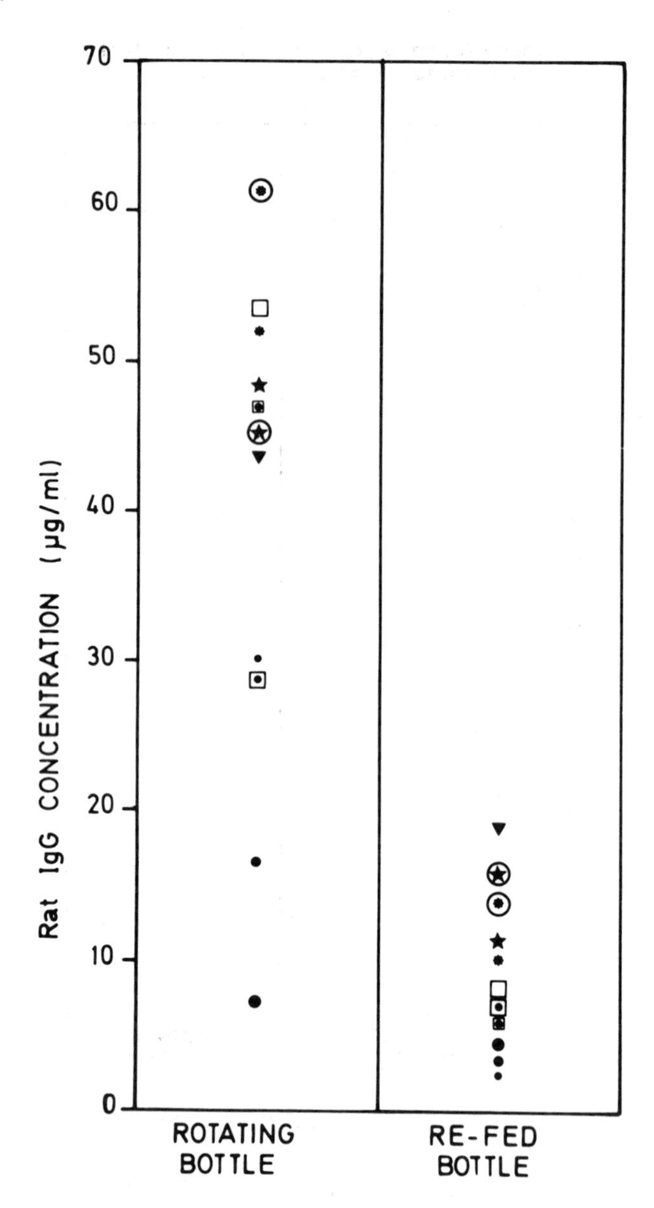

FIGURE 4. Rotating bottle system has been compared with the usual re-fed bottle technique. Eleven different hybridomas have been studied by these two systems; in the rotating system, the yield is two to eight times better.

any further manipulation. Moreover, recovery of the roller bottle supernatant can be made at any time after the 2 weeks since the Ig and antibody concentrations remain at the maximum level.

Finally it must be emphasized that the *in vitro* production of a MAb supplies the investigator with a material which is highly suitable for purification by immunoaffinity chromatography,[16] this point is discussed elsewhere in this volume (Chapter 15).

In conclusion, the usual cell culture in re-fed bottle can provide enough MAb for a restricted research program when micrograms are needed. Fermenters and similar industrial devices are available for those who consider industrial application and production of grams or even kilograms of material. In between, the stationary bottle, or better the roller bottle procedure is an easy and inexpensive way to produce intermediate amounts in the range of the milligram.

FIGURE 5. The Bellco cell production roller apparatus.

REFERENCES

1. **Köhler, G. and Milstein, C.,** Continuous cultures of fused cells secreting antibody of predetermined specificity, *Nature (London),* 256, 495, 1975.
2. **Galfré, G., Milstein, C., and Wright, B.,** Rat × rat hybrid myelomas and a monoclonal anti-Fd portion of mouse IgG, *Nature (London),* 277, 131, 1979.
3. **Nowinski, R. C., Tam, M. R., Goldstein, L. C., Stong, L., Kuo, C., Corey, L., Stamm, W. E., Handsfield, H., Knapp, J. S., Holmes, K. K., et al.,** Monoclonal antibodies for diagnosis of infectious diseases in humans, *Science,* 219, 637, 1983.
4. **Fuccillo, D. A. and Sever, J. L.,** Monoclonal antibodies as reagents, in *Concepts in Viral Pathogenesis II,* Notkins, A. L. and Oldstone, M. B. A., Eds., Springer Verlag, New York, 1986, 324.
5. **Bazin, H.,** Louvain rat immunocytomas, *Adv. Cancer Res.,* 50, 279, 1987.
6. **Bazin, H., Cormont, F., and De Clercq, L.,** Rat monoclonal antibodies. II. A rapid and efficient method of purification from ascitic fluid or serum, *J. Immunol. Methods,* 71, 9, 1984a.
7. **Ber, R. and Lanir, M.,** Fusion of plasmacytoma and host cells *in vivo:* selection of proliferating and nonproliferating cultures, *J. Natl. Cancer Inst.,* 72, 403, 1984.
8. **Hirsh, F., Vendeville, B., De Clercq, L., Bazin, H., and Druet, P.,** Rat monoclonal antibodies. III. A simple method for facilitation of hybridoma cell growth *in vivo, J. Immunol. Methods,* 78, 103, 1985.
9. **Lydersen, B. K., Pugh, G. G., Paris, M. S., Sharma, B. P., and Noll, L. A.,** Ceramic matrix for large scale animal cell culture, *Biotechnology,* 63, 1985.
10. **Freshney, R.I.,** in *Culture of Animal Cells. A Manual of Basic Technique,* 2nd ed., Allan R. Liss, New York, 1987.

11. **Adamson, S. R., Fitzpatrick, S. L., Behie, L. A., Gaucher, G. M., and Lesser, B. H.,** *In vitro* production of high titre monoclonal antibody by hybridoma cells in dialysis culture, *Biotechnol. Lett.*, 9, 573, 1983.
12. **Bodeus, M., Burtonboy, G., and Bazin, H.,** Rat monoclonal antibodies. IV. Easy method for *in vitro* production, *J. Immunol. Methods,* 79, 1, 1985.
13. **Burtonboy, G., Bazin, H., and Delferrière, N.,** Rat hybridoma antibodies against canine Parvovirus, *Arch. Virol.,* 71, 291, 1982.
14. **Paul, John,** in *Cell and Tissue Culture,* Churchill Livingstone, Edinburgh.
15. **Mancini, G., Carbonara, O. H., and Heremans, J. F.,** Immunochemical quantitation of antigens by single radial immunodiffusion, *Immunochemistry,* 2, 235, 1965.
16. **Bazin, H., Xhurdebise, L. M., Burtonboy, G., Lebacq, A. M., De Clercq, L., and Cormont, F.,** Rat monoclonal antibodies. I. Rapid purification from *in vitro* culture supernatants, *J. Immunol. Methods,* 66, 261, 1984b.

Chapter 14

IN VIVO PRODUCTION OF RAT MONOCLONAL ANTIBODIES

P. Manouvriez, J. P. Kints, A. De Cremer, and H. Bazin

TABLE OF CONTENTS

I. OPTIMALIZATION OF ASCITES PRODUCTION

The LOU/C rat easily develops ascites after intraperitoneal injection of syngeneic immunocytoma or hybridoma cells without any priming.[1,2] A rat is ten times as large as a mouse. A greater production per animal is therefore obtained from the former species. But in unprimed rats, the ascites production is proportionally lower than in oil-primed mice. The antibody production of hybridomas is also generally much lower than that of immunocytomas. For large scale MAb production the obtainment of ascites has thus to be optimalized in rats.

Plasmacytoma and hybridoma cells develop generally as a solid tumor and produce very little ascites after transplantation by the intraperitoneal route in unprimed mice. The yield of ascitic fluid is greatly enhanced by a pretreatment with 2, 6, 10, 14-tetramethylpentadecane also called pristane.[3,4] Incomplete Freund's adjuvant (IFA) is also successfully used.[5,6] These two substances and their v/v mixture, called PIFA, were compared in rats for their enhancing effect after intraperitoneal injection at the moment of cell transfer. As shown in Figure 1, PIFA gave the best results in terms of MAb production per rat. In some cases (not shown here) incomplete Freund adjuvant has shown itself to be as effective. Thus, the two substances, IFA and PIFA, should be tried before mass production. In case of equal efficacy, the former one will be chosen for its cheapness.

To optimalize the ascites production which is enhanced by incomplete Freund ajuvant or PIFA, different volumes of oil to be injected were tried, and so were different waiting periods between priming and hybridoma transfer. The most effective but also the most time-saving protocol was found to be the intraperitoneal injection of 2 ml PIFA or IFA at the moment of the tumor passage.[7]

A. CHOICE OF RATS

Hybridomas grow more easily in young rats, and particularly in 6- and 10-week-old rats. In older rats the proportion of unsuccessful tumor grafts increases and the volumes of ascites which are collected are not greater than in 8-week-old animals.

Perfectly histocompatible animals have to be chosen; if not tumor growth and MAb secretion are seriously hampered. If the fusion partner lymphocytes are from LOU rats, ascites will preferentially be produced in LOU/C rats or in the congenic LOU/C.IgK-1b(OKA) rats (see Chapter 2). The latter strain produces Kappa light chain immunoglobulins of the Kappa-1b allotype.[8] IgK-1a MAb produced on these animals can be purified specifically by a one-step immuno-purification with MARK-3, a mouse anti-rat IgK-1a allotype monoclonal antibody.[9] Conversely, IgK-1b MAb produced in rats of the IgK-1a allotype can be purified in one step by immunoaffinity chromatography on LO-RK1b-2, a rat MAb directed against rat IgK-1b allotype light chains (see Chapter 15). If the lymphocytes used as fusion partners are from strains other than LOU/C or LOU/C.IgK-1b(OKA), F1 hybrids have to be used.

The adaptation of hybridoma cells to *in vivo* culture has been discussed in Chapter 11.

B. REAGENTS AND MATERIAL

1. 6- to 10-week-old rats which are histocompatible with the hybridoma cells
2. Bell jar and ether
3. 5-ml syringes with a 18 G1 1/2 needle for oil injection and with a 22- G1 1/4 needle for cell injection
4. 10^7 to 5×10^7 hybridoma cells or 2 ml ascites diluted five times in PBS or physiologic water per rat to be injected
5. Incomplete Freund adjuvant (IFA) made by mixing one volume of m[illegible]ide monooleate (Arlacel No. A8009, Sigma) with nine volumes of Marcol (Exxon).
6. PIFA made with one volume IFA and 1 volume Pristane (2, 6, 10, 14-tetramethylpentadecane, Janssen fine chemicals, Belgium)

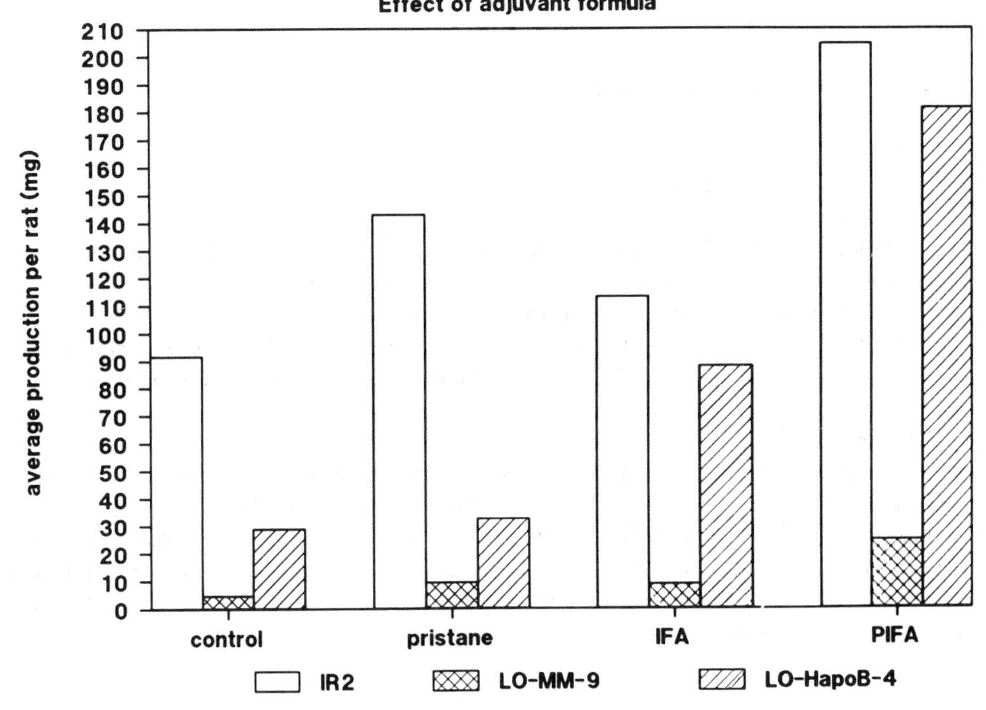

FIGURE 1. Average MAb production obtained in five rats receiving 2 ml pristane, or 2 ml incomplete Freund's adjuvant (IFA), or 2 ml v/v mixture of pristane and IFA (PIFA) at the moment of cell transfer. Control rats received cells alone.

7. Sterile dissection instruments (scissors and toothed forceps), a 20-ml syringe, a 18G1 1/2 needle, 100- or 250-ml flasks for ascites collection
8. 70% ethanol

C. PROCEDURE

Oil and cells are prepared in separate syringes. Never mix them because cells are killed in the emulsion. After being anesthetized, 2 ml adjuvant is injected intraperitoneally through the right flank and cells are also injected intraperitoneally, but through the left flank to avoid accidental intrasplenic inoculation. The abdomen is gently massaged to spread the cells and oil through the whole peritoneal cavity.

The rats are kept in standard conditions with food and water *ad libidum*. They are checked three times a week for ascitic tumor growth. The antibody concentration of the ascites is at its highest when tumor growth reaches its maximum development. We usually take ascites once at the moment of maximum tumor development. The rat is killed in an ether bell jar and first punctured with a 18G-needle, and ascites are collected in a clean flask. After swabbing with alcohol, the peritoneal cavity is opened on the left side of the abdomen. When no more liquid passes through the needle, the remaining ascitic fluid is aspirated with a 20-ml syringe without a needle. The best point to do this is on the side of the liver, where the syringe is not obturated by the intestine or the seminal vesicles.

II. STORAGE OF ASCITES

Ascites generally cannot be used in the state they are collected. Before use or storage they need a minimum treatment to ensure preservation of the antibody activity. The principal problems which occur are clotting after thawing, bacterial proliferation, and antibody inactivation by enzymes. To avoid these problems, the ascites are left to clot, and bactericides and enzyme inhibitors are added before freezing.

A. REAGENTS AND MATERIAL

1. 37°C water bath
2. Centrifuge and centrifuge tubes
3. Micropipettes and tips
4. Thrombin at 100 NIH units/ml (for example, Topostasine, Roche, Belgium)
5. Alpha-toluene sulfonyl fluoride (Janssen Chimica, Belgium No. 21.574.40, PMSF Merck No. 7349) at 40 mg/ml in ethanol stored at –20°C is used as an enzyme inhibitor[10,11]
6. Thimerosal (sodium ethylmercurithiosalicylate, Sigma No. T5125) is used as a bacteriostatic agent which is stored as a 0.1% solution in water at room temperature.

B. PROCEDURE

For clotting, 30 µl thrombin is added per milliliter of ascites or blood which is put in a 37°C water bath. After a few minutes the clot is detached from the walls of the flask and ascites or blood are left for 30 min to retract more. The clot is removed by 10 min centrifugation at 25,000 *g*. The supernatant is collected and 20 µl PMSF and 10 µl Thimerosal are added per milliliter of ascites or serum; 100 µl are kept for quality control and the remainder is eventually aliquoted for –20 or –70°C storage.

REFERENCES

1. **Bazin, H., Deckers, C., Beckers, A., and Heremans, J. F.,** Transplantable immunoglobulin-secreting tumours in rats. I. General features in LOU/Wsl strain rat immunocytomas and their monoclonal proteins, *Int. J. Cancer,* 10, 568, 1972.
2. **Bazin, H.,** Louvain rat immunocytomas, *Adv. Cancer Res.,* 50, 279, 1987.
3. **Potter, M., Pumphrey, J. G., and Walters, J. L.,** Growth of primary plasmacytomas in the mineral oil-conditioned peritoneal environment, *J. Natl. Cancer Inst.,* 49, 305, 1972.
4. **Hoogenraad, N., Helman, T., and Hoogenraad, J.,** The effect of pre-injection of mice with pristane on ascites tumour formation and monoclonal antibody production, *J. Immunol. Methods,* 61, 317, 1983.
5. **Camero, M. and Potter, M.,** The requirement of an adherent cell substratum for the growth of developing plasmacytoma cells *in vivo, J. Exp. Med.,* 144, 1554, 1976.
6. **Mueller, V. W., Hawes, C. S., and Jones, W. R.,** Monoclonal antibody production by hybridoma growth in Freund's adjuvant primed mice, *J. Immunol. Methods,* 87, 193, 1986.
7. **Kints, J. P., Manouvriez, P., and Bazin, H.,** Rat monoclonal antibodies. VII. Enhancement of ascites and monoclonal antibody production in rats by pretreatment with pristane and Freund adjuvant, *J. Immunol. Methods,* 1989, in press.
8. **Beckers, A., Querinjean, P., and Bazin, H.,** Allotypes of rat immunoglobulins. II. Distribution of the allotypes of kappa and alpha chain loci in different inbred strains of rat, *Immunochemistry,* 11, 605, 1974.
9. **Bazin, H., Cormont, F., and De Clercq, L.,** Rat monoclonal antibodies. II. Rapid and efficient method of purification from ascitic fluid or serum, *J. Immunol. Methods,* 71, 9, 1984.
10. **Moss, D. E. and Fahrney, D. E.,** Kinetic analysis of differences in brain acetylcholinesterase from fish or mamalian sources, *Biochem. Pharmacol.,* 27, 2693, 1978.
11. **Turini, P., Kurooka, S., Steer, M., Corbascio, A. N., and Singer, T. P.,** Action of phenylmethylsuflonyl fluoride on human acetylcholinesterase, chymotrypsin and trypsin, *J. Pharmacol. Exp. Ther.,* 167, 98, 1969.

Chapter 15

PURIFICATION OF RAT MONOCLONAL ANTIBODIES FROM ASCITIC FLUID, SERUM OR CULTURE SUPERNATANT

H. Bazin, J. M. Malache, F. Nisol, and T. Delaunay

TABLE OF CONTENTS

I. PURIFICATION OF RAT MONOCLONAL ANTIBODIES FROM ASCITIC FLUID OR SERUM

When considering the problem of purification of a rat monoclonal antibody (MAb), the first element that should be known is its concentration in the ascitic fluid or in the serum, as well as the concentrations of the other polyclonal immunoglobulins (Igs) and serum proteins. The second element which is necessary is a minimum knowledge of its physicochemical nature: isotype of its heavy chain, type of its light chain, isoelectric point, stability, etc.

A. PROBLEMS TO BE CONSIDERED BEFORE ANY PURIFICATION

1. Linked to the Host

The main contaminant of MAbs produced *in vivo* is generally the host polyclonal Ig. Its concentrations are highly dependent on the environment interactions and especially on bacterial, viral, and parasitic infections.

At birth, rats have very low serum Ig concentrations. They increase with the lactation and then slowly with the development of the immune system. Rat Ig concentrations reach a "normal" level when they are about 2 1/2 months. But this level is quite variable and not stable. Germ-free (also called axenic) rats have very low Ig serum levels which slowly increase with the antigenic aggressions due to the food intake, and perhaps with autoantigens. When 6- to 8-months old, germ-free animals are difficult to distinguish from their conventional counterparts kept in specific pathogen-free conditions. Rats bred in poor conditions and especially parasitized by natural or experimental helminths infections (syphacia are very common) may have extremely high serum Ig levels (see Chapter 1).

On the other hand, the concentration of MAb produced by a hybridoma can be quite variable depending on parameters which are poorly known. Thus, depending on the hybridoma's own secreting properties, on its size (more exactly, the number of living and secreting cells), on its localization (subcutaneous or in the peritoneal cavity), and on the half-life of the secreted MAbs, the serum concentration of MAb will be low, medium, or high, ranging from 1 mg/ml (or less) to 10 mg/ml and even more.

Various situations are given in Figures 1A and B, 2A and B, and 3A and B in function of the MAb serum concentrations and of the polyclonal Ig serum levels of the host. A rapid calculation shows that the percentage of MAb can considerably vary from about 5 to 70% of total Ig. Obviously, the difficulties encountered during the purification of a MAb will depend on many parameters, its own concentration and that of the host polyclonal Ig being of great importance. In Figure 4, some examples are given of rat MAbs produced in young LOU/C. IgK-1b rats (two to four) by comparison to a 6-month old normal rat serum.[1]

2. Linked to the MAb Itself

Some difficulties are linked to the isotype of the MAb, and others lie in the individual properties of the MAb.

a. Euglobulin Property

Insolubility in a low ionic strength buffer or even in distilled water can be used as a technique of purification, but denaturation often occurs and greatly limits the application of the technique. Moreover, the MAbs which disclose strong euglobulin properties cannot be purified by DEAE-chromatography as they are already insoluble in the buffer used to equilibrate the column.

Rat IgM and IgG2c MAbs must always be suspected of euglobulin properties.

b. Isoelectric Point or Electrophoretic Mobility

This property can be accurately tested by isofocalization electrophoresis. By agarose electrophoresis, it is easy to know whether a MAb is slow or very slow moving and can be easily purified by DEAE-chromatography. IgG2a (and more rarely IgG2b) rat MAbs sometimes present this characteristic. IgG2c frequently presents this slow moving property, but often being an euglobulin, it cannot be purified by DEAE chromatography as it precipitates at low ionic power.

c. Stability

The stability of a MAb is always interesting to know. Stability in function of time, at various temperatures from –70 to 37°C, and in various physical conditions: liquid phase with or without glycerol and purified or not, after lyophilization, after an acidic shock with a glycin-HCl buffer

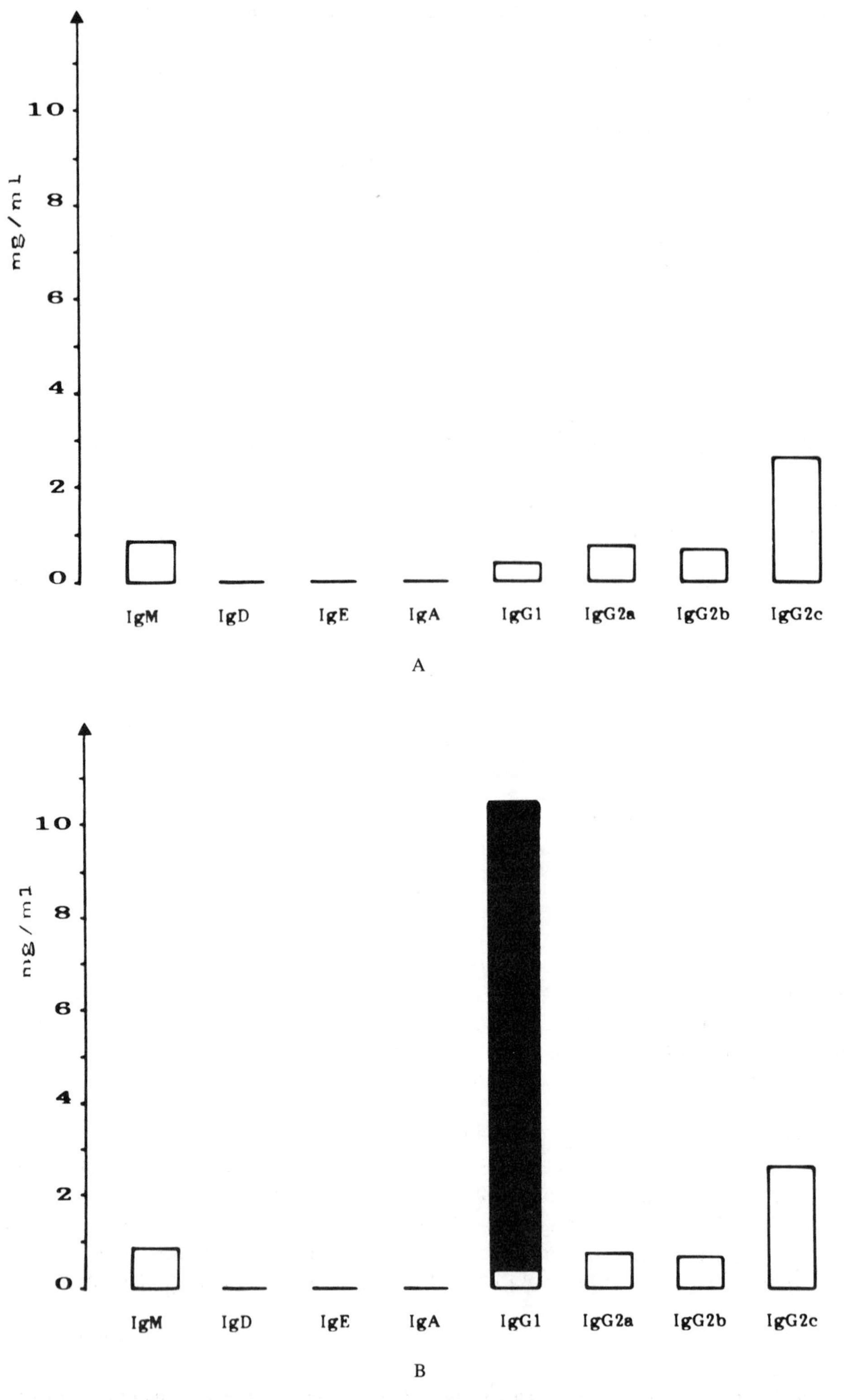

FIGURE 1. (A) Ig serum levels in specific pathogen-free (SPF) LOU/C.IgK-1b rats. Note the relatively high value of both IgM and IgG2c isotypes which are in axenic or SPF animals. (B) The same values and in black, the serum concentration of a high IgG1 secreting rat hybridoma. Quite clearly, the purification of such a MAb will generally be easy, the proportion of host polyclonal immunoglobulin being rather low (inferior to 30%).

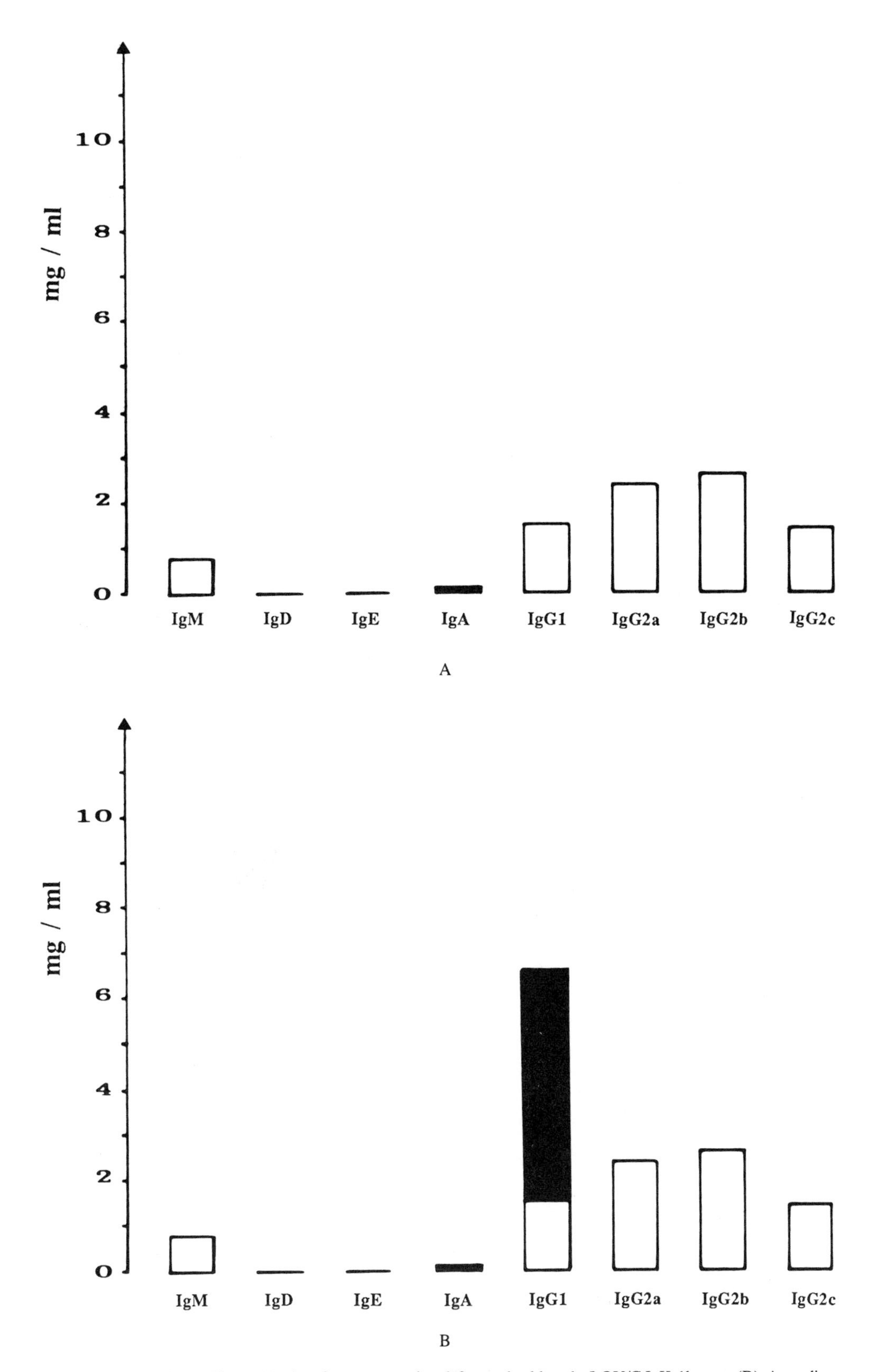

FIGURE 2. (A) Idem Figure 1A, but from conventional 3-month-old male LOU/C.IgK-1b rats. (B) A medium secreting IgG1 rat hybridoma. The difficulty of purifying will be normal, the percentage of host polyclonal IgG being about 50%.

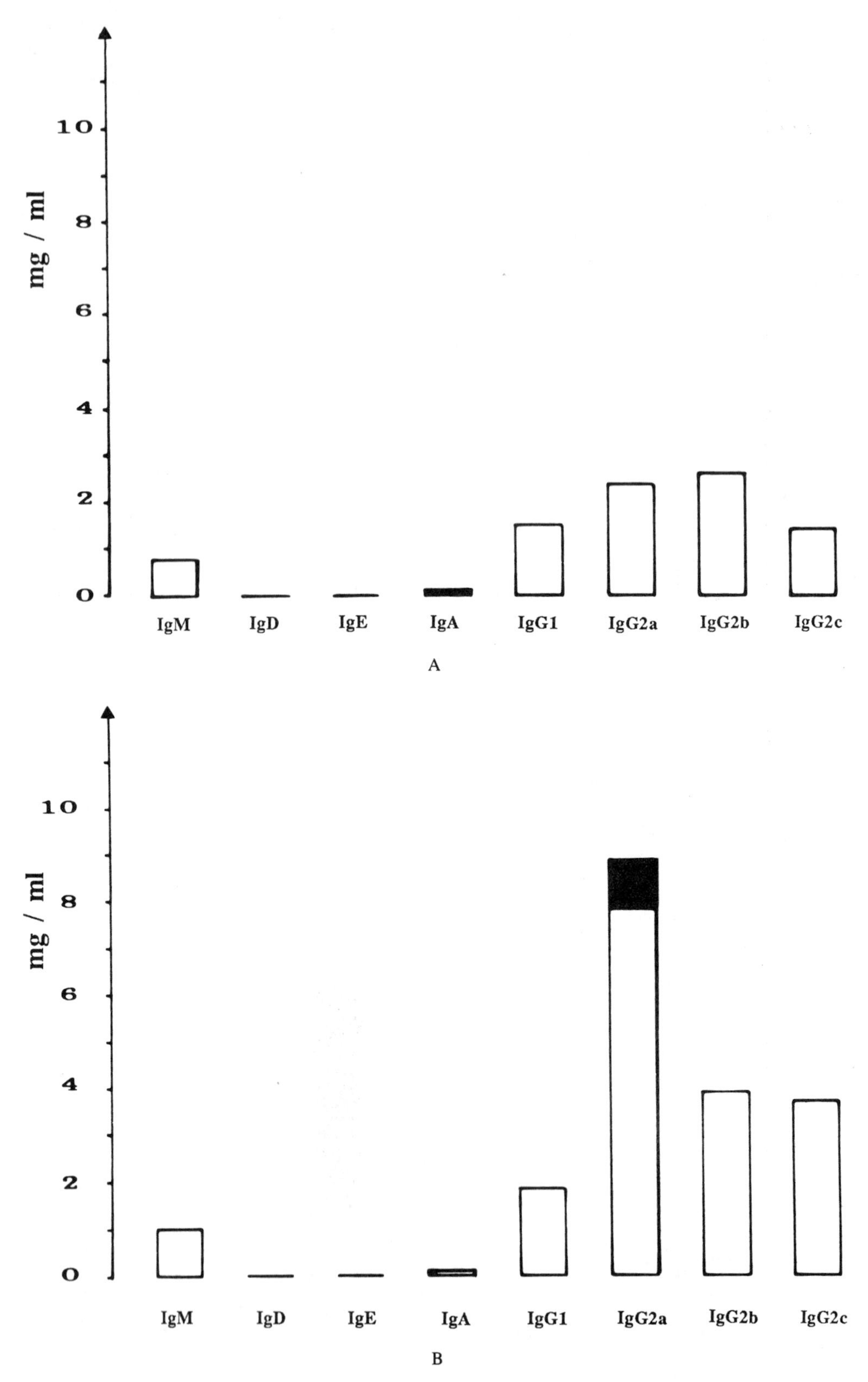

FIGURE 3. (A) Same as Figure 1, but old conventional LOU/C.IgK-1b rats. (B) A low secreting IgG2a rat hybridoma. The difficulty of purifying will be much higher than in the two previous examples, the percentage of host polyclonal IgG being superior to 90%.

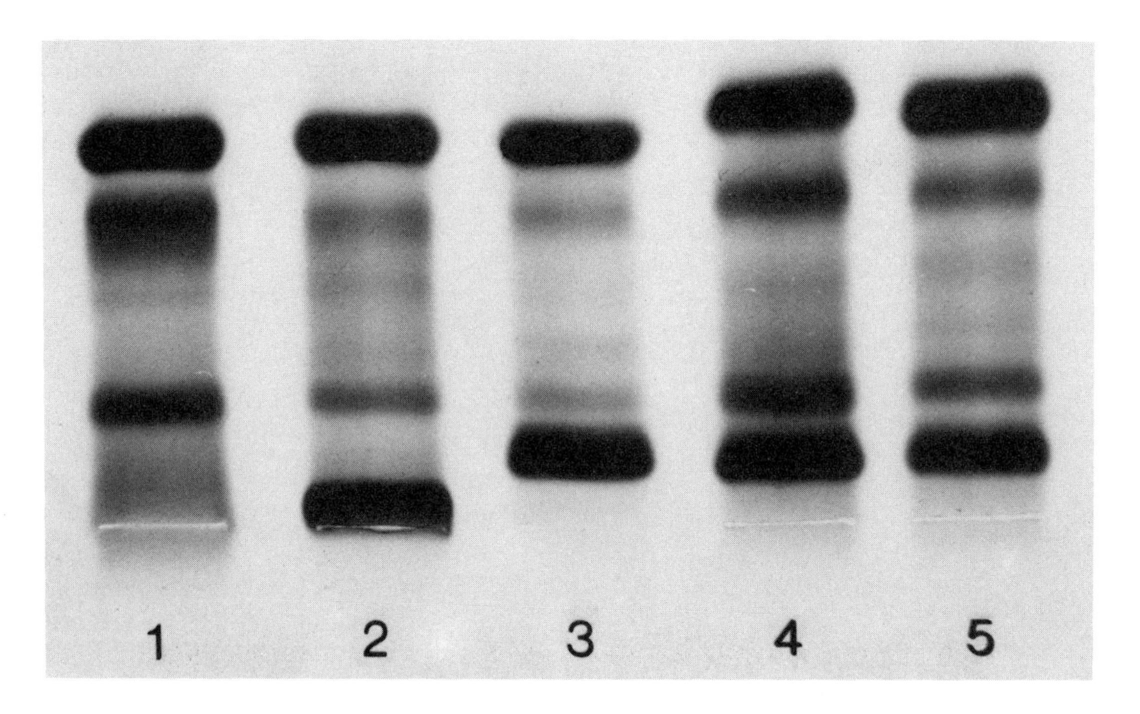

FIGURE 4. Agarose gel electrophoresis of (1) 6-month-old normal rat serum; (2) serum of a LOU/C.IgK-1b rat carrying the LO-HM-7, IgM-kappa secreting hybridoma; (3) serum of a LOU/C.IgK-1b rat carrying the LO-DNP-1, IgG1-kappa secreting hybridoma; (4) serum of a LOU/C.IgK-1b rat bearing the LO-RG-1, IgG2a-kappa secreting hybridoma; and (5) serum of a LOU/C.IgK-1b rat bearing the LO-HApoA1-8, IgG2b-kappa secreting hybridoma.

(0.01 *M;* pH 2.8) rapidly neutralized by a glycin-NaOH buffer (0.1 *M;* pH 8.6). The antigen-binding capacity of the MAb must be checked after these treatments.

B. CONVENTIONAL TECHNIQUES OF RAT Ig PURIFICATION

Many techniques have been proposed to purify either polyclonal or monoclonal rat immunoglobulin. The same basic procedures are commonly used and will be further discussed. Depending on the required purity, one or more steps are necessary. The technique of immunoaffinity chromatography will be described in the next section.

1. Description of the Conventional Techniques used for Rat Ig Purifications

a. Precipitation

i. Ammonium Sulfate

Materials

The materials used include saturated ammonium sulfate, saline, centrifuge, magnetic stirrer, tubes, flasks, etc. Saturated ammonium sulfate is made by dissolving ammonium sulfate in distilled water (approximately 1000 g/l to have a 4.05 *M* solution) at about 50 to 60°C and allow it to cool overnight at room temperature. Adjust it at pH 7.4 with NH_4OH. Saline consists of distilled water containing 0.9% of NaCl, adjusted at pH 7.4 with NaOH.

Techniques

Described below are the manipulations performed with the results.

Serum or ascitic fluid is diluted twice in saline with the result being diluted serum. The diluted

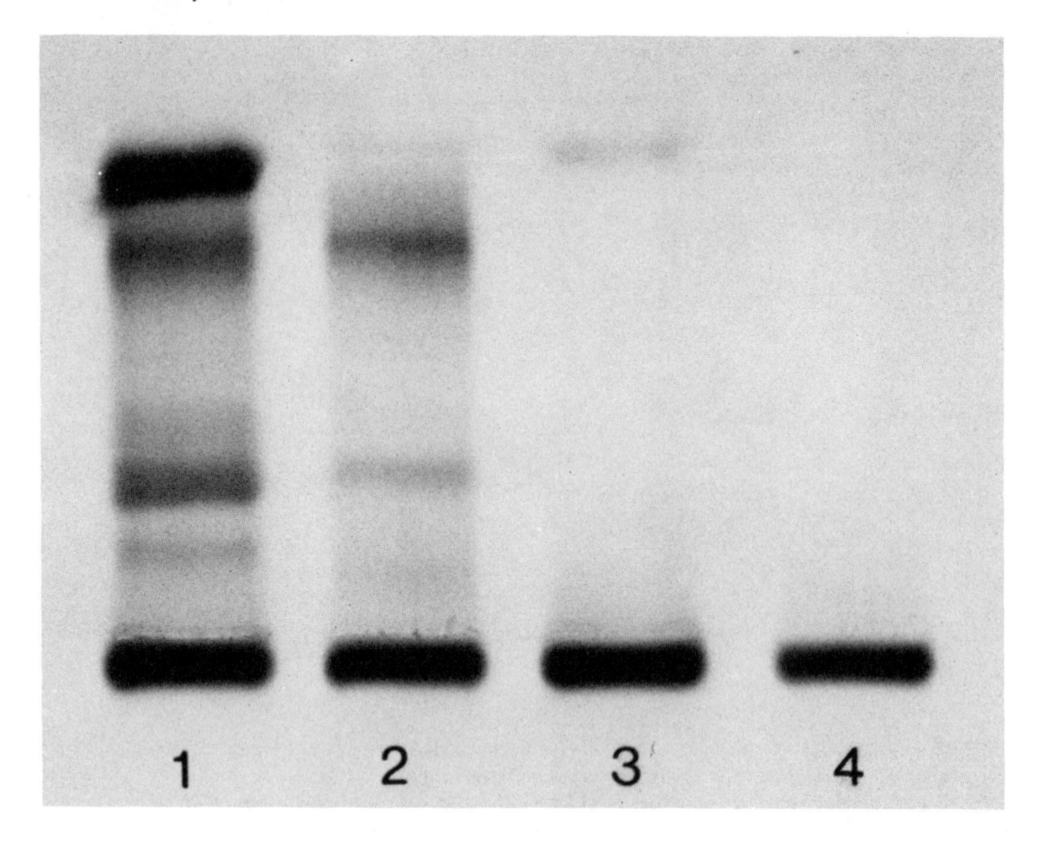

FIGURE 5. Agarose electrophoresis of (1) the serum of a rat carrying the LO-DNP-57 hybridoma secreting an IgG2b-kappa rat MAb anti-DNP hapten; (2) the same LO-DNP-57 serum after precipitation by ammonium sulfate as described in Section I.B.1.a.i (3) the same LO-DNP-57 serum after precipitation by caprilic acid as described in Section I.B.1.a.i and (4) the same LO-DNP-57 serum after caprilic acid and ammonium sulfate precipitation. The purified fractions have been corrected in order to have a volume identical to the one used at the beginning of the purification.

serum or ascitic fluid at 4°C is precipitated by ammonium sulfate adjusted to pH 7.4 at 40 to 50% final concentration depending on the various isotypes and centrifuged at 10,000 *g* for30 min at 4°C, resulting in a precipitate. The precipitate is washed twice again with 40 to 50% final concentration, depending on the various isotypes, saturated ammonium sulfate resulting in a washed precipitate. The washed precipitate is dissolved in approximately one third or half of the original serum or ascitic fluid volume, in saline or a buffer and thoroughly dialyzed 2 or 3 d against the same buffer at 4°C with continuous agitation and with two changes of saline or buffer everyday. Dialysis tubing such as Visking (Medicall Ltd, GB) can be used, having a 45,000 Da molecular weight cutoff. Alternatively, the dissolved precipitate can be desalted through Sephadex G25 equilibrated in saline or any other convenient buffer. The result is an ammonium sulfate precipitate serum.

Remarks

Ammonium sulfate must be slowly added (drop by drop) to the diluted serum which must be continuously stirred. Once the ammonium sulfate is totally added, the mixture must be kept 2 or 3 h or even overnight, with a gentle stir before being centrifuged.

Temperature is important. Room temperature is convenient and good results may be obtained at 4°C, but the results will be slightly different from those obtained at room temperature.

TABLE 1
Examples of Recovering of Rat MAbs After Purification by Caprilic Acid and Concentration by Millipore CX30, Ammonium Sulfate Precipitation or Both. No Concentration is Necessary after Ammonium Sulfate Precipitation

			Recovery (%)		
MAbs	Isotypes	Concentration of MAb (mg/ml) in the ascitic fluid	After precipitation by caprilic acid and concentration	After precipitation by $(NH_4)_2$	After precipitation by caprilic acid and $(NH_4)_2$ SO_4
LO-MG1-2	IgG1-kappa	1.8	10	93	14
LO-MG2b-2	IgG1-kappa	4.9	39	85	49
LO-MG2a-2	IgG2a-kappa	2.8	14	96	39
LO-MG2a-3	IgG2a-kappa	3.6	45	88	45
LO-DNP-57	IgG2b-kappa	12.7	68	86	52
LO-Tact-1	IgG2b-kappa	6.2	60	92	73

Results

The results show that the purity of the MAb purified by ammonium sulfate precipitation is rather poor (Figure 5). However, the percentage of recovery is high, around 85 to 95% (Table 1). Because the denaturation of the MAb being limited and sometimes giving a valuable concentration, this technique is interesting to perform. But, it must be noticed that the polyclonal Ig as well as the MAb are both purified, at the same time.

ii. Euglobulin

Material

One material used is a low ionic strength buffer, Tris HCl 0.005 *M* at pH 8.0.It is obtained by dissolving 0.6 g of Tris in 1 l of distilled water and then adjust to pH 8.0 with concentrated HCl. Also used is distilled water adjusted to pH 6.0 with 1 *N* NaOH, a centrifuge, magnetic stirrer, tubes, and flasks.

Techniques

The following are manipulations performed and the results.

Serum or a fractionated serum is dialyzed against a large quantity of buffer of low ionic strength (Tris-HCl 0.005 M, pH 8.0) and the result is euglobulin precipitate. Another method consists in diluting 1 volume of the serum in 19 volumes of distilled water adjusted to pH 6.0. The result is euglobulin precipitate. The precipitate is centrifuged at 15,000 *g* for 15 min and then redissolved in a convenient buffer resulting in euglobulin fraction

Remark

The euglobulin precipitation is a convenient and simple technique, but the precipitate is often difficult to redissolve, owing to the denaturation. Nevertheless, when no or little denaturation occurs, it appears to be a useful and very simple technique of purification with nearly 100% of recovery.

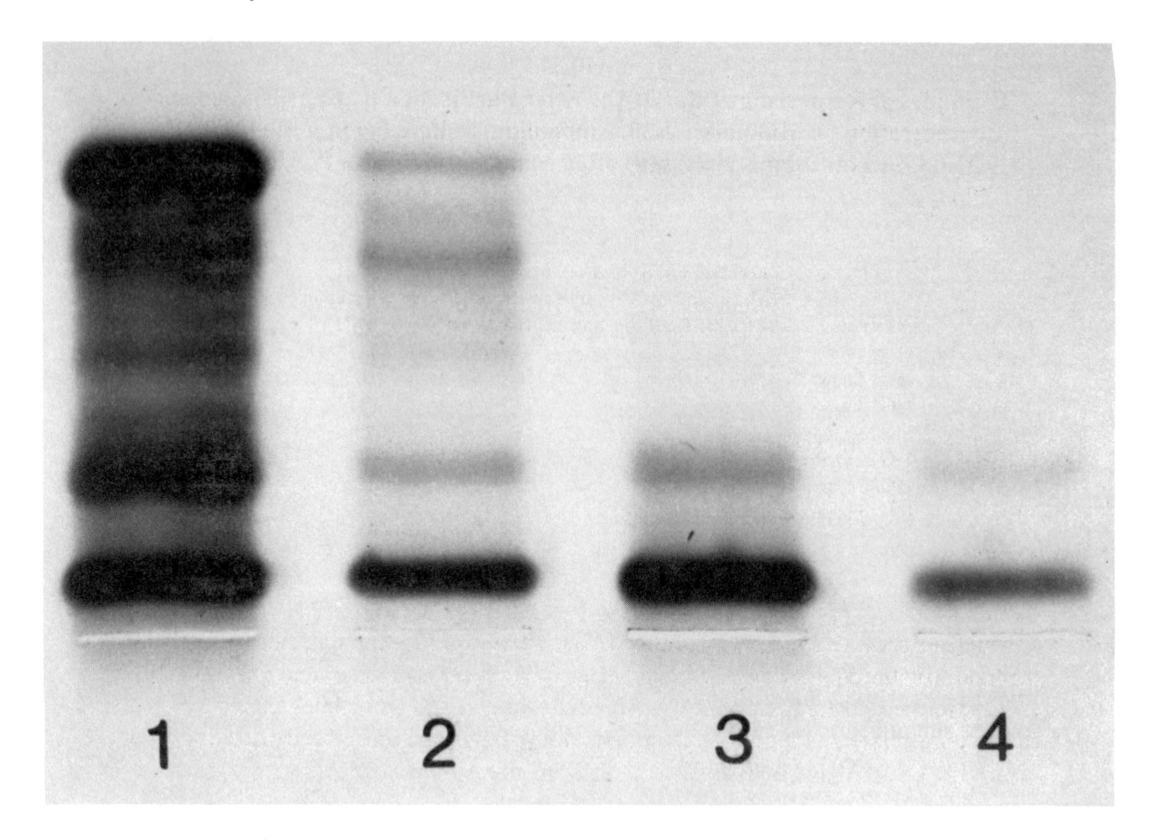

FIGURE 6. Agarose electrophoresis of (1) a serum of a LOU/C rat carrying the IR595 IgG1-kappa secreting immunocytoma; (2) The same serum after ammonium sulfate precipitation (see Section I.B.1.a.i.); (3) the same serum after ammonium sulfate precipitation, rivanol precipitation, and ammonium precipitation; and (4) The same serum after ammonium sulfate precipitation, rivanol precipitation, ammonium precipitation, and removal of the host polyclonal Ig by a LO-RK-1b-2 immunoaffinity column (see Section I.C.2.a.iv).

Results

They are quite variable, depending on the physicochemical properties of each MAb. It is difficult to give results which could be considered of general interest.

iii. Rivanol

We have used the method of precipitation by the rivanol described by Hardie and Van Regenmortel.[1]

Material and Reagents

The materials and reagents used are Rivanol 0.4% in H_2O (Rivanol, Ethacridine Lactate - Fedora SV, Brussels, Belgium), saturated KBr solution (5.35 g of KBr in 10 ml), centrifuge, and tubes.

Protocol

Mix 4 volumes of Rivanol solution with 1 volume of rat serum or ascitic serum and wait 10 min. Centrifuge at 1000 to 1500 *g* for 10 min. Add some drops of saturated KBr solution to precipitate the Rivanol. Centrifuge at 1000 to 1500 *g* for 5 min. Repeat the last centrifugation, if necessary. Thereafter, ammonium sulfate precipitation can be proceeded with.

Results

The results are interesting, but not as interesting as results obtained by caprilic acid precipitation, at least at the level of purity. Moreover, the MAbs are purified with all other polyclonal Ig. The recovery ranges from 40 to 60% of the MAbs (Figures 5 and 6).

iv. Caprilic Acid

McKinney and Parkinson[2] have proposed to use Steinbuch and Andran's[3] caprilic acid precipitation procedure associated with the ammonium sulfate precipitation in order to purify rabbit, sheep, goat, horse, mouse, and rat IgG and mouse MAbs.

The caprilic acid method proposed by McKinney and Parkinson[2] is the following:

1. One volume of serum (or ascitic fluid) is diluted with four volumes of acetate buffer (60 m*M*, pH 4.0) and the pH is adjusted to 4.5 with 0.1 *N* NaOH.
2. Caprilic acid (Sigma Chemical Co., St Louis, U.S.) is slowly added with thorough mixing, at a concentration of 25 ml of diluted serum and at room temperature. The solution is stirred for 30 min and the insoluble precipitate is removed by centrifugation at 10,000 *g* for 30 min.
3. The supernatant can be filtered through a prefilter (Millipore, France) to remove fine particles. The pH is adjusted to 7.4 with NaOH. This procedure can be followed by an ammonium sulfate precipitation (Figures 5 and 6).

The results are given in Table 1. Caprilic acid and ammonium sulfate precipitations are particularly interesting for some MAbs. Polyclonal Igs from the host which are the main contaminants can be removed by immunoaffinity chromatography using the kappa allotype system (see Chapter 1,C,2,a and this chapter, Section I.C.2.a.iv.).

b. Ion-Exchange Chromatography

i. DEAE-Chromatography

Materials

The materials include DEAE-cellulose (Whatman DE 32 No. 24322 - BALSTON - Modified cellulose, GB), conical flasks, syringes, or disposable column.

Techniques

Dialysis the serum or the fraction of serum against the starting buffer which has been used to equilibrate the column. Phosphate buffer could be employed, but Tris-HCl (0.05 *M*, pH 8.0) buffer is often used. The proteins are eluted with a linear gradient of increasing molarity (final buffer: 0.01 *M*, Tris HCl, 0.5 *M* NaCl, pH 8.0).

Remarks

Many DEAE-chromatography supports exist and they give similar results.

The method can be applied to small or large quantities of protein respectively in column or in batch; the method is not denaturating.

DEAE-chromatography can be performed at room temperature. However, air is less soluble at room temperature than at 4°C. The gel and the buffers, if they have been kept at 4°C, must be degassed by application of vacuum or allowed to reach room temperature before use.

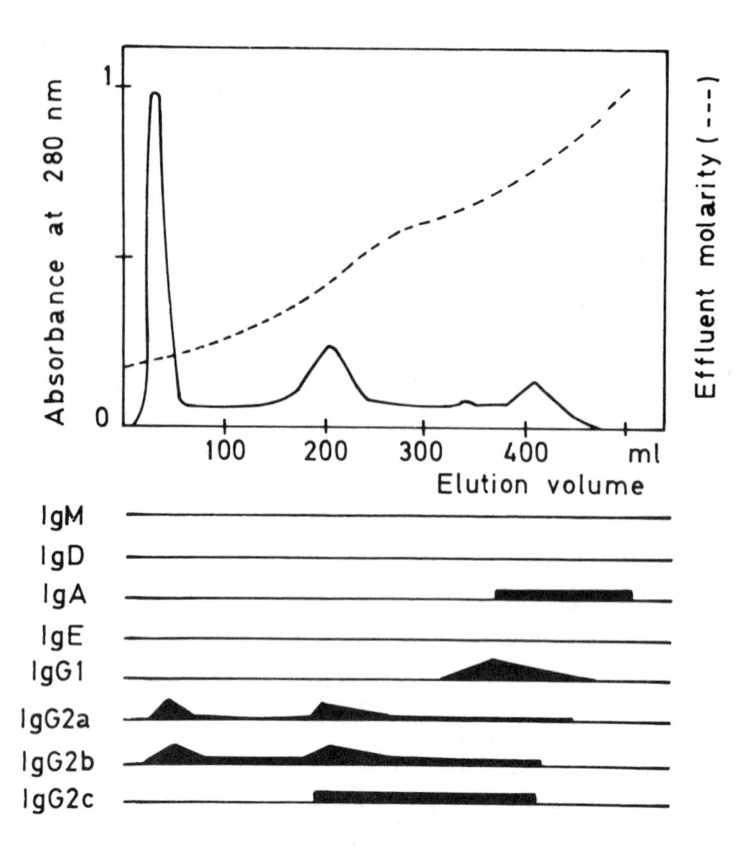

FIGURE 7. Distribution of the rat immunoglobulin classes or subclasses in the elution pattern of DEAE-cellulose chromatography. Column: 2.5 × 10 cm; sample: 8 ml of normal rat serum precipitated three times with cold 40% saturated ammonium sulfate and dialyzed against 0.05 *M* Tris-HCl buffer, pH 8.0. Elution buffer: linear gradient, at flow rate 70 ml/h, of Tris-HCl buffer pH 8.0, 0.05 to 0.05 *M* plus 0.5 *M* NaCl.

Results

Results obtained with a rat serum are given in Figure 7. The DEAE-chromatography is especially useful in case MAb has a pH of 7 to 8. This value can be easily determined by isofocalization analysis. More easily, agarose electrophoresis can be used to have an approximation value. MAbs or myeloma Ig of very low mobility will be first eluted in DEAE-chromatography. With a selected ionic strength of the buffer, only the monoclonal immunoglobulin or antibody to be purified will pass through the column, most other proteins being bound to the support. In such a case, the DEAE-chromatography is a very effective method. So, using a 0.005 *M* phosphate buffer at pH 7.5, McGhee et al.[4] have obtained pure polyclonal IgG2a molecules in the break-through peak. No other immunoglobulin class can be directly purified by DEAE-chromatography.

ii. Carboxylmethyl (CM) Chromatography

Materials

CM-cellulose (Whatman CM-32 No. 22322, Wand R. Balston, modified cellulose, GB).

TABLE 2
Binding of Rat IgG Subclasses to Protein A or Protein G and Comparison of Binding

	IgG1	IgG2a	IgG2b	IgG2c
Protein A	weak	no	no	strong
Protein G	weak	strong	weak	strong

TABLE 3
Rat IgG Subclass Content of Fractions Obtained by Chromatography on Protein A-Sepharose of 40% Ammonium Sulphate Precipitate of Normal Rat Serum

Fraction	Eluting buffer at pH	IgG1 (mg)	IgG2a (mg)	IgG2b (mg)	IgG2c (mg)
1	8.0	0.50	10.70	n.d.	n.d.
2	8.0	2.41	0.50	2.22	n.d.
3	7.0	4.31	n.d.	n.d.	n.d.
4	6.0	0.75	n.d.	n.d.	n.d.
5	5.0	0.39	n.d.	n.d.	0.27
6	4.0	n.d.[a]	n.d.	n.d.	1.80
7	3.0	n.d.	n.d.	n.d.	n.d.
Total recovered (mg)		8.36	11.20	2.22	2.07
Total applied (mg)		13.60	14.70	2.40	4.25
Recovery in percentage		61	76	92	49

[a] Not detectable by radial immunodiffusion (less than 0.05 mg).

Techniques

The serum or its fraction is dialyzed against a sodium acetate 0.2 *M*, 0.1 NaCl buffer, pH 4.2 and put on a CM-cellulose column equilibrated with the same buffer. The chromatography is performed with a sodium acetate 0.2 *M* buffer pH 4.2, 0.1 *M* sodium acetate with a linear gradient of NaCl up to 0.2 *M*.

Results

The technique is very rarely used, except for IgG2c purification. It is in general of low interest.

iii. Fast Liquid Chromatography

This is a very interesting method which can give excellent purification of mouse MAbs. Unfortunately, this technique requires expensive equipment (Pharmacia-LKB, Sweden) or Biorad (U.S.). Descriptions of results essentially devoted to mouse MAbs can also be obtained from these firms. Very probably, they could be transposed to rat MAbs.

c. Affinity Chromatography

These techniques will be described separately owing to their large range of applications.

i. Protein A Chromatography

Protein A is a membrane component of many strains of *Staphylococcus aureus*. Identified by Crov et al.[5] and Oeding et al.,[6] it has been purified and characterized by Shodahl.[7,8] Its interaction with IgG molecules has first been described by Forsgren and Sjöquist.[9]

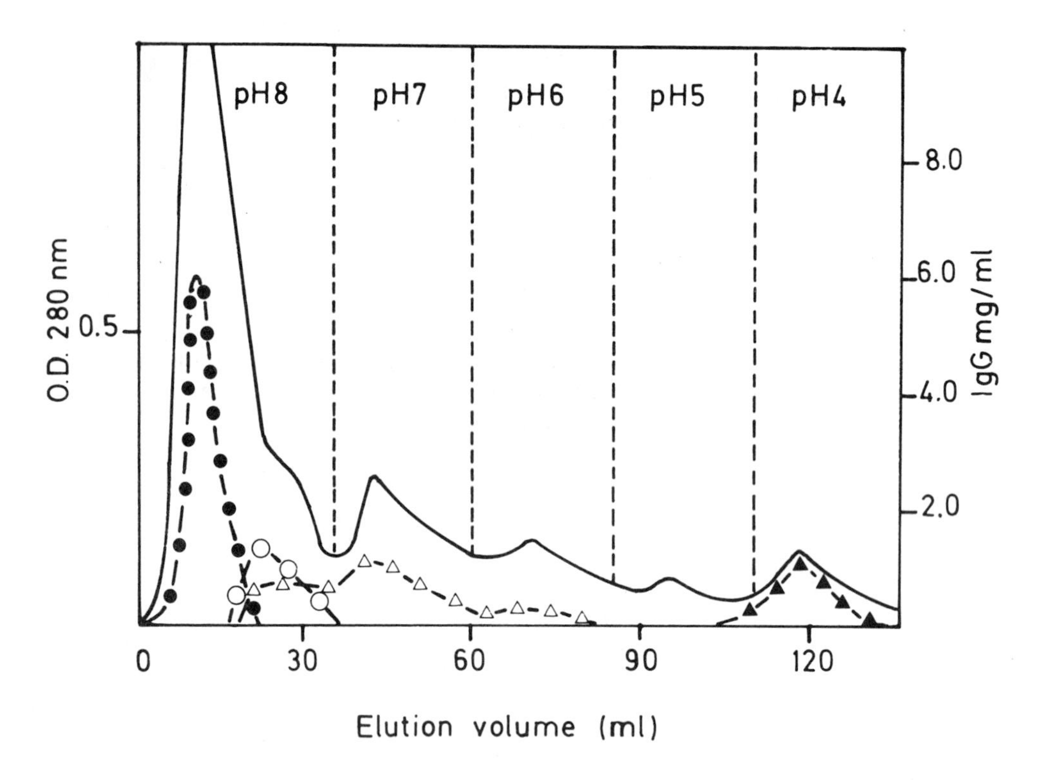

FIGURE 8. Elution at 15 ml/h of rat immunoglobulins from a 100 mg precipitate at 40% ammonium sulfate saturation of normal rat serum on a 6 ml protein A-Sepharose column. Sequential elution with 0.14 *M* Na-phosphate, pH 8.0 and 7.0 and 0.1 *M* Na-citrate, pH 6.0, 5.0, and 4.0. Broken lines represent the elution profile of the various IgG subclasses as determined by radial immunodiffusion in each fraction. -Δ-Δ- IgG1, ●-●- IgG2a, -O-O- IgG2b, -▲-▲- IgG2c (modified from Rousseaux et al., 1981).

In the rat species, many studies[10-16] have shown the binding of protein A to molecules of the IgG1 and IgG2c, and a very weak binding to IgG2b molecules (Table 2).

Material and Methods[14]

Staphylococcal protein A covalently linked to Sepharose CL-4B (Pharmacia Fine Chemicals, Sweden). One and half gram was swollen in phosphate buffered saline (PBS) at pH 8.0 containing 0.02% sodium azide, and packed into a 1 × 15 cm glass column.

One hundred milligrams of proteins from a precipitate at 40% ammonium sulphate saturation of normal rat serum, containing approximately 30 to 35 mg of rat IgG were applied to the column. Sequential elution with 0.14 *M* Na-phosphate, pH 8.0 and 7.0, then 0.1 *M* Na-citrate, pH 6.0, 5.0, 4.0, 3.0 was performed. Table 3 and Figure 6 give the results.

Conclusions

Differences in affinity between rat IgG subclasses to staphylococcal protein A were found to be related in part to the subclass with the order: IgG2c, good; IgG1, weak; IgG2b, very weak; and IgG2a, no affinity for protein A. However, differences in affinity exist between IgG2c monoclonal proteins, some being eluted at pH 4.0 to 3.0 and others at pH 6.0, the monoclonal IgG2c proteins being the more basic (having the slowest electrophoretic mobility) having the highest affinity for protein A. Such heterogeneity was also found for IgG1 monoclonal proteins which were eluted from pH 8.0 to pH 6.0.[14] Some molecules (15 to 20%) of rat IgM molecules can bind to protein-A-Sepharose.[10]

In practice, affinity chromatography on protein A-Sepharose can be used as a technique of purification of rat immunoglobulins or rat MAbs, especially for IgG2c MAbs which are rather rare and some MAbs of the IgG1 subclass.[10,12,14]

However, as shown in Figure 8, there are some IgG1 rat MAbs which will be eluted with the pool of polyclonal IgG2b. Moreover, most rat MAbs of the other classes or subclasses, especially those of the IgG2a and IgG2b subclasses, cannot be purified by protein A affinity chromatography.

ii. Protein G Chromatography

Protein G is a bacterial cell wall protein isolated from group c and b Streptococci. Its apparent molecular weight is 30 Kd. It can bind to many IgG from various species. Protein G binds mainly to the Fc region of the IgG molecules.[17,18] Table 2 gives the results compiled from the literature concerning protein G binding to rat IgG subclasses. Unfortunately, the binding of protein G to rat IgG1 and IgG2b immunoglobulins is weak. Protein G can be purchased from Perstorp Biolytica Sölvegatan 41, S-223 70 Lund, Sweden, and Zymed Laboratorie, Inc., 52 South Linden Avenue, South San Francisco, CA 94080, U.S., and Janssen Biochemica, Belgium, or Pharmacia, Sweden. Methods of IgG subclasses purification using protein G are described in booklets published by these commercial firms.

In the case of ascitic fluid, the purified rat MAbs will be contaminated with the polyclonal IgG from the host animal.

d. Gel Filtration

Gel filtration is a cheap and convenient system of purification, but its separation power is limited. However, it can be used in some circumstances:

1. Separation of immunoglobulins having different molecular weights: IgM from IgG and partially IgA from IgG and/or IgM
2. Purification of immunoglobulins from contaminants of different molecular weights, for example, IgG from Transferrin or albumin
3. Gel filtration can also be used to determine the molecular weight of a molecule, but the method has been supplanted by polyacrylamide gel or SDS polyacrylamide gel electrophoresis

Materials

Many supports for gel filtration are available. Sephadex G100 or G200 (Pharmacia, Sweden) is commonly used. Ultrogel (IBF, France) or Sephacryl (Pharmacia, Sweden) are very convenient for immunoglobulin purification. Sepharose (Pharmacia, Sweden) can also be used.

The length of the column is important. Depending on the quantities of proteins to separate, columns of 50 cm in length with a 1.0 cm diameter or 200 cm with a 2.5 cm diameter are convenient, but it is recommended to use a pump and to inject the proteins and the buffer from the bottom to the top. For large quantities of proteins, columns of 50 or 100 mm in diameter can be used. They must be as long as possible, 100 cm for example.

Techniques

The volume of the serum or the serum fraction to be applied on the column must be as small as possible (no more than about 1% of the gel volume). The volume of the fractions to be collected must be relatively limited and correlated to the degree of precision which is required for the purification.

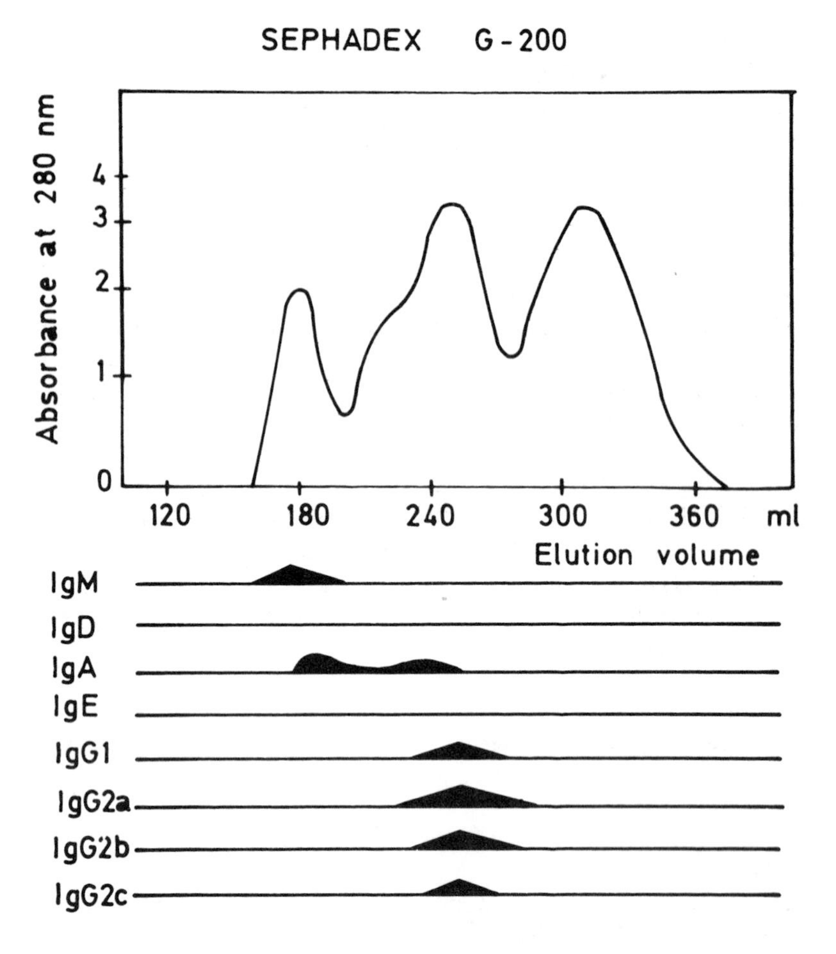

FIGURE 9. Distribution of the various immunoglobulin classes and subclasses in the elution pattern of G200 chromatography of normal rat serum. Column: 2.5 × 90 cm; sample: 2 ml of rat serum; buffer: 0.1 *M*, pH 8.0, Tris-HCl buffer containing 0.15 *M* NaCl; flow rate: 12 ml/h.

Results

Large molecules are totally excluded from the beads and emerge from the filtration after a volume equal to Vo, also known as the void volume. It is equal to about the space which exists between the beads. On the contrary, the small molecules emerge at a volume equal to the total volume of the column or Vt. The best separations are obtained when the molecules occur between Vo and Vt and are not excluded from the gel.

For the rat Igs, IgG subclasses, IgA, IgD, and IgE are well separated from the other serum proteins by ultrogel AcA34 and IgM by ultrogel AcA22 (IBF, France). Sephadex G200 is still considered as a reference gel for filtration but its resolution is much less efficient. Details concerning the technique to pour and to run the gel filtration can be found in booklets distributed by Pharmacia or IBF.

Figures 9 and 10 give the results obtained for gel filtration of normal rat serum on G200 and AcA34 respectively.

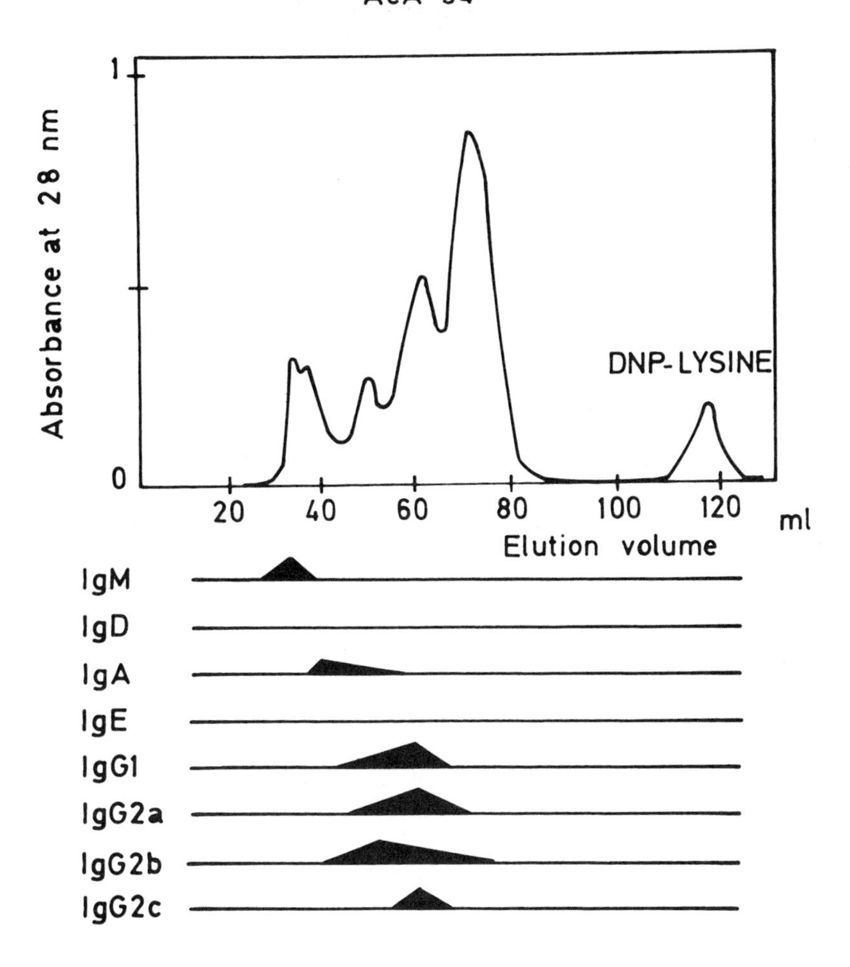

FIGURE 10. Distribution of the different immunoglobulin classes and subclasses in the elution pattern of AcA34 chromatography of normal rat serum. Column: 2.5 × 90 cm; sample: 2 ml rat serum; buffer: 0.1 *M* Tris-HCl, pH 8.0 containing 0.15 *M* NaCl; flow rate 15 ml/h.

Remarks

A minimum of salt concentration must be used in the buffer in order to avoid nonspecific absorption of proteins on the beads.

The sodium azide or another preservative must also be included in the buffer to inhibit microbial growth.

e. Zone Electrophoresis

Zone electrophoresis as a preparative method is no longer frequently used for Ig purification. However, for peculiar reasons, like immunization, agarose or acrylamide agarose electrophoresis of small quantities of Ig can be used.

f. Ultracentrifugation

Preparative ultracentrifugation is no longer used for Ig purification. It can be substituted by gel filtration.

2. Purification of Rat Monoclonal Ig Conventional Techniques

The serum or ascitic fluid must be centrifuged at 27,000 *g* for 10 min and filtered on a prefilter.

a. Rat IgM MAb

Various techniques of rat IgM purification have been described.[4,19-23] These authors all consider the molecular weight and/or the euglobulin properties of this Ig isotype.

- Step 1 Precipitation by ammonium sulfate at 45 to 50% final concentration.
- Step 2 Euglobulin precipitation in a Tris-HCl buffer at 0.005 *M*, pH 8.0.
- Step 3 Gel filtration in AcA22.

Remark

The recovery of the final product will mainly depend on the euglobulin precipitation step and more exactly on the resuspension of the precipitate which could be difficult. In that case, the second step must be avoided.

The purity of the IgM will rarely be perfect; traces of alpha2 macroglobulin will be present.

b. Rat IgD MAb

Until now, to our knowledge, no rat IgD MAb has been detected. It is certainly rare. Purification procedure used for immunocytoma IgD can probably be adapted for rat IgD MAb.

Briefly, as described by Bazin et al.[24]

- Step 1: Ascitic serum or serum containing a monoclonal IgD is submitted to a gel filtration on AcA22 (or AcA34) ultrogel at 4°C, by using a 0.1 M Tris-HCl buffer, pH 8.0, containing 0.15 M NaCl and 0.001 *M* epsilon-amino caprilic acid. The presence of IgD in the eluate can be checked by Ouchterlony test.
- Step 2: Concentration by ultrafiltration and repeated runs on ultrogel AcA34 column.
- Step 3: Preparative agarose or Pevikon electrophoresis.

Remark

Quite clearly, it is easier to use an immunoaffinity chromatography to prepare monoclonal IgD, but rat IgD molecules are rather labile.

c. Rat IgG1 MAb

- Step 1: 45 to 50% ammonium sulfate precipitation, at least two or three times.
- Step 2: DEAE-chromatography in a Tris-HCl buffer (0.05 *M*, pH 8.0). Elution is performed with a linear gradient of NaCl (up to 0.6 *M*) in the same buffer.
- Step 3: A further purification can be obtained by gel filtration chromatography on AcA34.

d. Rat IgG2a or IgG2b MAb

MAbs of these subclasses can be classified in function of their electrophoretic mobilities either gamma 1 or gamma 2 (respectively, the faster and the lower in electrophoretic analysis). Gamma 1 IgG2a or IgG2b MAb can be purified as described for those of IgG1 subclass. Gamma 2 IgG2a or IgG2b MAb can be purified by:

- Step 1: 45 to 50% ammonium sulfate precipitation, two or three times.
- Step 2: DEAE-chromatography; starting buffer: Tris-HCl buffer 0.05 *M*, pH 8.0; eluting buffer: linear gradient up to the final buffer 0.05 M Tris-HCl + 0.05 *M* NaCl, pH 8.0. Very often, the first peak will be constituted by the MAb.

e. Rat IgG2c MAb

MAb of this subclass can be difficult to purify owing to their euglobulin properties and their rather frequent poor stability. Some of them can be purified by:

- Step 1: Euglobulin precipitation and redissolved in acetate buffer at pH 5.5.
- Step 2: Gel filtration on AcA34.

Another possibility[23] is:

- Step 1: 45 to 50% ammonium sulfate precipitation, two or three times.
- Step 2: Dialysis of the precipitate in a Na-acetate buffer 0.2 *M*, pH 4.2 with 0.1 *M* NaCl and CM-chromatography with an linear gradient for elution up to 0.2 *M* Na-acetate, 0.2 *M* NaCl buffer at pH 4.2.

f. Rat IgE MAb

Rat IgE MAbs can be purified as described for immunocytoma IgE.[25]

- Step 1: 50% ammonium precipitation, two or three times.
- Step 2: DEAE-chromatography, equilibrated with a 0.05 *M* Tris-HCl buffer at pH 8.0. Elution with a convex gradient, starting buffer identical to that used for equilibration, limit buffer 0.1 *M* Tris-HCl, 0.4 NaCl, pH 8.0. The MAbs are detected by Ouchterlony test and concentrated by ultrafiltration.
- Step 3: The IgE fraction is dialyzed against a 0.1 *M* Tris-HCl, 0.15 *M* NaCl at pH 8.0 and applied to a AcA34 column.

The last chromatography gives three peaks: the first is monoclonal IgE, the second is IgG, and the third, transferrin. The purity of the IgE MAb can be assessed by agarose electrophoresis, SDS-PAGE or Ouchterlony tests.

Remark

IgE polyclonal or monoclonal antibodies are rather labile.

g. Rat IgA MAb

IgA immunocytoma proteins have been purified by repeated gel filtration on Sephadex G200, preparative electrophoresis on Pevikon blocks, and subsequent purification on Sepharose-6B of the IgA fraction.[26] Owing to the frequent heterogenity of the rat IgA, with regard to molecular weight and electrophoretic mobility, the percentage of recovery obtained with that procedure is relatively small.

C. PURIFICATION OF RAT MABS BY IMMUNOAFFINITY CHROMATOGRAPHY

Most rat MAbs are stable at acidic pH at least for a short period of time such as 3 to 5 min. They can be purified by immunoaffinity chromatography.

1. General Methodologies

a. Preparation of Serum or Ascitic Fluid: Clarification Step

It is always interesting to prepare the serum or the ascitic fluid mainly in order to avoid a mechanical blockage of the column by lipids or clots. This method can be applied to rat or mouse MAbs.

TABLE 4
Coupling MAb to Sepharose-4B[a]

Steps	Remarks
1. Weight the required quantity of Sepharose-4B	1 g of dry gel gives 3.5 ml of wet gel
2. Wash the Sepharose-4B on on sintered-glass funnel	Use HCl 0.1 *M* (200 ml/g)
3. Dialysis of MAb in the coupling buffer	Buffer $NaHCO_3$ (0.1 *M*, pH 8.3) with 0.5 *M* NaCl, about 5 to 10 mg/ml of gel
4. Mix the MAb solution with the Sepharose-4B	Use a mechanical stirrer for 3 h at room temperature (do not use a magnetic stirrer, which will destroy the gel)
5. Block the active groups which are still unblocked	Put the gel in the blocking buffer (0.2 *M* glycin at pH 8.0) for 16 h at 4°C
6. Wash the uncoupled MAb	Three cycles of acetate buffer pH 4.0 and borate buffer pH 8.0

[a] Rat or mouse MAb or Ig.

1. Ascitic fluid or serum is centrifuged at 22,000 *g* for 10 min.
2. Dilipidation at room temperature by adding one volume of Trichlorotrifluoroethane (Frigen 113 TR-T, Hoechst, FRG) for one volume of serum or ascitic fluid. Stir gently for 30 min.
3. Centrifugation at 2500 *g* for 10 min at 4°C. The serum (or ascitic fluid) is collected at the top of the tube.
4. Addition of 10 μl of ethylenediamine tetra-acetic acid tetrasodium salt (EDTA) (BDH, GB) at 100 g/l/ml of serum (or ascitic fluid).
5. Filtration on Millipore prefilter (AP25).

b. Coupling Mouse or Rat MAbs (or Igs) to Sepharose-4B

Material

The materials used include a blocking buffer, glycine 0.2 *M* pH 8.0; Na-acetate buffer, 0.1 *M* pH 4.0 with NaCl 0.5 *M;* borate buffer, 0.1 *M* pH 8.0 plus NaCl 0.5 *M;* MAb in the coupling buffer, and activated Sepharose-4B.

Procedure

Sepharose-4B is rehydrated for 15 min in HCl solution (Table 4) and washed with the same solution (about 200 ml/g of dry Sepharose-4B). One gram of dry Sepharose-4B gives 3.5 ml of wet gel. Then the gel is washed in the coupling buffer (about 5 ml/g of dry gel) and immediately dropped in the solution of MAb. It is very interesting to quantify the uncoupled MAb which can be found in the last washing buffer to know how much MAb has been coupled with the Sepharose-4B. Econo-column (BIORAD, U.S.) or IBF column of about 10 to 15 cm high and 2.3 cm diameter are the most convenient to use.

c. General Procedure of MAb Purification Using Immunoaffinity Column

The general procedure is the following:

- Step 1: Ascitic or serum is applied on the column and the column is immediately washed with phosphate buffered saline (PBS).

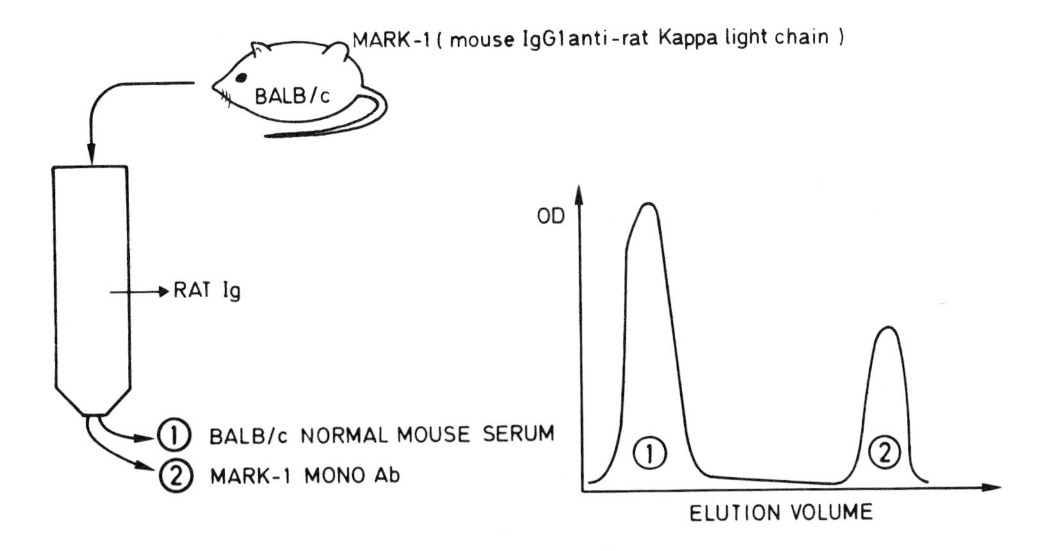

FIGURE 11. Purification method used to purify mouse MAb (MARK-1 or MARK-3) to rat kappa light chain. (From *J. Immunol. Methods,* 1984, 66, 261. With permission.)

- Result of step 1: A first peak is collected. It is composed of the ascitic fluid (or serum) devoid of the molecules bound to the Sepharose-4B-MAb.
- Step 2: Washing of the column with 2.5 *M* NaCl in PBS. This step is necessary to wash molecules bound nonspecifically to the Sepharose. Rarely, can it be sufficient to elute the bound molecules. Then, the column is washed with PBS to return back its salinity to normal.
- Result of step 2: Washings, to be discarded.
- Step 3: Elution with glycin-HCl buffer 0.1 *M* NaCl. The pH of the buffer must be chosen by preliminary experiment. pH between 4.5 and 2.8 are generally used. The column is then washed with PBS.
- Result of step 3: Second peak containing the immunoaffinity purified molecules which must be rapidly neutralized with an appropriate volume of glycin-NaOH buffer (0.1 *M*, pH 8.6).

d. Preparation of Sepharose-4B-Mouse MAb Anti-Rat Ig Immunoaffinity Column

Mouse MAbs anti-rat Ig can be prepared by immunoaffinity purification or by conventional techniques.

i. Anti-Rat Kappa Light Chains

MARK-1 MAb is most generally used. MARK-1 antibodies can be prepared from ascitic fluid or serum of BALB/c mice bearing the MARK-1 hybridoma. An amount of 0.5 ml of Pristane (2-, 6-, 10-, and 14-tetramethylpentadecane) (Janssen Pharmaceutica, Belgium) injected intraperitoneally a few days prior to the injection of the hybridoma cells is necessary in order to obtain the highest quantities of ascitic fluid and MAb.[27] Ascitic fluid is centrifuged and can be directly applied to a rat Ig-Sepharose column or clarified as described in Section I.C.1.a if necessary.

- Step 1: Ascitic fluid is applied to a rat Ig Sepharose column. MARK-1 MAb contained in 10 to 12 ml at a concentration of 5 to 10 mg/ml can be purified on a column of 100 mg (Figure 11).
- Result of step 1: First peak is composed of normal mouse serum devoid of MAb.
- Step 2: Washing of the column with 20 to 30 ml of phosphate buffered saline (PBS)

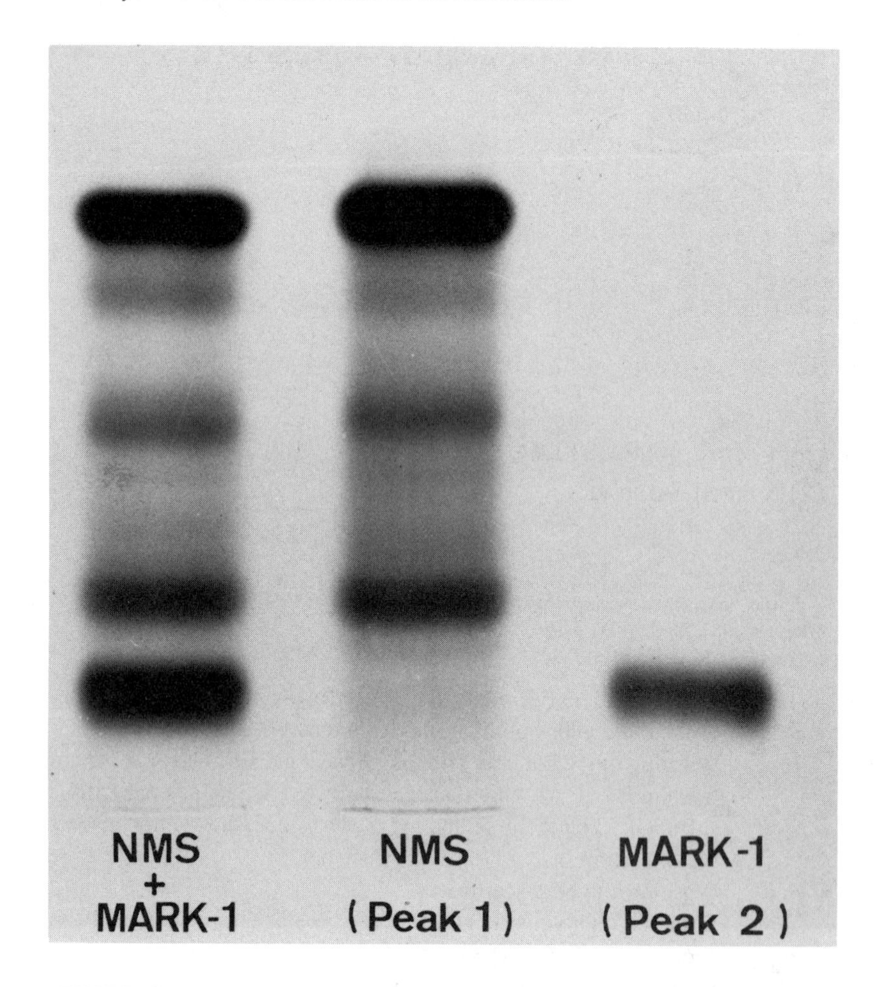

FIGURE 12. Agarose gel electrophoresis of purified MARK-1 (mouse MAb anti-rat kappa light chain) on a rat Ig Sepharose-4B column. From left to right: ascitic fluid containing MARK-1, peak one: rat serum without MARK-1, peak two: MARK-1. (From *J. Immunol. Methods,* 1984, 66, 261. With permission.)

and then the same quantity of 2.5 *M* NaCl in PBS followed by a 20 ml of PBS to return back to the normal salinity.

- Result of step 2: Washing buffer, to be discarded.
- Step 3: Elution with an acidic buffer (glycin-HCl, pH 2.8, 0.1 *M* + 0.15 *M* NaCl) up to the end of the elution peak.
- Result of step 3: Second peak containing the MAb.

This fraction must be neutralized with a glycin NaOH (0.1 *M,* pH 8.6) buffer. After concentration by ultrafiltration (Immersible CX-30, Millipore), the purity of the MAb can be checked by electrophoresis (Figure 12). The MARK-1 MAb is then coupled to Sepharose-4B as indicated in Section I.C.1.b.

ii. Anti-Rat Kappa Allotype Light Chains

The procedure is identical to the one described in the preceding paragraph, but using the correct kappa allotype of the rat Ig column, MARK-3 MAb must be purified on a rat IgK-1a allotype Ig column and LO-RK-1b-1 (or LO-RK1b-2) must be purified on a rat IgK-1b allotype Ig column.

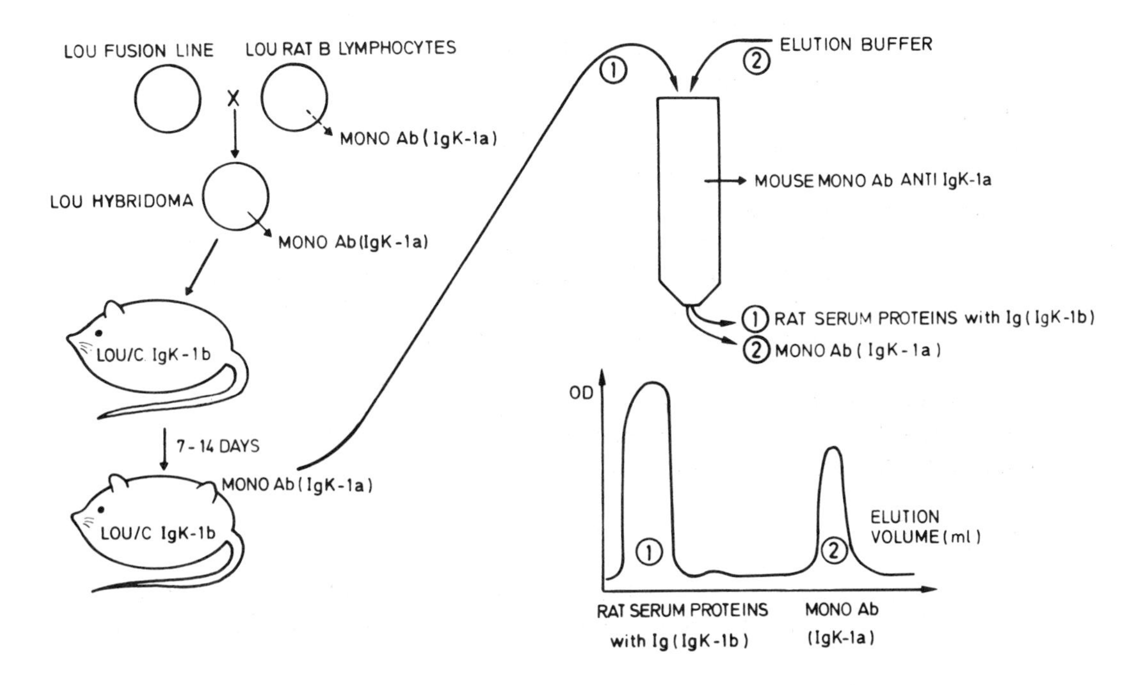

FIGURE 13. Purification method used to purify the LOU monoclonal antibodies from the host serum proteins including its own Ig. (From *J. Immunol. Methods,* 1984, 71, 9. With permission.)

iii. Anti-Rat Lambda Column

MARL-4 (or MARL-5) MAb can be purified by conventional techniques or by immunoaffinity on a column of rat polyclonal Ig (about 5% of lambda for 95% of kappa isotype) or on a column of rat monoclonal lambda Ig, like those synthesized by the IR31 immunocytoma.

iv. Anti-Rat Heavy Chain Isotypes

Mouse MAbs anti-rat Ig isotypes can be purified by conventional techniques. They can also be purified by immunoaffinity on a Sepharose-4B column containing enough rat Ig or a monoclonal rat Ig of the given isotype.

2. Purification of Rat MAbs Using Immunoaffinity Column of Sepharose-4B-Mouse MAb Anti-Rat Ig Light Chains

a. Purification of Rat MAb from Ascitic Fluid or Serum by Immunoaffinity with a Mouse MAb Anti-Rat Kappa Light Chains

Coupled with 1 g of activated Sepharose-4B is 20 to 25 mg of MARK-1. Columns of 150 to 200 mg of MARK-1 are usually used, but smaller MARK-1 columns can also be prepared. As recommended by Goding,[28] it is advisable to "precycle" the immunoaffinity column before each use of the column with the eluting buffer at least when the column has not been used for days or weeks.

- Step 1: A number of milliliters of serum or ascitic fluid of rat serum containing kappa type MAb is applied to a MARK-1 column at a rate of 2 ml/min, at room temperature(a column of 100 mg of MARK-1 can bind about 60 to 70 mg of rat IgG).
- Result of step 1: First peak, rat "normal" serum.

- Step 2: The column is washed with PBS until the chart recorder reaches the baseline and then it is washed again with about half the quantity of PBS containing 2.5 *M* sodium chloride and returned to normal salinity with the same quantity of PBS.
- Result of step 2: Washings.
- Step 3: The pH of the column is decreased by running an acidic buffer (0.1 *M* glycin-HCl + 0.15 *M* NaCl, at pH 2.8).
- Result of step 3: Second peak with the MAb.
- Step 4: The pH of the MAb solution must be returned to neutrality as soon as possible with drop(s) of Tris 1 *M* or by addition of a quantity of glycin-NaOH (0.1 *M*, pH 8.6) buffer. Likewise the pH of the immunoaffinity column must be returned to normal with PBS.

The purity of the MAb is far from perfect, as nearly 95% of the host polyclonal Ig will also be recovered in the second peak. However, the technique is perfectly convenient for mouse-rat hybridoma (of the kappa type) raised in nude mice.

Remark

The antigen binding capacity of the MARK-1 MAb has been found to be equal to 0.62 mg of rat kappa Ig/mg of MARK-1 coupled with Sepharose-4B after ten runs of the studied column.

b. Purification of Rat MAb Using Immunoaffinity with Anti-Kappa Allotype Mouse MAbs

In the rat species, a major Ig allotype is located on the constant part of the kappa light chain.[29-31] Two alleles have been described, the IgK-1a being carried by the LOU/C rat strain and the IgK-1b allotype by the DA or the OKA strains.[30] Hybridomas made by fusing cells of the nonsecreting LOU/C 983 cells[32] and immune cells from LOU/C rats are fully histocompatible with the LOUVAIN (LOU/C or LOU/M) rats. Transplantation of a LOU/C hybridoma into LOU/C.IgK-1b(OKA) or LOU/C.K1b (see Chapter 2) leads to the synthesis of MAb of the IgK-1a allotype in the serum (or ascitic fluid) of the host, which itself secretes polyclonal Ig of the IgK-1b allotype. As described in Chapter 2, LOU/C and LOU/C.K1b are two congenic strains having, respectively, the kappa light chain locus of the IgK-1a and IgK-1b allotype on the same genetic LOU/C background.[30]

As shown in Figure 13, the IgK-1a MAb, or the IgK-1a monoclonal protein synthesized by a LOU/C immunocytoma (see Chapter 3), can be separated from the normal host serum proteins including their own IgK-1b by affinity chromatography with a MAb against rat IgK-1a allotype, immobilized in a Sepharose-4B column. Such an antibody can be produced by the MARK-3 (Mouse Anti-Rat Kappa IgK-1a allele number 3) hybridoma.

- Step 1: 8 to 10 ml of ascitic fluid or serum from LOU/C.K1b rat carrying a LOU/C.IgK-1a hybridoma are applied to a column of 7.5 g of Sepharose-4B on which about 100 mg of MARK-3 MAb had been immobilized, at a rate of 2 ml/min at room temperature.
- Result of step 1: Peak one, host serum proteins.
- Step 2: The column is washed with about 50 to 80 ml of phosphate buffered saline (PBS), then with 50 ml of PBS containing 2.5 *M* NaCl and returned to normal salinity with 50 ml of PBS.
- Result of step 2: Washings, to be discarded.
- Step 3: The elution is obtained by decreasing the pH of the column with glycin-HCl 0.1 *M* plus 0.15 *M* NaCl buffer at pH 2.8.
- Result of step 3: Peak 2, purified rat MAbs.

Peak 2, which contains the MAb, is returned to neutral as early as possible after elution with the help of a glycin-NaOH buffer, 0.1 *M*, pH 8.6. Likewise, the column is washed and returned to neutrality with about 20 ml of PBS (Figure 14).

The same method can be used to purify rat IgK-1a MAb synthesized by a hybridoma transplanted in a nude (rnu/rnu) rat of the IgK-1b allotype or in a nude (nu/nu) mouse.

This methodology can be used with a LOU/C.K1b hybridoma transplanted in a LOU/C rat (having the IgK-1a allotype). In that case, the MAb can be purified on an immunoaffinity column where a MAb anti-rat IgK-1b has been coupled, such as the LO-RK1b-1 or LO-RK1b-2.

It can also be interesting to use the reverse system (Figure 15) and to remove the host polyclonal Ig by an immunoaffinity chromatography which leaves the MAb in the first peak with the host serum proteins except their Ig. The MAb is purified from the other proteins by a conventional method such as a DEAE-chromatography especially when the MAb has a low electrophoretic mobility at pH 8.6, and then will go across the column. This method is interesting when the MAb is rather unstable at the acidic pH necessary for elution during the affinity chromatography. Evidently, these reverse techniques of purification do not consider the 5% lambda Ig, which are contaminants of the first peak, but can be eliminated by the conventional technique of purification if necessary.

Remark

The antigen-binding capacity of MARK-3 coupled with Sepharose-4B has been found to be 0.66 mg of rat IgK-1a immunoglobulins per milligrams of MAb.

c. Purification of Rat MAbs Using Immunoaffinity Column of Sepharose-4B-Mouse MAb Anti-Rat Lambda Light Chains

Rat hybridomas synthesizing MAb carrying lambda light chains are rather rare, but they exist. They can be purified by immunoaffinity chromatography using mouse MAb anti-rat lambda chains such as MARL-4 or MARL-5. Using a direct method, the technique will give about 5% contaminants which are the polyclonal lambda Ig from the host. A reverse method can eliminate 95% of the host Ig (those carrying the kappa light chains) and leave the lambda MAb in the first peak, without acidic shock, to be purified to host non-Ig serum proteins by conventional techniques.

3. Techniques Using Anti-Ig Heavy Chain MAbs

Immunoaffinity chromatography has been proposed as a step in purifying human IgM or IgA from normal human serum.[33] Such a method could perfectly be used for rat MAb using polyclonal antibodies anti-rat Ig isotypes. However, the broad spectrum of avidies of such antibodies is an important disadvantage leading to a very large peak of elution, diluting the MAb in a great volume of acidic buffer which must be neutralized.

It is much easier to use a mouse MAb anti-rat Ig isotype as available in many commercial firms or in cell collections (see Chapter 33).

4. Percentage of Recovery Using Immunoaffinity Methods

The percentages of recovery given in Table 5 have been established by comparing the quantities of MAbs recovered in the concentrated (by Millipore CX30 or by ammonium sulfate precipitation) eluates of the chromatography purification. The three methods give a purity which is approximately identical and clearly superior to 95%. In fact, no contaminant can be detected by agarose or PAGE electrophoresis when the immunoaffinity column is still in good shape. With time and repeated uses, the column can bind and release contaminants, and in that case the column must be discarded. In the case of the column "antigen", the Sepharose-4B was coupled with mouse myeloma IgG2b and IgG2a, respectively. The second method using a column of

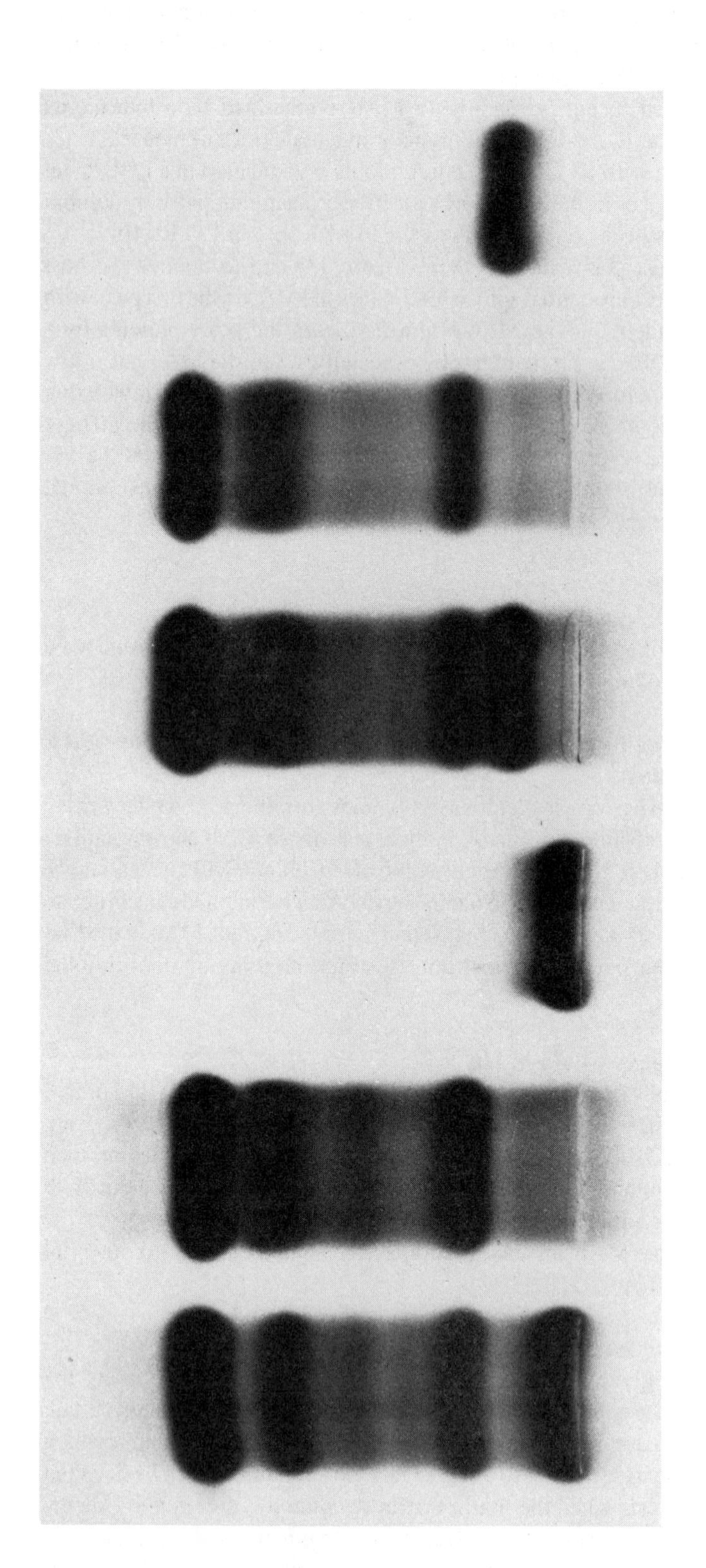

FIGURE 14. Agarose electrophoresis of immunoaffinity chromatography purified rat IR731 IgD myeloma protein (see Chapter 3) and rat LO-DNP-2 MAb (rat IgG1 anti-DNP); from left to right: serum of IR731 carrying LOU/C.IgK-1b rat; peak 1: rat serum; peak 2: IR731 protein; serum of a LO-DNP-2 (rat IgG1 MAb anti-DNP) carrying LOU/C.IgK-1b rat; peak 1: rat serum; peak 2: LO-DNP-2 MAb. (Reprint, *J. Immunol. Methods*, 1984, 71, 9. With permission.)

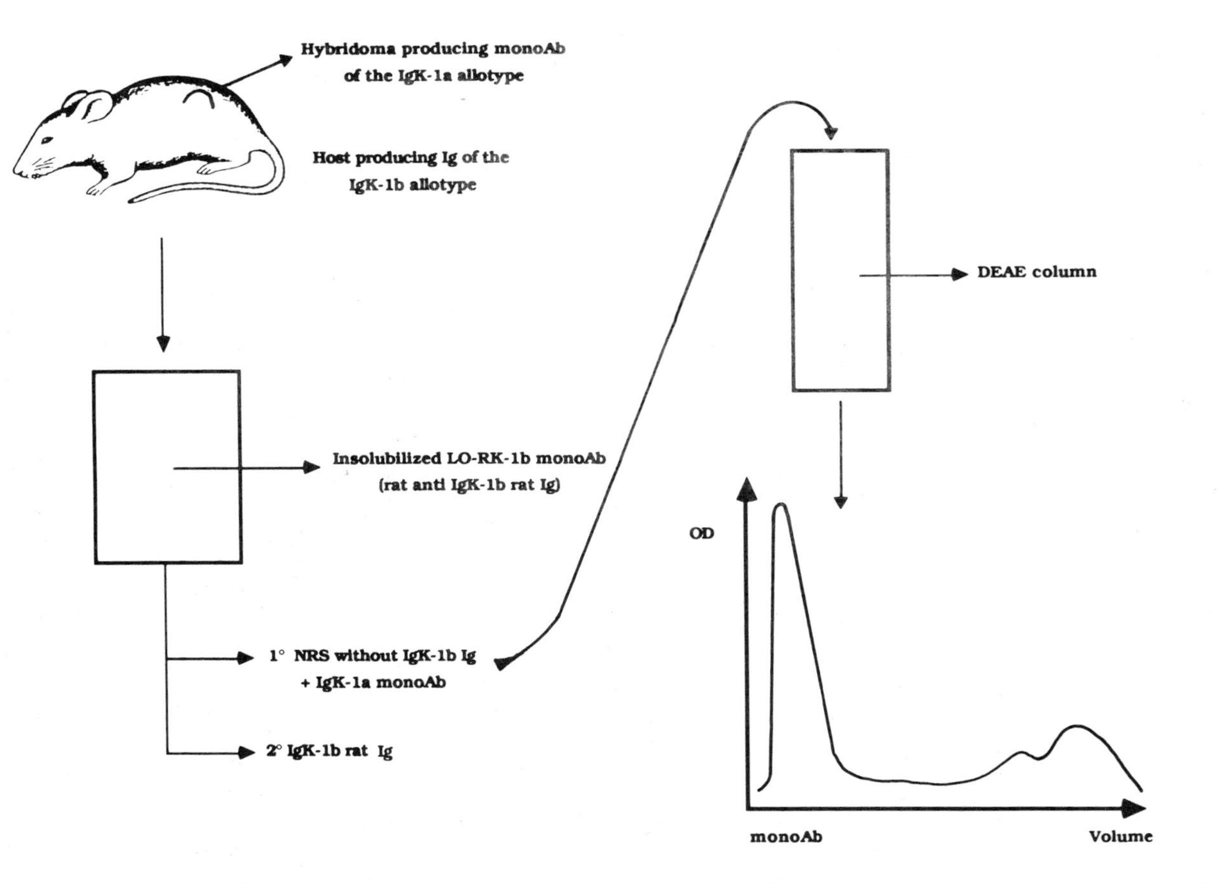

FIGURE 15. Purification method to purify LOU monoclonal antibodies by reverse immunoaffinity chromatography. The first step removes the host Ig from the bulk of the other proteins. By a conventional technique such as DEAE-chromatography, it is possible to purify the MAb from them.

TABLE 5
Percentage of Recoveries of Purified Rat MAbs by Immunoaffinity Chromatography

			Immunoaffinity chromatography on Sepharose-4B[c]		
MAbs	Isotype	Antigen	Anti-rat Ig isotope + anti-host kappa allotype[a]	Kappa allotype (MARK-3 column)[a]	Kappa allotype (MARK-3 column)[b]
LO-MG2b-2	IgG1-kappa	63	63	78	93
LO-MG2a-2	IgG2a-kappa	76	ND	73	80
LO-DNP-57	IgG2b-kappa	ND	59	73	ND

[a] Concentrated by Millipore CX30.
[b] Concentrated by ammonium sulfate precipitation and dialysis.
[c] See text for explanation.

Sepharose-4B coupled with MAb anti-rat IgG1 and IgG2b, respectively, followed by an affinity chromatography using the LO-RK1b-2 MAb to eliminate the host polyclonal Ig. The third method was the classical "kappa allotype" technique. The last two methods need a production of the MAb in a kappa congenic LOUVAIN rats (see Chapter 2).

It is certain that these values obtained in a limited number of experiments are only indicative. They include the concentrations of the eluates (Millipore AX30 or by ammonium sulfate precipitation) in volumes identical to those of the various MAbs in the ascitic fluid.

5. Automated Techniques

A technique to automate the immunoaffinity chromatography technique has been described by Bazin and Malache.[34] It operates on a time basis and considers only two parameters: state of the reservoir containing the sample and measurements of the optical density (at 280 nm) of the immunoaffinity column eluate. Each run includes different steps: first wash, injection of the serum, second wash, special high ionic strength washing, third wash, elution, and last wash.

In collaboration with Fabelec (Rue du Peigne d'Or 6, B-5988 Nethen, Belgium), a Belgian firm specialized in electronics, we were able to set up a "programmable multi-source sequencer" which can operate the fluid network of the system with electrovalves (LKB No. 11300 or Pharmacia No. 19.5164.01, Sweden) on a time basis and in function of the UV absorption levels (Uvicord SII and level sensor, LKB, Sweden) related to both the beginning and the end of each peak collection. The details of the technique are given in Bazin and Malache.[34]

With this apparatus, already in use for 3 years, we have been able to run about 2000 to 3000 cycles, without any problem except at the beginning when we had clot formation in the tubes or the valves. Such problems have been discarded by carefully preparing the serum fluid before use. The results are remarkably constant and can be repeated many times without any change (Figure 16).

6. "Batch" Purification of Rat MAbs

In order to quickly purify large amounts of rat MAb (up to 2 g), immunoaffinity technology applied to a "batch" purification can be performed. Mouse MAb anti-rat kappa light allotype 1a (MARK-3) coupled with CNBr-activated Sepharose-4B is used as specific immunosorbent.

Technique

The amount of 800 mg of MARK-3 coupled with CNBr-activated Sepharose-4B are first washed with PBS on a sintered-glass funnel. Then, 250 ml of prepared rat ascitic fluid (1 mg of rat MAb per millilitre of ascitic fluid) are incubated, under constant agitation, with the MARK-

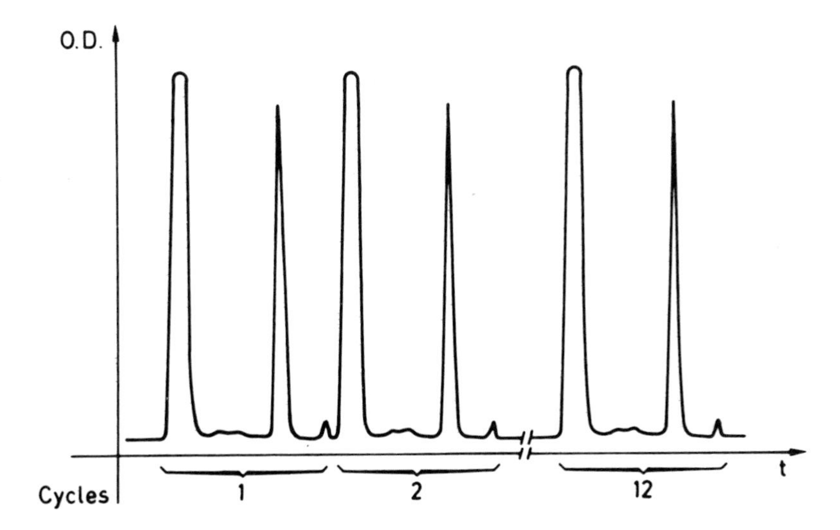

FIGURE 16. Cycles of rat MAb immunopurification, using the allotype MARK-3 method and our automated technique. As an example, the first, the second, and the twelfth cycles of purification of a LO-MG1-2 (rat IgG1-kappa anti-mouse IgG1) MAb.

3 immunosorbent. The normal ascitic proteins and IgK-1b immunoglobulins are eliminated 1 h later by washing with PBS and PBS-NaCl 2.5 *M*. The MAb retained is eluted with a glycine HCl 0.1 *M* NaCl 0.15 *M* pH 2.8 buffer and collected in a neutralizing buffer.

Results

From 1 to 1.5 g of rat MAb per day can be purified according to this methodology and the achievement of such a technique is similar to that obtained by immunoaffinity column (Table 6).

7. Longevity of the Immunoaffinity MAb-Matrix Complex

Clogging of columns by clots and lipids are the main problem of immunoaffinity column survival. Slow denaturation of the MAbs during the elution step at acidic pH must also be taken into account (Figure 17). Ascitic fluid or sera containing high quantities of biliary pigments, hemoglobulin, or lipids are the most dangerous. Attention must be paid to delipidate and to centrifuge these sera (see Section I.C.1). They can also be precipitated by caprilic acid, which is an efficient method, but will decrease the recovery coefficient of the technique.

Columns could be cleaned from nonspecific binding by guanine, 5 *M* or NH_4 SCN 1.2 or 3 *M*. However, these two treatments will, in any case, diminish the antigen-binding capacity of the column.

Moreover, with good and well-clarified sera, the use of the column will slowly give rise to a conglomerate of matrix beads which will also impair the column capacity. A possibility to restore such an immunoaffinity column is to remove the gel from the column into a backer, wash the matrix in NaCl 2.5 *M* two or three times, and then leave in contact for a night with the same solution. Eventually gently stir with glass beads (about 2 mm diameter) for some minutes in order to destroy the aggregates.

Immunoaffinity columns of antigen are very often rather stable. In the case of MAb, the stabilities of the column are certainly lower, but still interesting. There does not seem to be a difference in stability between rat and mouse MAb at the level of species origin, but much more at the level of individual properties of each MAb.

TABLE 6
Examples of Number of Runs Made with Immunoaffinity Columns of Rat Ig or MAbs Anti-Rat Ig

Antigens coupled to Sepharose-4B	Number of runs already done with the column	Columns still in use
LOU kappa Ig	130	no
LOU kappa Ig	132	yes
OKA kappa Ig	58	yes
IgM (IR202)	40	yes
IgG1 (IR595)	39	yes
IgG2b (IR863)	32	yes
MAbs coupled to Sepharose-4B		
MARK-1	69, 216, 170	no
MARK-3	50, 113, 57, 88, 65, 176, 174	no
MARK-3	54	yes
LO-RK-1b-1	109	yes
LO-RK-1b-2	171	no
MARM-4	53	yes
MARG1-1	38	yes
MARG2b-3	30	yes

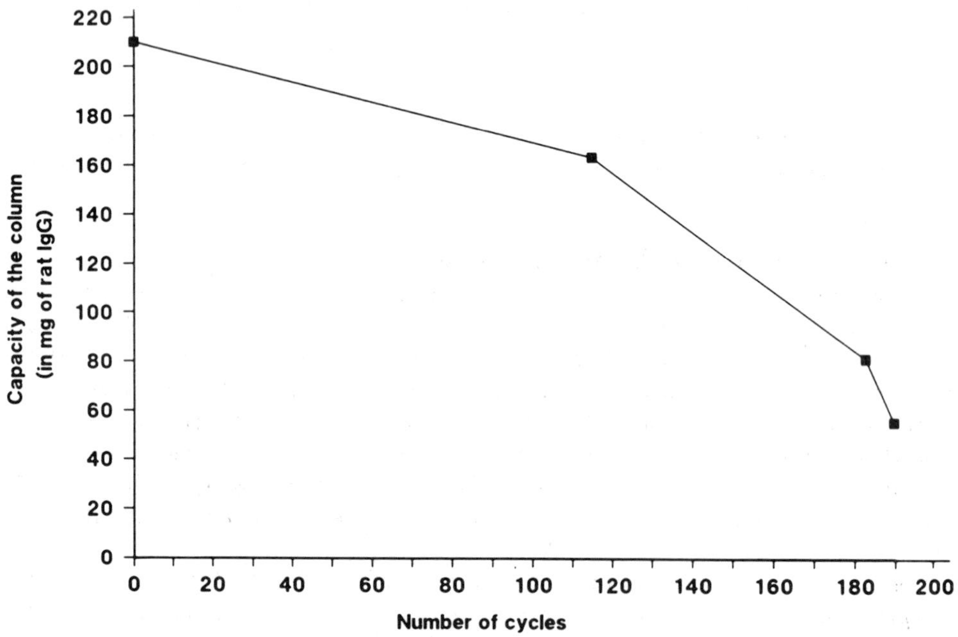

FIGURE 17. Capacity of rat IgG molecules (mg) in a column of Sepharose 4B-MARK-1 monoclonal antibodies (280 mg), as a function of the number of cycles performed. (From *J. Immunol. Methods,* 1986, 88, 19. With permission.)

D. GENERAL STRATEGIES

There is no method of purification that can be used in every circumstance. Clearly the immunoaffinity techniques are very efficient and can be used in most of the cases.

In Tables 7 and 8, some possibilities are considered and can help to find an efficient and well-adapted technique.

TABLE 7
General Strategies of Rat Monoclonal Antibody Purification by Immunoaffinity Chromatography

Immunoaffinity column	MAb or host Ig	Direct or reverse methods	Contaminants	Can be eliminated	Remark
Anti-kappa isotype	Kappa MAb	Direct	Kappa host Ig (95% of the total)	Majority, by conventional technique	—
Anti-kappa allotype	MAb of the correct allotype	Direct	no	—	—
Anti-kappa allotype	Anti-host allotype	Reverse	Lambda host Ig (5% of the total host Ig) + host serum proteins	Majority, by conventional technique	no acidic shock
Anti-lambda	Lambda MAb	Direct	Polyclonal lambda Ig (about 5% of the total host Ig)	Possible	—
Anti-isotype	—	Direct	Polyclonal kappa Ig of the same isotype	Majority, by conventional technique	—

II. PURIFICATION OF RAT MABS FROM *IN VITRO* CULTURE SUPERNATANTS

A. PROBLEMS TO BE CONSIDERED BEFORE ANY PURIFICATION

1. Linked to the Culture Media

The main problems to solve in order to purify rat MAb from *in vitro* culture supernatant are linked to the proteins which are in solution in the medium and especially of calf or horse immunoglobulins when fetal calf serum, newborn calf serum, or horse serum are used. Even after concentration by ultrafiltration or by ammonium sulfate precipitation, it is difficult to purify rat MAb from polyclonal calf or horse immunoglobulins by conventional procedures.

2. Linked to the MAb Itself

See Section I.A.2.

B. PURIFICATION BY CONVENTIONAL PROCEDURES

There are two major possibilities which can be used.

1. DEAE-Chromatography

Ion exchange chromatography is a useful technique for purifying rat monoclonal antibody from *in vitro* culture supernatant, especially IgG MAb. The technique is described in many excellent books and reviews[28] or in the instructions of the manufacturers. However, the method is not suitable for IgM MAb. Moreover, the method does not seem very efficient when calf or horse polyclonal Ig must be carefully removed from the MAb preparation.

2. Protein A Affinity Chromatography

As described for purification of rat MAb from ascitic fluid or serum, protein A affinity chromatography can be used for IgG2c and most of the IgG1 MAb, but not for IgM, IgG2a, or IgG2b MAbs. Moreover, the technique cannot distinguish the MAb from the polyclonal calf or horse immunoglobulins.

The isolation of rat IgG MAbs by protein G affinity chromatography from *in vitro* culture

TABLE 8
General Conclusions; Purification of Rat MAbs from Ascitic Fluids or Sera and Percentage of Recovery and Purity

				Anti light chain				
	Ammonium sulfate precipitation	**Caprilic acid precipitation**	**Both precipitations**	**Kappa isotype**	**Kappa allotype**	**lambda**	**Anti isotype**	**Anti antigen**
Theoretical percentage of recovery of MAb	100%	100%	100%	100%	100%	100%	100%	100%
Theoretical purity	MAb + 100% poly[a]	MAb + 100% poly[a]	MAb + 100% poly[a]	MAb + 95% poly[a]	MAb	MAb + 5% poly[a]	MAb + poly[b]	MAb
Experimental percentage of recovery of MAb	84—87%[c]	42—69	34—52	—	—	—	—	—
Experimental purity	MAb + poly[a]	MAb + poly[a]	MAb + poly[a]	MAb + 95% poly[a]	MAb	MAb + 5% poly[a]	MAb + poly[b]	MAb

[a] Poly: polyclonal Ig from the host.
[b] Depending on the isotype.
[c] Range.

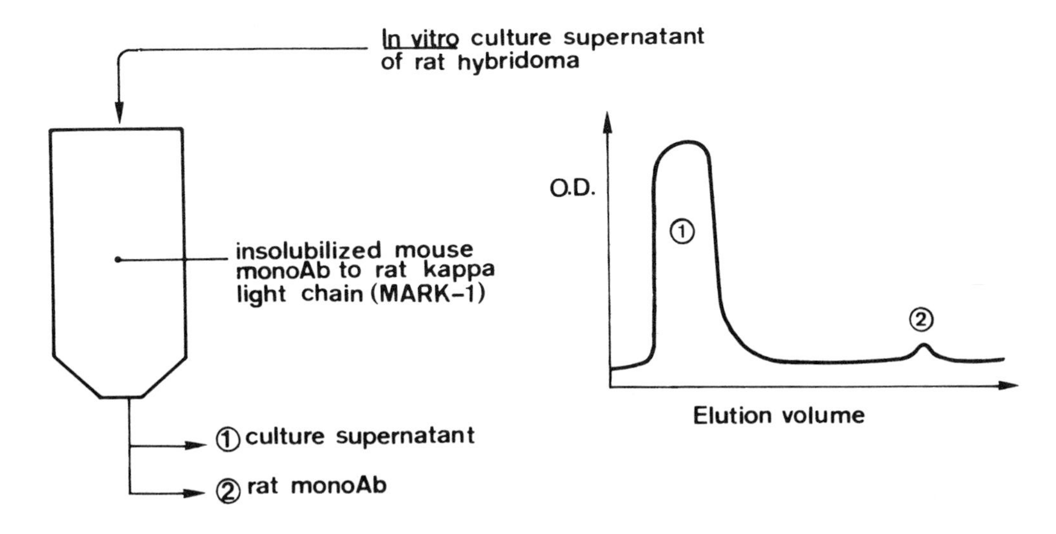

FIGURE 18. Method used to purify rat MAb from *in vitro* culture supernatant of rat hybridoma.

supernatants is possible. However, as protein G binds to bovine IgG1 and IgG2 and horse IgG, bovine or horse Ig will contaminate the purified rat MAbs, except if the cell culture has been performed in a synthetic medium.

C. PURIFICATION BY IMMUNOAFFINITY CHROMATOGRAPHY

Rat MAb can be purified from *in vitro* culture supernatant after concentration by ultrafiltration or even directly. Rat MAb carrying lambda light chains can be purified by using mouse MAb anti-rat lambda chain such as MARL-4 or MARL-5. Rat MAb of the various isotypes can be purified with the corresponding mouse MAb. Rat MAb carrying kappa light chains, i.e., about 95% of them can be purified with MAb anti-rat kappa light chain such as MARK-1 (Figure 18).

1. Procedure Using MAb Anti-Rat Kappa Light Chain[35]

- Step 1: 50 to 500 ml of *in vitro* culture supernatant are applied to a MARK-1 column (50 to 150 mg of MAb) at a rate of approximately 2 ml/min at room temperature.
- Result of step 1: Culture supernatant.
- Step 2: The column is then washed with PBS until the chart recorder (Uvicord S, LKB, Sweden) reaches the baseline, and then with 50 to 100 ml of PBS containing 2.5 *M* NaCl and returned to normal salinity with 100 ml of PBS.
- Result of step 2: Washings to be discarded.
- Step 3: The elution of the MAb can be obtained by decreasing the pH of the column by running an acidic buffer (0.1 *M* glycin-HCl + 0.15 *M* NaCl at pH 2.8). The pH of the eluate is returned to normal pH by adding drops of Tris-HCl buffer (0.1 *M* at pH 8.0) or with a glycin-NaOH buffer (0.1 *M*, pH 8.6). The pH of the column is also neutralized by washing the column with PBS.
- Result of step 3: Purified MAb.

The technique is easy to perform and to reproduce. A MARK-1 column can be used for many runs without extensive loss of the antigen-binding capacity (Figure 17). Many MAb anti-rat kappa light chains (isotype or allotype) can be used as given in the Chapter 33. The quantity of the MAb which is bound to the column will greatly influence the percentage of recovery of the MAb. An important advantage of the technique is its selectivity of binding which gives the possibility to purify the MAb without calf or horse polyclonal immunoglobulins.

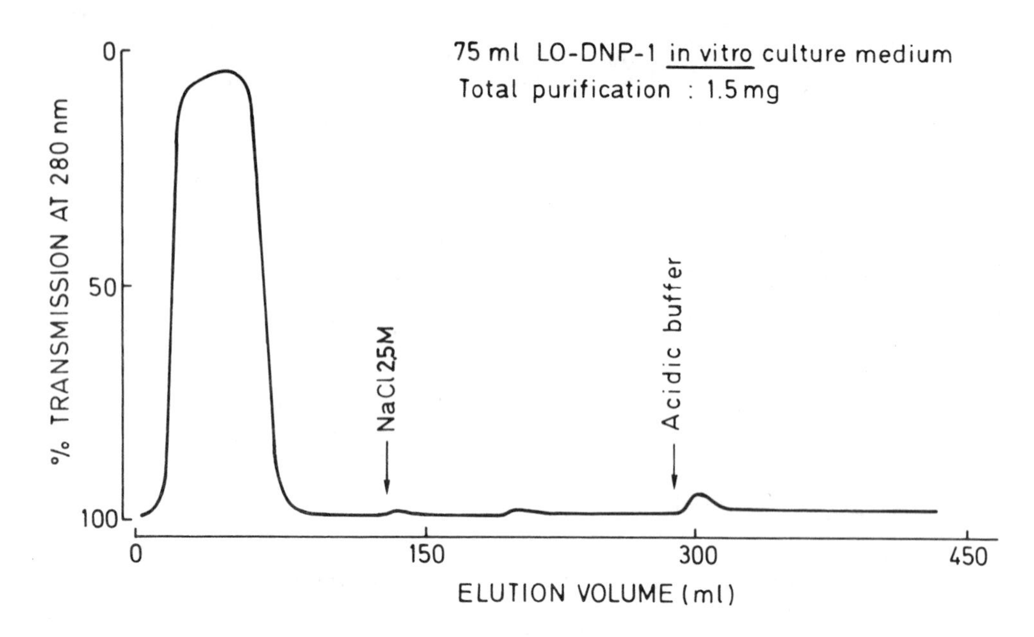

FIGURE 19. Elution pattern of LO-DNP-1 rat IgG1 MAb from an immunoaffinity column of MARK-1-Sepharose-4B.

Figure 19 gives an example of elution pattern of a rat MAb anti-dinitrophenyl hapten. In Table 9, some examples of rat MAb purification are given. The supernatants have been produced by the roller bottle system described in Chapter 13.

2. Procedure Using a Mouse MAb Anti-Rat Lambda Light Chain

When the monoclonal antibody is of the lambda type, it is possible to purify it with an anti-lambda immunoaffinity column. MARL-4 monoclonal antibodies can be used for such purposes. They can themselves be purified on lambda type immunocytoma proteins produced by LOU/C tumors such as IR31, RH58, or RH68 lambda type hybridomas (Lefebvre, Burtonboy, Bodeus and Bazin, unpublished results). When coupled with Sepharose-4B, MARL-4 allows the purification of the lambda type MAb with a great efficiency.

3. Procedure Using a Mouse MAb Anti-Rat Ig Isotypes

Mouse MAb anti-rat Ig isotypes can be used to purify rat MAb of a given isotype. Most mouse MAb anti-rat Ig isotypes do not cross-react with calf or horse Ig; however, it is advisable to check this possibility.

4. Automated Techniques

See Section I.C.4.

III. CONCLUSIONS

Each MAb must be studied before purification in order to know a minimum of its physicochemical properties. Depending on them, a technique must be chosen, keeping in mind what the purpose of the purifications and the various physicochemical properties of the MAb are.

TABLE 9
***In Vitro* Production of Anti-canine Parvovirus Monoclonal Antibodies**

Rat hybridomas	Concentration in the supernatant (μg/ml)	Total production (mg) (300 ml)
RH29	52.4	15.72
RH35	43.3	13.00
RH40	53.2	15.96
RH54	7.1	2.13
RH58	61.2	18.36
RH61	47.0	14.10
RH62	16.5	4.95
RH64	47.8	14.34
RH67	29.9	8.97
RH68	44.5	13.35
RH71	28.6	8.58
		Mean ± SE: 11.77 ± 1.50

Results given by Drs. G. Burtonboy and M. Bodeus (Virology Unit, University of Louvain, Brussels, Belgium).

REFERENCES

1. **Hardie, G. and Van Regenmortel, M. H.,** Isolation of specific antibody under conditions of low ionic strength, *J. Immunol. Methods,* 15, 305, 1977.
2. **McKinney, M. M. and Parkinson, A.,** A simple, non-chromatographic procedure to purify immunoglobulins from serum and ascites fluid, *J. Immunol. Methods,* 96, 271, 1987.
3. **Steinbuch, M. and Audran, R.,** The isolation of IgG from mammalian sera with the aid of caprilic acid, *Arch. Biochem.,* 134, 279, 1969.
4. **McGhee, J. R., Michalek, S. M., and Ghanta, V. K.,** Rat immunoglobulins in serum and secretions: purification of rat IgM, IgA and IgG and their quantitation in serum, colostrum, milk and saliva, *Immunochemistry,* 12, 817, 1975.
5. **Crov, A., Myklestad, B., and Oeding, P.,** Immunochemical studies on antigen preparation from *Staphylococcus aureus.* I. Isolation and chemical characterization of antigen A., *Acta Pathol. Microbiol. Scand.,* 61, 588, 1964.
6. **Oeding, P., Grov, A., and Myklestad, B.,** Immunochemical studies on antigen preparation from *Staphylococcus aureus.* II. Precipitating and erythrocyte-sensitizing properties of protein A (antigen A), *Acta Pathol. Microbiol. Scand.,* 62, 117, 1964.
7. **Sjodahl, J.,** Repetitive sequences in protein A from *Staphylococcus aureus.* Arrangement of five regions within the protein, four being highly homologous and Fc binding, *Eur. J. Biochem.,* 73, 343, 1977.
8. **Sjodahl, J.,** Structural studies on the four repetitive Fc-binding regions in protein A from *Staphylococcus aureus, Eur. J. Biochem.,* 78, 471, 1977.
9. **Forsgren, A. and Sjoquist, J.,** Protein A from *Staphylococcus aureus.* I. Pseudoimmune reaction with human gammaglobulin, *J. Immunol.,* 97, 822, 1966.
10. **Medgyesi, G. A., Fust, G., Gergely, J., and Bazin, H.,** Classes and subclasses of rat immunoglobulins: interaction with the complement system and with staphylococcal protein A, *Immunochemistry,* 15, 125, 1978.
11. **Goding, J. W.,** Use of staphylococcal protein A as an immunological reagent, *J. Immunol. Methods,* 20, 241, 1978.
12. **Ledbetter, J. A. and Herzenberg, L. A.,** Xenogeneic monoclonal antibodies to mouse lymphoid differentiation antigens, *Immunol. Rev.,* 47, 63, 1979.
13. **Fust, G., Medgyesi, G. A., Bazin, H., and Gergely, J.,** Differences in the ability of rat IgG subclasses to consume complement in homologous and heterologous serum, *Immunol.Lett.,* 1, 249, 1980.
14. **Rousseaux, J., Picque, M. T., Bazin, H., and Biserte, G.,** Rat IgG subclasses: differences in affinity to a protein A-Sepharose, *Mol. Immunol.,* 18, 639, 1981.

15. **Nillson, R., Myhre, R., and Kronwall,** Fractionation of rat IgG subclasses and screening for IgG Fc-binding to bacteria, *Mol. Immunol.,* 19, 119, 1982.
16. **Lindmark, R., Thoren-Tolling, K., and Sjöquist, J.,** Binding of immunoglobulins to protein-A and immunoglobulin levels in mammalian sera, J. Immunol. Methods, 62, 1, 1983.
17. **Björck, L. and Kronvall, G.,** Purification and some properties of streptococcal protein G., a novel IgG-binding reagent, *J. Immunol.,* 133, 969, 1984.
18. **Akerström, B., Brodin, T., Reis, K., and Björck, L.,** Protein G: A powerful tool for binding and detection of monoclonal and polyclonal antibodies, *J. Immunol.,* 135, 2592, 1985.
19. **Bazin, H., Beckers, A., and Querinjean, P.,** Three classes and four subclasses of rat immunoglobulins: IgM, IgA, IgE and IgG1, IgG2a, IgG2b, IgG2c, *Eur. J. Immunol.,* 4, 44, 1974.
20. **Carter, P. and Bazin, H.,** Immunology, in *The Laboratory Rat,* Vol. II, Academic Press, New York, 1980, 182.
21. **Cremer, N. E., Taylor, D. O. N., Lennette, E. H., and Hagens, S. J.,** IgM production in rats infected with Moloney leukemia virus, *J. Natl. Cancer Inst.,* 51, 905, 1973.
22. **Oriol, R., Binaghi, R., and Coltorti, E.,** Valence and association constant of rat macroglobulin antibody, *J. Immunol.,* 104, 932, 1971.
23. **Rousseaux, J. and Bazin, H.,** Rat immunoglobulins, *Vet. Immunol. Immunopathol.,* 1, 61, 1979.
24. **Bazin, H., Beckers, A., Urbain-Vansanten, G., Pauwels, R., Bruyns, C., Tilkin, A. F., Platteau, B., and Urbain, J.,** Transplantable IgD immunoglobulin-secreting tumours in rats, *J. Immunol.,* 121, 2077, 1978.
25. **Bazin, H. and Beckers, A.,** IgE myeloma in rats, Nobel Symposium No. 33 *"Molecular and Biological Aspects of the Acute Allergic Reaction",* Stockholm, Johansson, S. G. O., Stranberg, K., and Uvnas, B. Eds., Plenum Company, New York, 1976, 125.
26. **Bazin, H., Beckers, A., Vaerman, J. P., and Heremans, J. F.,** Allotypes of rat immunoglobulins. I. An allotype at the alpha-chain locus, *J. Immunol.,* 112, 1035, 1974.
27. **Hoogenraad, N., Helman, T., and Hoogenaard, J.,** The effect of pre-injection of mice with pristane on ascites tumour formation and monoclonal antibody production, *J. Immunol. Methods,* 61, 317, 1983.
28. **Goding, J. W.,** In *Monoclonal Antibodies: Principle and Practice,* Academic Press, London, 1983, 276.
29. **Wistar, R., Jr,** Immunoglobulin allotype in the rat. Localization of the specificity to the light chain, *Immunology,* 17, 23, 1969.
30. **Beckers, A., Querinjean, P., and Bazin, H.,** Allotypes of rat immunoglobulins. II. Distribution of the allotypes of kappa and alpha chain loci in different inbred strains of rat, *Immunochemistry,* 11, 605, 1974.
31. **Gutman, G. A., Bazin, H., Rockhlin, C. V., and Nezlin, R. S.,** A standard nomenclature for rat immunoglobulin allotypes, *Transplant Proc.,* 15, 1685, 1983.
32. **Bazin, H.,** Production of rat monoclonal antibodies with the LOU rat non secreting IR983F myeloma cell line, in *Protides Biological Fluids, Proc. 29th Colloq.,* 1981, Peeters, E., Ed., Pergramon Press, Oxford, 1982, 615.
33. **Cripps, A. W., Neoh, S. H., and Smart, I. J.,** Isolation of human IgA and IgM from normal serum using polyethylene glycol precipitation and affinity chromatography, *J. Immunol. Methods,* 57, 197, 1983.
34. **Bazin, H. and Malache, J. M.,** Rat (and mouse) monoclonal antibodies. V. A simple automated technique of antigen purification by immunoaffinity chromatography, *J. Immunol. Methods,* 88, 19, 1986.
35. **Bazin, H., Xhurdebise, L. M., Burtonboy, G., Lebacq, A. M., De Clercq, L., and Cormont, F.,** Rat monoclonal antibodies. I. Rapid purification from *in vitro* culture supernatants, *J. Immunol. Methods,* 66, 261, 1984.

Chapter 16

QUALITY CONTROL OF PURIFIED ANTIBODIES

P. Manouvriez

TABLE OF CONTENTS

I. INTRODUCTION

Quality controls must be performed at each step of monoclonal antibody (MAb) production and purification. Agarose electrophoresis and immunoprecipitation tests are routinely performed to detect the contaminants possibly present in a purified antibody solution. When very pure antibody preparations are needed, the more sensitive polyacrylamide gel electrophoresis and ELISA assays are also performed. Finally, comparison with native ascites or culture supernatants of known antibody content is useful for evaluating the degree of activity of the purified antibody.

II. PURITY TESTS

A. AGAROSE ELECTROPHORESIS

Compared with electrophoresis on cellulose acetate membranes or on polyacrylamide gel, the agar gel electrophoresis is relatively less sensitive and resolving. However, it is cheap and very easy to perform. It permits the visualization of the major components of sera or ascites fluids: albumin, alpha2-macroglobulin, transferrin, normal immunoglobulins (Igs), and the band of MAb.[1] In some cases the electrophoretic migration of the MAb is the same as that of transferrin and thus cannot be discriminated by agarose electrophoresis. After each purification step an agarose electrophoresis is performed. The choice of the next purification step depends on the results of this test.

For most uses a MAb preparation which seems pure by agarose electrophoresis is pure enough. Since the lower detection limit of a contaminant is approximately 0.05 mg/ml, and if the MAb concentration is low, a relative high proportion of undetected contaminants may be present. Figure 1 shows agarose electrophoresis performed with normal rat serum, LO-Tact-1 ascites and purified MAbs. After three clonings, LO-Tact-1 still shows two bands in agarose electrophoresis. This phenomenon has also been described by others with mouse MAbs.[2,3]

1. Preparation of the Gel

a. Reagents and Material

1. 0.075 *M* veronal buffer, pH 8.6, 5-diethylbarbituric acid 10.55 g, 5,5-diethylbarbituric sodium salt 65.7 g, Ca lactate $2H_2O$ 2.88 g, dissolve in 5 l
2. Indubiose (agarose EEQ = -0.017, IBF, France)
3. NaN_3
4. 2-l conical flask
5. Magnetic stirrer heater
6. Bunsen
7. 56°C water bath
8. 15-ml tubes

b. Procedure

One liter veronal buffer is brought to 80°C on a magnetic stirrer heater; during stirring 10 g Indubiose are very progressively added. When the gel has acquired a homogeneous consistency, the conical flask is left for 1 h at 56°C, then 1 g NaN_3 is added and the gel is aliquoted in 15-ml tubes.

Once the gel is solidified, the tubes are stored at 4°C until further use.

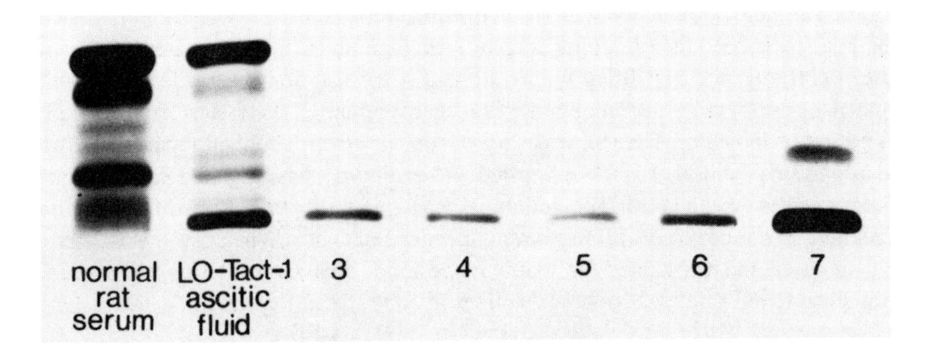

FIGURE 1. Agarose electrophoresis of purified LO-Tact-1 (rat MAb anti-human IL2 receptor) compared with normal rat serum and LO-Tact-1 ascitic fluid. LO-Tact-1 is a double-banded MAb (cloned three times) of the IgG2b subclass with IgK-1a light chain. MAb of lanes 3 to 6 were at 2 mg/ml. Lane 3: first peak of DEAE ion exchange chromatography of immunoaffinity purified LO-Tact-1 from batch of lanes 6 and 7. Lane 4: LO-Tact-1 purified by double precipitation with octanoic (caprilic) acid and ammonium sulfate followed by removal of host immunoglobulins on LO-RK1b-2 (rat MAb anti-rat IgK-1b allotype). Lane 5: LO-Tact-1, same as in lane 6, but lyophylized and reconstituted. Lane 6: LO-Tact-1 specifically purified by immunoaffinity chromatography on MARK-3 (mouse MAb anti-rat IgK1a). Lane 7: same batch than for lane 6, but 15 times more concentrated, i.e., at 30 mg/ml. In this lane, the second MAb band is more easily seen, but also contaminants which are not visible at a lower MAb concentration. The purification techniques are detailed in Chapters 15 and 25.I.

2. Slab Gel Preparation and Electrophoresis

a. Reagents and Material

1. 1% agar in veronal buffer pH 8.6
2. Veronal buffer pH 8.6, 0.1% azide
3. Bromophenol blue
4. Fixative solution; 15 g picric acid in 1 l distilled water is left to stir overnight at room temperature and filtered on grade 1 Whatmann paper. 800 ml filtrate is mixed with 200 ml acetic acid
5. Staining solution; glacial acetic acid 60 ml, Na acetate 8.2 g, Amido-Schwartz 2 g, adjusted to 2 l with distilled water
6. Destaining solution; glacial acetic acid 100 ml, distilled water 1900 ml
7. Filter paper
8. 10.5 × 20 cm clean glass slides
9. Moulding set with a 1-mm-thick gasket
10. Water refrigerated electrophoresis apparatus with power supply
11. Fixing, staining, and de-staining baths
12. Boiling water baths
13. Microsyringe
14. 1-cm-wide filter papers
15. Mask for nine 1-cm slots

b. Procedure

Agar is melted in a boiling water bath, then the cassette is filled and left to gelify for 10 to 15 min until opacification. The cassette is set for 10 min at 4°C and then the cassette is withdrawn. The slots are punched with the help of the mask and 1-cm-wide filter papers. When the buffer has moistened the filter papers up to 1-cm high, the filter papers are withdrawn. The slots are slightly opened and filled with 6 µl to test the solution. One reference containing bromophenol blue is added. Bromophenol blue stains albumin, the fastest migrating protein, used as migration indicator. The electrodes are put on, the electrophoresis apparatus is switched on, and a constant potential of 100 V is applied. When albumin has migrated (approximately 50 min) the current is switched off. The gel, on its glass-plate support, is fixed for 20 min in the fixative solution. It is covered with wet filter paper and left to dry. When dry, it is washed under running tap water, and it is stained for 30 min in the amido-Schwartz solution and destained for a few minutes. Dried agar electrophoresis slides may be stored for years.

Remarks

Electrode tanks are filled with 0.075 *M* veronal pH 8.6 buffer. Polarity is inverted at each run and the cathode applied on the slot side of the slides.

Wicks may be made with 4% agar in the veronal buffer and this buffer may be used for storage.

B. POLYACRYLAMIDE GEL ELECTROPHORESIS (PAGE)

The polyacrylamide gel electrophoresis is a technique that has been used for many years. There are many systems in use; they differ widely in their gel molding techniques and sample application procedures.

For IgG, a conventional electrophoresis in a 7.5% acrylamide gel with 3% cross-linking and in nondenaturing conditions allows running normal serum proteins in their native form. SDS-polyacrylamide gel electrophoresis[4] (10 to 12% acrylamide in presence of SDS and reducing agents like mercaptoethanol or dithiothreitol) gives a good resolution for the light chains (22,000 to 25,000 Da) and the heavy chains (50,000 to 60,000 Da).

Conventional polyacrylamide gel electrophoresis performed in nondenaturing conditions with native proteins produces after staining, results very similar to those obtained by agarose electrophoresis.

After 5 min disulfide bond reduction at 100°C with 2% mercaptoethanol the light and heavy chains are perfectly separated in SDS-PAGE. If some denaturation has occurred during purification procedures, one of the two chains can be entirely denatured and multiple bands of lower molecular weight can appear.

If Coomassie brilliant blue staining is used, PAGE is at least ten times more sensitive than agarose electrophoresis performed as described above. The silver staining (for example, Biorad silver stain Cat. No. 161-0443) has the same sensitivity as autoradiography without the risks inherent in the use of radioisotopes. Less than 1% contaminants can thus be detected in a purified MAb solution using these electropheretic techniques.

C. ISOELECTROFOCUSING

Isoelectrofocusing is a widely used high resolution technique for separating proteins and peptides according to their isoelectric point. MAbs show generally multiple isoelectric points. This has been explained by Staines[5] as the result of variations in glycosylation. Each MAb has its typical isoelectrophoretic pattern. This technique can thus be used to ascertain the identification of the MAb.

D. IMMUNOPRECIPITATION ASSAYS

The sensitivity of the immunoprecipitation techniques in flat agar gel depends on the quality of the antisera used. Antigens can be detected in a 2 µg/ml to 2 mg/ml range. They have been described in detail by Crowle.[6]

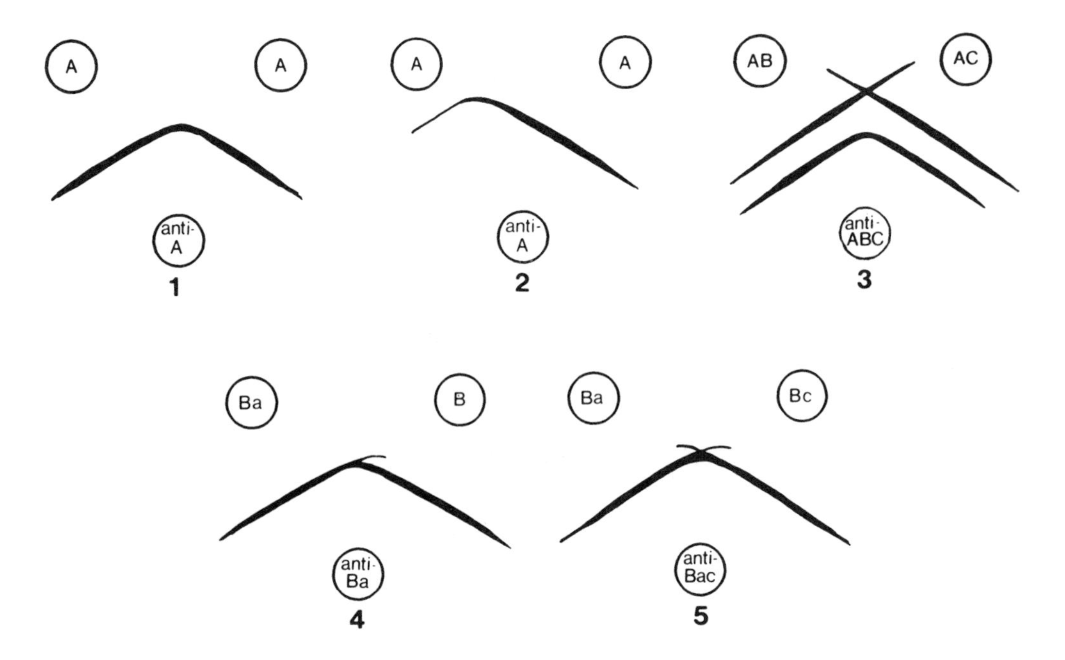

FIGURE 2. Precipitation patterns commonly observed in double radial immunoprecipitation tests in which two antigen solutions are compared using an antiserum as an analytical agent. Pattern 1 (of identity) is observed when the compared antigens (A) are identical serologically and are used in equal concentrations against their specific antiserum (anti-A). A skewed pattern results when one of the two same antigens is less concentrated (2). When two different antigens are compared using an antiserum which contains antibodies against both of them, each antigen-antibody system precipitates independently of the other, so that the resulting precipitin bands cross in the pattern of nonidentity. Often, identity and nonidentity patterns are present simultaneously as multiple precipitin bands (3). An antigen which is similar enough to another to be capable of precipitating some antibodies in the antiserum against the latter (the homologous antigen) will form a precipitin band which is arrested at its juncture with the band formed by the homologous antigen. The latter band, however, continues to grow, forming a "spur" (pattern of partial identity or partial intersection), whose length is inversely proportional to how closely related the cross-reacting antigen is to the homologous antigen and whose curvature and faintness are directly related to this relationship (4). A double spur forms if two antigens are compared which are different but are related to a third antigen, and antiserum to this third antigen is employed (5). Modified from Crowle, A. J., *Immunodiffusion,* 1961, 333.

1. Double Radial Immunodiffusion

Double radial immunodiffusion is the simplest test to carry out. Agar gel (1 to 2% in pH 8.5 buffer) is poured on a slide and wells are punched. One well is filled with the antigenic solution, another with the antiserum, and a third is filled with a standard. If MAbs are used as antiserum, PEG 6000 is added to the gel to enhance the appearance of precipitation lines after overnight incubation at room temperature.

Multiple precipitation lines appear when multiple proteins are present in the antigenic solution. Their interpretation requires in each test the use of adjacent wells filled with solutions of known compositions to interpret the precipitation lines as shown in Figure 2.

The materials and procedure of double radial immunodiffusion have been described in Chapter 10. The diameters of the wells will be adapted to the antigen and/or the antibody concentrations. We mostly use 2.5-, 3.8-, or 5-mm diameter wells. The most sensitive tests can be carried out when the concentrations of test solutions and the diameters of wells have been adapted so that precipitation lines appear approximately at mid-distance between the antigen and the antibody wells.

2. Immunoelectrophoresis

The immunoelectrophoresis technique, invented by Grabar and Williams[7] combines the advantages of gel electrophoresis with those of immunoprecipitation in gel.

Electrophoresis is performed on the MAb solution to be tested. The groove is filled with a polyclonal antiserum directed against whole rat serum or directed against specific rat Ig isotypes. The results of such a test are shown in Figure 3 and the typical pattern of rat Igs in normal serum is shown in Figure 2 of Chapter 1.

E. ELISA OR RIA

ELISA or RIA are very sensitive tests which can detect trace contaminations at concentrations lower than 10 ng/ml. Purity of this order is practically impossible to obtain in one step with antibodies purified from ascites fluid or serum. Even with culture supernatants, some contaminants may come from the horse or fetal calf serum used. These problems will be eliminated with entirely synthetic culture media.

In each case, the test will be adapted to the contaminations which have to be detected.

III. SPECIFICITY TESTS

At any time, a hybridoma may have mutated and its specificity or that of a subclone may have been modified. Moreover, labeling errors can also occur and then the wrong hybridoma could be processed. The minimal specificity test is to be performed against the antigen. This is also the case for activity tests as described. If there is any doubt about an unexpected band in electrophoresis or immunoprecipitation among others, but also at regular intervals a whole battery of specificity tests should be used. This battery is not different from that used for the screening of newly developed hybridomas, therefore it will not be described in this section.

IV. ACTIVITY TESTS

The minimal activity test to perform consists in establishing the optimal use concentration. Activity is, however, usefully checked when purification procedure of a new MAb is developed. Indeed, some treatments, such as acidic pH or euglobulin precipitation, can be harmless for some MAb, but can inactivate others to a lesser or greater extent.

Remaining activity of new MAb will also be tested after labeling repeated thaw-freezing cycles and long-term storage of culture supernatant, ascites fluid, and purified antibody.

The use concentration of an antibody depends on its antigenic specificity and on the test in which it is used. Therefore, each user will have to adapt an activity test to his own requirements to optimalize his assays.

Only, general guidelines are given here. The reader should refer to Parts IV and V of this book or to the specialized literature of his particular field to choose the appropriate activity test.

Activity will never be expressed in mass units per milliliter. A titer will be found as the result of the comparison with the activity of a standard. This standard will generally be used in a native form, i.e., in culture supernatant, ascites fluid, or serum. They must be aliquoted and preserved at -70°C until further use. The antibody concentration of the standard will be measured as described in Chapter 26. The reduction of activity caused by one thaw-freezing cycle will also be quantified to give a correction factor to the real native activity of the standard. A thawed aliquot of standard will never be frozen again. Repeated thaw-freezing cycles are very harmful to the antibodies. The choice of the standard is very important. An inappropriate choice will result in artifacts and over- or under-estimation of the antibody activity. Ascites fluid will be taken from healthy animals and will be clear (not hemorrhagic or charged with billirubin). Culture supernatant must not be yellowish. This ascites fluid or these culture media will probably not contain great concentrations of MAb, but the proportion of denatured antibodies will be very low.

The MAb content of the standards must be determined. It can be measured by chromato-affinity purification or by competitive-binding assay with MARK-3, a mouse anti-rat kappa-1a allotype MAb (see Chapters 15 and 25).

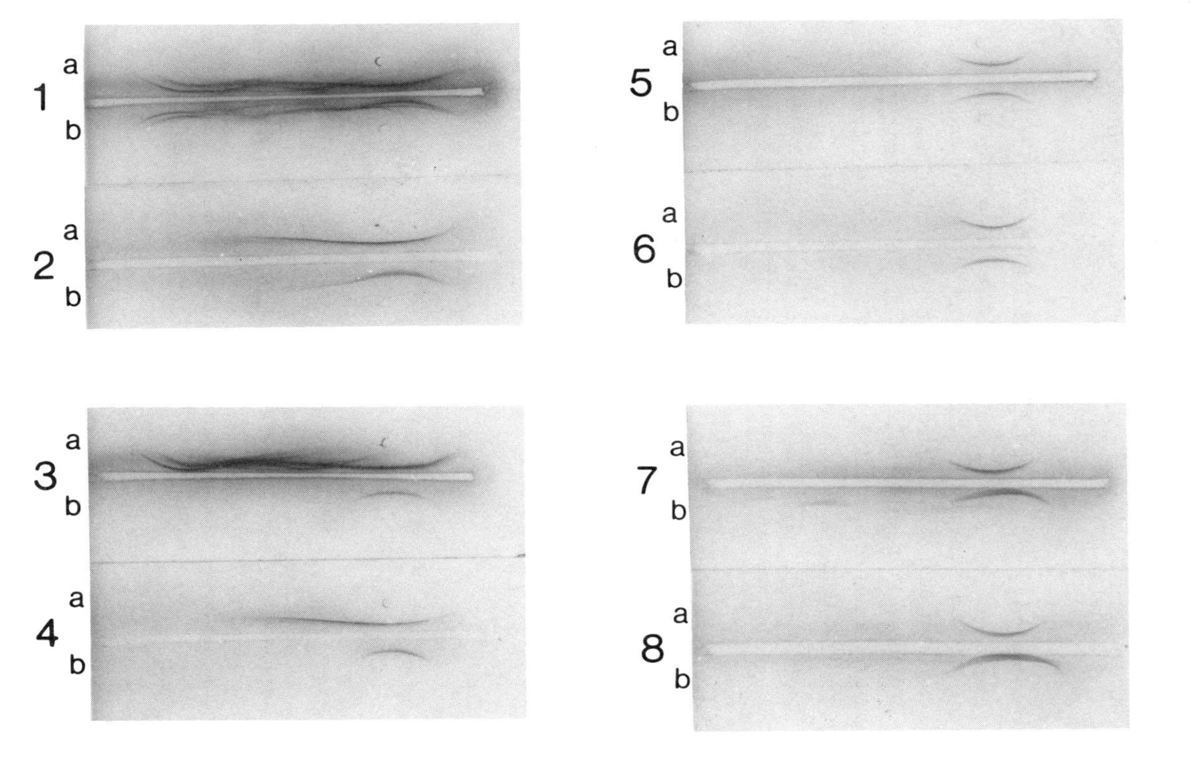

FIGURE 3. Immunoelectrophoretic analysis of normal rat serum and of different purified preparations of LO-Tact-1 (see Figure 1 for the agarose electrophoresis of the same samples and details of purification). Each sample has been tested against two different antisera, goat anti-rat whole serum (1, 3, 5, 7) and against rabbit anti-rat IgG2b (2, 4, 6, 8). 1a,2a,3a,4a: normal rat serum; 1b, 2b: LO-Tact-1 ascites; 3b, 4b: first peak of DEAE ion exchange chromatography of immunoaffinity purified LO-Tact-1 of slides 7 and 8; 5a, 6a: LO-Tact-1 purified by double precipitation with octanoic acid and ammonium sulfate followed by removal of the immunoglobulins of the host on LO-RK-1b-2 (rat MAb anti-rat IgK-1b allotype); 5b, 6b: immunoaffinity purified LO-Tact-1, same as in slides 7 and 8, but lyophylized and recontituted; 7a, 8a: LO-Tact-1specificallypurified by immunoaffinity chromatography on MARK-3 (mouse MAb anti-rat IgK-1a); 7b, 8b: same batch than for 7a and 8a, but 15 times more concentrated, i.e., at 30 mg/ml. At this concentration, some contaminants are visible which were not at a lower MAb concentration. The purification techniques are detailed in Chapters 15 and 25.I.

A. TITRATION OF THE ACTIVITY OF ANTIBODIES DIRECTED AGAINST SOLUBLE ANTIGENS

96-well microtitration plates are coated with the antigen and are incubated with serial dilutions of determined concentrations of purified antibody and of a standard. The amount of bound antibodies is measured with labeled mouse anti-rat light chain MAbs or with labeled polyclonal anti-rat Ig antisera which can be obtained from several manufacturers, but their specificity will be checked before use to see if there is a risk of cross-reaction with the antigen. The comparison of the standard and test curves gives the antibody activity (may be in percentage).

B. TITRATION OF THE ACTIVITY OF ANTI-CELL MEMBRANE ANTIGEN ANTIBODIES

Fixed numbers of target cells are incubated in serial dilutions of standard and purified antibodies. Washed cells are incubated with labeled polyclonal anti-rat Ig antibodies (a heterologous serum is used to improve the positive signal). The labeled cells can then be studied. When RBA (radiobinding assay) or ELISA are used only the total amount of bound antibodies can be measured. The immunofluorescence labeling permits an analytical study by microscope observation, but cytofluorography allows a more detailed study of the percentages of labeled cells and the mean fluorescence intensity, which are two useful parameters.

V. SPECIAL POINTS TO CONSIDER FOR INJECTABLE PRODUCTS

Hoffman[8] has described in detail several points to consider in the manufacture of injectable MAbs. All precautions will be taken for purity, absence of pathogens, toxins, and pyrogens. In addition, rat MAb preparations will also be screened for the absence of Hantaan virus which can provoke from asymptomatic infections to lethal diseases in humans.[9]

REFERENCES

1. **Weber, K. and Osborn, M. J.,** Reliability of molecular weight determination by dodecyl sulfate-polyacrylamide-gel electrophoresis, *Biol. Chem.*, 244, 4406, 1969.
2. **Chaffotte, A. F., Djavadi-Ohaniance, L., and Goldberg, M. E.,** Does a monospecific hybridoma always secrete homogeneous immunoglobulins?, *Biochimie*, 67, 77, 1985.
3. **Lee, S. M.,** Affinity purification of monoclonal antibody from tissue culture supernatant using Protein A-Sepharose CL-4B, in *Commercial Production of Monoclonal Antibodies, A Guide for Scale-Up*, Seaver, S. S., Ed., Marcel Dekker, New York, 1987, 199.
4. **Laemmli, U. K.,** Cleavage of structural proteins during the assembly of the head of bacteriophage T4, *Nature*, 227, 680, 1970.
5. **Staines, N. A.,** Monoclonal antibodies, in *Biochemical Research Techniques*, Wrigglesworth, J. M., Ed., John Wiley & Sons, New York, 1983, 177.
6. **Crowle, A. J.,** *Immunodiffusion*, Academic Press, New York, 1961.
7. **Grabar P. and Williams, C.,** Méthode permettant l'étude conjugée des propriétés electrophorétiques et immunochimiques d'un mélange de protéines. Application au sérum sanguin, *Biochem. Biophys.Acta*, 10, 193, 1953.
8. **Hoffman, T.,** Regulatory issues surrounding therapeutic use of monoclonal antibodies, in *Methods of Hybridoma Formation*, Bartal, A. H. and Hirshaut, Y., Eds., Humana Press, Clifton, New Jersey, 1987, 447.
9. **Lee, H. W., French, G. R., Lee, P. W., Back, L. J., Tsucheya, K., and Foulke, R. S.,** Observations on natural and laboratory infection of rodents with the etiologic agent of Korean haemorrhagic fever, *Am. J. Trop. Med. Hyg.*, 30, 1477, 1981.

Chapter 17

FRAGMENTATION OF RAT MONOCLONAL ANTIBODIES

J. Rousseaux and R. Rousseaux-Prévost

TABLE OF CONTENTS

I. INTRODUCTION

Since the first report of the production of a rat monoclonal antibody,[1] many other antibodies with different specificity and from various immunoglobulin classes and subclasses have been obtained. For use in immunochemical assays as well as for therapeutic purposes, the preparation of fragments F(ab′)2 or Fab is often necessary, for example, to avoid the nonspecific binding of antibodies to cell surface receptors. In the present chapter, procedures for the preparation of monovalent fragments, Fab or Fab-like, and bivalent fragments, F(ab′)2 or F(ab)2-like, are described for rat monoclonal antibodies (MAbs). Methods and optimal conditions will mainly be focussed on IgG subclasses. Preparation of proteolytic fragments from monoclonal rat IgE will also be reported.

II. PREPARATION OF PROTEOLYTIC FRAGMENTS FROM RAT IGG SUBCLASSES

A. FAB, FAB(T), AND F(AB′) FRAGMENTS

Papain digestion is the most common method to produce Fab and Fc fragments. However, the optimal conditions vary according to IgG subclass. Digestion with trypsin for MAbs of the IgG2b and IgG2c subclasses, and incubation with pepsin for IgG2c subclass, have also been shown to be efficient methods for preparation of monovalent fragments. Procedures for enzymatic cleavage have already been reported.[2-4] Accent will be made on some new methods used for purification of the fragments.

1. Enzymatic Digestions

a. Papain

Mercuripapain as a suspension in 70% ethanol (Sigma Chemical Co) is the enzymatic preparation in our hands, which gives the most reproducible results. The suspension is vigorously stirred and an aliquot is centrifuged at 4°C (3000 rpm, 10 min). The pellet is dissolved, at a final concentration 1 mg/ml, in digestion buffer: 0.075 *M* sodium phosphate, 0.075 *M* NaCl, pH 7.0 containing 0.01 *M* cysteine. IgG has been previously dissolved or equilibrated against the same buffer at a final concentration of 10 mg/ml, and incubated 30 min at 37°C before the addition of papain. One volume of papain solution is added to 10 volumes of IgG solution, i.e., a final enzyme to protein ratio 1% (w/w). After a 2 to 4 h incubation at 37°C, the digestion is stopped with the addition of iodoacetamide (final concentration 0.015 *M*).

b. Trypsin

IgG, 10 mg/ml in 0.1 *M* Tris-HCl, 0.02 *M* $CaCl_2$ buffer, pH 7.8, is incubated for 30 min at 37°C before digestion with trypsin. Trypsin, TPCK treated (Worthington Biochemicals), is dissolved in the same buffer at a concentration of 2 mg/ml. One volume of trypsin solution is added to 10 volumes of IgG solution (enzyme to protein ratio 2%, w/w). Incubation at 37°C for 4 h is usually sufficient and digestion is stopped with a 2 mg/ml solution (in digestion buffer) of soybean trypsin inhibitor (Sigma Chemical Co) added as 2 to 10 volumes of protein solution.

c. Pepsin

IgG, 10 mg/ml in 0.1 *M* sodium acetate buffer, pH 4.5, is incubated 30 min at 37°C before digestion. Pepsin (Sigma Chemical Co) is dissolved in distilled water at a concentration of 1 mg/ml. One volume of pepsin solution is added to 10 volumes of protein solution (enzyme to protein ratio: 1%, w/w). Incubation is achieved for 18 h at 37°C and stopped by raising the pH to 8 (either with 0.1 *M* NaOH or with 1 *M* Na-phosphate buffer).

TABLE 1
Molecular Weights of the Monovalent Fragments Released by Papain, Trypsin, and Pepsin Digestion of Rat IgG Subclasses (Determination by Polyacrylamide Gel Electrophoresis in SDS)

	Fab	Fc[a]	Fab(t)[b]	Fc(t)[b]	Fab′[b]
IgG1	50 KDa	28 KDa	—	—	—
IgG2a	50 KDa	27 KDa	—	—	—
IgG2b	50 KDa	56 KDa 27 KDa	50 KDa	56 KDa	50 KDa
IgG2c	50 KDa	52 KDa	50 KDa	52 KDa	50 KDa

[a] The molecular weight of Fc from IgG1 and IgG2a corresponds to one C gamma 2-C gamma 3 chain, as for these two subclasses, a noncovalent Fc is produced. In the case of IgG2b a covalent Fc (MW 56 KDa) and a noncovalent Fc (MW 27 KDa) are obtained.
[b] Fab(t) and Fc(t) are not released from trypsin digests of IgG1 and IgG2a. Fab′ is not obtained from pepsin digests of IgG1 and IgG2a.

2. Optimal Conditions According to IgG Subclass

An assay with a small amount of purified MAb is always necessary before preparation of proteolytic fragments from a larger quantity. This assay is performed with various times of digestions from 1 to 18 h. Progress of the fragmentation can be followed by polyacrylamide gel electrophoresis in SDS (PAGE-SDS). The molecular weight of the fragments released by papain (Fab and Fc), trypsin (Fab(t) and Fc(t)), and pepsin (Fab′) is given in Table 1. Identification of the Fab or Fab-like fragments may be completed by immunoblotting using an antiserum specific for rat kappa light chains (such as mouse MAb MARK-1, see below). PAGE-SDS will also show products of incomplete digestion.

As a guide, digestion with papain for 2 h is usually sufficient to cleave IgG2a and IgG2b MAbs into Fab and Fc fragments. A longer time (4 h) is better for digestion of IgG1 and IgG2b proteins (though sometimes incomplete for IgG1). Cleavage with trypsin is an easy method to cleave IgG2b and IgG2c into Fab(t) and Fc(t) fragments (time of digestion 4 h at 37°C). Large amounts of a Fab′ fragment are only obtained from monoclonal IgG2c (18 h of digestion at 37°C are necessary). Then pepsin digestion is a good method for production of a monovalent fragment with antibody activity, from proteins of IgG2c subclass.

3. Purification of Proteolytic Fragments

Purification of Fab or Fab(t), and of Fc fragments released by papain or trypsin cleavage, needs removal of any undigested or partially digested proteins. This may be performed by gel filtration on Ultrogel AcA44 column (IBF, France). Alternatively, small amounts of material (up to 2 mg) may be fractionated on HPLC size exclusion chromatography columns such as TSK G3000 SW (7.5 × 600 mm, Pharmacia-LKB). Conventional methods for fractionation of Fab and Fc fragments are ion-exchange chromatography on DEAE cellulose (Whatman) or DEAE Trisacryl (IBF) columns. Alternative method for IgG2c subclass is the use of protein A Sepharose chromatography.[4]

Nevertheless, the development of a MAb to rat kappa chains named MARK-1,[5] led us to reconsider methods of purification. The following procedure has been found the most convenient.

An immunoaffinity column is prepared by coupling purified MARK-1 MAb to CNBr activated Sepharose 4B (Pharmacia-LKB). For example, 100 mg of purified MAb[5] are coupled to 4 g of activated gel, according to manufacturer instructions. Such an affinity matrice (approximate volume 14 ml) is able to retain up to 20 mg of IgG or proteolytic fragments containing kappa chains, (Fab, Fab(t), and F(ab′)2). A proteolytic digest of 30 mg of IgG (in 6 ml PBS, pH 7.4) is loaded onto the column containing MARK-1 MAb and left in contact during

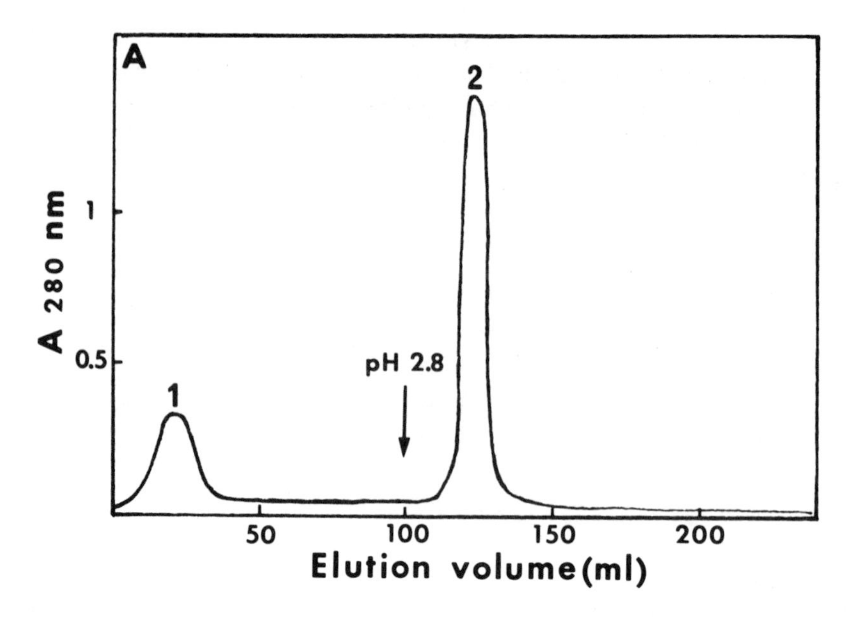

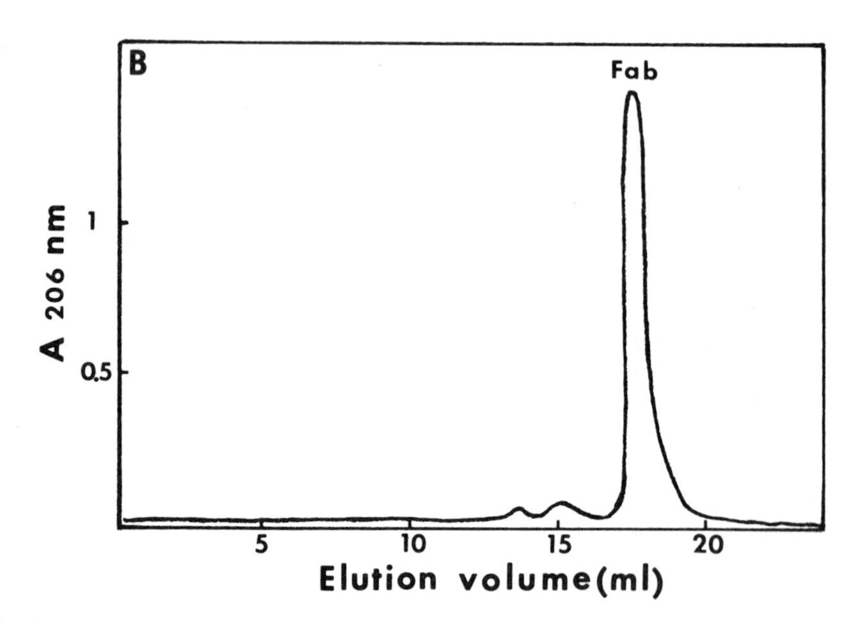

FIGURE 1. Purification of Fab fragment released by papain digestion of rat monoclonal IgG1 IR27 (20 mg). (A) Chromatography on an immunosorbent column of MARK-1 MAb coupled to Sepharose 4B (about 14 ml). Elution with PBS, pH 7.4 containing 0.5 *M* NaCl, followed by elution (arrow) with 0.1 *M* glycine, 0.15 *M* NaCl buffer pH 2.8 (flow rate: 20 ml/h). Peak 1: Fc; peak 2: Fab with some undigested or partially digested IgG. (B) Chromatography of peak 2 (1 mg) on a TSK G3000 SW column (7.5 × 600 mm, Pharmacia-LKB). Flow rate 0.5 ml/min.

2 h at room temperature or overnight at +4°C. The column is then washed (flow rate 20 ml/h) with PBS containing 0.5 *M* NaCl (about 100 ml). The effluent contains essentially intact Fc fragments and some Fc subfragments. The column is eluted with 0.1 *M* glycine-HCl, 0.15 *M* NaCl buffer, pH 2.8 (50 ml). Fractions of 5 ml are collected, with each fraction collector containing 0.5 ml of 1 *M* Tris-HCl pH 8.5 (Figure 1A). The eluate contains Fab, (or Fab(t)), as well as some undigested or partially digested IgG. This contaminating material is eliminated by gel filtration chromatography. A column of Ultrogel Aca44 (1.6 × 90 cm) equilibrated in PBS

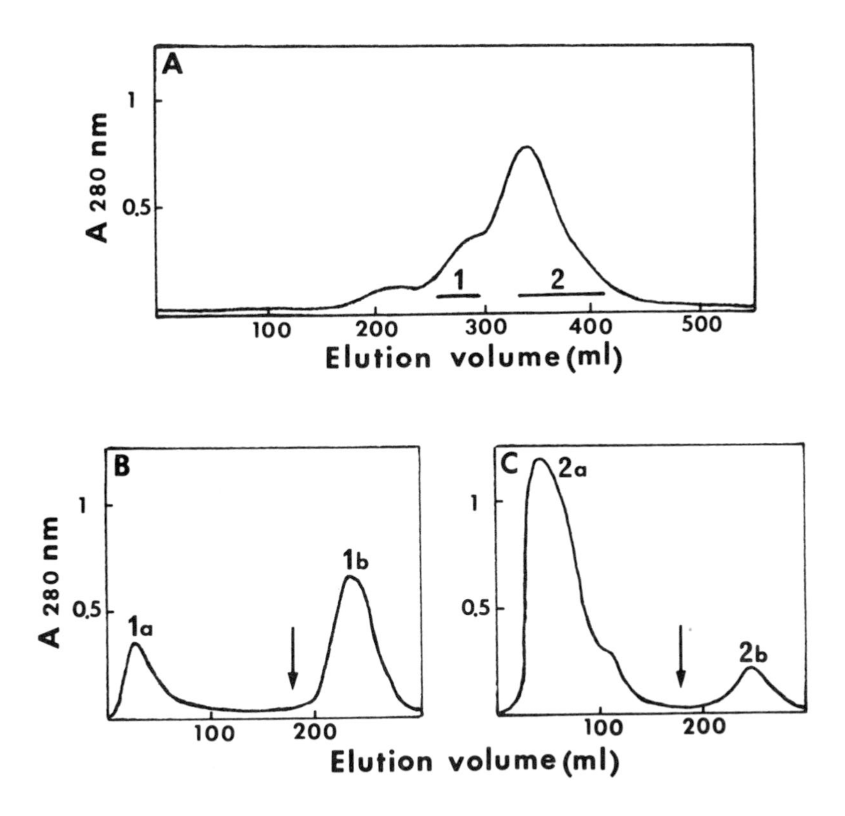

FIGURE 2. Purification of Fab fragment released by papain digestion of rat monoclonal IgG2a IR 418 (50 mg). (A) Gel filtration on Ultrogel AcA44 column (2.6 × 90 cm, flow rate 15 ml/h) in PBS, pH 7.4. (B) Purification of peak 1 by DEAE cellulose chromatography (column 2 × 15 cm; flow rate 15 ml/h). Elution with 0.01 *M* Na-phosphate buffer pH 7.8, followed by 0.2 *M* Na-phosphate buffer, pH 6.8 (arrow). Peak 1a: Fab; peak 1b: Fc. (C) Purification of peak 2 on a DEAE cellulose column (experimental conditions as in B). Peak 2a: Fab; peak 2b: Fc.

pH 7.4 may be used. The column is loaded with the material eluted from the immunoaffinity column concentrated to a volume of 2 to 3 ml. Elution is performed with PBS, pH 7.4 at a flow rate 6 ml/h. Alternatively, small amounts of material eluted from the immunoaffinity column may be purified by HPLC gel filtration chromatography on a TSK G3000 SW column (7.5 × 600 mm, Pharmacia-LKB). Up to 2 mg (in 0.2 ml) are injected into the column and fractionation is performed at a flow rate 0.5 ml/min. Fractions of 0.5 ml are collected (Figure 1B).

If the immunoaffinity column of MARK-1 MAb is not available, gel filtration chromatography followed by ion-exchange chromatography may be used. Such a procedure has already been described.[4] Compared to the method described above it gives lower yields of material. Briefly, IgG (50 mg) digested with papain or trypsin is loaded on an Ultrogel AcA44 column (2.6 × 90 cm). Undigested or partially digested IgG is eluted before Fab and Fc. This gel filtration step gives a partial separation of Fc or Fc(t) and Fab or Fab(t) fragments; Fc being eluted before Fab (Figure 2). Fractions corresponding to Fab are pooled, concentrated, and equilibrated in 0.01 *M* Na-phosphate buffer pH 7.8. Ion-exchange chromatography is performed on a DEAE-Trisacryl (IBF) column (2 × 15 cm), and equilibrated in 0.01 *M* Na-phosphate buffer, pH 7.8. The following buffers are used for elution: equilibration buffer, then 0.01 *M* Na-phosphate pH 6.8, 0.05 *M* Na-phosphate pH 6.8, and 0.2 *M* Na-phosphate pH 6.8. Fab or Fab(t) are usually eluted with the first or second buffer (Figure 2).

For monoclonal IgG2c, which have high affinity for protein A, the Fab from a papain digest can be purified after gel filtration on Ultrogel AcA44 by chromatography on a protein-A Sepharose column (6 ml) equilibrated in 0.14 *M* Na-phosphate buffer pH 8. Fab is eluted with equilibration buffer, while Fc is eluted with 0.1 *M* Na-citrate buffer, pH 3.

The purification of Fab′ fragment released by pepsin digestion of IgG2c is performed by gel filtration, either on a conventional column such as Ultrogel AcA44 or an HPLC support such as TSK G3000 SW. This chromatographic procedure is sufficient to separate Fab′ fragment from remaining F(ab′)2 or undigested IgG.

B. F(AB′)2 AND F(AB)2 FRAGMENTS

F(ab′)2 fragments are usually prepared by pepsin digestion. However, in the case of IgG2b monoclonal antibodies, cleavage with *Staphylococcus aureus* V8 proteinase is the best method to produce a F(ab)2 fragment.

1. Enzymatic Digestions

a. Pepsin

The most important feature for pepsin digestion of rat IgG subclasses is the resistance of IgG2a and IgG1 MAbs to pepsin cleavage. This difficulty can be overcome if MAbs of these two subclasses are first incubated at acid pH before digestion with pepsin.[4] The following procedure is used: IgG (about 20 mg/ml) in 0.01 *M* Na-phosphate, 0.15 *M* NaCl buffer, pH 7.4 is equilibrated at pH 2.8 by dialysis at 4°C against 0.1 *M* Na-formate buffer (three changes of 100 volumes of buffer each, for 1 volume of IgG solution). Alternatively, the solution may be adjusted to pH 2.8 by chromatography on a Trisacryl GF05 column equilibrated in Na-formate buffer, pH 2.8. After incubation at pH 2.8, the solution is equilibrated at pH 4.5, by dialysis against 0.1 *M* Na-acetate buffer, pH 4.5. The concentration of the IgG solution is adjusted to 10 mg/ml and digestion is performed for 4 h at 37°C (enzyme to protein ratio, 1% w/w). Such a procedure is not necessary for IgG2b and IgG2c monoclonal proteins. Optimal conditions are enzyme to protein ratio 5%, 18 h at 37°C for IgG2b, and enzyme to protein ratio 1%, 4 hr at 37°C in the case of IgG2c.

b. **Staphyloccus aureus** *V8 Proteinase*[6]

Monoclonal IgG2b are cleaved by this enzyme into F(ab)2 and Fc-like fragments (with some additional Fab-like fragments). IgG (10 mg/ml) in 0.1 *M* Na-phosphate, 0.002 *M* EDTA, pH 7.8 is incubated at 37°C for 30 min, then *S. aureus* V8 proteinase (Miles Laboratories) (1 mg in 0.3 ml buffer) is added as 1 volume of enzyme to 10 volumes of IgG (i.e., enzyme to protein ratio 1:30, w/w). After 4 h at 37°C, the digestion is stopped by quickly freeezing the solution. The material is stored at –20°C.

2. Controls of Enzymatic Digestion

Controls are performed by polyacrylamide gel electrophoresis in SDS. The molecular weight of F(ab′)2 is about 100,000 Da. IgG2b cleaved with *Staphylococcus aureus* V8 proteinase gives fragments F(ab′)2 (MW 105,000) and Fc (MW 27,000). A pFc′ fragment (MW 12,000) is released from pepsin digestion of IgG2a, IgG2b, IgG2c, but not of IgG1.

3. Purification of Proteolytic Fragments

As for Fab or Fab-like fragment, the use of an immunosorbent with MARK-1 MAb is the most easy method to purify F(ab′)2 or F(ab)2-like fragments. This method gives excellent yields (more than 70% recovery). It must be followed by gel filtration on Ultrogel AcA44 column or HPLC TSK G3000 SW column to remove either undigested IgG or any Fab′ or Fab-like fragment.

Alternatively gel filtration either on conventional or on HPLC support may be used. Figure 3 shows an example of separation of proteolytic fragments released from monoclonal IgG2c: F(ab′)2 or Fab′ (according to the protocol of digestion with pepsin).

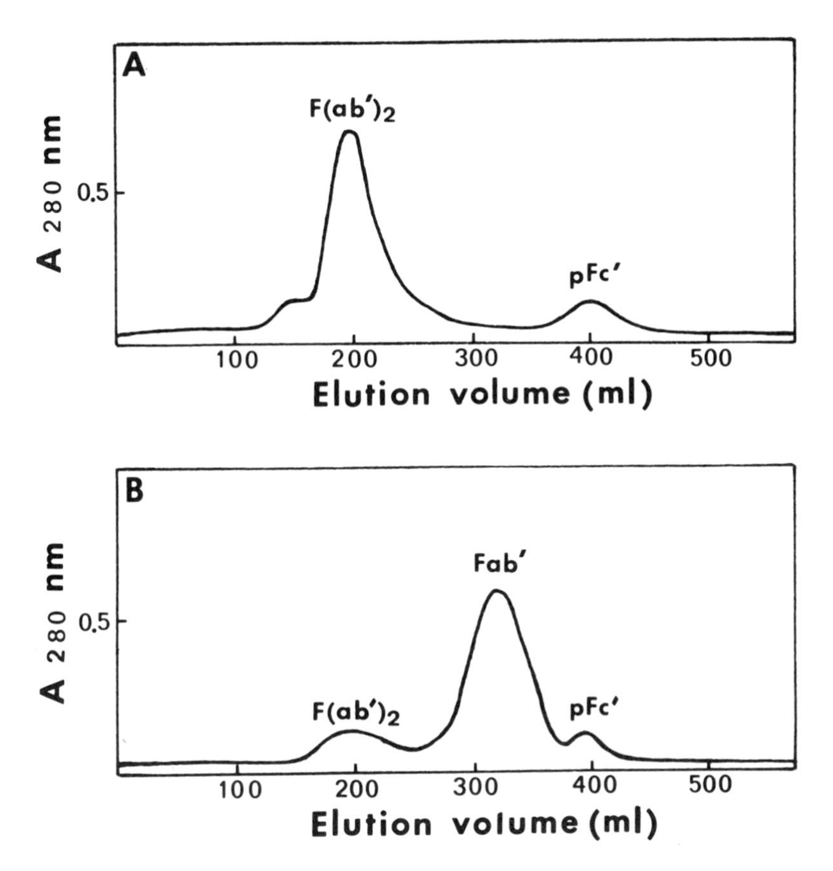

FIGURE 3. Purification of F(ab′)2 and Fab′ fragments released by pepsin digestion of monoclonal IgG2c IR 304. Chromatography on Ultrogel AcA44 column (2.6 × 90 cm; flow rate 15 ml/h) in PBS, pH 7.4. (A) Fractionation of IR 304 after digestion with pepsin (enzyme to protein ratio 1%) for 4 h at 37°C. (B) Fractionation of IR 304 after digestion with pepsin (enzyme to protein ratio 5%) for 18 h at 37°C.

III. PREPARATION OF PROTEOLYTIC FRAGMENTS FROM RAT IGE

Proteolytic cleavage of rat IgE was mainly performed to produce fragments with binding properties to rat mast cell receptors as well as receptors from other cells (lymphocytes, macrophages, eosinophiles, platelets). Unfortunately, several attempts to produce Fc or other fragments with affinity to various cells were reported unsuccessful.[7-9] Recently, we described the characterization of proteolytic fragments obtained by short time digestion with papain.[10] Details for the preparation of F(ab′)2-epsilon and C epsilon 4 fragments are given in the present chapter. In addition, high yield production of a Fab′-epsilon fragment from heat denaturated IgE is presented (Rousseaux-Prévost et al., manuscript in preparation).

1. Enzymatic Digestions

Production of a fragment F(ab′)2-epsilon and of a fragment related to the C epsilon 4 domains is obtained by short time digestion (10 to 30 min) with papain or trypsin. A fragment Fab′-epsilon is released from IgE previously treated at 56°C for 2 h.

a. Papain

An aliquot of mercuripapain as a suspension in 70% ethanol is centrifuged at 4°C (2500 rpm,

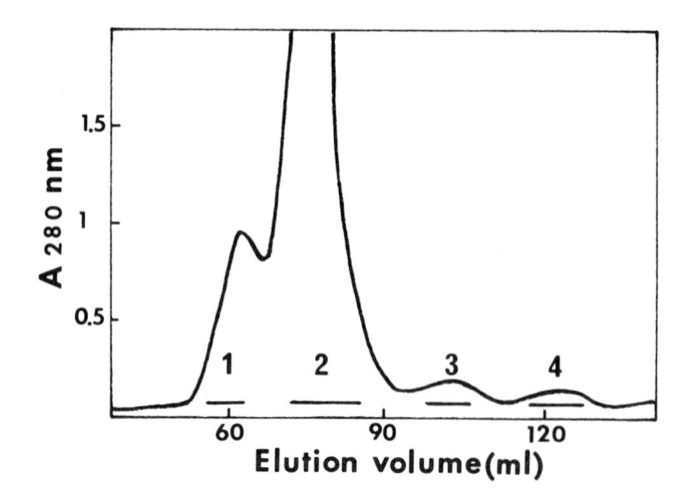

FIGURE 4. Purification of fragments released by papain digestion (30 min, enzyme to protein ratio 1%) of IgE IR 162. Chromatography on Ultrogel AcA44 column (1.6 × 90 cm) in PBS, pH 7.4 (flow rate: 6 ml/h). Peak 1: undigested IgE and IgE deleted of one C epsilon 3 domain; peak 2: F(ab′)2-epsilon; peak 3: Fab′-epsilon; peak 4: dimer of C epsilon 4 domains.

10 min) and the pellet is dissolved in phosphate buffered saline (PBS), pH 7.4 containing 0.01 *M* cysteine (at a final concentration 1 mg/ml). IgE (10 mg/ml) in PBS, pH 7.4 is incubated 10 min at 37°C before addition of papain solution (1 volume to 10 volumes of IgE solution, i.e., an enzyme to protein ratio 1%, w/w). The digestion is performed for 10 to 30 min at 37°C and stopped by addition of iodoacetamide (final concentration 0.015 *M*). In order to obtain only a F(ab′)2-epsilon fragment the digestion is performed for 3 or 4 h at 37°C.

b. Trypsin

Experimental conditions described above for IgG are used for tryptic digestion of IgE (enzyme to protein ratio 2%, w/w). After 30 min at 37°C, the reaction is stopped by addition of soybean trypsin inhibitor.

c. Enzymatic Digestion of Heat Denaturated IgE

Preparation of heat denaturated IgE is performed as follows: IgE (10 mg/ml) in PBS, pH 7.4, is incubated for 2 h at 56°C. After cooling at 4°C, the IgE solution is digested either with papain or with trypsin, as described above.

2. Purification and Characterization of Proteolytic Fragments

Gel filtration on an Ultrogel AcA44 column equilibrated in PBS, pH 7.4 is used to separate proteolytic fragments from IgE. Approximately 40 mg of enzymatic digest is loaded on a 1.6 × 90 cm column. The elution profile of a papain digest (30 min, 37°C, enzyme to protein ratio 1%, w/w) is shown in Figure 4. Peak 1 contains undigested IgE together with a fragment related to IgE deleted of one C epsilon 3 domain; peaks 2, 3, and 4 contain, respectively, F(ab′)2-epsilon, Fab′-epsilon, and a fragment related to a dimer of C epsilon 4 domains.[10]

When IgE is previously heated for 2 h (or more) at 56°C, the main fragment released from papain or trypsin digestion is Fab′-epsilon. This fragment may be purified by chromatography on conventional supports, such as Ultrogel AcA44, or by HPLC on TSK G3000 SW column (Rousseaux-Prévost et al., manuscript in preparation).

REFERENCES

1. **Galfré, G., Milstein, C., and Wright, B.,** Rat × rat hybrid myelomas and a monoclonal anti-Fd portion of mouse IgG, *Nature,* 277, 131, 1979.
2. **Rousseaux, J., Biserte, G., and Bazin, H.,** The differential enzyme sensitivity of rat immunoglobulin G subclasses to papain and pepsin, *Mol. Immunol.,* 17, 469, 1980.
3. **Rousseaux, J., Rousseaux-Prévost, R., Bazin, H., and Biserte, G.,** Tryptic cleavage of rat IgG: a comparative study between subclasses, *Immunol. Lett.,* 3, 93, 1981.
4. **Rousseaux, J., Rousseaux-Prévost, R., and Bazin, H.,** Optimal conditions for the preparation of Fab and F(ab′)2 fragments from monoclonal IgG belonging to the different rat IgG subclasses, *J. Immunol. Methods,* 64, 141, 1983.
5. **Bazin, H., Cormont, F., and De Clercq, L.,** Rat monoclonal antibodies. II. Rapid and efficient technique of purification from ascitic fluid or sera, *J. Immunol. Methods,* 71, 9, 1984.
6. **Rousseaux, J., Rousseaux-Prévost, R., Bazin, H., and Biserte, G.,** Proteolysis of rat IgG subclasses by *Staphylococcus aureus* V8 proteinase, *Biochim. Biophys. Acta,* 748, 205, 1983.
7. **Bennich, H. H., Ellerson, J. R., and Karlsson, T.,** Evaluation of basic serum IgE levels and the IgE antibody response in the rat by radioimmunoassay, *Immunol. Rev.,* 41, 261, 1978.
8. **Karlsson, T., Ellerson, J. R., Dahlbom, I., and Bennich, H.,** Analysis of the serum IgE levels in non-immunized rats of various strains by a radioimmunoassay, *Scand. J. Immunol.,* 9, 217, 1979.
9. **Perez-Montfort, R. and Metzger, H.,** Proteolysis of soluble IgE-receptor complexes: localization of sites on IgE which interact with the Fc receptor, *Mol. Immunol.,* 19, 1113, 1982.
10. **Rousseaux-Prévost, R., Rousseaux, J., and Bazin, H.,** Studies of the IgE binding sites to rat mast cell receptor with proteolytic fragments and with a monoclonal antibody directed against epsilon heavy chain: evidence that the combining sites are located in the C epsilon 3 domain, *Mol. Immunol.,* 24, 187, 1987.

Chapter 18

LABELING OF MONOCLONAL ANTIBODIES

P. Manouvriez

TABLE OF CONTENTS

I. LABELING WITH FLUOROCHROMES

When a single immunofluorescence labeling is needed, fluorescein isothiocyanate (FITC) is usually chosen. FITC has a high quantum yield, is relatively hydrophilic, and is inexpensive. The major problem occurring with the use of FITC is its rapid fade-down under the excitation of the high intensity ultraviolet emission of modern mercury-vapor lamps.

For double immunofluorescence experiments TRITC (tetramethylrhodamine-isothiocyanate) is mostly used as second fluorochrome. The intense red fluorescence of TRITC is stronger than that of rhodamine isothiocyanate, which has a quantum yield equivalent to FITC.

TRITC and XRITC or "Texas Red", which is a sulfonyl chloride derivative of rhodamine, are not adequately used in cytofluorometers equipped with argon lasers (488 and 514 nm emission lines), but may be used with the 568 nm line of krypton lasers.[1]

More recently phycobiliproteins from red algae have been introduced as very efficient fluorochromes.[2] They possess multiple bilin chromophores and have high quantum yields. One of them, R-phycoerythrin, is very suitable for use with an argon laser: it is efficiently excited with the 488 nm line and has an optimal emission at 580 nm. These phycobiliproteins can thus be used with the same filter set as for FITC to allow the visualization of two different labelings at the same time in microscope fluorescence.

Until now phycoerythrin has only been commercially available from Becton Dickinson as a biotin-avidin conjugate.[2] However, remaining biotin-binding sites are not very numerous. Techniques have to be improved to conjugate directly phycoerythrin to antigens or antibodies.

A. OPTIMAL FLUOROCHROME-PROTEIN RATIO

Molar F/P ratio can only be calculated for FITC derivatives with the formula proposed by The and Feltkamp.[3,4]

$$F/P = \frac{2.87 \times \text{OD } 495 \text{ mm}}{\text{OD } 280 \text{ mm} - (0.35 \times \text{OD } 495 \text{ mm})}$$

$$\text{mg IgG} = \frac{\text{OD } 280 \text{ mm} - (0.35 \times \text{OD } 495 \text{ mm})}{1.38}$$

FITC conjugation lowers the isoelectric point of the antibody. In case of over-conjugation, highly acidic molecules will be produced and will possibly stick nonspecifically to the cells. Optimal F/P ratios have been proven to be of 2:3 for fixed cells and 4:6 for live cells.[5] Our experience in light microscopy and flow cytofluorometry with the conjugates used at a concentration of less than 40 μg/ml shows that the artifact problems occur very rarely for membrane fluorescence with a F/P ratio as high as 8 and with a F/P ratio of 4:5 for cytoplasmic fluorescence. Self-quenching due to over-conjugation is not a problem with FITC.

Proteins must not be labeled too heavily with TRITC because of high self-quenching, which severely reduces the fluorescence intensity. Moreover, TRITC is highly hydrophobic, and heavily conjugated antibodies will precipitate because of hydrophobic interactions or produce artifactual adsorption of labeled antibodies by hydrophobic interactions between proteins and TRITC.

B. BINDING WITH FITC

The quality of FITC varies widely among manufacturers and even among batches of the same manufacturer. Moreover, some MAbs are easily overlabeled and others underlabeled, using the same conditions for conjugation. The optimal protein to fluorochrome concentration ratios used for the labeling reaction will therefore be investigated experimentally for each MAb and each FITC batch.

TABLE 1
FITC-Hydroxysuccinimide Ester Concentration to Use with Different Concentrations of MAbs

MAb concentration	FITC-hydroxysuccinimide per mg MAb
15—20 mg/ml	15 μg
<15 mg/ml	30 μg
<10 mg/ml	60 μg
< 5 mg/ml	90 μg

Note: The obtained FITC/protein molar ratio varies with the MAb and is between 6 and 9.

Antibodies preferentially at a concentration of 10 to 15 mg/ml are dialyzed in a carbonate-bicarbonate or borate buffer at pH 9.5 (with eventually 0.15 *M* NaCl). The FITC succinimide ester concentration has to be adapted to protein concentration as shown in Table 1. These conditions allow a F/P ratio of 5:7 with most MAb. This F/P ratio gives an accurate conjugate for immunofluorescence assays on live cells. To lower the F/P ratio, the quantity of FITC added has to be reduced down to fourfold.

1. FITC — Labeling of Purified Antibodies

a. Reagents and Material

1. 0.1 *M* carbonate-bicarbonate buffer pH 9.5
2. Fluorescein isothiocyanate isomer 1 (Sigma or Nordic, The Netherlands) has to be used in microgram quantities. To facilitate weighing, FITC is carefully mixed at 1:100 (w/w) in Pevikon beads (C870, Serva) used as inert charge. FITC-Pevikon can be stored dry at 4°C for long periods.
3. Phosphate buffered saline pH 7.2
4. A 25-cm G25 medium (Pharmacia, Sweden) column equilibrated with PBS
5. 10-ml tubes
6. Centrifuge tubes
7. Centrifuge which permits a 3000 *g* centrifugation
8. Dialysis membrane
9. Erlenmeyer
10. A spectrophotometer

b. Procedure

The purified MAbs are dialyzed against carbonate-bicarbonate buffer for 24 h at 4°C. The immunoglobulin concentration is assessed by measuring the optic density at 280 mm and assuming an extinction coefficient of 1.38 (see Chapter 1) Ig concentration in mg/ml = D.O./1.38. If possible the protein concentration is adjusted at 10 to 15 mg/ml. The quantity FITC-Pevikon added will be adapted to the immunoglobulin concentration, the batch of FITC and the MAb to label. Average conditions are given in Table 1. Before the opening of the FITC-Pevikon vial, it will be left to reach room temperature. Otherwise, water condensation would denaturate the activated fluorochrome by hydrolysis. While vortexing, FITC-Pevikon is dissolved directly in the protein solution. The solution is incubated, in obscurity, and gently shaken for 2 h at room temperature or 1 h at room temperature and over night at 4°C. The insoluble Pevikon is precipitated by gravity or by 5 min centrifugation at 3000 *g*. Free FITC is separated from the conjugated protein on a 25 cm high G25 (Pharmacia-LKB) column with PBS and the first peak is collected.

c. Remarks

The conjugation is strongly favored at protein concentrations >10 mg/ml.

Repeated thaw-freezing cycles may seriously damage the antibody activity of the conjugates. They would be best aliquoted before storage or stored at –20°C in 50% glycerol.

2. Fast FITC Conjugation of Antibodies in Ascitic Fluid

a. Reagents and Material

1. Fluorescein isothiocyanate isomer I (FITC) (Sigma)
2. Anhydrous dimethylsufloxide (DMSO)
3. 0.29 *M* bicarbonate buffer pH 9.3
4. Saturated sodium sulfate pH 7.4
5. Ascitic fluid containing MAbs
6. PBS, 0.1% BSA, 0.1% azide
7. Vortex
8. Microfuge
9. Microfuge tubes

b. Procedure

At room temperature, 250 µl freshly filtered saturated sodium sulfate is added to 250 µl ascitic fluid in a microfuge tube. After 2 h incubation at room temperature, the tube is centrifuged, the supernatant is discarded, and the pellet is redissolved in 100 µl bicarbonate buffer. While vortexing, 15 µl of a 1 mg/ml FITC in DMSO is added. After 1 h at room temperature 885 µl PBS-BSA are added to consume excess FITC. The solution can be stored at 4°C or at –20°C in 50% glycerol.

c. Appreciation

This labeling technique of crude MAb in ascites fluid has been adapted by Ledbetter from Goding[5] and published by Clark and Einfeld.[6] It gives very good results without any modification. With saturated ammonium sulfate instead of sodium sulfate the precipitation occurs in a much shorter time. The main advantage of this technique is that it does not require long preliminary purification procedures of the MAb. However, owing to the high total labeled-protein concentration there is a risk of nonspecific labeling by protein-protein interaction. If the MAb concentration is low, the proportion of irrelevant immunoglobulins will be high and high dilutions impossible to be used. Binding of labeled immunoglobulins to Fc receptors will thus often occur and will also give false positive tests, but if artifacts are easily distinguished and if direct FITC-labeling is a selection criterium for the obtained MAbs, it is worth trying the FITC-labeling technique of Ledbetter.

C. BINDING WITH TRITC

1. Peculiarities of TRITC

The antigen concentration and the pH of reaction are the same as for the conjugation with FITC. However, the highly hydrophobic TRITC is very difficult to dissolve directly in buffer. Before mixing with the protein solution, it should first be dissolved in anhydrous dimethylsulfoxide (DMSO). The desired amount of solubilized TRITC will then be added dropwise to the antibody solution while stirring. If the protein concentration is low, the amount of TRITC added must be increased. In this case the pH at the beginning of reaction will be checked and eventually adjusted at pH 9.5 with NaOH.

For the separation of free fluorochrome a G50 column (Pharmacia) will be preferred, because it is very difficult to wash out a G25 column of some precipitated TRITC - protein conjugates.

The F/P ratio is generally heterogenous; some species will thus have a poor fluorescence efficiency due to over- or under-labeling. It will thus often be useful to separate them by ion exchange chromatography and to use only the fraction with the best fluorescence efficiency.

2. Rhodamine Conjugation of Purified Antibodies

a. Reagents and Material

1. Tetramethylrhodamine isothiocyanate isomer R (Sigma or Nordic)
2. Anhydrous DMSO
3. 0.1 *M* carbonate-bicarbonate buffer pH 9.5
4. 0.15 *M* PBS pH 7.2
5. NaOH 1 *M*
6. Glycerol
7. Purified antibody at preferably >10 mg/ml
8. Dialysis membrane
9. 30 cm G50 medium (Pharmacia) column
10. Vortex
11. Tubes

b. Procedure

The antibody is dialyzed against a carbonate-bicarbonate buffer, and then 1 mg TMRITC is dissolved in 1 ml anhydrous DMSO. While stirring, 10 µl TMRITC solution are added dropwise per mg protein. During the first 10 min of incubation pH will be monitored and adjusted between 9 and 9.5 with NaOH. The incubation is performed for 2 h at room temperature and shielded from light. The solution is centrifuged for 10 min at 2000 *g* to remove aggregates. Free fluorochrome is separated from the conjugate on a 30 cm G50 column with PBS as elution buffer. The first colored peak contains the conjugate. The conjugate is best stored at –20°C in 50% glycerol.

c. Remarks

The conjugation is strongly favored at protein concentrations >10 mg/ml. For lower protein concentrations it would be useful to increase the quantity of TMRITC added.

Over conjugation with TMRITC results in a quenching of the fluorescence. According to Goding,[7] the best O.D. 550 nm/O.D. 280 nm ratio would be of 0.2:0.3.

The protein concentration can be measured as follows: protein in mg/ml = O.D. 280 nm – 0.385 × O.D. 515 nm/1.38

D. SPECIAL OBSERVATION

Fluorochrome conjugation is often the first labeling performed on a newly obtained MAb. Our general experiment has shown that MAbs inactivated by this labeling are generally also denaturated by biotin-labeling or ^{125}I-labeling. These MAbs are also often those which are difficult to purify without inactivation. However, they can be stable in ascites fluid and are then to be used in indirect labeling techniques with second layer antibodies.

II. THE BIOTIN-AVIDIN SYSTEM

The biotin-avidin system is one of the most attractive of immunology. Biotin is easily coupled under mild conditions with little amounts of antibody. The tetramer avidin which has an exceptionally great affinity for biotin (Keq = 10^{15} M^{-1}) can be used as a quasi universal second layer reagent. Indeed, avidin is commercially available and labeled with all the current fluorochromes, radio-isotopes, and enzymes.

Two tests can be performed on the same sample with two different biotin-labeled antibodies. The sample is incubated with the first conjugated antibody which is revealed with labeled avidin.

After elimination of any free avidin, the sample can be incubated with a second biotin-conjugated antibody which is revealed by differently labeled avidin. The very strong and stable avidin-biotin binding avoids any cross-reaction with the first biotin-labeled antibody which is already saturated with avidin.

Biotin has to be activated before conjugation. The easiest coupling method uses the N-hydroxysuccinimide ester which is commercially available. This ester allows very stable amide bonds and the conjugation procedure is very mild. Immunoglobulin biotinilation is very fast and easy to perform. The conjugates can be revealed by fluorochrome-, enzyme-, or radio-labeled avidin.

A. REAGENTS AND MATERIAL

1. Biotin N-hydroxysuccinimide esther (Sigma; IBF, France) N-dimethylformamide, anhydrous
2. 0.2 *M* carbonate-bicarbonate buffer pH 8.8, plus if necessary 0.15 *M* NaCl
3. Phosphate buffered saline pH 7.2: NaCl 8 g/l, KCl 0.2 g/l, Na_2HPO_4 0.2 g/l, and $KH2PO_4$ 0.2 g/l, in 1 l distilled water. It is convenient to make a ×10 stock solution for storage.
4. NH_4Cl 1 *M*, pH 6
5. Glycerol
6. Dialysis tubing
7. Erlenmeyer

B. PROCEDURE

The protein is adjusted at 1 mg/ml in the carbonate-bicarbonate buffer. A solution of 2 mg biotin N-hydroxysuccinimide esther is made in 500 µl anhydrous N-dimethylformamide. While stirring happens, 20 µl of this biotin solution are added per milliliter of protein solution. The conjugation is performed for 5 to 15 min at room temperature (20 to 25°C). The reaction is stopped by addition of 100 µl NH_4Cl. The excess of biotin is eliminated by dialyzing the protein for 24 h at 4°C against PBS. The biotinilated protein is best stored at –20°C in 50% glycerol. Over-labeling with biotin can inactivate a proportion of the antibodies. This occurs if the reaction is continued for a long time and if the excess of biotin-N-hydroxysuccinimide is not inactivated with NH_4Cl before dialysis.

III. RADIOLABELING OF MONOCLONAL ANTIBODIES

A. IODINATION

Protein iodination is mainly done using the lactoperoxidase, the chloramine-T, or the iodogen methods.

The first one is very gentle when bead-immobilized enzyme is used.[8] The preparation of the labeling is, however, more technically demanding than the chloramine-T procedure. The latter can be as mild as lactoperoxidase when optimized.[9] The iodogen procedure can also be used for the labeling of MAb.[10] All these techniques are detailed in Chard.[11]

The controlled chloramine-T iodination, modified by Manouvriez et al.[12] from De Meyts,[13] is the labeling technique used in the laboratory. A maximum of a two molar excess of $Na^{125}I$ is added to the protein in PBS. The reaction is controlled by a limiting quantity of chloramine-T, and ^{125}I incorporation is checked by trichloracetic precipitation. In case of insufficient incorporation of iodine, the labeling may be enhanced by the addition of a limited quantity of chloramine-T. Free iodine is separated by gel filtration.

By this technique the antibodies are easily labeled stoichiometrically with very little or no loss of activity.

Heavy iodinations may result in antibody inactivation. Another cause of damage is the use of high concentrations of chloramine-T, which will oxidize the protein. In any case, when a high proportion of the iodinated proteins is denaturated, the tests will be performed with high concentrations of heavily labeled antibodies. The possible benefit which would be obtained by a higher positive signal is then generally entirely masked by a great increase in the background signal.

1. Quantity of $Na^{125}I$ to Use

Iodine is used as Na ^{125}I at 14 mCi $^{125}I/\mu g$; 0.5 mCi = 275×10^{-9} mole = 5 µl at 100 mCi/ml, 400 µg rat IgG = 250×10^{-9} mole.

If 0.5 mCi is reacted with 400 µg antibody 45 to 100% of the iodine must be incorporated. Too heavy labeling will result in radiolysis of the antibody.

The following protocol permits a controlled and mild iodination of immunoglobulins.

2. Reagents and Material

1. $NaH_2PO_4.H_20$ 138 mg/5 ml distilled water
2. Chloramine $T.3H_2O$ at 40 µg/ml in PBS must be prepared extemporaneously and shielded from light
3. Sodium metabisulfite 1.6 mg/ml
4. KI 1 mg/ml
5. PBS
6. PBS 5% serum or PBS-1% bovine serum albumin grade V (BSA) (Sigma)
7. 20% trichloracetic acid (TCA)
8. $Na^{125}I$ at 14 mCi/µg and 100 mCi/ml (IMS30 Amersham)
9. 400 µg proteins in 200 to 400 µl PBS without azide
10. Fume board
11. Gas mask
12. 20 µl syringe
13. 50 µl pipette
14. Cryotube with screwstopper 43 × 12.5 (Nunc 3-66656) or any tight- stopped tube for the iodination reaction.
15. 10-ml syringe filled with 10-ml G25 fine or a PD-10 column (Pharmacia)
16. Microfuge tubes filled with 500 µl PBS-5% serum
17. Microfuge
18. Disposable tubes for collection of iodinated protein
19. Gammacounter
20. Microfuge

3. Procedure

Step 1—The amount of 5 µl $Na^{125}I$ is taken and put in the screw cap reaction tube, and then 5 µl NaH_2PO_4 is added to neutralize NaOH used as antioxidation agent in the $Na^{125}I$ solution.

Finally, the protein solution and 50 µl chrloramine-T at 40 µg/ml are also added to the tube.

Step 2—The reaction is continued for 5 min while stirring. A small drop sample is taken with the point of a Pasteur Pipette and transferred into a 500 µl PBS-serum tube. It is mixed and one drop is transferred to a second 500 µl PBS-serum tube. Then a drop is put from this tube into a third containing 500 µl PBS-serum. After mixing 500 µl, 20% TCA is added and the tube is centrifuged for 1 min. The supernatant is put in a 5-ml tube. The bottom of the microfuge tube containing the protein pellet is cut and put in another 5-ml tube. Pellet and supernatant are counted in the gamma counter. The proportions of bound (in the pellet) and unbound (in the supernatant) ^{125}I can then be assessed. If more than 50% of ^{125}I is incorporated, go to step 3. If 10 to 40% of ^{125}I is incorporated, 20 µl chrloramine-T is added and step 2 is repeated. If less than 10% of ^{125}I is incorporated, 50 µl chrloramine-T is added and step 2 is repeated.

Step 3— 50 µl KI is added and the mixture laid on a G25 column. Elution is made with PBS. In 5-ml tubes, 1- to 1.5-ml fractions are collected (12 tubes are generally enough). The column is stopped and discarded with radioactive waste. From each tube 20 µl are taken and counted. The first peak contains the iodinated protein, the second contains free ^{125}I.

Remark

If the two peaks are not very distinct, the first peak must be dialyzed for 48 h and 4 changes with PBS to eliminate free ^{125}I.

B. LABELING WITH TRITIUM

The labeling with ^{125}I has several disadvantages: radiolysis with heavy labelings and a half-life of 60 d. Labeling with tritium allows very heavy labeling without denaturation of the MAb and has an isotope half-life of 12 years which permits the use of the same batch over a very long time. We succeeded in labeling monoclonal rat IgE without a detectable loss of activity,[12] using the technique described by Means and Feeney.[14] The tritiated protein was stored in aliquots for 3 years at –70°C.

IV. PEROXIDASE LABELING

Two main labeling techniques are used: the coupling with glutaraldehyde as a bifunctional reagent and the method of Nakane and Kawaoi.[15]

A. THE METHOD OF AVRAMEAS

The conjugation can be performed in one step. The enzyme and the protein to label are mixed together and glutaraldehyde is added.[16] However, the reaction products are heterogenous and not well-defined. The two-step method[17] allows a better control of the products. The enzyme is first activated with glutaraldehyde. After the excess of glutaraldehyde has been removed, the activated enzyme is mixed with the protein to label.

These techniques will not be described here in detail.

B. THE METHOD OF NAKANE AND KAWAOI

Horseradish peroxidase possesses several carbohydrate groups. Sodium periodate can oxidize vicinal hydroxyl groups. The carbohydrate chain is then broken and two aldehyde groups are formed. These may react with the amino groups of the protein added and will form Schiff's bases at pH 9. To prevent the formation of peroxidase homopolymers by the same way, the amino groups of the enzyme must first be blocked by dinitrofluorobenzene. The procedure described here is adapted from the original procedure; it differs by the final stabilization step which we do not perform.

1. Reagents and Material

1. Horseradish peroxidase (Boehringer, Germany, grade I code 108090)
2. Antibody to be labeled at 10 mg/ml in carbonate-bicarbonate buffer 0.01 *M* pH 9.5
3. A solution of 2,4 fluorodinitrobenzene (FDNB) at 0.1 g/l in ethanol (10 mg/10 ml, diluted 1/10: 200 µl in 1.8 ml ethanol)
4. 85.5 mg sodium periodate ($NaIO_4$, Sigma 1878) prepared extemporaneously in 5 ml distilled water
5. 0.1 *M* sodium phosphate buffer, pH 6.8; $NaH_2PO_4.H_2O$: 7.01 g, and $Na_2HPO_4.2H_2O$: 8.76 g dissolved in distilled water and adjusted at 1 l
6. 0.1 *M* sodium phosphate buffer, pH 7.4; $NaH_2PO_4.H_2O$: 2.7 g, and $Na_2HPO_4.2H_2O$: 14.31 g dissolved in distilled water and adjusted at 1 l

7. 0.3 *M* sodium bicarbonate
8. 0.01 *M* carbonate-bicarbonate buffer, pH 9.5
9. Anhydrous glycerol

2. Precautions

The glassware must be very clean and free from any trace of detergent. All glassware will be washed with sulfochromic acid and thoroughly rinsed with distilled water. Peroxidase is very sensitive to light, thus all the incubations will be done shielded from light. Do not use azide, which is a very strong inhibitor of peroxidase. The activated PO is unstable. The conjugation has to be done without delay once PO is activated.

3. Protocol

The protocol is given for 5 mg PO and 10 mg/ml protein to label in a 0.01 *M* carbonate-bicarbonate buffer, pH 9.5. The success of the labeling is linked to this pH which will never be lower than 9.5.

a. Activation of the Enzyme

The amount of 5 mg PO is dissolved in 1 ml 0.1 *M* sodium phosphate buffer pH 6.8. The enzyme solution is mixed with 1 ml 0.3 *M* sodium carbonate and 0.1 ml of the 2,4 FDNB solution at 0.1 mg/ml. Incubation is performed for 2 h at room temperature and in obscurity, then 1 ml of 0.08 *M* sodium periodate is added. The solution is incubated further for 1 h at room temperature and in obscurity. The reaction is stopped with 0.2 ml glycerol. The activated enzyme is dialyzed against 0.01 *M* carbonate-bicarbonate buffer, pH 9.5 (4×15 min in 500 ml buffer). After dialysis the pH must be checked and eventually adjusted to 9.5.

b. Conjugation with the MAb

The activated enzyme is mixed with 10 mg MAb in 1 ml 0.01 *M* carbonate-bicarbonate buffer, pH 9.5. Reaction is brought for 3 h at room temperature and obscurity. The mixture is dialyzed overnight at 4°C against 0.1 *M* phosphate buffer, pH 7.4.

Remark

The concentration of the MAb can be reduced down to 2.5 mg/ml without reduction in the labeling efficiency. The amount of activated enzyme will be adapted.

c. Storage

If the presence of free peroxidase is undesirable, it can be separated by gel filtration from the MAb-peroxidase conjugate. The conjugate can be stored in a dark tube in 50% glycerol at –20°C for at least two years. Long-term storage can also be done at 4°C as a precipitate in 50% saturated ammonium sulfate. By these two storage techniques, Montoya and Castell[18] estimated the half-life of the conjugated antibodies to be of 8 years. The conjugate can also be kept frozen but repeated thaw-freezing cycles must be avoided. Once thawed the sample may be kept for a few weeks at 4°C. Lyophilization will be avoided at any time because only less than 30% of the conjugate activity will be recovered after reconstitution (Montoya and Castell[18] and Manouvriez and Malache, unpublished results).

C. COMPARISON OF THE TWO HORSERADISH PEROXIDASE LABELING METHODS

Hagenaars et al.[19] have compared the molecular weight and the activities of the products obtained after conjugation.

The Nakane-conjugate had a high molecular weight (MW 300,000). The conjugation was almost complete. The Avrameas-conjugate was less polymerized and contained more free IgG and peroxidase.

The comparison of the activity in ELISA also showed the superiority of the Nakane-conjugate which was chosen in our laboratory. Only 1 out of about 50 MAbs, which have been labeled until now, has been inactivated by this treatment. The labeled MAb can be used at concentrations of less than 1 µg/ml and can be stored for more than 2 years at –20°C in 50% glycerol without any detectable loss of activity.

REFERENCES

1. **Titus, J. A., Haugland, R., Sharrow, S. O., and Segal, D. M.,** Texas red a hydrophilic, red-emitting fluorophore for use with fluorescein in dual parameter flow microfluorometric and fluorescence microscopic studies, *J. Immunol. Methods,* 50, 193, 1982.
2. **Oi, V. T., Glazer, A. N., and Stryer, L.,** Fluorescent phycobiliprotein conjugate for analysis of cells and molecules, *J. Cell Biol.,* 93, 981, 1982.
3. **The, T. H. and Feltkamp, T. E.W .,** Conjugation of fluorescein isothiocyanate to antibodies. I. Experiments on the conditions of conjugation, *Immunology,* 18, 865, 1970a.
4. **The, T. H. and Feltkamp, T. E. W.,** Conjugation of fluorescein isothiocyanate to antibodies. II. A reproductible method, *Immunology,* 18, 875, 1970b.
5. **Goding, J. W.,** Conjugation of antibodies with fluorochromes: modifications to the standard methods, *J. Immunol. Methods,* 13, 215, 1976.
6. **Clark, E. A. and Einfeld, D.,** Human B cell surface molecules defined by an international workshop panel of monoclonal antibodies, in *Leukocyte Typing II, Vol. 2, Human B Lymphocytes,* Reinherz, E. L., Haynes, B. F., Nadler, L. M., and Bernstein, I. D., Eds., Springer Verlag, New York,1984, 156.
7. **Goding, J. W.,** Monoclonal antibodies principles and practice, Academic Press (London), 1983.
8. **Karonen, S. L., Mörsky, P., Siren, M., and Seuderling, U.,** An enzymatic solid-phase method for trace iodination of proteins and peptides with 125-iodine, *Anal. Biochem.,* 67, 1, 1975.
9. **Heber, D., Odell, W. D., Schedewie, H., and Wolfsen, A. R.,** Improved iodination of peptides for radioimmunoassay and membrane radioreceptor assay, *Clin. Chem.,* 24, 796, 1978.
10. **Ballou, B., Levine, G., Hakala, T. R., and Solter, D.,** Tumor location detected with radioactively labeled monoclonal antibody and external scintigraphy, *Science,* 206, 844, 1979.
11. **Chard, T.,** *An Introduction to Radioimmunoassay and Related Techniques,* Elsevier Biomedical Press, 1982.
12. **Manouvriez, P., Ravoet, A. M., and Bazin, H.,** Fc epsilon receptors on rat B lymphocytes: specificity and binding kinetics, *Mol. Immunol.,* 22, 1201, 1985.
13. **De Meyts, P.,** Insulin and growth hormone receptors in human cultured lymphocytes and peripheral blood monocytes, In *Methods in Receptor Research, Part I,* Blecher, M., Ed., Marcel Dekker, New York, 1980, 301.
14. **Means, G. E. and Feeney, R. E.,** Reductive alkylation of amino groups in proteins, *Biochemistry,* 7, 2192, 1968.
15. **Nakane, P. J. and Kawaoi, A.,** Peroxidase-labeled antibody: a new method of conjugation, *J. Histochem. Cytochem.,* 22, 1084, 1974.
16. **Avrameas, S.,** Coupling of enzyme to proteins with glutaraldehyde. Use of the conjugates for the detection of antigens and antibodies, *Immunochem.,* 6, 43, 1969.
17. **Avrameas, S. and Ternynck, T.,** Couplage des anticorps aux enzymes à l'aide du glutaraldéhyde, in *Les Techniques de l'Immunofluorescence et les Réactions Immunoenzymatiques,* Farne M., Dupouchy, P., and Morelec, M. J., Eds., 1977, 153.
18. **Montoya, A. and Casrtell, J. V.,** Long term storage of peroxidase-labeled immunoglobulins for use in enzyme immunoassays, *J. Immunol. Methods,* 99, 13, 1987.
19. **Hagenaars, A. M., Kuipers, A. J., and Nagel, J.,** Preparation of enzyme-antibody conjugates, In *Immunoenzymatic Assay Techniques,* Malvano, R., Ed., Martinus Nijhoff, The Hague, The Netherlands, 1980, 16.

Part IV
Examples of Production and Characterization of Rat-Rat Hybridomas Against Various Antigens

Chapter 19.I

RAT MONOCLONAL ANTIBODIES AGAINST MURINE IMMUNOGLOBULINS

M. Lefebvre, C. Vincenzotto, C. Digneffe, F. Cormont, C. Genart, and H. Bazin

TABLE OF CONTENTS

I. INTEREST OF ANTI-MOUSE IMMUNOGLOBULIN REAGENTS

Since the advent of the hybridoma technology, most monoclonal antibodies (MAbs) have been derived from the mouse species. Among the screening assays which have been developed to identify the mouse MAbs in culture supernatants, the common method utilizes the detection of antibodies capable of binding to the fixed antigen, using labeled anti-immunoglobulins (anti-Igs) as second antibodies. Nowadays, such anti-Igs reagents are important since the knowledge of Ig class and subclass is required to select the mouse MAbs for specific purposes, as complement fixation or Fc-receptors binding,[1-4] and to devise the strategy of their purification.

On the other hand, specific anti-mouse Ig reagents have broad applications in studies of the murine humoral immune system. They allow immuno-modulation experiments by depleting populations of mouse B lymphocytes (see Chapter 25) and by transduction of either stimulatory or inhibitory signals.[5-7]

The anti-Ig antibodies have either polyclonal[8] or monoclonal origins. MAbs offer definite advantages over polyclonal anti-mouse Ig antibodies, including a virtually unlimited supply and a restricted specificity. MAbs can be produced within the murine species and then isotype-associated allotypes serve as an appropriate target for the generation of Mabs. Although easier to produce, allogeneic MAbs seem to have been poorly used,[9-12] xenogenic anti-Ig MAbs are much more common. But, it is amazing that almost all hybridomas generated until this time have been of mouse × rat origin,[13-20] and thus required *in vitro* exploitation or *in vivo* production on "nude" mice. Clearly, rat × rat fusion is advantageous for making xenogenic anti-mouse Ig MAbs[20-23] and allows for efficient large scale production.

II. PREPARATION OF RAT MAbS ANTI-MOUSE Ig ISOTYPIC DETERMINANTS

A. IMMUNOGENS AND IMMUNIZATION SCHEDULE

Rat immunization techniques can be found in Chapter 5. Specific remarks on anti-mouse Ig isotype immunizations are given below.

Hyperimmune mouse sera contain substantial, but heterogeneous Ig quantities and cannot be easily fractionated. Sera and ascites of mice bearing either myelomas or hybridomas are convenient starting materials for Ig purification.[24]

Specific immunization for anti-mouse heavy chain isotype requires extensive Ig class and subclass purification. On the contrary, anti-light chain isotype immune responses can be expected with Ig preparations which have only been purified by ammonium sulfate precipitation.

The optimal Ig dose per animal injection averages 100 μg. The murine Igs are soluble proteins and require adjuvant to be immunogenic in rats (see Chapter 5).

As the stimulation is made with monoclonal Ig proteins, care must be taken to alternate different idiotype-associated Ig in order to avoid massive anti-idiotype response.

For immunization, 8 to 12-week-old female or male LOU/C rats are adequate. No difference has been found between LOU/C, DA, or BN inbred rats.

The intraperitoneal route is commonly used for 2 to 4 priming injections given at 1 to 2 week intervals. Animals are bled by retro-orbital sinus ponctions 8 to 15 d after the last injection. Stimulation effectiveness is checked by Ouchterlony immunoelectrophoresis or by an ELISA. The usual rest period is 4 to 6 months and the final boost injection in saline is intravenously given 3 d prior to fusion.

B. ISOLATION OF SPECIFIC HYBRIDOMAS

The fusions were performed according to the experimental design described in Chapters 6 and 7. Screening assays were initiated 8 to 12 d after fusion. Each supernatant of growing colonies was first screened by ELISA for its differential binding to the mouse Ig, which is desired, and further characterized for its reactivity with other isotypes (immunodot, Ouchterlony). Selected hybridomas are subcloned twice in microculture plates by limiting dilution (see Chapter 8).

To date, out of 20 fusion experiments we have selected about 60 hybridoma cell lines that produce rat MAbs specifically reactive with isotypic determinants of mouse IgM, total IgG, IgG1, IgG2a, IgG2b, IgG3, IgE, IgA classes, and kappa light chains.

III. CHARACTERIZATION OF RAT ANTI-MOUSE Ig MAbS

A. SPECIFICITY

It has been reported that anti-mouse Ig MAbs can react with determinants expressed by other isotypes within the murine species or even with Ig of other species.[12] Some of these cross-reactions may be attributed to an epitope identity between Ig proteins, which present some degrees of structural homology (partial common amino acid sequences or carbohydrate moieties). On the other hand, it is clear that the conformational structure of Ig proteins is also preponderant and can be modified in the assay. The specificity of rat anti-mouse Ig MAbs needs then to be investigated in the context of their potential uses (see Chapter 25).

It may be anticipated that future optimal preparations of anti-mouse Ig MAbs will be pools of several different MAbs, such that their nominal specificity will be maintained in all circumstances.[26]

Some specificity profiles of rat anti-mouse Ig MAbs produced in our laboratory are given in Chapter 33, Section II. Among them, the anti-kappa light chain MAbs (LO-MK) are particularly useful as second reagents since a large proportion (>95%)[27] of mouse Igs contain kappa light chains.

B. ISOTYPES

Determimation of the class and subclass is performed according to the procedures described in Chapter 10.

C. AVIDITY

Kinetics studies provide an insight into the process of antigen-antibody binding and into the fine specificity of anti-Ig reagents. The method of Van Heyningen et al.[28] is employed to perform the anti-mouse Ig MAb avidity. In this technique, the coated mouse-Ig concentration is kept constant while the MAb concentrations are varied. The avidity is correlated to the MAb concentration, which achieves a 50% binding on an identical quantity of antigen coupled to a solid-phase (e.g., different MAb avidity values are listed in Table 1, Chapter 24.I). The fact that most selected MAbs bound their corresponding antigen with relatively high avidity is relevant to their selections.

REFERENCES

1. **Kipps, T. J., Parham, P., Punt, J., and Herzenberg, L. A.,** Importance of immunoglobulin isotype in human antibody-dependent, cell-mediated cytoxicity directed by murine monoclonal antibodies, *J. Exp. Med.*, 161, 1, 1985.
2. **Neuberger, M. S. and Rajewsky, K.,** Activation of mouse complement by monoclonal mouse antibodies, *Eur. J. Immunol.*, 11, 1012, 1981.
3. **Parham, P., Kipps, T. J., Ward, F. E., and Herzenberg, L. A.,** Isolation of heavy chain class switch variants of a monoclonal anti-DC1 hybridoma cell line: effective conversion of non-cytotoxic IgG1 antibodies to cytotoxic IgG2 antibodies, *Hum. Immunol.*, 8, 141, 1983.

4. **Bargellesi, A., Spira, G., Fishberg, E., Cosulich, E., and Sharff, M. D.,** Isolation of class and subclass switch variants SIB-selection and ELISA, in *Proc. Int. Symp. Impact of Biotechnology on Diagnostics,* Koprowski, H., Ferrone, S., and Albertini, A., Eds., Elsevier Science, Amsterdam, 1985, 17.
5. **Parker, D. C.,** Induction and suppression of polyclonal antibody responses by anti-Ig reagents and antigen-non specific helper factors: a comparison of the effects of anti-Fab, anti-IgM and anti-IgD on murine B cells, *Immunol. Rev.,* 52, 115, 1980.
6. **Sieckmann, D. G.,** The use of anti-Ig to induce a signal for cell division in B lymphocytes via their membrane IgM and IgD, *Immunol. Rev.,* 52, 181, 1980.
7. **Cambier, C., Heusser, C. H., and Julius, M. H.,** Abortive activation of B lymphocytes by monoclonal anti-immunoglobulin antibodies, *J. Immunol.,* 136, 3140, 1986.
8. **Spalding, D. M. and Koopman, W. J.,** Antiisotypic antibodies, in *Methods in Enzymology,* Vol. 116, Colowick, S. P. and Kaplan, N. O., Eds., Academic Press, London, 1985, 146.
9. **Oi, V. T., Jones, P. P., Goding, J. W., Herzenberg, L. A., and Herzenberg, L. A.,** Properties of monoclonal antibodies to mouse Ig allotypes, H-2 and Ia antigens, in *Current Topics in Microbiology and Immunology,* Vol. 81, Melchers, F., Potter, M., and Warner, N. L., Eds., Springer-Verlag, New York, 1978, 115.
10. **Huang, C., Parsons, M., Oi, V T., and Herzenberg, L. A.,** Genetic characterization of mouse immunoglobulin allotypic determinants defined by monoclonal antibodies, *Immunogenetics,* 18, 311, 1983.
11. **Goroff, D. K., Stall, A., Mond, J. J., Finkelman, F. D.,** *In vitro* and *in vivo* B lymphocyte-activating properties of monoclonal anti-delta antibodies, *J. Immunol.,* 136, 2382, 1986.
12. **Luo, S. and Bankert, R. B.,** Monoclonal antibodies specific for the mouse IgG1 allotypic determinants: reactivity with inbred and outbred mice, *Hybridoma,* 4, 319, 1985.
13. **Kincade, P. W., Lee, G., Sun, L., and Watanabe, T.,** Monoclonal rat antibodies to murine IgM determinants, *J. Immunol. Methods,* 42, 17, 1981.
14. **Yelton, D. E., Desaymard, C., and Scharff, M E.,** Use of monoclonal anti-mouse immunoglobulin to detect mouse antibodies, *Hybridoma,* 1, 5, 1981.
15. **Ware, C. F., Reade, J. L., and Der, L. C.,** A rat anti-mouse kappa chain specific monoclonal antibody, 187.1.10: purification, immunochemical properties and its utility as general second-antibody reagent, *J. Immunol. Methods,* 74, 93, 1984.
16. **Weiss, S., Lehmann, K., and Cohn, M.,** Monoclonal antibodies to murine immunoglobulin isotypes, *Hybridoma,* 2, 49, 1983.
17. **Yagawa, K. and Vitetta, E. S.,** Induction of proliferation and differenciation of murine B cells bearing surface Ig lambda by rat monoclonal antibody to lambda chain, *Hybridoma,* 2, 169, 1983.
18. **Julius, M. H., Heusser, C. H., and Hartmann, K. U.,** Induction of resting B cells to DNA synthesis by soluble monoclonal anti-immunoglobulin, *Eur. J. Immunol.,* 14, 753, 1984.
19. **Leptin, M.,** Monoclonal antibodies specific for murine IgM: activation of B lymphocytes by monoclonal antibodies specific for the four constant domains of IgM, *Eur. J. Immunol.,* 15, 131, 1985.
20. **Labit, C. and Pierres, M.,** Rat monoclonal antibodies to mouse IgG1, IgG2a, IgG2b, and IgG3 subclasses, and kappa chain isotypic determinants, *Hybridoma,* 3, 163, 1984.
21. **Galfre, G., Milstein, C., and Whright, B.,** Rat × rat hybrid myelomas and a monoclonal anti-Fd portion of mouse IgG, *Nature,* 277, 131, 1979.
22. **Haaijman, J. J.,** Production of monoclonal antibodies for the analysis of the ontogeny of the murine lymphoid system by flow cytometry, in *Immuno Fluorescence Technology,* Wick et al., Eds., Elsevier Science, Amsterdam, 1982, chap. 8.
23. **Cormont, F., Manouvriez, P., De Clercq, L., and Bazin, H.,** The use of rat monoclonal antibodies to characterize, quantify, and purify polyclonal or monoclonal mouse IgM, in *Methods in Enzymology,* Vol. 121, Colowick, S. P. and Kaplan, N. O., Eds., Academic Press, London, 1986, 622.
24. **Smith-Gill, S. J., Finkelman, F. D., and Potter, M.,** Plasmacytoma and murine immunoglobulins, in *Methods in Enzymology,* Vol. 116 Colowick, S. P. and Kaplan, N. O., Eds., Academic Press, London, 1985, 121.
25. **Tijsen, P.,** *Practice & Theory of Enzyme Immunoassays,* Burdon, R. H. and Knippenberg, P. H., Eds., Elsevier Science, Amsterdam, 1986, chap. 13.
26. **Goding, J. W.,** *Monoclonal Antibodies: Principles and Practice,* Academic Press, London, 1983, chap. 2.
27. **Hood, L., Gray, W. R., Sanders, B. G., and Dreyer, W. T.,** Light chain evolution, *Proc. Cold Spring Harbor Symp. Quantitative Biology,* 32, 133, 1967.
28. **Van Heyningen, V., Brock, D J., and Van Heyningen, S.,** A simple method for ranking the affinities of monoclonal antibodies, *J. Immunol. Methods,* 62, 147, 1983.

Chapter 19.II

ALLOTYPE SPECIFIC MONOCLONAL ANTIBODIES TO RAT KAPPA LIGHT CHAIN

M. Lefebvre and H. Bazin

TABLE OF CONTENTS

I. INTRODUCTION

The proportion of kappa-bearing immunoglobulins (Igs) in the laboratory rats, Rattus Norvegicus, ranges from 87% (LOU/C strain) to 98.5% (ACI strain) of the total.[1]

Nowadays, the structure and the genetics of the rat kappa light chain are well known.[2-6] Two allelic variants have been found: the IgK-1a, being carried by the LOU/C strain and the IgK-1b allotype being carried by the DA strain. Antigenic allotypic differences depend on 11 amino-acid substitutions in several sites distributed along the light chain C-region[4] (see Chapter 2).

To date, a dozen monoclonal antibodies (MAbs) reacting against different determinants of IgK-1a and IgK-1b allotypes have been described.[7-12] Their derivation has been restricted either to the mouse × mouse or to the mouse × rat system, which should impair their large scale *in vivo* production. The homogeneous rat × rat production has proved interesting in that matter and we have taken advantage of the 983 cell line availability to generate hybridomas secreting rat MAbs specific either of the IgK-1b or of the IgK-1a rat allotypes.

II. ISOLATION OF HYBRIDOMAS

For the anti-IgK-1b immunizations, 8-week-old female LOU/C rats were injected intraperitoneally (i.p.) with IgK-1b bearing Ig, which were purified by immunoaffinity chromatography from sera of OKA inbred animals (see Chapter 15 for procedures). But for the anti-IgK-1a immunizations, we primed with purified LOU/C Ig proteins some LOU/C.IgK-1b rats,[2] which are congenic to the LOU/C strain, but bear the kappa light chain locus from the OKA strain in the LOU/C background. In both cases, eight antigenic injections were performed with 200 μg of the corresponding antigens mixed in complete Freund's adjuvant. After rest periods of 5 and 8 months, respectively, and just before the spleen harvesting, 3 boost i.p. injections were given daily with the same antigen doses in saline.

Fusion and hybrid culture maintenance were performed as described in Chapters 6 and 7.

Specific MAb products were selected for their ability to agglutinate sheep red blood cells sensitized by the chromium chloride method with purified polyclonal or monoclonal Igs either of LOU/C or of DA strain origin (see Chapter 9).

Titrations were carried out in microtiter plates with phosphate buffer plus 10% horse serum as diluent. MARK-1 (mouse MAb anti-rat kappa light chain)[8] was used at the initial concentration of 1 μg/ml as a positive control.

After cloning, the individual reaction profiles of the selected MAb were investigated by the same method using purified rat and mouse monoclonal Igs from different isotypes and allotypes.

Thus from two fusion experiments, we managed to isolate two hybridoma cell lines secreting IgG1 MAbs against the rat IgK-1b allotype (LO-RK-1b-1, LO-RK-1b-2) and 1 IgM MAb reacting against the IgK-1a allotype (LO-RK-1a-1). The three hybridomas were exploited by *in vivo* production. Unfortunately LO-RK-1a-1, which has an unstable *in vivo* behavior, could only be produced *in vitro*. On the other hand, we obtained good yield with LO-RK-1b-1 and LO-RK-1b-2, which were used successfully for purification of IgK-1b bearing rat Ig by immunoaffinity chromatography, as well as labeled with biotin, horseradish peroxidase or fluorescent marker for the characterization of IgK-1b membrane Ig on dispersed cells or on tissue sections (see Chapter 26).

REFERENCES

1. **Zhou, D., Leslie, G. A., Guo, K., and Gutman, G. A.,** Expression of Immunoglobulin lambda chains in the laboratory rat, *J. Immunogenet.,* 13, 299, 1986.
2. **Beckers, A., Querinjean, P., and Bazin, H.,** Distribution of the allotypes of kappa and alpha chain loci in different inbred strains of rats, *Immunochemistry,* 11, 605, 1974.
3. **Nezlin, R. S. and Rokhlin, O. V.,** Allotypes of light chains of rat immunoglobulins and their application to the study of antibody biosynthesis, *Contemp. Top. Mol. Immunol.,* 5, 161, 1976.
4. **Gutman, G. A.,** Genetic and structural studies on rat kappa chain allotypes, *Transplant. Proc.,* 13, 1483, 1981.
5. **Gutman, G. A., Bazin, H., Rokhlin, O. V., and Nezlin, R. S.,** A standard nomenclature of rat immunoglobulin allotypes, *Transplant. Proc.,* 15, 1685, 1983.
6. **Barton Frank, M., Besta, R. M., Baverstock, P. R., and Gutman, G. A.,** The structure and evolution of immunoglobulin kappa chain constant region gene in the genus Rattus, *Mol. Immunol.,* 24, 953, 1987.
7. **Gutman, G. A.,** Rat kappa chain allotypes. IV. Monoclonal antibodies to distinct RI-1b specificities, *Hybridoma,* 1, 133, 1982.
8. **Bazin, H., Xhurdebise, L. M., Burtonboy, G., Lebacq, A. M., De Clercq, L., and Cormont, F.,** Rat monoclonal antibodies. I. Rapid purification from *in vitro* culture supernatants, *J. Immunol. Methods,* 66, 261, 1984.
9. **Cherapakhin, V. V., Chervonsky, A. V., Filatov, A. V., and Rokhlin, O. V.,** Monoclonal antibodies against common and IgK-1a allotypic determinants of rat immunoglobulin kappa chain constant domain, *Immunol. Lett.,* 10, 217,1985.
10. **Lanier, L. L., Gutman, G. A., Lewis, D. E., Griswold, S. T., and Warner, N. L.,** Monoclonal antibodies against rat immunoglobulin kappa chains, *Hybridoma,* 1, 125, 1982.
11. **Springer, T. A., Bhattacharya, A., Cardoza, J. T., and Sanchez Madrid, F.,** Monoclonal antibodies specific for rat IgG1, IgG2a and IgG2b subclasses and kappa chain isotypic and allotypic determinants. Reagents for use with rat monoclonal antibodies, *Hybridoma,* 1, 257, 1982.
12. **Gray, D., MacLennan, I. C. M., and Lane, P. J. L.,** Virgin B cell recruitment and the lifespan of memory clones during antibody responses to 2,4, dinitrophenyl hemocyanin, *Eur. J. Immunol.,* 16, 641, 1986.

Chapter 19.III

APPLICATION OF RAT HYBRIDOMAS TO STUDIES ON OCULAR IMMUNOBIOLOGY

J. V. Peppard and P. C. Montgomery

TABLE OF CONTENTS

I. INTRODUCTION

As part of an ongoing effort to define parameters regulating the induction and expression of IgA antibodies at mucosal surfaces, particularly in the eye, we have utilized rat hybridomas and monoclonal antibodies (MAbs) and the rat experimental animal model. The relationship between some ocular subcompartments and the mucosal immune system has been explored through studies described here, on the sources of IgA in rat tears, the production of B cell hybridomas from rat lacrimal glands, and the production of monoclonal anti-idiotypic antibodies to probe the expression of antibody subpopulations at diverse mucosal sites.

II. SOURCE OF THE IgA IN RAT TEARS

As in other mucosal-associated secretions in rats, such as saliva or intestinal fluids, the predominant immunoglobulin (Ig) in tears is secretory IgA (SIgA), which is present at concentrations of about 200 μg/ml.[1] Polymeric IgA reaches secretions by traveling across epithelia via a receptor-mediated transport system[2] using the poly-Ig receptor (membrane bound secretory component) synthesized by epithelial cells. Many IgA-secreting cells underly the secretory epithelium of the lacrimal gland and secretory component has been demonstrated on epithelial cells of its acini, ducts, and tubules.[3] Thus the polymeric IgA (pIgA) produced in the gland is well-positioned to gain access to the tears. However, it was not clear that local production of IgA was the sole source of this Ig in tears. To assess the contribution of plasma-derived pIgA to tears, we made use of the 91c[4,5] rat hybridoma line, which produces IgA antibody to horseradish peroxidase and grows well *in vivo*. The pIgA produced by this hybridoma *in vivo* served to overcome the extremely effective uptake and excretion of pIgA by the liver in rats,[6] thus raising the plasma concentration of IgA with a known antibody activity. Tears and saliva were obtained from rats in which this hybridoma was growing subcutaneously, and the concentration of 91c IgA in these fluids was compared to that in the bloodstream using a radioimmunoassay (RIA). We found that 91c IgA in its polymeric form and combined with secretory component, did become detectable in tears and saliva once the concentration of plasma IgA rose sufficiently.[7] I+ was estimated from these experiments that the concentration of IgA in tear samples was about 0.3% of those in matched serum samples. Thus a plasma contribution is made to the tears but >99% of tear IgA may be assumed to be produced locally. A similar estimate was found to be applicable in the case of salivary IgA.

III. PRODUCTION OF HYBRIDOMAS FROM RAT LACRIMAL GLANDS

Past studies[4] have shown that the mesenteric lymph node (MLN) is a good source of cells to use for fusion if B cell hybridomas producing IgG or IgA antibody are required. In recent efforts to produce anti-DNP hybridomas we used MLN cells for this purpose (see Section IV). However, the immunization route found most effective for MLN stimulation has been found to be the injection of particulate antigen into Peyer's patches of the rats, which necessitates surgery for each immunization time. We have been studying routes of immunization to induce IgA antibody in tears,[8] and have found that ocular-topical application of particulate antigen, heat-killed Pneumococcus coupled to DNP (DNP-Pn), was an effective method to elicit IgA anti-DNP both in tears and bile with IgG and IgM anti-DNP responses occurring in serum. As mentioned in Section II, the lacrimal gland is a good source of IgA secreting cells. Therefore, we questioned whether the lacrimal gland might be used much as the MLN has been used in the past to provide cells for fusion purposes.

As a preliminary experiment, we prepared cells from the exorbital lacrimal glands of rats and used them in a fusion to study whether the whole process was feasible using this tissue, and the distribution of antibody isotypes.

TABLE 1
Production of Immunoglobulin by Lacrimal Gland Hybridoma Cells

Fusion	Number of wells positive for: IgM	IgG	IgA	Total wells positive for Ig
1	0	12	23	35/380
2	13	97	16	126/384

Lacrimal glands from 12 rats were used to provide cells. The rats were killed and submerged briefly in 90% alcohol. The glands were removed, minced finely, and the tissue was digested twice for 45 min at 37°C with dispase (1.5 mg/ml in 20 ml Joklik medium containing 10% fetal calf serum). The released cells were washed and the lymphocytes further purified using Lymphocyte M. A total of 1×10^8 gland cells (containing about 10% B cells) and 2×10^7 IR983F cells[9] were mixed and fusion was brought about as described previously.[4] The cells were plated out in HAT selective medium into 4×96-well plates. These contained feeder layers of confluent fibroblasts, derived from rat *Xiphisternae,* which had been irradiated (30 Gy) and the medium removed just before the fusion was carried out. The cultures were observed over the following month.

At first (from around 6 d), there was evidence of an outgrowth of fibroblasts from the gland. These proved to be no major problem in the long run, except that there was some necessity to add the medium somewhat more frequently than in an equivalent spleen or MLN fusion, and so the use of 24-well plates at this stage might have been an improvement. The presence of these cells would obviously require cloning of hybridomas at a comparatively early stage to eliminate them. A few days later an apparently increasing population of large cells, which appeared to contain lipid droplets, was evident. Although obtrusive in appearance, these cells decreased in numbers after another week or so, and again proved to be of no major importance. Two weeks after the fusion, hybridoma colonies started to be evident. The supernatant medium from each well was screened at three weeks using PVC plates coated with specifically purified rabbit anti-rat L chain. Supernatant was incubated overnight in the wells of triplicate plates, the plates were washed and the isotype of Ig present was determined using ^{125}I-labeled anti-rat mu, gamma, or alpha.

The results of two separate fusions are shown in Table 1. The lacrimal gland resembles the MLN rather than the spleen with respect to the relatively high incidence of IgA producers and low incidence of IgM producers. While further studies are required to determine the optimal ocular-topical immunization protocol, the use of the lacrimal gland is thus feasible for the production of hybridomas if IgA producers are required.

IV. PREPARATION OF MONOCLONAL ANTI-IDIOTYPIC ANTIBODIES TO PROBE ANTIBODY EXPRESSION

A. PRODUCTION OF ANTI-DNP HYBRIDOMAS AND MONOCLONAL IgA ANTI-DNP ANTIBODIES

In order to study the mucosal immune response in the tears and other secretions of rats immunized by various routes, we wished to produce anti-idiotypic probes. The initial stage of this project involved producing a series of anti-DNP hybridoma lines. This was carried out by immunizing Fischer 344 rats twice in the Peyer's patches with 0.2 mg DNP-Pn followed by an i.p. injection of 0.1 mg DNP-Pn 3 d before the fusion. The MLN was used as a source of cells, and fusion with IR 983F cells was carried out as already described.[4] The hybrids were screened by RIA using PVC plates coated with DNP-bovine serum albumin (DNP-BSA), and ^{125}I-labeled anti-mu, gamma, or alpha. Eleven clones secreting IgA and five secreting IgG were selected and cloned twice by limiting dilution. Five of the IgA clones were grown up, adapted for growth in Dulbecco's Modified Eagle's Medium containing 4% fetal calf serum, and several liters of culture supernatant were collected.

The IgA antibodies produced by these clones were found to be sensitive to the use of at least some chaotropic agents and thus not suitable candidates for affinity chromatography. Therefore, for screening of anti-idiotypic clones produced in the next stage, it was simplest to prepare IgA by lectin chromatography using a column of Con A-Sepharose (Pharmacia). Culture supernatant, containing about 15 μg/ml IgA, was concentrated by ammonium sulfate precipitation, dialyzed against 0.01 *M* Tris-HCl containing 0.15 *M* NaCl, 2 m*M* $CaCl_2$, 2 m*M* $MgCl_2$, and 0.02% NaN_3, pH 7, and passed through the column in the same buffer. After extensive washing of the column, bound protein was eluted in the same buffer containing 2% Methyl alpha-Mannopyranoside. Analysis of the Con A eluates by PAGE and immunoelectrophoresis showed that some bovine protein contaminants still remained; for a single step purification, however, the yield of IgA was high. In one case, the IgA content of the eluate was assessed by passing a small amount through a DNP-BSA/AH-Sepharose column: about 65% of the total protein was removed. Analysis of another Con A eluate, from CB1 hybridoma supernatant, by gel filtration on an analytical HPLC column (BioSil TSK 400, BioRad) coupled with an RIA analysis[7] (Figure 1) showed that both polymeric and monomeric IgA had been eluted.

B. PRODUCTION OF RAT MONOCLONAL ANTI-IDIOTYPIC ANTIBODIES

In order to screen hybrids for production of anti-idiotypic antibody, IgA-secreting hybridoma cell clones were used to immunize rats. None of the clones used have been found to grow *in vivo* in unprimed rats and this approach was used simply as a method of giving antigen, in this case IgA, in particulate form, which has been shown to generate good anti-idiotypic responses.[10] About 10^7 viable hybridoma cells were injected i.p. into Fischer 344 rats; two or three injections were given about 1 month apart and 3 d before the fusion, a final injection was given. The spleen was taken and fusion with IR983F cells was performed as described above. The screening assay was set up using as antigen the semipurified IgA (described above) from the same cell line, and anti-idiotypic antibodies of IgG, and IgM isotypes were screened for using a mixture of radiolabeled anti-gamma and anti-mu antibody. The cell lines proving positive in this first screen were cultured further and subjected to differential screens using IgA from other anti-DNP clones, 91c anti-HRP IgA, or polyclonal SIgA prepared from bile, again with radiolabeled anti-mu/gamma to detect positive binding. To date, two fusions have been carried out using rats immunized with two different cell lines. One yielded 10 positive clones, all IgM; the other gave 25, which were almost equally distributed between IgM and IgG. Most of these clones bind to IgA from all sources; however, only two or three exhibited specific binding for the immunizing IgA. It is not yet known whether the cross-reaction represents idiotype or allotype reactivity. Further fusions are planned using immunized F1 offspring from LOUVAIN and Fischer 344 rats in an attempt to increase the frequencies of anti-idiotypic antibody-producing clones. After characterization, the monoclonal anti-idiotypic antibodies will be used to probe the idiotype distribution of antibodies in secretions of animals during the course of immune responses.

V. CONCLUSION

Knowledge of the functions and properties of Ig isotypes has been derived mainly from studies of the homogenous proteins formed by plasmacytomas of mouse and man, and in recent years, mouse hybridomas. However, these species are not suited for all investigative purposes. Mice are restrictive for certain types of surgery and the amounts of body fluids that can be obtained. Humans, apart from being inconveniently outbred, present limitations with respect to experimental manipulation. With the availability of plasmacytomas,[11] and more recently

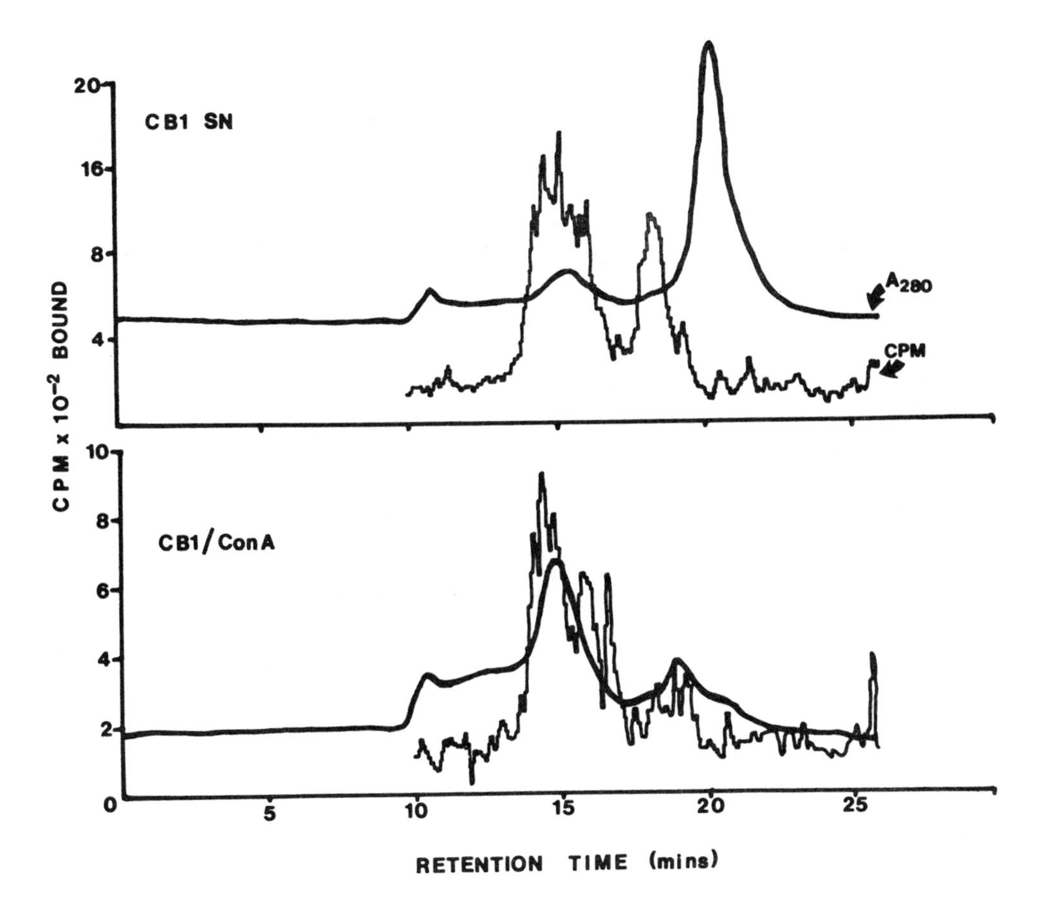

FIGURE 1. Analysis of a 20 µl of 22× concentrated supernatant from the CB1 hybridoma cell line (CB1 SN), compared with 5 µl of the protein (1 mg/ml) eluted from Con A-Sepharose (CB1/Con A) after adsorption of the same supernatant concentrate. Protein was separated by size using a Bio Sil TSK400 1 × 60-cm column running at 1 ml/min. Fractions (4 drops) were collected into 96-well PVC plates coated with DNP-BSA, from 10 min into the run and stopped when the 144th well was filled; the stop time was recorded. The plates were incubated overnight, and IgA binding was detected using ^{125}I-labeled anti-rat alpha. The integrator trace obtained from the column monitor set at 280 nm (thick line) was overlayed onto the same plot as the RIA results (thin line), to indicate the size distribution of IgA in the protein mixtures applied to the HPLC column. The integrator attenuation was the same for both runs. From previous runs using purified IgA[7] the elution times of tetrameric, trimeric, dimeric and monomeric IgA has been found to be approximately 15.5, 16.3, 17.1, and 19.3 min respectively.

hybridomas,[12,13] the versatility of the rat model has been enhanced and some relatively inaccessible components of the immune system, such as the eye, are somewhat more amenable to study. Aside from technical advantages it is important that studies are simultaneously carried out on the immune systems of several species, and the many differences even between mouse and rat serve to underscore the danger of extrapolation from one system to another.

REFERENCES

1. **Sulivan, D. A. and Allansmith, M. R.,** Source of IgA in tears of rats, *Immunology,* 53, 791, 1984.
2. **Solari, R. and Kraehenbuhl, J. P.,** The biosynthesis of secretory component and its role in the transepithelial transport of IgA dimer, *Immunology Today,* 6, 17, 1985.

3. **Franklin, R. M., Kenyon, K. R., and Tomasi, T. B.,** Immunohistologic studies of human lacrimal glands: localization of immunoglobulins, secretory component and lactoferrin, *J. Immunol.,* 110, 894, 1973.
4. **Dean, C. J., Gyure, L. A., Hall, J. G., and Styles, J. M.,** Production of IgA-secreting rat × rat hybridomas, in *Methods in Enzymology,* Vol. 121, Langone, J. J. and Van Vunakis, H., Eds., Academic Press, Orlando, Florida, 1986, 52.
5. **Peppard, J. V., Jackson, L. E., Hall, J. G., and Robertson, D.,** The transfer of immune complexes from the lumen of the small intestine to the blood in sucking rats, *Immunology,* 53, 385, 1984.
6. **Orlans, E., Peppard, J. V., Payne, A. W. R., Fitzharris, B. M., Mullock, B. M., Hinton, R. H., and Hall, J. G.,** Comparative aspects of the hepatobiliary transport of IgA, *Ann. N.Y. Acad. Sci.,* 409, 411, 1983.
7. **Peppard, J. V. and Montgomery, P. C.,** Studies on the origin and composition of IgA in rat tears, *Immunology,* 62, 193, 1987.
8. **Peppard, J. V., Mann, R. V., and Montgomery, P. C.,** Antibody production in rats following ocular-topical or gastrointestinal immunization: kinetics of local and systemic antibody production, *Curr. Eye Res.,* 7, 471, 1988.
9. **De Clercq, L., Cormont, F., and Bazin, H.,** Generation of rat-rat hybridomas with the use of the LOU IR983F non secreting fusion cell lines, in *Methods in Enzymology,* Vol. 121, Langone, J. J. and Van Vunakis, H., Eds., Academic Press, Orlando, Florida, 1986, 234.
10. **Weiler, I. J.,** Neonatal and maternally induced idiotypic suppression, in *Lymphocytic Regulation by Antibodies,* Bona, C. and Cazenave, P.A., Eds., John Wiley & Sons, New York, 1981.
11. **Bazin, H., Deckers, C., Beckers, A., and Heremans, J.F.,** Transplantable immunoglobulin-secreting tumours in rats. I. General features of LOU/Wsl strain rat immunocytomas and their monoclonal proteins, *Int. J. Cancer,* 10, 586, 1972.
12. **Bazin, H.,** Production of rat monoclonal antibodies with the LOU rat non secreting IR983F myeloma cell line, in *Protides of the Biological Fluids,* Peeters, H., Ed., Pergamon Press, Oxford, 1982, 615.
13. **Galfré, G., Milstein, C., and Wright, B.,** Rat × rat hybrid myelomas and a monoclonal anti-Fd portion of mouse IgG, *Nature,* 277, 131, 1979.

Chapter 19.IV

RAT MONOCLONAL ANTIBODIES AGAINST RABBIT IMMUNOGLOBULIN

C. Digneffe and H. Bazin

Three-month old LOU/C rats were given two injections of 500 μg of rabbit immunoglobulin mixed with 200 μl of complete Freund's adjuvant by the intraperitoneal route. After a rest period of 4 months, the boost of 500 μg of rabbit Ig was done 4 d before the fusion by the intravenous route (fusion made by L. De Clercq). One rat hybridoma secreting monoclonal antibodies (MAbs) against rabbit IgG was selected for its binding activity and its secretion. This MAb (LO-RG-1) is an IgG2a, and can be used to purify rabbit immunoglobulin. After peroxidase labeling, it can be very useful as second layer in immunochemistry (see Figure 5 in Chapter 24.VI).

Chapter 19.V

PREPARATION OF ANTI HUMAN IgG MONOCLONAL ANTIBODIES USING RAT × RAT HYBRIDOMA SYSTEM

X. Zuoling and P. Zhiming

TABLE OF CONTENTS

I. INTRODUCTION

Since the introduction of the cell fusion technique between a suitable myeloma cell and a lymphoid cell from an immunized animal,[1] specific monoclonal antibodies (MAb) have been produced in numerous laboratories. The most common animal model is the mouse × mouse model. The first report of rat × rat hybridoma was published in 1979 by Galfré et al.[2] Compared with the mouse × mouse model, the rat × rat hybridoma system has some advantages for the preparation of antigen-specific MAb, (1) the antibody-generating repertoire of the rat is different from that of the mouse; (2) rats are approximately ten times bigger than mice, so that they can give a mean production of 50 to 150 mg of MAb per animal, whereas the production in mice is only about 25 mg per animal; (3) the rat MAb is easy to purify from the ascitic fluid or serum by immunoaffinity chromatography; and (4) rat antibodies of IgG1, IgG2a, and IgG2b isotypes can easily fix the human complement. However, long immunization schedules for soluble antigens are usually needed. The best results are obtained with a rest time of 4 to 8 months between the primary immunization and the booster before fusion.

In this chapter the results show that we could obtain rat MAb against human IgG by using low dosage of soluble antigen and short-term immunization through the combination of intradermal and intrasplenic injections.

II. MATERIAL AND METHODS

Cell lines — Rat immunocytoma IR983F,[3] and mouse anti-rat Ig kappa light chain hybridoma cell line, MARK-1, which had been given by H. Bazin, were grown in RPMI-1640 (Gibco) supplemented with 10% fetal calf serum, 10% horse serum, gentamycin (50 μg/ml), and sodium bicarbonate at 37°C in a humid atmosphere of 5% CO_2 in the air. The cells were maintained in exponential growth.

Immunization — For fusion purpose, the LOU/C rats were immunized as following: the antigen, 0.1 mg human IgG, was mixed thoroughly with 0.2 ml complete Freund's adjuvant, then injected intradermally into multi-points (about 30 points) of the rats' back skin. The second and third injections were carried out at an interval of 2 weeks using 0.1 mg antigen mixed with 0.2 ml incomplete adjuvant in the same place. Two weeks later, a booster injection of 45 μg of purified human IgG in 0.1 ml of saline was given 3 d before fusion by direct intrasplenic injection.

Fusion and screening — According to the methods described in Chapter 4, fusion cells 983 were fused with the spleen cells from the immunized rats. A feeder layer of mouse peritoneal cells was prepared in HAT medium the day before fusion. The hybrid cells appeared on day 4 and grew fast by the end of the first week. The screening of supernatants began on day 15 after fusion. The positive clones were selected by ELISA, subcloned, and frozen. The subclasses of rat immunoglobulin were identified by immunodiffusion using rabbit anti isotypes of rat Ig given by H. Bazin. The analyses of chromosomes were carried out by conventional method.

III. RESULTS

The growth of hybridomas was observed on day 4 after fusion. They grew fast in HAT medium. On day 11, HAT medium was changed for HT medium, and then changed to conventional medium on day 15. From 288 wells, growing hybridoma cells appeared in 44 wells. The secretion of rat immunoglobulins (Igs) in 20 wells could be detected by ELISA. There were 3 wells which showed positive reaction with human IgG. By selection, one of them, H2, had higher specificity, titer, and secreting stability. The doubling time of hybridoma cell H2 was 12 hours (Figure 1). The amount of rat MAb secreted by H2 in culture supernatants was 0.7 μg/10^6 cells/24 h. The rat hybridoma, H2, was also inoculated subcutaneously into one LOU/C.IgK-1b

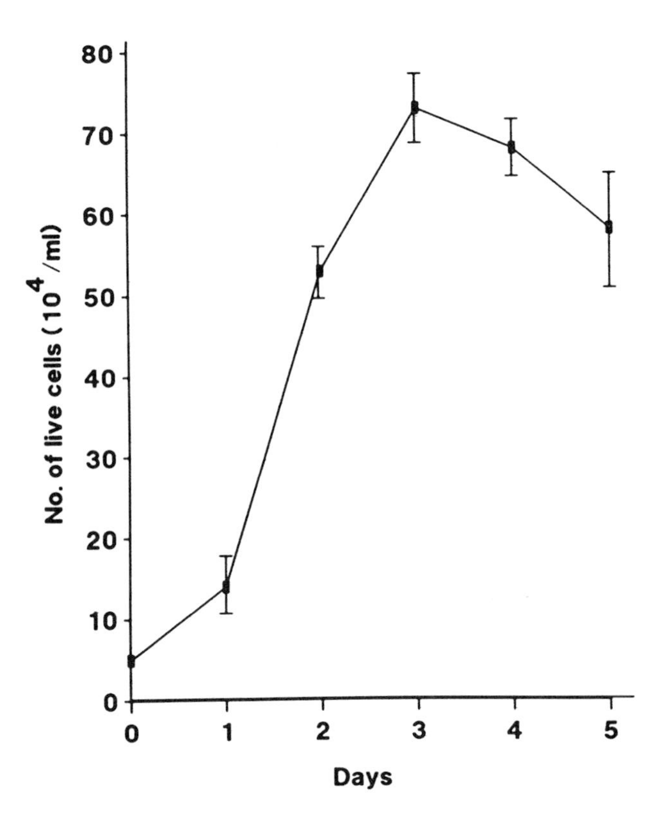

FIGURE 1. Growth curve of H_2 · 5 × 10^5 hybrid cells were seeded per culture flask.

TABLE 1
Titer of Rat MAb Against Human IgG (ELISA)

	Negative control	1:2	1:4	1:8	1:16	1:32	1:64	1:128	1:256
O.D.value	0.0055	1.033	0.724	0.469	0.299	0.209	0.158	0.127	0.115

Note: 10 μg/ml human IgG was coated on the plate, then added the different dilutions of rat MAb, MARK-1, and enzyme-labeled rabbit anti-mouse IgG antibody.

(abbreviated to LOU/C.K1b) rat, congenic rat of LOU/C; then the hybrid cells from the tumor mass were transplanted (5 × 10^6 cells per rat) into the peritoneal cavity of LOU/C-K1b rats without treatment by pristane. Ascitic fluid (250 ml) was harvested from these four rats 2 weeks later. The concentration of rat MAb in ascitic fluid was about 1.1 mg/ml. The average production of rat MAb in ascitic fluid was 70 mg per rat.

The specificity of MAb secreted by H2 was identified by the response to different kinds of serum, such as the serum from sheep, horse, calf, rabbit, and mouse, and different human Ig subclasses, such as human IgM, IgG, and IgA. The MAb only showed the positive reaction with human IgG and negative reactions with all other kinds of antigens.

The results of inhibition test indicated that the MAb activity in the supernatants could be blocked by purified human IgG (results not shown). Positive detections of MAb bound to the antigen were observed in ELISA even at a dilution of 1:256 of the culture supernatant. The O.D. values of positive wells detected by photospectrometer was 20 times higher than that of the control group (Table 1).

The subclass of the MAb is IgG2a which was determined by the sera of rabbit anti-different classes of rat Igs. The hybridoma cell line kept continuously secreting after 2-time subclones. So far there has not been only change in its secreting capacity and its titer. The analyses of the model of chromosomes have revealed that it is a pseudotetraploid.

IV. DISCUSSION

Three hybrid cell lines which are of specificity against human IgG were established successfully within 3 months. So far one of them has kept continuously secreting MAb with high titer after two-time subclones. The specificity of the MAb was identified by an inhibition test and binding with different sources of sera. Ascitic fluid (250 ml)was obtained from four rats. The average production of MAb per rat was 70 mg. Obviously, the quantity of ascitic fluid and the production of MAb per animal were much higher than those of mice. Compared with the yield of MAb from mice, the MAb from rats has greater economic benefits.

For fusion purpose, there are two ways to immunize rats by using soluble antigens. One method consists of immunizing the rat in the peritoneum or subcutaneously twice with antigen and adjuvant at two week intervals, and then boosting it intravenously 4 to 8 months later.[4] This method is effective, but the disadvantages are that it takes a long time and consumes large amounts of antigens. The other method consists of injecting the antigen intrasplenically once.[5,6] It takes a short time and needs fewer amounts of antigen, but the use of this method often produces IgM secreting MAb. We used the first method. We injected 100 μg antigen with adjuvant intradermally; then a booster was given by intrasplenic inoculation with 45 μg antigen. It is well-known that there is a large amount of Langerhans cells in the skin. This kind of cell has important functions in antigen processing and presentation. Using this method, we were able to obtain the MAb in a shorter time and with smaller amounts of antigen. However, the results also showed that the fusion frequency is low. It might be correlated with the procedure of immunization, as well as with the batches of fetal calf serum or the culture conditions of fusion partners.

ACKNOWLEDGMENTS

The author would like to thank Prof. H. Bazin for kindly offering the rat hybridoma system and his advice. We would also like to thank the Committee of National Science of China and the European Communities-People's Republic of China Cooperation Programme for their support. Special thanks are also due to Dr X.H. He and Z.X. Lin for language corrections.

REFERENCES

1. **Köhler, G. and Milstein, C.,** Continuous cultures of fused cells secreting antibody of predefined specificity, *Nature,* 256, 495, 1975.
2. **Galfré, G., Milstein, C., and Wright, B.,** Rat × rat hybrid myelomas and a monoclonal anti-Fc portion of mouse IgG, *Nature,* 277, 131, 1979.
3. **Bazin, H.,** Production of rat monoclonal antibodies with the LOU rat non-secreting IR983F myeloma cell line, in *Protides of the Biological Fluids,* Proceedings of the 29th Colloquium, Peeters, H., Ed., New York, 1982, 615.
4. **Bazin, H., Xhurdebise, L. M., Burtonboy, G., Lebacq, A. M., De Clercq, L., and Cormont, F.,** Rat monoclonal antibodies. I. Rapid purification from *in vitro* culture supernatants, *J. Immunol. Methods,* 66, 261, 1984.
5. **Spitz, M., Spitz, L., Thorpe, R., and Engui, E.,** Intrasplenic primary immunization for the production of monoclonal antibodies, *J. Immunol. Methods,* 70, 39, 1984.
6. **Gearing, A. J. H., Thorpe, R., Spitz, L., and Spitz, M.,** Use of single shot intrasplenic immunization for production of monoclonal antibodies specific for human IgM, *J. Immunol. Methods,* 76, 337, 1985.

Chapter 19.VI

RAT MONOCLONAL ANTIBODIES AGAINST HUMAN IMMUNOGLOBULIN

C. Digneffe, M. Jiang, C. Vincenzotto, and H. Bazin

TABLE OF CONTENTS

I. INTRODUCTION

Monoclonal antibodies (MAbs) reacting with human immunoglobulin (Ig) may be used for detection or quantitation of Ig in serum and biological fluids. In addition to this, these MAbs make the identification and isolation of anti-viral or anti-bacterial antibodies of a given class possible. The purification of human antibodies or Ig can also be resolved by using the immunoaffinity chromatography with these rat MAbs.

II. IMMUNIZATION SCHEDULE

Female 10- to 12-week-old LOU/C rats were immunized with 500 μg of human IgM, IgD, IgG, IgE, IgA, Ig kappa, and Ig lambda of polyclonal or monoclonal origin emulsified with complete Freund's adjuvant. For every antigenic preparation, five rats were used. Two or three intraperitoneal injections were made at days 0, 10, and sometimes 20. The rats' blood samples were checked for their serum antibody levels by immunoelectrophoresis and Ouchterlony against purified human polyclonal or monoclonal Ig. If no strength precipitation reaction was obtained, the rats were immunized again with antigen and alumine hydroxide (6 mg/ml Amphogel, Wyeth). Alternatively, human monoclonal of various origins or polyclonal Igs, when possible, were used in order to avoid anti-idiotype immune responses. The rest period was 4 to 6 months and the LOU/C rat chosen for fusion was boosted by the intravenous route and by inoculation of 500 μg of Ig at day 4 before fusion.

III. FUSIONS AND SCREENING

Splenocytes from a boosted LOU/C rat were fused with 983 fusion cells according to standard procedures (see Chapter 6).

Antibody production by the growing clones was assessed by hemagglutination or by ELISA (see Chapter 9) with different isotypes of human Igs.

The class or subclass of antibody was identified by the Ouchterlony test (see Chapter 10).

IV. RESULTS

Some properties of most MAbs used, obtained by this technique, are described in Table 1.

V. UTILIZATION

Rat MAbs anti-human Ig can be used for many purposes. Some examples are given below.

A. PURIFICATION OF HUMAN Igs

LO-HM-8 and LO-HG-7 have been used for purification of human MAbs by affinity chromatography. Actually, human-human hybridomas or cells lines transformed by Epstein Barr virus produce human MAbs of the IgM or IgG isotypes. Their purifications by affinity chromatography can easily be developed. They allow easy elimination of calf or horse serum immunoglobulin used in culture medium. They also permit excellent recovery of the MAbs even in diluted solution as they are in culture supernatants.

B. SELECTION OF Ig ISOTYPE FOR ELISA TEST SPECIFIC OF CLASS ANTIBODIES

LO-HM-8, LO-HG-7, and LO-HK-3 can be useful to test patient serum with viral infection, the rat MAbs being used to provide the isotype specificities in the tests.

TABLE 1
Some Properties of Selected Rat MAbs Developed in Our Laboratory Against Human Igs

Code	HDB number[a]	Isotype	Specific antibody properties[b]	Precipitation[c]	Immunoaffinity	Avidity ($\times M^{-1}$)[d]	Labeling[e]		
							PO	FITC	Biotin
LO-HM-2	7121	IgG1, kappa	Human IgM	yes	yes	—[f]	—	yes	—
LO-HM-7	—	IgM, kappa	Human IgM	yes	yes	—	—	yes	—
LO-HM-8	—	IgM, kappa	Human IgM	yes	yes	—	yes	yes	—
LO-HD-1	7125	IgG1, kappa	Human IgD	no	—	—	—	—	—
LO-HD-3	7126	IgM, kappa	Human IgD	no	—	—	—	—	—
LO-HG-1	7131	IgM, —	Human IgG (4 subclasses)	yes	yes	—	—	—	—
LO-HG-3	—	IgG2c, kappa	Human IgG (4 subclasses)	yes	yes	—	—	—	—
LO-HG-6	7134	IgM, kappa	Human IgG (4 subclasses)	yes	yes	—	—	—	—
LO-HG-7	7135	IgM, kappa	Human IgG (4 subclasses)	yes	no	5.3×10^9	—	—	—
LO-HG-19	—	IgM, kappa	Human IgG (4 subclasses)	yes	no	9.0×10^7	—	—	—
LO-HG-20	—	IgM, kappa	Human IgG (4 subclasses)	yes	yes	1.4×10^9	—	—	—
LO-HG-21	—	— kappa	Human IgG (4 subclasses)	—	yes	6.4×10^8	—	—	—
LO-HG-22	—	IgG, kappa	Human IgG (4 subclasses)	yes	—	2.7×10^8	—	—	—
LO-HG-23	—	— kappa	Human IgG (4 subclasses)	—	no	1.1×10^9	—	—	—
LO-HE-7	7130	IgG1, kappa	Human IgE	no	no	—	—	—	—
LO-HA-1	—	IgG1, kappa	Human IgA1	yes	—	—	—	—	—
LO-HA-2	—	IgG1, kappa	Human IgA1	yes	—	—	—	—	—
LO-HA-3	—	IgG2a, kappa	Human IgA1 and IgA2	yes	—	—	yes	—	—
LO-HA-4	—	IgG2a, kappa	Human IgA1 and IgA2	yes	yes	1.2×10^9	—	—	—
LO-HA-5	—	IgG, kappa	Human IgA1 and IgA2	yes	no	1.4×10^9	—	—	—
LO-HA-6	—	IgG2a, kappa	Human IgA1 and IgA2	yes	yes	1.7×10^9	yes	yes	yes
LO-HK-1	7136	IgG2a, kappa	Human kappa light chain	yes	yes	—	—	—	—
LO-HK-3	—	IgG1, kappa	Human kappa light chain	yes	yes	—	—	yes	yes

TABLE 1 (continued)
Some Properties of Selected Rat MAbs Developed in Our Laboratory Against Human Igs

Code	HDB number[a]	Isotype	Specific antibody properties[b]	Precipitation[c]	Immunoaffinity	Avidity ($\times M^{-1}$)[d]	Labeling[e]		
							PO	FITC	Biotin
LO-HL-1	—	IgG, kappa	Human lambda light chain	no	—	1.9×10^{9}	—	—	—
LO-HL-2	—	IgG1, kappa	Human lambda light chain	yes	—	1.5×10^{10}	yes	yes	yes
LO-HL-3	—	IgG2a, lambda	Human lambda light chain	no	—	2.8×10^{7}	—	—	—
LO-HL-4	—	— kappa	Human lambda light chain	no	—	3.6×10^{9}	—	—	—

[a] HDB: Hybridoma Data Bank (see Chapter 32).
[b] Specificity was tested by ELISA, RIA, or Hemagglutination.
[c] Precipitating properties were determined by the Ouchterlony test (see Chapter 10).
[d] Avidity was calculated by the Van Heimingen method.[1]
[e] Labeling with peroxidase, FITC and biotin (see Chapter 18).
[f] Means not tested.

C. SECOND ANTIBODY AFTER LABELING WITH PEROXIDASE OR FLUORESCEIN OR BIOTIN

The rat MAb LO-HA-6 has been labeled with peroxidase, biotin, and fluorescein and can be used in ELISA test and for immunopurification of human IgA.

ACKNOWLEDGMENTS

The authors wish to thank Dr. J.L. Preud'Homme (Laboratoire d'Immunologie et d'Immunopathologie, Centre Universitaire de Poitiers, France) and Dr J.P. Vaerman (Institute of Cellular Pathology, University of Louvain, Brussels, Belgium) for helping in the characterization of the specificity of LO-HG, LO-HL, and LO-HA, respectively.

REFERENCES

1. **Van Heiningen, V., Brock, D. J., and Van Heiningen, S.,** A simple method for ranking the affinities of monoclonal antibodies, *J. Immunol. Methods,* 62, 147, 1983.

Chapter 19.VII

PRODUCTION AND CHARACTERIZATION OF RAT MONOCLONAL ANTIBODIES DIRECTED AGAINST HUMAN APOLIPOPROTEIN B

T. Delaunay, H. Bazin, C. Durieux, C. Fievet, P. J. Volle, and J. C. Fruchart

TABLE OF CONTENTS

I. INTRODUCTION

Applications of monoclonal antibodies (MAbs) to the field of serum lipoproteins started only to 6 years ago, and already new information has been gained which may not have been possible otherwise.

Monoclonal antibodies to human plasma apolipoproteins B have been used for the mapping of apolipoprotein B epitopes,[1] for studying the apolipoprotein B expression in various lipoproteins,[2,3] for exploring the apolipoprotein B-lipid interactions,[4] and for quantifying of apolipoprotein B.[5,6]

All the anti-apolipoprotein B monoclonal antibodies developed at present have been raised in the mouse system because the hybridoma technology was first introduced by Köhler and Milstein.[7] In this chapter, we describe the generation and the characterization of six cloned and stable LOU rat hybridoma lines which produce antibodies specific for human apolipoprotein B in large quantities. These monoclonal antibodies have affinity constants in the range of 109 M-1 and are specific for all apolipoprotein-B-containing lipoproteins. Comparatively with the mouse model, no differences in immune response, efficiency of fusion procedure, and chemical quality of rat monoclonal antibodies were observed.

II. MATERIALS AND METHODS

A. ISOLATION OF Apob-CONTAINING LIPOPROTEINS, IMMUNIZATION

From pooled fresh normal plasma preserved with proteinase inhibitor, antioxydants and antibiotics, we isolated VLDL (d < 1.006 kg/l), IDL (d = 1.006 – 1.019 kg/l), LDL (d = 1.0025 – 1.063 kg/l) and LpB (d = 1.040 – 1.053 kg/l), using standard ultracentrifugation techniques. From a normal subject after a fatty meal, we isolated chylomicrons by ultracentrifugation at 120,000 *g* for 2 h.

Female LOU rats at an age of 7- to 8-weeks-old (having the kappa light chain allotype IgK-1a) were immunized with 3 intraperitoneal injections of 100 to 200 μg ApoB at a 2 week interval. The first two injections included 50% of complete Freund's adjuvant (Difco, Pasture, Belgium) and the last had 50% of incomplete Freund's adjuvant. The immune responses were quantified by a direct ELISA assay. The immunized serum was kept as positive control for the screening of hybridoma cell supernatants.

B. FUSION AND CULTURE

A booster of 100 to 200 μg LDL in saline was given intraperitoneally 6 months later, 4 d before the fusion. Splenocytes from an immunized LOU rat were fused with nonsecreting 983 myeloma cells[8] according to the Köhler and Milstein's procedure[7] adapted to the rat-rat hybridoma technology, as described in Chapter 4. The hybridoma cells obtained were cultured on rat feeder layer peritoneal cells, in a selective hypoxanthine-aminopterin-thymidine Dubelco's modified eagle medium supplemented with heat-inactivated fetal calf serum and horse serum.

As cell proliferation approached confluency (2 weeks), supernatants were screened for the presence of anti-LDL by an ELISA test developed in the laboratory.

C. SCREENING OF SUPERNATANTS

Microtiter plates (Flow Laboratories, Cat No. 76-381-04) were coated with LDL (20 μg/ml, 50 μl/well) in PBS at 4°C overnight. Then, the free binding sites were blocked with 10% serum in PBS. After washing, cell culture supernatants were added (50 μl/well) and incubated for 2 h at 37°C. The wells were washed in PBS and 50 μl of peroxidase conjugated mouse monoclonal antibodies anti-rat kappa light chain (MARK-1 + MARK-3) diluted 1/4000 in PBS were incubated for 2 h at 37°C. This was followed by three washes in PBS and by the addition of the peroxidase substrate (O-phenylenediamine, Sigma P 1526, 0.4 mg/ml and H_2O_2 0.003% in citrate phosphate buffer 0.1 *M* pH 5.5, 50 µl/well). The reaction was stopped 15 min later by the addition of 50 µl HCl 1 *N*. A reading of the optical absorbance at 495 nm was made, using Behring ELISA reader.

D. PRODUCTION OF RAT MAbS

The method is described in Chapter 12. Briefly, the first passage on animals was always carried out subcutaneously. Hybridoma cells were injected in LOU congenic rats having the kappa light chain allotype IgK-1b. Two weeks later, the teased solid tumor was injected intraperitoneally. The ascitic fluid was collected 2 weeks later and assayed by agarose electrophoresis. The MAb was then purified on a mouse anti-rat kappa light chain (MARK-3) affinity column, as explained in Chapters 13 and 14.

Finally, the isotype of the rat MAb was confirmed by immunoelectropheresis, ELISA, or immunodot assay.

E. CHEMICAL CHARACTERIZATION

1. Iodination of Human Lipoprotein

LDL were radio-iodinated with $Na^{125}I$ according to a modification of McFarlane's procedure,[9] as previously described.[10] The specific activity of ^{125}I· LDL was 455 cpm/ng.

2. Immunoblotting Technique

LDL, VLDL, and chylomicrons-ApoB subspecies were separated by analytical sodium dodecyl sulfate (SDS)-polyacrylamide gel electrophoresis in a vertical slab gel apparatus (70X80X0.75 mm, Mini Protean II, Biorad). Electrophoresis was performed in 3 to 6% acrylamide (w/v) slab gels gradient containing 0.1% SDS (w/v). The lipoproteins were delipidated by boiling for 3 min in electrophoresis sample buffer that contained 2.5% (SDS) (w/v), 5% beta-mercaptoetanol (v/v), 10% glycerol (v/v), 0.01% bromophenol blue (v/v) and 20 to 50 µg of protein per 10 µl. The gels were electrophoresed at constant voltage (150 V) for 45 min.

After migration, the separated protein bands were transferred electrophoretically (Mini Trans Blot, Biorad) at a constant current (150 mA, 1 h at 4°C) to nitrocellulose paper in 25 m*M* Tris, 192 m*M* glycine, 20% (v/v) methanol pH 8.3. The remaining active binding sites on the nitrocellulose were saturated by soaking 90 min in 20 m*M* Tris HCl buffer pH 8, 150 m*M* NaCl, 0.1% Tween 20 (v/v) and 15% heat-inactivated serum (v/v). The nitrocellulose paper was then cut into strips and each strip was placed individually into a solution of each MAb in the Tris HCl buffer 5% serum (v/v) for 1 h.

After washing the strips with the same buffer, we incubated them in a 2000-fold dilution of MARK-1 + MARK-3 peroxidase conjugated for 1 h. After two washes with the Tris HCl buffer and one wash with phosphate-buffered isotonic saline buffer (PBS), the color was developed in the strips by staining with 4-chloro-1-naphtol (Sigma, C8890) 0.4 mg/ml in PBS, in presence of 0.03% (w/v) H_2O_2.

3. Fluid Phase Radioimmunoassay

The competitive radioimmunoassay was carried out under equilibrium conditions, at an antibody concentration sufficient to bind 50% of labeled LDL in absence of unlabeled LDL.

We used Microcentrifuge test tube (1.5 ml, Cat. No. 9624949001, Treff Lab., Switzerland). Each sample contained, in a total volume of 150 µl of buffer (PBS, 3% gammaglobulins, Sigma No. G-5009), 50,000 cpm I^{125}LDL, rat MAb at a determined concentration and varying amounts of unlabeled LDL (ranging from 0.1 to 15 µg). The reaction mixture was incubated for 18 to 20 hours at 4°C. Then, we added 100 µl of MARK-1 + MARK-3 coupled with Sepharose-4B (25 mg/g of dried gel) in each test tube to separate unbound I^{125}LDL. One hour later, 50 µl of PBS, 3% gammaglobulins were added and the samples were centrifuged for 1 min at 15,000 v/min.

The supernatants of each sample were kept and pooled with a second washing supernatant (500 µl PBS 3% gammaglobulins).

The radioactivity in 1150 µl was counted in a scintillation gamma counter. The inhibition of labeled LDL antibody binding was calculated according to the formula given by Müller:[11]

$$\text{Inhibition } (\%) = \frac{\text{cpm } 1 - \text{cpm } 2}{\text{cpm } 3 - \text{cpm } 4}$$

where cpm 1 = cpm in the pooled supernatants of inhibitor containing sample (test sample),.cpm 2 = cpm in the pooled supernatants of sample without inhibitor (0% inhition = maximal binding), and cpm 3 = cpm in the pooled supernatants of sample without antibody (100% inhibition = nonspecific binding, typical nonspecific binding was 5% of total ^{125}I-LDL added).

All assays were performed in triplicate. For each antibody, the inhibition labeled LDL antibody binding was plotted against the unlabeled LDL concentration and the 50% inhibition values (It) were determined. The antibody affinity constants were calculated from the It values according to the equation given by Müller.[11]

4. Solid Phase Radioimmunometric Assay

To determine whether the MAbs bind all forms of the antigen with the same affinity, and to test the role of lipids in the expression of antigenic determinants recognized by our antibodies, we compared the competitive displacement curve obtained with different competitors: VLDL, IDL, LDL, and two types of plasma (normo and hypertriglyceridermic). Three delipidation procedures were performed on the latter. In the first delipidation, we treated the plasma with alipolytic enzyme (pancreatic triglyceride lipase, EC3.1.1.3), in the second with mixture of n-butanol-diisoprophylether (40/60 v/v), and in the third with a detergent solution as SDS.

Microtiter wells were coated with 100 µl of each monoclonal antibody solution (20 µg/ml of PBS) overnight at room temperature. After having been emptied, the wells were then incubated with PBS containing 10 g of bovine serum albumine (BSA) per liter (PBS-BSA) for 1 h at room temperature to saturate extra binding sites on the plastic. After removal of this blocking solution, increasing amounts of each competitor (expressed as protein ApoB content) were added in 50 µl PBS-BSA together with a constant amount of ^{125}I-labeled LDL (45,500 cpm corresponding to about 100 ng of protein). After incubation overnight at room temperature, the wells were emptied and the contents were washed three times with PBS, sliced, and counted in a gamma counter. The results were plotted as competitive displacement curves. Logit transformation of data produced linear plots which permitted statistical analysis of their slopes.[12,13]

5. Immunoprecipitation

Double immunodiffusion in 1% (w/v) Indubiose A-37 (IBF) in veronal-Tris buffer pH 9.2, 0.05 *M* was carried out by the technique of Ouchterlony.[14]

LDL and MAbs (corresponding to about 10 µg of protein) were applied to wells in 10 µl aliquots and left to diffusion for 18 h at room temperature in a damp chamber.

III. RESULTS

A. PRODUCTION OF RAT MAbs

With the method described above, six cloned hybridoma lines producing antibodies to human LDL were selected and characterized. They are listed in Table 1 and are called LO-HApoB, followed by a chronological number (abbreviation of "LOUVAIN" anti-"human" "apolipoprotein B"). The usual recovery of ascitic fluid which was generated from intraperitoneal growth of each cloned hybridoma line amounted to 30 to 50 ml per rat. Depending on the hybridoma

TABLE 1
Apoprotein-B Specific Rat Hybridoma Antibodies

Name of cell line	Chain type	Immunoreactivities to Apo-B species	Affinity constant (Ka) for LDL (4°C L/mol × 10^9)
LOH-ApoB1	IgG1, kappa-1a	B-100, -74[a], -48[a], -26	0.36
LOH-ApoB2	IgG1, kappa-1a	B-100, -74	0.14
LOH-ApoB3	IgG1, kappa-1a	B-100, -74	0.16
LOH-ApoB4	IgG1, kappa-1a	B-100, -74[a], -48	0.46
LOH-ApoB5	IgG1, kappa-1a	B-100, -74, -48	2.8
LOH-ApoB6	IgG2a, kappa-1a	B100, -74	0.19

[a] Weak reaction.

TABLE 2
Comparison of the Expression of LDL B Epitopes on VLDL, IDL and Two Plasma

Antibody	Homologous competitor (LDL)	VLDL	IDL	Slope analysis of heterologous competitor	
				Normolipemic plasma	Hyperlipemic plasma
B1	–2	–0.75[a]	–1.80[b]	–0.86[a]	–1[a]
B2	–2.5	–1.76[a]	–2.57b	–2.12[b]	–2.04[c]
B3	–2.06	–1.49[a]	–2.30[d]	–2.15[b]	–2.13[b]
B4	–3.7	–2.99[b]	–3.43[b]	–2.30[a]	–2.59[b]
B5	–1.94	–2.04[b]	–2.41[b]	–2.02[b]	–1.70[b]
B6	–1.29	–0.99[a]	–1.32[b]	–1.00[a]	–1.02[a]

[a] $p < 0.001$
[b] not significant
[c] $p < 0.05$
[d] $p < 0.01$

cells, 1 to 5 mg of antibody per milliliter of ascitic fluid could be purified, or 30 to 250 mg of MAb per rat.

B. CHEMICAL CHARACTERIZATION

1. Western Blot Analysis

The results, listed in Table 1, show that all six MAbs reacted with species B 100 (MW = 550 kDa) and B74 of ApoB. LO-HApoB-4 and LO-HApoB-5 reacted strongly with B-48. Only LO-HApoB-1 recognized B-26.

2. Antibody Affinities

As shown in Table 1, all six MAbs had high affinity constants for ApoB epitopes on intact LDL, ranging from 0.14 to 2.8 × 10^9 l/M.

3. Solid Phase Radioimmunoassay

Apolipoprotein B is present in varying quantities in plasma, VLDL, IDL, and LDL lipoproteins. Complete competition was observed with each antibody at high concentration of competitors, which indicates that each of ApoB epitopes was expressed by the competitors. For each MAb, the slopes of the dose titration regression lines were compared statistically (Table 2). For a given MAb, a significant difference between the two slopes indicated that the ApoB

TABLE 3
Slopes of Displacement Curves in Competitive Immunoassays Using ^{125}I-LDL as Ligant and a Hyperlipemic Plasma in the Presence or Absence of a Lipolytic Enzyme

Antibody	Slope analysis	
	Plasma without lipase	Plasma incubated with lipase
B1	−1.00	−1.04[a]
B2	−2.04	−2.06[a]
B3	−2.13	−2.37[a]
B4	−2.59	−2.37[a]
B5	−1.70	−1.89[a]
B6	−1.02	−1.01[a]

[a] Not significant.

epitope recognized by this antibody on different ApoB lipoprotein particles was not identical to the ApoB epitope expressed on LDL. Only LO-HApoB-5 showed regression lines for the different competitors with similar slopes, thus indicating that this antibody recognized one ApoB epitope expressed equally by all ApoB containing lipoprotein particles which we had tested.

As shown in Table 3, no statistical differences between the slopes of regression lines with lipase-pretreated plasma and normal plasma were obtained, whatever the MAb was; this suggests that lipids are not located in the antigenic site.

4. Immunoprecipitation of ApoB by MAbs

LO-HApoB-1 and LO-HApoB-6, studied separately, did not present a precipitation line with LDL. But a clear precipitation line was observed in the reaction of LDL with separate LO-HApoB-2,3,4,5. This finding is interesting, and to our knowledge such MAbs have not yet been produced.

IV. DISCUSSION

Six stable rat hybridoma cell lines were developed, producing high affinity MAbs to human plasma apolipoprotein B. These antibodies showed no cross reactivity with other human plasma apolipoproteins.

Among the antibodies described, only LO-HApoB-5 showed identical binding affinities for all major apolipoprotein B-containing lipoproteins density classes. Antibodies LO-HApoB-1 and LO-HApoB-6 exhibited regression lines for LDL and IDL that had similar slopes, whereas the slopes of the VLDL and the two plasma dose titration were different. Therefore, the epitopes recognized by these two monoclonal antibodies were expressed on a similar manner only on IDL and LDL.

It has been shown by MAbs that the expression of apolipoprotein B in plasma lipoproteins varied and depended on interaction of the apolipoprotein with lipids. Some regions of apolipoprotein B can be masked or conformationally changed in the presence of lipids. Results of this study suggested that the six MAbs reacted with epitopes confined to a stable domain of apolipoprotein B that is not altered by the interaction with the lipids.

Conventional polyclonal anti-LDL antibodies are a heterogeneous population that can precipitate LDL. In general, a single monoclonal antibody cannot precipitate LDL but we have four monoclonal antibodies that are able to precipitate purified LDL. An immunoprecipitation for ApoB determination in human plasma is under development, using a mixture of different MAbs.

ACKNOWLEDGMENTS

This work was supported by the Commission of the European Communities in the frame work of the Biomolecular Engineering Program. We would like to acknowledge our thanks to J.P. Kints, J.M. Malache, and F. Nisol. We also wish to thank Hoechst-Behring (France) for providing materials for investigation. F. Bolle is acknowledged for excellent preparation and editing of the manuscript.

REFERENCES

1. **Marcel, Y. L., Weech, P. K., Wilthorp, P., Terce, F., Vezina, C., and Milne, R. W.,** Monoclonal antibodies and the characterization of apolipoprotein structure and function, *Prog. in Lipid Res.*, 23, 169, 1985.
2. **Tikkanen, M. J., Cole, T. G., Hahm, K. S., Krul, E. S., and Schonfeld, G.,** Expression of Apolipoprotein B epitopes in very low density lipoprotein subfractions. Studies with monoclonal antibodies, *Arteriosclerosis,* 4, 138, 1984.
3. **Cardin, A. D., Price, C. A., Hirose, N., Krivawek, M. A., Blankenship, D. T., Chao, J., and Mao, S. J . T.,** Structural organization of apolipoprotein B-100 of human plasma low density lipoproteins, *J. Biol. Chem.*, 261, 1674, 1986.
4. **Marcel, Y. L., Hogue, M., Weech, P. K., and Milne, R. W.,** Mapping of antigenic determinants of human apolipoprotein B using monoclonal antibodies against low density lipoproteins, *J. Biol. Chem.*, 259, 6952, 1982.
5. **Marcovina, S., France, D., Phillips, R. A., and Mao, S. J. T.,** Monoclonal antibodies can precipitate low density lipoprotein. I. Characterization and use in determining apolipoprotein B, *Clin. Chem.*, 31, 1654, 1985.
6. **Slater, H. R., Tindle, R. W., Packard, C. J., and Shepherd, J.,** A monoclonal-antibody-based immunoradiometric assay for apoprotein B in low-density lipoprotein, *Clin. Chem.*, 31, 841, 1985.
7. **Köhler, G. and Milstein, C.,** Continuous culture of fused cells secreting antibody of predefined specificity, *Nature,* 256, 495, 1975.
8. **Bazin, H.,** Production of rat monoclonal antibodies with the LOU rat non secreting IR983F myeloma cell line, in *Protides of the Biological Fluids,* Peeters, H., Ed., Pergamon, Oxford, 1982, 615.
9. **McFarlane, H. S.,** *Mammalian Protein Metabolism,* Munro, M. W. and Allison, J. B., Eds., Academic Press, New York, 1964, 331.
10. **Mao, S. J. T., Kazmar, R. E., and Silverfield, J. C.,** Immunochemical properties of human low density lipoproteins as explored by monoclonal antibodies binding characteristics distinct from those of conventional serum antibodies, *Biochim. Biophys. Acta,* 743, 365, 1982.
11. **Müller, R.,** Calculation of average antibody affinity in anti-hapten sera from data obtained by competitive radioimmunoassay, *J. Immunol. Methods,* 34, 345, 1980.
12. **Rodbard, D.,** Statistical quality control and routine data-processing for radioimmunoassay and immunoradiometric assays, *Clin. Chem.*, 20, 1255, 1974.
13. **Zettner, A.,** Principles of competitive binding assays (saturation analysis).I.Equilibrium techniques, *Clin. Chem.*, 19, 699, 1973.
14. **Ouchterlony, O.,** Diffusion in gel method for immunological analysis, *Prog. Allergy,* 5, 1, 1958.

Chapter 19.VIII

STUDIES ON RAT-RAT HYBRIDOMA TECHNIQUE AND ITS APPLICATION TO OBTAIN RAT MONOCLONAL ANTIBODIES ANTI-HORSERADISH PEROXIDASE

S. L. Sun, X. X. Zhang, and Q. Zhao

TABLE OF CONTENTS

I. INTRODUCTION

The rat-rat hybridoma technique was introduced by Bazin and his colleagues in 1986 in China and it has been successfully established for the first time at our laboratory of Beijing Institute for Cancer Research. As Galfré and Bazin indicated, the rat hybridoma system is a very helpful tool for biomedical research.[1,2] For this research, a large amount of ascitic monoclonal antibodies (MAbs) against horseradish peroxidase (HRP) was produced and it was used to prepare peroxidase anti-peroxidase (PAP complex). Their characteristics were studied. The purpose of this research is to establish a more sensitive and specific unlabeled antibody-enzyme (PAP method) of rat for widespread application of the rat hybridoma technique.

II. PRODUCTION OF RAT HYBRIDOMAS

A. IMMUNIZATION AND FUSION

A LOU/M rat (3-month-old female) was immunized with 500 μg horseradish peroxidase (HRP) (Type VI, RZ = 3.03, Biochemistry Institute, Shanghai) emulsified in complete Freund's adjuvant administered by the intraperitoneal and subcutaneous routes. Secondary immunization was performed 2 weeks later under the same conditions. The hyperimmune rat was boosted with HRP 500 μg in 0.9% NaCl solution intravenously 2 months after the initial immunization and the fusion was performed 3 d later. Splenocytes from this rat were fused with LOU rat non-secreting myeloma cell line IR983F developed by Bazin.[3]

B. SCREENING OF MAb AND HYBRIDOMA CLONING

The resulting hybridomas were screened for antibody production by ELISA as described by Burdon[4] (Figure 1). The positive culture wells were identified and cloned by limiting dilution. Six hybridomas (GC8, GD1, IA7, IG8, BA5, and EA10) secreting anti-HRP MAb were obtained after subcloning five times.

III. CHARACTERIZATION OF MAb

A. COMMON CHARACTERIZATION OF MAb

According to the titer of rat MAb (supernatant) and the amount of ascitic fluid produced by *in vivo* transplantation, three hybridomas (GC8, GD1, IG8) were selected and characterized. The results are shown in Table 1.

B. PURIFICATION OF MONOCLONAL ANTIBODIES

MAbs GC8, GD1, and IG8 were purified from ascitic fluid by immunoaffinity chromatography (Figure 2) as described in Bazin et al.[6,7]

C. SPECIFICITY OF MONOCLONAL ANTIBODY

Specification of rat anti-HRP MAbs GC8 and GD1 were identified by ELISA.[8] ELISA microplates were coated overnight at 4°C with specific antigens (HRP Type VI, Sigma, RZ = 2.83; South Africa, RZ = 3; Biochemistry Institute, Shanghai, RZ = 3.03; HRP Type I, Sigma, RZ = 1.12; HRP crude, Sigma, RZ = 0.4) and unrelated protein antigen (BSA, HSA, Normal rabbit serum, Normal mouse IgG, beta-galactosidase, Sigma) in 50 m*M* sodium carbonate buffer pH 9.5 (20 μg/ml, 100 μl/well). After blocking, ascites were added in serial dilution and the plates were then washed. Alkaline phosphatase-conjugated MARK-1[9] (50 μl/well) diluted 1:40 with PBS-Tween was added and then the plates were washed. Finally, chromogen solution (P-Nitrophenyl phosphate 1 mg/ml in 50 m*M* sodium carbonate buffer pH 9.5 containing 1 m*M* $MgCl_2$, 50 μl/well) was added. The reaction was stopped by the addition of 2 *N* sodium hydroxide (50 μl/well) and then the OD was determined, using the ELISA autoreader at 405 nm. OD 405

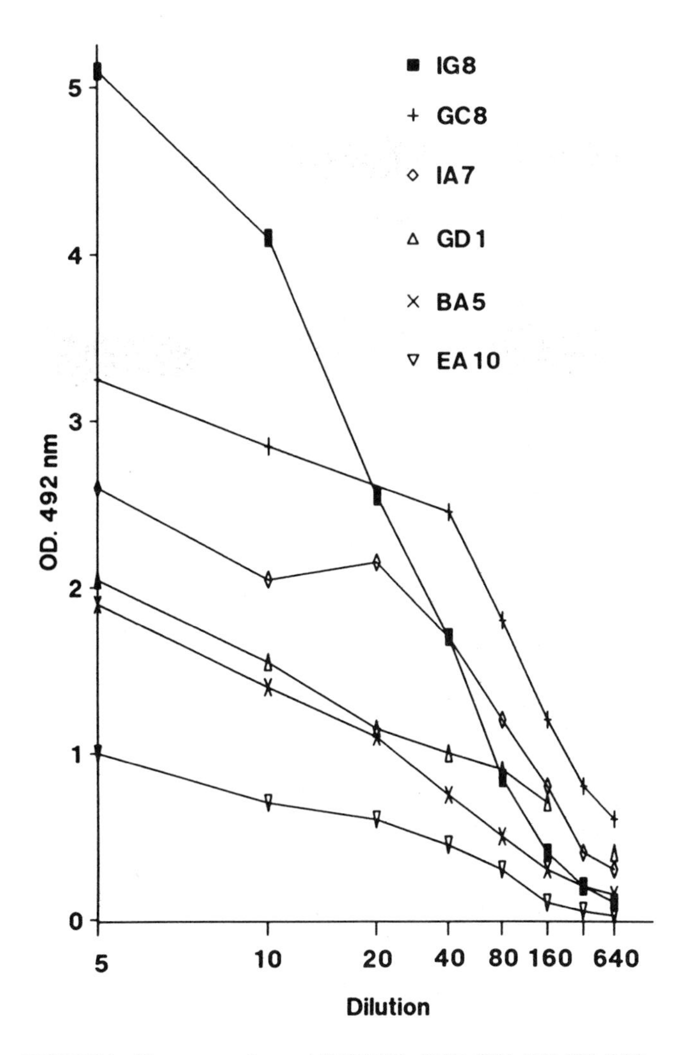

FIGURE 1. Titer curves of rat anti-HRP MAbs (GC8, GD1, IA7, IG8, BA5, and EA10). An ELISA plate was coated with MARK-1 overnight at 4°C; it was blocked with blocking solution for 1 h, incubated with serial dilutions of MAb supernatants (GC8, GD1, IA7, IG8, BA5, and EA10) for 1 hour, washed with PBS-Tween; and then an HRP solution (50 μg/ml) was added. The plates were washed again and revealed with OPD for 30 min. The wells were evaluated with OD492 nm on an ELISA reader.

TABLE 1
Characterization of Rat Anti-HRP MAbs

Characteristic hybridomas	Doubling time (h)	Titer[a]		Vol (ml) of ascitis fluid obtained per animal (range)	Isotypes	Number of chromosomes ($X \pm DS$)
		Supernatant	Ascites			
CG8	14	10^3	ND[b]	60—98	IgG1	69.9 ± 7.2
GD1	17	10^3	10^7	50—61	IgG2a	71.3 ± 5.4
IG8	18	10^2	ND	40—49	IgM	70.4 ± 10.8

[a] Titer in ELISA: maximum dilution giving a positive test.
[b] Not determined.

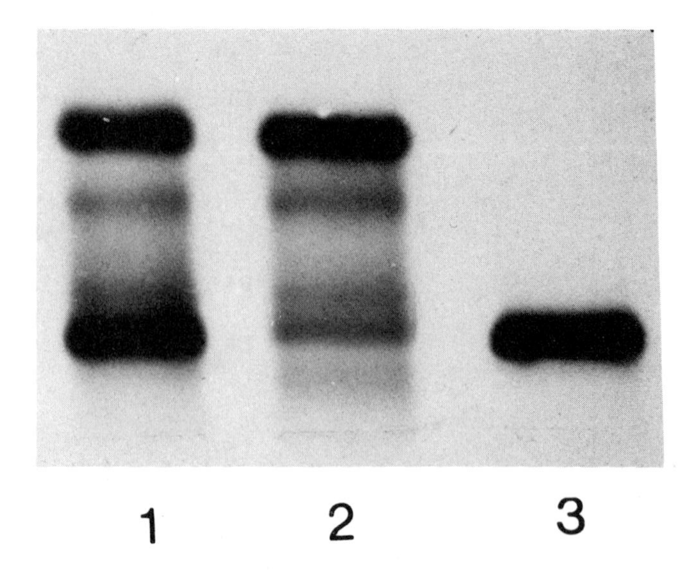

FIGURE 2. Agarose gel electrophoresis after purification of rat MAb against HRP (GD1) by immunoaffinity chromatography. From left to right: (1) ascites of LOU/C.IgK1b(OKA) rat carrying MAb; (2) first peak, LOU/CIgK1b(OKA) rat ascites; and (3) second peak, purified MAb.

higher than 0.2 was considered positive and below 0.2, negative. Other procedures were carried out as described in Burdon.[4] The two rat MAbs GC8 and GD1 were positive with the various HRP and negative with the unrelated antigens (data not shown).

IV. EFFECT OF PAP

Peroxidase-anti-peroxidase (PAP complex) was made with GC8, GD1, IG8, or a mixture of GC8.GD1 (PAP-GC8, PAP-GD1, PAP-IG8, and PAP-mix). Its effects were observed in the screening of rat MAb anti-gastric cancer cell line M85 with the micro dot immunohistochemistry[4] and cell ELISA assay. The second antibody (bridge antibody) was rabbit anti-rat serum (RArG) in this test system. Simultaneously, HRP-labeled RArG is compared with the PAP. The OD492 of PAP-GD1 and PAP-GC8 at the dilution of 1:64000 was 0.8 to 0.9; the OD492 of PAP-IG8 at the dilution of 1:64000 was 0.3 in ELISA method, and the sensitivity of PAP was 20 times higher than that of HRP-RArG. The reaction of PAP-GD1 (1:4000) turned a brown color ++ +++ (Figure 3), 1:500 of HRP-RArG provoked weak staining + ++ with the micro dot immunohistochemistry method. The background obtained with the PAP staining was clear, but the effect of PAP was obviously stronger than that of HRP-RArG alone. Staining effects of PAP-GD1 and PAP-mix was similar. This fact indicates that when GD1 was mixed with GC8, there was no enhancing effect. On the contrary, PAP-GD1 and PAP-IG8 staining effects were different. This might be due to the immunoglobulin subclasses of GD1 and IG8, which are respectively, IgG2a and IgM.

V. CONCLUSION

The rat-rat hybridoma technique has been successfully established in China. Six hybridomas, GC8, GD1, IG8, IA7, BA5, and EA10 have been obtained, stably secreting MAbs against HRP.

A large amount of MAb (ascites) has been produced by the hybridomas. Generally, 50 to 60 ml of ascites were harvested from each rat, with the highest harvest of ascites being 98 ml from one rat.

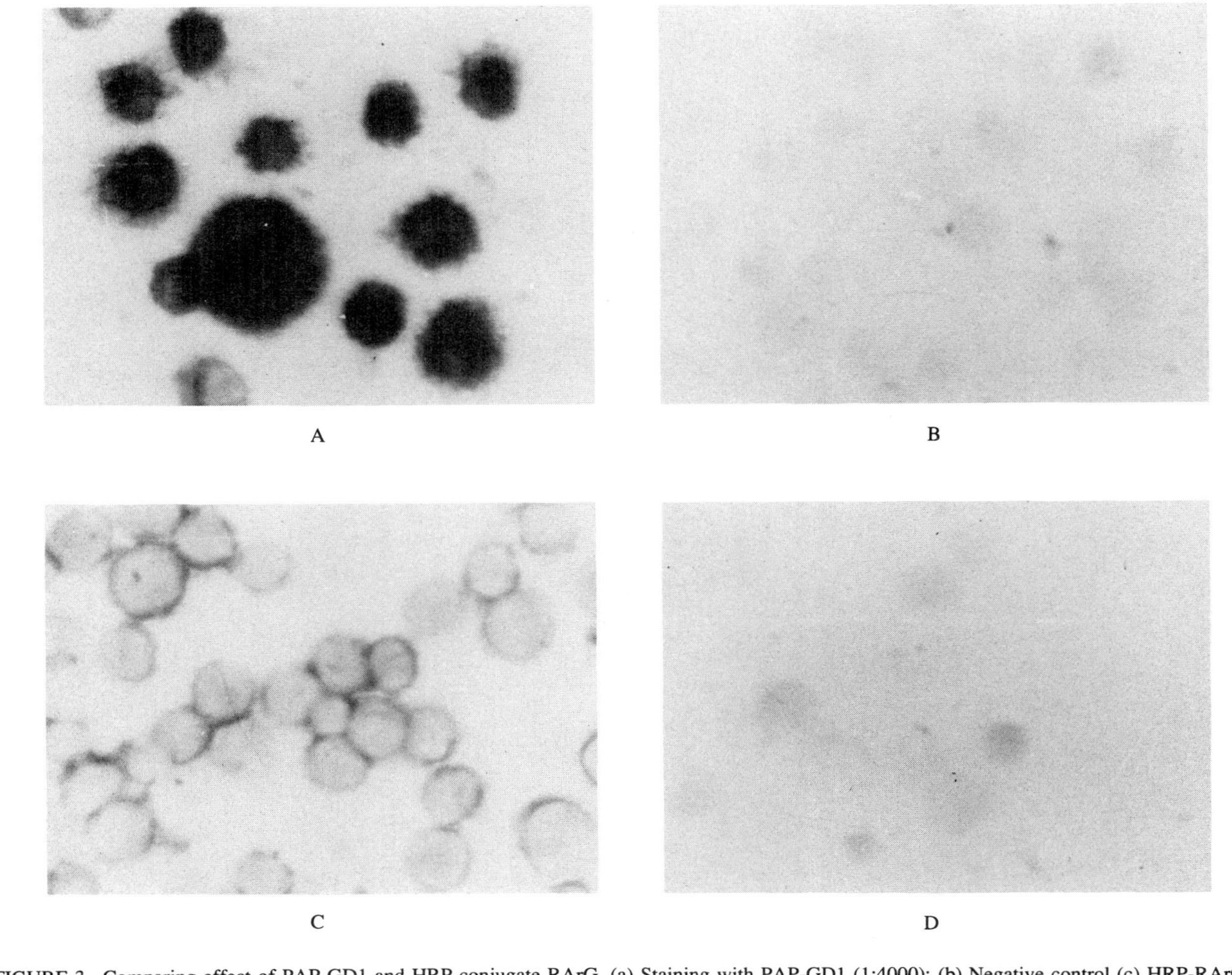

FIGURE 3. Comparing effect of PAP-GD1 and HRP conjugate RArG. (a) Staining with PAP-GD1 (1:4000); (b) Negative control (c) HRP-RArG (1:500); and (d) Negative control.

The rat MAbs, GD1, GC8, against HRP reacted with HRP type VI (RZ=3). No cross reaction with nonspecific antigen (such as type I, crude HRP, BSA, HSA, rabbit serum, and mouse serum) was observed. Unexpectedly, we found the GC8 had a strong reaction with beta-galactosidase, additionally. It was difficult to explain this phenomenom. To identify the specification of rat anti-HRP MAbs, we change the ELISA protocol as follows: the ELISA plates were coated overnight at 4°C with MARK-1, blocked with blocking solution, incubated with serial dilutions of MAbs GD1 and GC8 for 1 h, washed and then the HRP (type VI, type I, and crude HRP) and beta-galactosidase (20 μg/ml) were added. After reacting, the plates were washed again and HRP and beta-galactodidase were revealed with OPD and O-Nitrophenyl-beta-galactopypanoside, respectively. Finally, the wells were evaluated with OD_{492} and OD_{414} nm on the ELISA reader. The results of this experiment showed that the MAbs GD1 and GC8 reacted with HRP (type VI, type I, and crude HRP) and the cross reaction of GC8 with beta-galactosidase disappeared.

The effect of PAP was perfect. This PAP will be used for research work and it will show great economic value.

ACKNOWLEDGMENTS

The authors would like to thank Prof. H. Bazin for having introduced the rat hybridoma technique and for having given cell lines IR983F, MARK-1, and MARK-3. We would also like to thank Dr. Dong Zi-wei for helpful suggestions and Dr. Liu Hua for language corrections.

REFERENCES

1. **Galfré, G., Milstein, C., and Wright, B.,** Rat × rat hybrid myelomas and a monoclonal anti-Fd portion of mouse. IgG, *Nature,* 277, 131, 1979.
2. **Bazin, H.,** Rat-rat hybridoma formation and rat monoclonal antibodies, in *Methods of Hybridoma Formation,* Bartal, A. H. and Hirshaut, Y., Eds., Humana Press, Clifton, NJ, 1987, 337.
3. **Bazin, H.,** Production of rat monoclonal antibodies with the LOU rat non secreting IR983F myeloma cell line, in *Protides of the Biological Fluids,* Peeters, H., Ed., Pergamon Press, Oxford, 1982, 615.
4. **Burdon, R. H.,** *Laboratory Techniques in Biochemistry and Molecular Biology,* Vol. 13, Complell, A. M., Ed., Elsevier, Amsterdam, 1984.
6. **Bazin, H., Xhurdebise, L. M., Burtonboy, G., Lebacq, A. M., De Clercq, L., and Cormont, F.,** Rat monoclonal antibodies. I. Rapid purification from *in vitro* culture supernatants, *J. Immunol. Methods,* 66, 261, 1984.
7. **Bazin, H., Cormont, F., and De Clercq, L.,** Rat monoclonal antibodies. II. Rapid and efficient method of purification from ascitic fluid or serum, *J. Immunol. Methods,* 71, 9, 1984.
8. **Voller, A. and Bidwell, D. E.,** A simple method for detecting antibodies to rubella, *Br. J. Exp. Pathol.,* 56, 338, 1975.
9. **Engvall, E. and Perlmann, P.,** Enzyme-linked immunosorbent assay, ELISA. III. Quantitation of specific antibodies by enzyme-labeled anti-immunoglobulin in antigen-coated tubes, *J. Immunol.,* 109, 129, 1972.

Chapter 19.IX

RAT-RAT AND RAT-MOUSE HYBRIDOMAS SECRETING RAT MONOCLONAL ANTIBODIES ANTI-HORSERADISH PEROXIDASE

M. Jiang, D. Wauters, and H. Bazin

TABLE OF CONTENTS

I. INTRODUCTION

For various reasons which we have already explained, we consider that rat-rat fusions are more interesting to perform than rat-mouse fusions. However, the latter are efficient and can be done perfectly. In this chapter, we shall describe such a rat-mouse fusion and its equivalent in the normal rat-rat technology. The chosen antigen was the horseradish peroxidase (HRP), which is a good immunogen for the rat.

II. IMMUNIZATION

Three-month-old female rats were immunized with 200 μg of HRP (Boehringer, Grad I) mixed with 50 μl of complete Freund's adjuvant (Difco, U.S.) inoculated by the peritoneal route (i.p.) on days 0, 15, and 22. They were tested 10 d later by Ouchterlony test against HRP at a concentration of 1 mg/ml and found immunized. Four months later, they received an intravenous booster of 500 μg of HRP in saline, three days before the fusion.

III. FUSION

A. FIRST FUSION

The fusion was made with the spleen cells with the technique described in Chapter 6, using the rat 983 or the mouse PAIO fusion cells, in order to have a comparison between the two systems. The spleen from the LOU/C immunized rat gave 250×10^6 cells. 120×10^6 spleen cells were fused, respectively, with 24×10^6 983 or PAIO fusion cells. After fusion, the cells were distributed in 96- or 24-well plates at a dose of 2×10^5 or 5×10^5 cells per well, respectively.

B. SECOND FUSION

The fusion was made with the 200×10^6 spleen cells with the technique described in Chapter 6, using the 983 fusion cells.

IV. RESULTS

The results of fusion 1 are given in Table 1. Quite clearly, no significant difference seems to exist between the two systems, which at least in this unique experiment, showed an identical potential. To our knowledge, these identical potentialities of the rat and mouse fusion lines to fuse with immunized rat spleen cells are generally obtained. The results of fusion 2 are also given in Table 1. In Table 2, there is a selected list of MAbs anti-HRP and some of their properties.

V. *IN VIVO* GROWTH OF RAT-MOUSE HYBRIDOMAS

With a subcutaneous dose of 2×10^6 hybrid cells per animal, which is a normal dose to inoculate histocompatible rats or mice with, out of seven cell lines tested, no mouse-rat hybridomas have grown either in LOU/C rats or in BALB/c mice. The same hybridomas, at the same dose and with the same route, have been injected in nude (nu/nu) BALB/c and in (rnu/rnu) LOU/C rats, which are a congenic strain of rats having the rnu allele (Bazin and Kints, unpublished results). The hybridomas have all grown very well in nude BALB/c mice, but not in nude LOU/C rats (Figure 1).

VI. *IN VIVO* PRODUCTION OF RAT MONOCLONAL ANTIBODIES SYNTHESIZED BY RAT-MOUSE HYBRIDOMAS

In nude mice, the rat-mouse hybridomas can easily grow, at least those that we tested, and

TABLE 1
Results Obtained with Spleen Cells from LOU/C Rats Immunized by HRP and Mouse PAIO or Rat 983 Fusion Cells

Fusion line	Number of feeded wells	Hybridoma positive wells	Tested wells[a]	Positive wells
Fusion 1				
BALB/c PAIO	288 + 24[b]	197	80	6
Fusion 1				
LOU/C 983	288 + 24[b]	183	90	5
Fusion 2				
LOU/C 983	960 + 0[b]	697	272	24

[a] Only wells containing the most vigorous clones have been tested.
[b] Small wells + large wells.

TABLE 2
Some Properties of Rat MAbs Developed in Our Laboratory Against Horseradish Peroxidase

Code	Mouse-rat	Rat-rat	Isotype	Avidity
LO-HRP-8	+		IgG2a	
LO-HRP-9	+		Ig2a, kappa	
LO-HRP-10	+		IgG2a, kappa	
LO-HRP-12		+	IgG2a, kappa	6.7×10^9
LO-HRP-13		+	IgG, kappa	5.1×10^9
LO-HRP-14		+	IgG1, kappa	1.7×10^{10}
LO-HRP-15		+	IgG1, kappa	1.0×10^{10}
LO-HRP-16		+	IgG, kappa	4.1×10^9
LO-HRP-17		+	IgM, kappa	1.1×10^9
LO-HRP-18		+	IgM, kappa	2.7×10^9
LO-HRP-19		+	IgG, kappa	$2.8 \text{ x } 10^9$

they can produce large quantities of rat MAb which can be purified by immunoaffinity chromatography, using anti-rat kappa light chain like MARK-1 or anti-rat isotype (Figure 2).

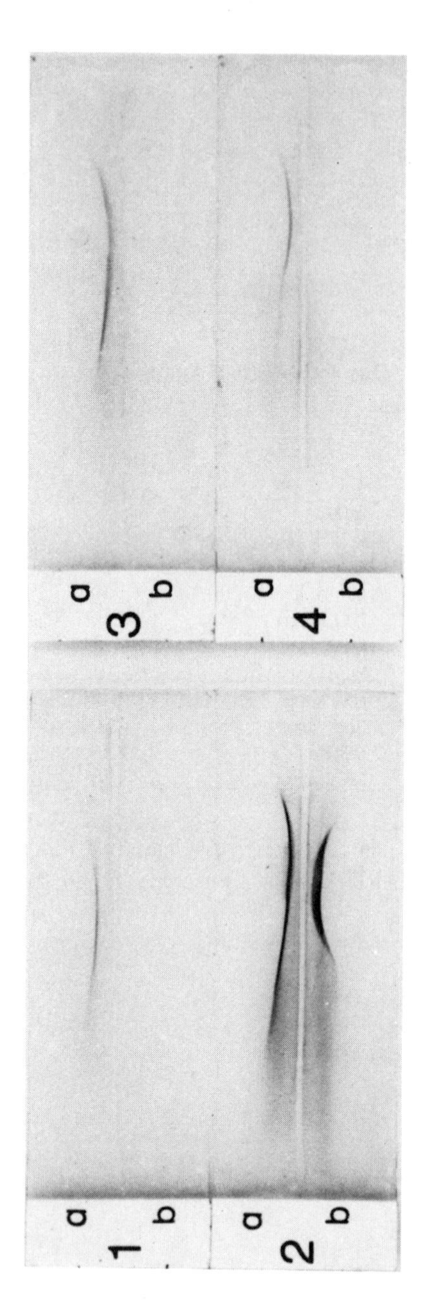

FIGURE 1. Immunoelectrophoresis analysis of top: normal rat serum; bottom: serum of a BALB/c nude mouse bearing the LO-HRP-8 (IgG2a-kappa) Rat-Mouse hybridoma. In the slots, rabbit serum anti-rat: (1) gamma1; (2) gamma2a; (3) gamma2b; and (4) gamma2c. Note the specific reaction obtained in gamma2b.

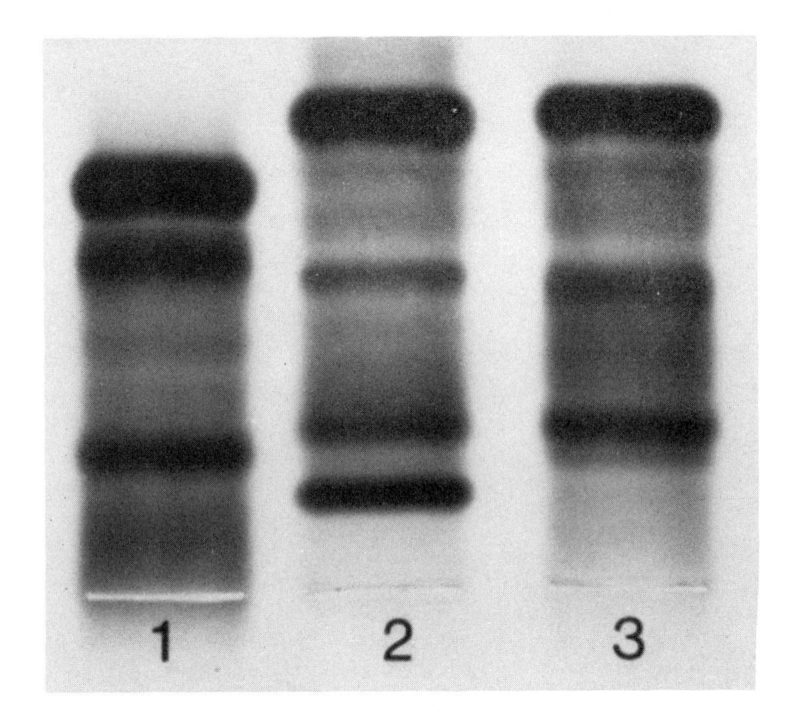

FIGURE 2. Agarose electrophoresis of various sera: (1) normal rat serum; (2) serum of a nude BALB/c mouse carrying a rat-mouse (LOU/C × BALB/c) hybridoma secreting a rat MAb anti-peroxidase; and (3) normal mouse serum.

Chapter 19.X

LOU RAT MONOCLONAL ANTIBODIES AGAINST *XENOPUS* VITELLOGENIN

P. Querinjean

TABLE OF CONTENTS

I. OBTAINMENT

LOU/C rat as wells as BALB/c mice were immunized with male *Xenopus* plasma from animals stimulated by 17-β-oestradiol. Both species reacted equally well after two i.p. inoculations of 0.4 mg plasma proteins with complete Freund adjuvant at a 1-week interval. The LOU rat chosen for fusion was boosted by i.v. inoculation of 0.4 mg plasma proteins at days 3, 2, and 1 before fusion.

Fusion was carried out according to standard procedures in the Experimental Immunology Unit of the University of Louvain.[1] From the 17 hybridomas selected for their growing properties and their positive response in an ELISA for vitellogenin, a single one was cloned and the others frozen for 2 years. After limiting dilution, this hybridoma gave a monoclonal antibody (MAb) producing hybridoma LO-VG1. This was typed as an IgG2a immunoglobulin.

After 2 years, two other hybridomas were rescued and provided two MAbs, LO-VG2 and LO-VG3. The three antibodies exhibited different electrophoretic mobilities.

The work with these MAbs was pursued in two directions: the purification of the MAbs and the setup of an indirect sandwich ELISA for vitellogenin and antibodies directed against this antigen.

II. PURIFICATION OF THE MAbS

LO-VG1 and LO-VG2 were satisfactorily purified with an immunoadsorbant of MARK-1 (mouse anti-rat kappa chain MAb) described elsewhere (Chapter 15), starting from ascites fluid or from a 40% ammonium sulfate precipitate of the same material.

In order to handle larger quantities of material, a nonimmunological method of purification was designed. The best setup was obtained for the LO-V62 MAb. A 1.6 *M* ammonium sulfate precipitate was fractionated on S-Sepharose FF (Pharmacia) and gave a minimum contamination by the Igs from the host.

Other combinations with DEAE-Trisacryl M, DEAE-Sepharose CL6B and S-Sepharose FF did not provide a satisfactory purification for LO-VG1 antibody.

The use of a strong anionic ion-exchanger like the QAE-Sepharose or an hydrophobic support should give the key to obtain a pure MAb. This work provided a good example of the need for the optimization of the purification procedure for individual MAbs and the possibility to reduce, if not exclude, the Igs from the host.

III. ELISA FOR VITELLOGENIN (Vg) AND ANTIBODIES AGAINST THIS ANTIGEN

A rather complex assay was set for the screening of hybridomas in order to use the same ELISA for the evaluation of vitellogenin and its antibodies.

The polystyrene microplate was coated with polyclonal rabbit anti-Vg and saturated with dry defatted evaporated milk at 5%. Vitellogenin was then added in PBS. Antibodies, polyclonal or monoclonal, were laid on vitellogenin, followed by rabbit anti-rat Ig conjugated with horseradish peroxidase (DAKO, Denmark). Peroxidase was revealed by hydrogen peroxide (8.8 m*M*) in presence of ABTS (1.9 m*M*) in a 0.1 *M* sodium citrate-dipotassium hydrogenophosphate buffer adjusted at pH 4.0. The absorbance readings were performed with a BIOTEK 307 BER ELISA reader at 405 nm after azide stopping.

Special attention was given to the optimization of the ELISA data evaluation using the BIOTEK REGRESS software.[2]

The raw data were first manually plotted and subsequently processed off-line with the help of the BIOTEK REGRESS software complemented with the LOTUS 1, 2, 3 software.

The data processing used a combination of the 1st, 2nd or 3rd order polynomial regression and the lin/lin, lin/log, or log/log plotting systems of the REGRESS software. The concentrations of the coating antibody, of the antigen (Vg), and of the primary antibody were optimized.

The most suitable amount of coating antibody was 100 μl of a 20 μg/ml solution (C.V. = 10%). Vitellogenin, 100 μl at 40 ng/ml (correlation coefficient r = 1.0), should be used to evaluate rat anti-Vg antibodies in the concentration range of 4 to 100 μg/ml. With 100 μl of a 40 μg/ml (r = 0.996) solution of rat anti-Vg primary antibody, the useful range of the Vg concentrations to be measured extended from 2 to 400 ng/ml.

Preliminary evalutions of the Vg content of *Xenopus borealis* untransformed (XB 693 C) and transformed (XB 693 T) cell extracts[3] as well as the detection of LOUVAIN rat anti-Vg MAbs[1] in hybridoma culture supernatants or in sera and ascites from rats carrying the LO-VG hybridomas were performed using the optimized ELISA.

The LO-VG MAbs have been assayed successfully to ascertain the sex of *Xenopus* animals and may be helpful to different sex fish species.

ACKNOWLEDGMENTS

This work results from the sum of several collaborations of A. Avno, C. Blum, J. Delcourt, C. Gilles, M. Guillaume, P. Herman, P. Hermand, M. Lefebvre, P. Lempereur, and P. Voet. All these students are here acknowledged for their contributions. This work was supported by grant no. 3.154.79 from the FRSM, Belgium. P.Q. is a research associate from the FNRS, Belgium.

REFERENCES

1. **Bazin, H.,** Production of rat monoclonal antibodies with the LOU rat non secreting IR983F myeloma cell line, in *Protides of the Biological Fluids,* Peeters, H., Ed., Pergamon Press, Oxford, 1982, 615.
2. **Gilles, C., Delcourt, J., and Querinjean, P.,** Optimalization of an ELISA for the evaluation of *Xenopus laevis* vitellogenin and a few monoclonal antibodies against this antigen, *Arch. Int. Physiol. Biochem,* 95, B, 1988.
3. **Picard, J. J., Afifi, A., and Pays, A.,** An oncogenic cell line inducing transplantable metastazing adenocarcinomas, in *Xenopus borealis, Carcinogenesis,* 4, 739, 1983.

Chapter 19.XI

PRODUCTION AND CHARACTERIZATION OF RAT-RAT HYBRIDOMAS AGAINST DNP-HAPTEN

B. Platteau, M. Rits, F. Cormont, D. Wauters, and H. Bazin

TABLE OF CONTENTS

I. INTRODUCTION

Rat anti-dinitrophenyl hapten (DNP) monoclonal antibodies (MAbs) of different isotypes were obtained by fusing 983 rat fusion cells with lymph node or spleen cells from appropriately immunized LOU/C rats.

Hybrids secreting monoclonal rat IgM-, IgG1-, IgG2a-, IgG2b-, IgG2c-, IgE-, or IgA-antibodies specific for the DNP-hapten were produced after immunizing 6 to 8-week-old LOU/C rats with the DNP-hapten conjugated to different soluble immunogens as ovoalbumin (OVA), hydroxyethyl starch (HES), or more or less insoluble antigens as *Nippostrongylus brasiliensis* extract (NP), *Ascaris suum* extract (ASC), or to a particulate antigen like *Salmonella typhimurium* (S).

Several different routes and schedules of immunization were tested, giving each a large number of hybridoma colonies producing anti-DNP antibodies.

II. MATERIALS AND METHODS

A. IMMUNOGENS AND ADJUVANT

1. *Nippostrongylus brasiliensis* (Nb) larvae were kindly provided by the Union Chimique Belge, UCB, Braine-L'Alleud, Belgium. A saline extract of *N. brasiliensis* was prepared according to the method described by Kojima et al.[1]. The DNP-Nb conjugate was prepared as described by Ovary and Benacerraf.[2]
2. Dinitrophenylated *Ascaris suum* extract (DNP-ASC) was kindly given by T. Tada (Tokyo University, Japan) and prepared by the method described by Tada and Okumara.[3]
3. Lightly dinitrophenylated egg albumin (DNP2-OVA) was prepared as described by Ishizaka and Okudaira.[4]
4. Dinitrophenylated bovine gammaglobulin (DNP28-BGG) was prepared by the same method and used for antibody titrations.
5. Dinitrophenylated hydroxyethyl starch (DNP-HES) was kindly given by J. H. Humphrey, Royal Postgraduate Medical School, London, and was prepared as described by Humphrey.[5]
6. Dinitrophenylate conjugated *Salmonella typhimurium* (DNP-S) was prepared as described by Rits et al.[6]
7. The adjuvant, when used by primary immunizations, was *Bordetella pertussis* (Bp), Perthydral vaccine containing 0.1% of alumin hydroxide (Institut Pasteur Production, Paris).

B. PREPARATION OF HAPTEN CONJUGATES

The haptenic reagent, 2,4-dinitrobenzene sulfonic acid, in the form of its salt (DNBS), was purchased from Eastman Kodak Company, Rochester, N.Y. The DNP-antigen conjugates were prepared by the reaction of sodium 2,4-dinitrobenzene sulfonate with antigens in 0.2 *M* Na_2CO_3 solution, pH 10.7. The uncoupled hapten was removed by gel filtration through a column of Sephadex G-25 equilibrated with a solution of phosphate-buffered saline (PBS) containing 0.15 *M* NaCl in 0.02 *M* phosphate buffer, pH 7.2.

The concentration of the reagent and the reaction time for each conjugate are described in the references.[2-6]

C. IMMUNIZATIONS

Different immunization protocols were tested, using different immunogens conjugated to the DNP-hapten with or without adjuvant. The rats were immunized by injecting the antigen in the footpads (i.f.), intraperitoneally (i.p.), intravenously in the lateral tail vein (i.v.), subcutaneously (s.c.), in the Peyer's patches (i.p.p.), or in the spleen (i.spl.) (Figure 1).

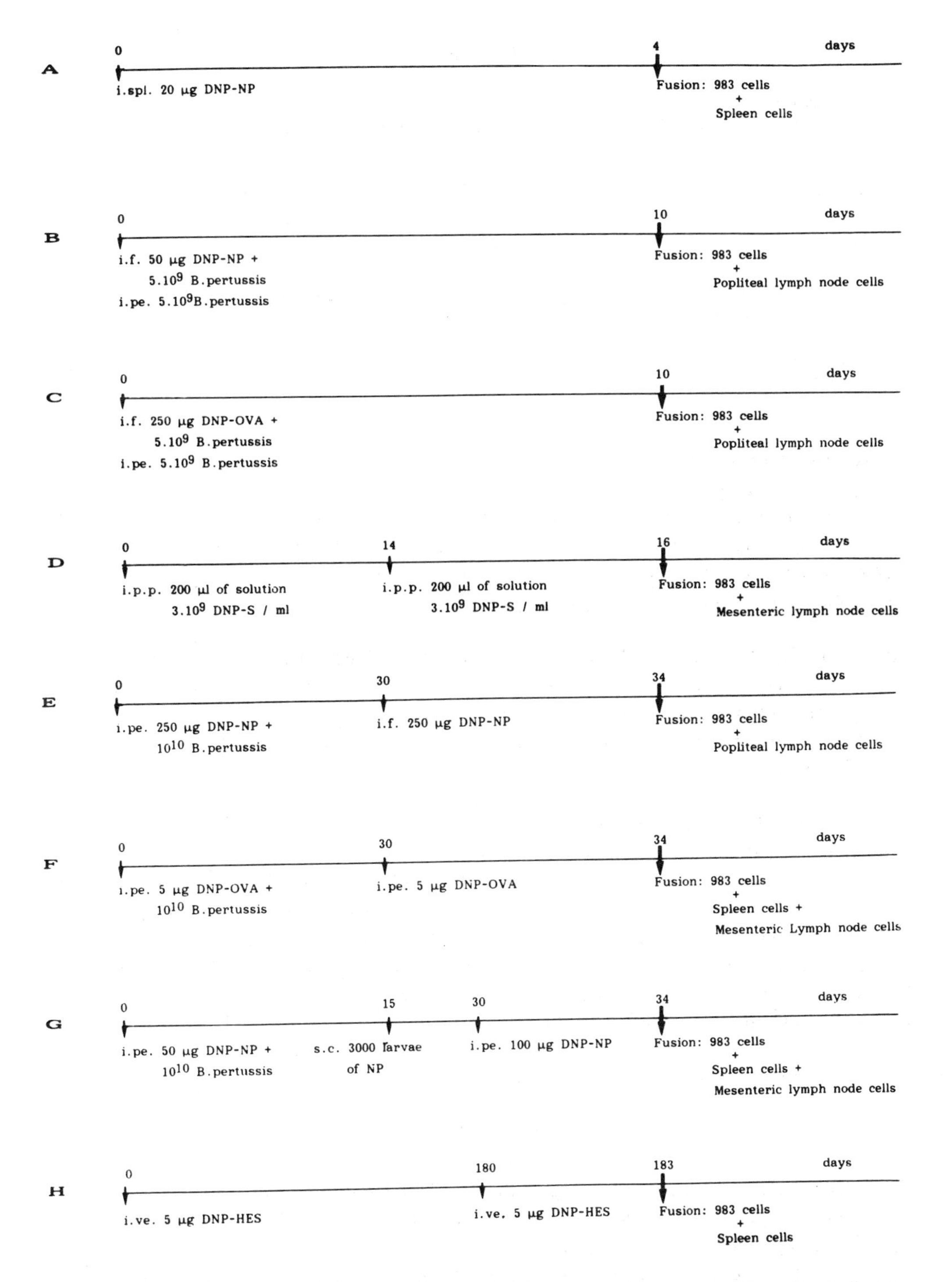

FIGURE 1. Immunization protocols. Rats were injected with dinitrophenylated immunogens with or without adjuvant. Abbreviations—immunogens: ovoalbumine (OVA), hydroxyethyl starch (HES), *Nippostrongylus brasiliensis* extract (NP), *ascaris suum* extract (ASC), Salmonella typhimurium (S); adjuvant: *Bordetella pertussis* (BP); injection routes: in the footpads (i.f.), intraperitoneally (i.p.), intravenously in the lateral tail vein (i.v.), subcutaneously (s.c.), in the Peyer's patches (i.p.p.), or in the spleen (i.spl.).

TABLE 1

Immunization schedule[a]	IgM	IgG1	IgG2a	IgG2b	IgG2c	IgA	IgE
A	+	–	–	–	–	–	–
B, C, E, F, G	+	+	+	+	+	–	+
D	+	+	NT[b]	+	NT	+	NT
H	+	–	–	–	+	–	+

Note: Isotypes of rat monoclonal antibodies against the DNP hapten usually resulting from the various models of immunization.

[a] See Figure 1.
[b] Not tested.

D. HYBRIDIZATIONS

Spleen cells and/or lymph node cells from immunized LOU/C donor rats were fused with the 983 rat myeloma cells in a 5:1 ratio. Culture media were screened for antibodies against DNP, and the isotype of anti-hapten antibodies was determined. After subcloning by limiting dilution the positive hybrid cell lines were either maintained *in vitro,* or alternatively maintained *in vivo* as ascitic tumors in LOU/C rats[7] (see Chapter 4).

E. SCREENING AND CHARACTERIZATION OF HYBRIDOMA SUPERNATANTS

Several methods were used to assay the hybridoma supernatants for anti-DNP antibodies.

Specific antibodies for DNP activity can be detected by a solid-phase radioimmunoassay (RIA) or by an enzyme-linked immunosorbent assay (ELISA), or even by Ouchterlony against DNP28-BGG.

The determination of the isotype of anti-hapten antibody was carried out by ELISA. Polystyrene wells were coated with the hapten on a carrier not used for immunization (DNP28-BGG), incubated with the hybridoma supernatant, washed and further incubated with peroxidase-labeled mouse MAbs specific for either rat IgM, IgG1, IgG2a, IgG2b IgG2c, IgE, or IgA.

Reaginic anti-DNP IgE antibodies were also detected by passive cutaneous anaphylaxis (PCA) as described by Platteau and Bazin.[8] Briefly, an intradermal injection of 0.1 ml of hybridoma supernatant was given on the shaved back of 250 g female WISTAR R rats. After a sensitization period of 48 to 72 h the rats were challenged by intravenous injection with a saline solution containing 2.5 mg DNP28 BGG and 5 mg of Evans Blue (Gurr, England). The skin reactions were examined after 20 min and a blueing reaction on the skin showed the presence of DNP reaginic antibodies in the supernatant.

III. RESULTS

A considerable number of immunoglobulin-secreting hybridomas with specificity for DNP has been derived from the fusions of 983 rat myeloma cells with spleen cells, mesenteric, or popliteal lymph node cells of rats immunized as described in Figure 1. IgA secreting hybridomas against DNP had only been obtained by immunizing the rats via the Peyer's patches with DNP-S as described by Rits et al.[6]

Table 1 shows that hybridomas secreting MAbs of different classes and subclasses have been obtained with the different immunization routes, as well as with the different DNP conjugated carriers.

IV. CONCLUSIONS

Quite clearly, these results as well as those shown in Chapter 1 (rat specific immune responses against DNP) demonstrate that it is important to choose correctly the immunization protocol according to the isotype of MAb required. However, many such protocols can give identical results, at least those giving IgM-IgG immune responses.

REFERENCES

1. **Kojima, S., Yokogawa, M., and Tada, T.,** Production and properties of reaginic antibodies in rabbits infected with clonorchis sinensis or Schistosoma japonicum, *Exp. Parasitol.,* 35, 141, 1974.
2. **Ovary, Z. and Benacerraf, B.,** Immunological specificity of the secondary response with Dinitrophenylated proteins, *Proc. Soc. Exp. Biol. Med.,* 114, 72, 1963.
3. **Tada, T. and Okumura, K.,** Regulation of homocytotropic antibody formation in the rat. I. Feed-back regulation by passively administered antibody, *J. Immunol.,* 108, 1002, 1971.
4. **Ishizaka, K. and Okudaira, H.,** Reaginic antibody formation in the mouse. II. Enhancement formation by priming with carrier, *J. Immunol.,* 110, 1067, 1973.
5. **Humphrey, J. H.,** Tolerogenic or immunogenic activity of hapten conjugated polysaccharides correlated with cellular localization, *Eur. J. Immunol.,* 11, 212, 1981.
6. **Rits, M., Cormont, F., Bazin, H., Meykens, R., and Vaerman, J. P.,** Rat monoclonal antibodies. VI. Production of IgA secreting hybridomas with specificity for the 2,4 dinitrophenyl (DNP) hapten, *J. Immunol. Methods,* 89, 81, 1986.
7. **Bazin, H., Xhurdebise, L. M., Burtonboy, G., Lebacq, A. M., De Clercq, L., and Cormont, F.,** Rat monoclonal antibodies. I. Rapid purification from *in vitro* culture supernatants, *J. Immunol. Methods,* 66, 261, 1984.
8. **Platteau, B. and Bazin, H.,** TB-mice: a model of high and long lasting reaginic responders, *J. Immunol.,* 20, 53, 1978.

Chapter 20.I

METHODS FOR ANALYSIS OF RAT MONOCLONAL ANTIBODIES DIRECTED AGAINST HUMAN LEUKOCYTE DIFFERENTIATION ANTIGENS

A. M. Ravoet, D. Latinne, J. Seghers, P. Manouvriez, J. Ninane, M. De Bruyère, H. Bazin, and G. Sokal

TABLE OF CONTENTS

I. INTRODUCTION

In this chapter methods for the analysis of the specificity of rat monoclonal antibodies (MAbs) against cell surface antigens are illustrated with examples of MAbs directed against human normal leukocytes, activated lymphocytes, or leukemic cells (cALL). Analysis includes:

1. Study of the reactivity of the MAbs with a large number of normal cells or leukemic/lymphoma cells
2. Biochemical analysis of the antigen
3. Competition/blocking assays using reference MAbs
4. Determination of the biological activity of the MAbs (see also Chapter 30).

Detailed procedures are given at the end of this chapter.

II. PRODUCTION AND FIRST SELECTION OF HYBRIDS

LOU (LOU/C or LOU/M) rats were immunized by injection of thymocytes, peripheral blood mononuclear cells contaminated with platelets, phytohemagglutinin(PHA)-activated lymphocytes, leukemic cells from patients with chronic lymphocytic leukemia (B-CLL), common acute lymphoblastic leukemia (cALL), or in lymphoid blast of chronic myelogenous leukemia (Table 1). From 2 to 5×10^7 cells were injected into the peritoneal cavity of the rat at a 3-week interval. Four days after the second injection, fusion of splenocytes with IR983F immunocytoma cells was performed according to the methods described in this volume (Chapters 6 and 7). From the 5th d after fusion, clonal growth was observed. Hybrids were transferred into 2-ml wells, 8 to 20 d after fusion. Once the clones had grown large enough (covering $^1/_3$ of the well surface and medium turning yellowish), hybrid supernatants were screened for antibody activity towards the target cells. In order to carry out a first selection, positive supernatants were screened on granulocytes. Cross-reacting MAb-secreting clones (about 90% of the positive clones) were discarded. The interesting hybrids were cloned, expanded and produced *in vivo*. Only after *in vivo* growth and cryopreservation of the tumor, was detailed analysis started.

III. REACTIVITY OF MAbS WITH DIFFERENT CELL TYPES

A first step in analysis of MAb against leukocyte differentiation antigens, is to determine the overall reactivity pattern of the MAbs with cells of different lineages and at different stages of differentiation. The panel therefore includes normal cells (complex mixtures and purified cell preparations), clinical samples of leukemia/lymphomas and the universally used leukemia cell lines. Serum or ascites of a hybridoma-bearing rat is used as a source of MAb. A 500 to 2000 times dilution of this monoclonal ascites generally provides enough MAb to saturate all available antigen, and reduces the interference of the polyclonal antibodies.

TABLE 1
Rat Monoclonal Antibodies Discussed in this Chapter

Immunogen	Name	Monoclonal antibodies: Specificity		Subclass
Thymocyte	LO-CD1-a	CD1	p45	IgG2a, NT
	LO-CD5-a	CD5	p68	IgG2a, kappa
Peripheral blood	LO-MO1	—	—	NT
mononuclear cells	LO-PL3-a	CDW41	106; 96; 23	IgG2a, kappa
+ platelets	LO-PL3-b	CDW41	106; 96; 23	IgG2a, kappa
	LO-PL4	CD9	p24	
PHA activated	LO-Tact-1	CD25	p63	IgG2b, kappa
lymphocytes	LO-Tact-2	CD25	p63	IgG2b, kappa
	LO-DRa	HLA-DR	p29; 33	IgG2b, kappa
Tonsil cells	LO-DQ-a	HLA-DQ	p27; 31	NT
	LO-CD21-a	CD21	p140	NT
B-CLL lymphocytes	LO-panB-a	—	—	IgG2b, kappa
	LO-D-b	MHCII	p29; 33	IgG2a, kappa
	LO-D-c	MHCII	p27; 31	IgG2b, kappa
	LO-D-d	MHCII	p27; 31	IgG2b, lambda
CML-blast crisis	AL1-a	CD24	p180, 220	IgG2c, NT
lymphoblasts[a]	AL6	CD9	p24	IgG2b, NT
cALL lymphoblasts[a]	AL2	CD10	p100	IgG2b, NT
	AL4	CD10	p100	IgM, NT

[a] These early experiments were described by Lebacq and Bazin[1] and Lebacq-Verheyden et al.[2]

Different techniques can be used for analysis: immunofluorescence, immunohistology, radiobinding assay, and ELISA. Whenever a cytofluorograph equipped with an electronic analyzing system is available, immunofluorescence is the most rapid and accurate method. Information about the percentage of labeled cells in a suspension and the intensity of labeling are provided. Moreover, analysis can be performed on a subpopulation of a cell sample. Even without the facility of a cytofluorograph, immunofluorescence or immunohistology should be preferred over ELISA or radiobinding assays. These two methods measure the total amount of antigen in the sample, but do not distinguish a subpopulation of brightly labeled cells from a whole population of dimly labeled cells (see also Chapter 9).

Reactivity patterns of our rat MAbs with normal cells are given in Table 2. Some specificities were very neatly delimited: LO-CD5-a recognized the T-lineage, LO-CD1-a reacted with thymocytes, LO-panB-a, LO-CD21-a, and LO-DQ-a reacted with B cells, LO-MO1 reacted only with mature monocytes, LO-PL3-b with platelets, and LO-Tact-1 with activated T lymphocytes. Other reactivity patterns were more complex: LO-DRa reacted with B cells, monocytes, and activated T cells. The data obtained for AL1a, AL2, and AL6 were not informative regarding the specificities; only the cross-reactivities with granulocytes (AL1a, AL2), platelets, and monocytes (AL6) were noticed.

Reactivities of the MAbs with leukemia/lymphoma derived cell lines (Table 3) and with leukemia/lymphoma clinical samples (Table 4) confirm and extend the specificities: LO-CD5-a reacted with T-ALL, T-CLL, and weakly with B-CLL and B-NHL (non-Hodgkin lymphoma); LO-CD1-a with some T-ALL; LO-DR-a with non T-ALL, B-CLL, B-lymphomas, and most ANLL. LO-panB-a strongly stained the B-NHL, much weaker the B-CLL, and pre-B-ALL. Staining with LO-CD21-a and LO-DQ-a was much weaker on leukemia/lymphoma clinical samples than on normal B cells. Neither LO-Tact1, nor LO-MO1 were found to stain the leukemia/lymphoma cell lines tested; LO-PL3-b was found on occasional ANLL (5/157), therefore classified as ANLL-M7. AL2 defined cALL and also reacted with some B-NHL; AL1-

TABLE 2
Reactivity Pattern of Rat MAbs Directed Against Human Leukocyte Differentiation Antigens with Normal Cell Types

Name of the MAb reactivity	LO-CD5-a CD5	LO-CD1-a CD1	LO-panB-a —	LO-CD21-a CD21	LO-DR-a HLA-DR	LO-DQ-a HLA-DQ	LO-Tact-1 CD25	LO-MO1 —	LO-PL3-b CD41	AL1a CD24	AL2 CD10	AL6 CD9
Blood cells												
Erythrocytes	–	–	–	–	–	–	–	–	–	–	–	–
Platelets	–	–	–	–	–	–	–	–	+++	–	–	++
Granulocytes	–	–	–	–	–	–	–	–	–	+	(+)	–
Monocytes(+)	– bg	– bg	–	++	–	–bg	+++	–	–bg	–	(+)	
Lymphocytes total	++s	–	+s	+s	+s	–	–<5%	–	–	–	–	+s
E+	+	–	–	NT	–	NT	–<5%	–	–	–	–	+s
Lymphocytes PHA act.	++	–	–	NT	(+)	NT	+++	–	–	NT	NT	NT
Tonsil cells total	+s	–	+s	+s	++s	+s	–	–<5%	–	(+)s	–<5%	+s
E-	–	–	+	+	+++	+	–	–	–	(+)	–	
E- SAC act.	–	NT	++	NT	+++	NT	–	–	NT	NT	NT	NT
Thymus total	++	++	– <5%	–	+s	–	–	–	–	–	–	+s
Bone marrow MN	– <5%	–	(+)s	NT	+s	NT	–	+s	–	+s	((+))<5%	+<5%

Note: Monocytes and granulocytes were purified by centrifugation on a Percoll gradient (Ulmer and Flad,[3] Mannoni et al.[4]). A purified platelet preparation was obtained by differential centrifugation. E- and E+ lymphocytes were separated on a Ficoll-Paque gradient after rosetting with 2-aminoethyllisothiouronium (AET)-treated sheep erythrocytes. Tonsil B cells were activated by a 4 d culture in the presence of formalin-fixed *Staphylococcus aureus* Cowan I (SAC) strain (1×10^7 bacteria/ml). Phytohemagglutinin (PHA)-activation of T lymphocytes was obtained after a 3 d culture in the presence of 3 μg phytohemagglutinin/ml. Results are expressed as -: negative; bg: background reactivity, i.e., same intensity and same percentage as in the control containing normal rat serum; (+): weakly labeled, hardly or not detected by microscopy, but clearly labeled when analyzed on EPICS; +: clearly; ++: bright; +++: very bright; s: subpopulation labeled >5%.

TABLE 3
Reactivity Pattern of Rat MAbs Directed against Human Leukocyte Differentiation Antigens with Human Leukemia/Lymphoma Lines

Name of the MAb reactivity	LO-CD5-a CD5	LO-CD1-a CD1	LO-panB-a —	LO-CD21-a CD21	LO-DR-a HLA-DR	LO-DQ-a HLA-DQ	LO-Tact1 CD25	LO-MO1 —	LO-PL3-b CD41	AL1a CD24	AL2 CD10	AL6 CD9
nTnB ALL												
KM3[5]	–	–	+	–	+++	–	–	–	NT	++	+	+
NALM-6[6]	–	–	–	NT	++	NT	–	–	–	++	++	+++
T-ALL												
HPB-ALL[7]	++	++	–	NT	–	NT	–	–	–	–	(+)	–
CCRF-CEM[8]	++	((+))	–	NT	–	NT	–	–	–	–	(+)	++
Jurkatt	(+)	(+)	–	NT	–	NT	–	–	–	–	–	–
CCRF-HSB2[9]	–	–	–	–	–	–	–	–	NT	((+))	–	–
B-lymphoma Raji[10]/Daudi[11]	–	–	+	+	++	+	–	–	NT	–	+	–
EBV-lymphoblastoid UC927-6	–	–	+++	++	+++	++	–	–	NT	–	–	–
Plasmablast												
ARH77[12]	–	–	+++	++	+++	++	–	–	NT	–	–	(+)
IM9	–	–	++	++	+++	++	–	–	NT	–	–	–
Myeloid												
K562[13]	–	–	–	–	–	–	–	–	–	++	–	+
HL60[14]	–	–	–	–	–	–	–	–	NT	–	–	–
U937[15]	–	–	((+))	–	++	–	–	–	NT	–	–	–

Note: Continuous cell lines were gifts from different laboratories. (Imperial Cancer Research Fund, London; Royal Free Hospital, London, Vrije Universiteit of Brussels) They were maintained in RPMI-10% FCS.

TABLE 4
Reactivities of Rat MAbs with Clinical Samples of Leukemias and Lymphomas

MAb	LO-CD5-a	LO-CD1-a	LO-panB-a	LO-CD21-a	LO-DR-a	LO-DQ-a	AL1a	AL2	AL6
Null AUL	NT	NT	NT	NT	NT	NT	2/9	0/9	9/9
cALL	0/5	0/5	8^w/9	0/4	9/9	3^w/4	38/48	48/48	47/50
T-ALL	6/6	1/6	0/6	5^w/6	0/6	0/6	0/6	0/6	2/6
T-CLL	1/1	0/1	0/1	0/1	0/1	0/1	0/1	0/1	0/1
B-NHL	4^w/4	0/4	4/4	2^w/2	4/4	2^w/2	11/11	2/11	0/9
B-CLL	4^w/4	0/4	12/14	3^w/4	14/14	4^w/4	10/10	0/10	4/10
Myeloma	NT	NT	0/1	NT	0/1	NT	0/1	0/1	NT

a labeled B-CLL, B-NHL, most cALL, and some null-ALL. Hence the AL1-a antigen appears earlier during B differentiation than common acute lymphoblastic leukemia antigen (CALLA) and is a useful marker of the early B-lineage.

Reactivity patterns of most of our MAbs correlated exactly with those of mouse reference MAbs: LO-CD5-a with anti-Leu-1 (Becton Dickinson), LO-CD1-a with OKT6 (Ortho Diagnostics), LO-DRa with OKIa (Ortho Diagnostics), AL2 with J5 (Coulter), and AL6 with BA2 (Hybritech). For other MAbs cell typing on a large number of clinical samples was needed before small differences with reference MAbs were noticed: AL1a labeled non T-ALL cells as brightly as BA1 (Hybritech) did. However, B-CLL's, B-lymphoma's, and normal B tonsil cells were stained much weaker with AL1a than with BA1; differences of labeling between LO-panB-a and B1 (Coulter) are noticed when cALL samples are analyzed.

In the case of LO-Tact1 and LO-Tact2, it was necessary to use different preparations of activated T cells in order to get some insight into the specificities of these MAbs (Table 5). Not detected on normal resting peripheral blood lymphocytes, the antigen of LO-Tact1 and LO-Tact2 appeared after activation. Polyclonal activation by lectins transformed all T cells into lymphoblasts, brightly labeled with LO-Tact1 and LO-Tact2. After a 5 to 8 d mixed lymphocyte culture, only a fraction of the lymphocytes were activated. Precisely this blast population was stained with LO-Tact1 and LO-Tact2. Expression of the LO-Tact1 and LO-Tact2 antigen paralleled the expression of the interleukin-2 (IL2) receptor, as defined by a mouse anti-IL2 receptor MAb (2A3) (Becton Dickinson).

IV. IMMUNOPRECIPITATION OF THE ANTIGEN AFTER CELL RADIOLABELING

A second step in the analysis of the MAbs is to get some information about the antigen recognized. The most straightforward technique in this respect is radiolabeling of cells, followed by purification of the antigen by immunoprecipitation, and analysis of the antigen by polyacrylamide gel-electrophoresis (PAGE).[16]

In order to detect, after purification, the minute amounts of antigen present in cells (for example, 10^6 Nalm 6 cells will contain around 30 ng of CALLA), antigens are radiolabeled before immunoprecipitation. The labeling must be nonaggressive. Hence, biosynthetic incorporation of labeled amino acids or sugars, or enzyme-catalyzed radiolabeling of surface proteins or carbohydrates are chosen.

The method most commonly used is ^{125}I-iodination of cell surface tyrosyl residues by lactoperoxidase using Na ^{125}I and H_2O_2.[17,18] *In situ* H_2O_2 is generated continuously by the glucose oxidase-glucose system. Using 1 mCi Na ^{125}I to label 10^7 cells, around 10^8 cpm/10^7 cells are currently incorporated, which means a high labeling of most antigens (Figures 1 and 4). Background labeling remains low.

Using PHA-activated lymphocytes as a source of antigen, LO-Tact1, LO-Tact2 as well as the

TABLE 5
Expression of the LO-Tact1 and LO-Tact2 Antigens after Activation of Lymphocytes

	Percentage of positive cells				
	LO-Tact1	**LO-Tact2**	**IL2 receptor**	**HLA-DR**	**Transferrin receptor**
Lymphocytes					
Resting	5	3	10	7	0
PHA activated	95	96	95	47	78
ConA activated	96	94	91	26	83
MLR day					
0	3	0	3	8	3
4	4	2	9	7	0
5	3(52)	1(48)	9(63)	9(44)	1(48)
6	12(61)	6(25)	21(74)	18(50)	19(49)
7	18(68)	14(47)	20(68)	17(53)	13(45)
8	30(71)	21(48)	38(78)	26(62)	24(47)

Note: Peripheral blood mononuclear cells (PBMN) (1×10^6/ml) were cultured for 3 d in the presence of 3 μg phytohemagglutinine/ml or 20 μg concanavalin A/ml in 24-well culture plates. For bidirectional mixed lymphocyte reaction (MLR), PBMN from two blood donors (1×10^6cells/ml) were cocultured for 4 to 8 days in tubes (4 ml/tube). Analysis was done on the total population of viable cells or on the large size cells only (numbers between brackets).

mouse anti-IL2 receptor MAb, 2A3 (Becton Dickinson) were shown to precipitate a 63 kDa antigen; LO-CD5-a and Leu-1 (Becton Dickinson) precipitated a 68 kDa antigen (Figure 1). Using HPB-ALL cells, LO-CD1-a precipitated an antigen with exactly the same molecular weight as OKT6, i.e., a 43 kDa protein. The molecular weight of the antigens precipitated by the other MAbs (Table 1) was determined by the same method, except for the antigen of AL1-a, which requires endogenous labeling.[19]

In another method of external labeling, surface galactosyl residues are oxidized by galactose oxidase, then reduced by NaB_3H_4. The efficiency of ^3H incorporation is enhanced by a neuraminidase pretreatment, which uncovers galactosyl residues in complex-type glycoproteins.[20] Incorporation of 5×10^6 cpm/10^7 cells allows detection of major glycoproteins.[2] This method is especially valuable when the antigen is a highly glycosylated protein and large amounts of cells are available. For instance, the platelet glycoprotein complexes Ib and IIb-IIIa are more clearly detected by this method than by iodination[21] (Figure 2). For biosynthetic incorporation of amino acids, ^{35}S-methionine and ^3H-leucine are the most commonly used. Culture media lacking these amino acids are commercially available. Although this method is easy to perform, it gives high background labeling and therefore requires multiple preclearing steps before immunoprecipitation. Some studies, including kinetics of biosynthesis, detection of large molecular weight precursors,[22] inhibition of N-glycosylation by tunicamycin, and effective demonstration of biosynthesis of an antigen by a certain cell type, require biosynthetic labeling. For example, when ^{35}S-methionine incorporation was performed in the presence of 25 μg tunicamycin/ml, the CALLA molecule immunoprecipitated by AL2 migrated with an apparent molecular weight of 86 kDa, instead of 100 kDa in untreated cells (Figure 3). Hence, CALLA is a N-glycosylated protein. After Braun et al.[23] reported that ^{125}I-labeled CALLA could be precipitated from granulocytes, some controversy arose as to whether granulocytes synthesized CALLA or adsorbed the molecule from the blood. An experiment described in Figure 3 shows that CALLA is indeed synthesized by granulocytes, but has a molecular weight slightly higher than in pre-B lymphoblasts.

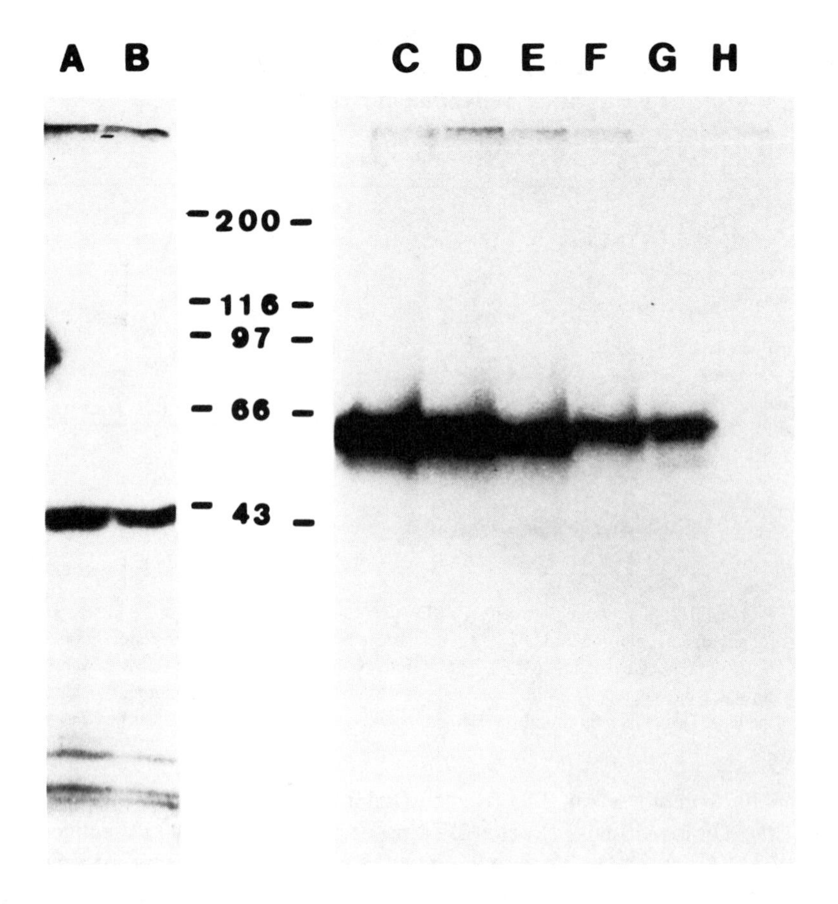

FIGURE 1. Immunoprecipitation from iodinated cell extracts of LO-CD1-a, LO-CD5-a antigens and IL2 receptor by rat MAbs. HPB-ALL cells (Lanes A,B) or PHA-activated lymphocytes (lanes C, D, E, F, G, H) were ^{125}I-iodinated and membrane proteins were extracted. Antigens were precipitated using LO-CD1-a (B), LO-Tact1 (D), LO-Tact2 (E), LO-CD5-a (F), or normal rat serum (H) followed SAC coated with rabbit anti-rat Ig antibodies, or OKT6 (A), and 3A2 (anti-IL2 receptor) (C) followed by SAC coated with rabbit anti-mouse Ig antibodies, or Leu-1 (G) followed by SAC.

Membranes are solubilized with a high concentration of a nonionic detergent: 0.5% Nonidet P40 or Triton X-100. We found no benefit in the addition of the anionic detergent, sodium deoxycholate. To inhibit serine-proteases, 2 m*M* phenyl methane sulfonyl fluoride (PMSF) is added immediately after addition of the detergent.

After nuclei and debris have been spun out and a preclearing step with *Staphylococcus aureus* Cowan I (SAC) has been performed, ascites containing from 1 to 5 μg MAb, is added. Immunoprecipitation with rat MAbs almost always requires an intermediate layer which brings about a bridge between the rat MAb and the SAC. This intermediate layer consists of affinity-purified rabbit anti-rat Ig antibodies. Direct binding to protein A and hence to SAC is observed with rat IgG2c (strongly bound), IgG1 (weakly bound), and with very few IgM's (see Chapter 15). We found Sepharose-Protein A to be less effective in promoting immunoprecipitation than formalin-fixed SAC, probably owing to the lower density of protein A on the beads. After washing of the immunoprecipitate, antigen and antibodies are eluted in sodium dodecyl sulfate (SDS) and submitted to an electrophoresis in a polyacrylamide gel. The radiolabeled antigen is detected by autoradiography (^{125}I, ^{35}S) or by fluorography (^{3}H).[24]

In order to demonstrate that molecules recognized by two MAbs are identical, depletion

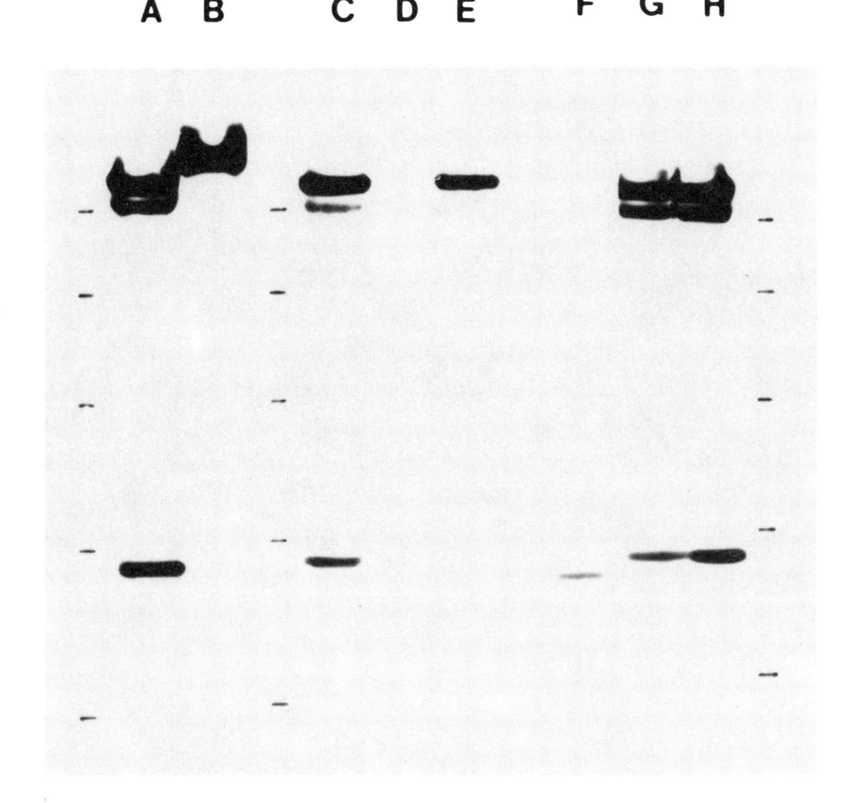

FIGURE 2. Immunoprecipitation of the platelet IIb-IIIa and Ib glycoprotein complexes and p24 protein, after tritium-labeling of platelets. MAbs used for immunoprecipitation were provided by the Third International Workshop on Human Leukocyte Differentiation Antigens or produced in our laboratories: lane A: 96-5-C3 (Villela), B: AN51 (McMichael), C: LO-PL3-a, D: normal rat serum, E: LO-PL3-b, F: LO-PL4, G: 111-2-B5, and H: 111-5-A1 (Villela).

experiments are performed. Three cycles of immunoprecipitation with a first MAb are usually necessary to deplete the extract of an antigen completely. Subsequent immunoprecipitation with the first MAb as a control and with other MAbs is performed. From the autoradiography, shown in Figure 4, we conclude that J5 and AL4 attach to the same molecule as AL2. In the experiment reported in Figure 5, 4 anti-MHC class II MAbs were compared. Depletion of the antigen precipitated by LO-D-d results in concomitant disappearance of the LO-D-c antigen, while the LO-D-b antigen and the antigen recognized by the 7-2 MAb (Martin, Second International Workshop on Human Leukocyte Differentiation Antigens) remain in the extract. This is in keeping with the small differences in molecular weights observed.

V. COMPETITIVE BINDING, BLOCKING ASSAYS

The identity or difference between epitopes defined by 2 MAbs is usually demonstrated by competitive binding or blocking experiments. In the former, competing and tracer MAbs are added together to the cells and the binding approaches an equilibrium state. In the latter, the blocking MAb is added first and allowed to bind to the cells, before the tracer MAb is added. The blocking assay is more sensitive than the competitive assay, but only the latter allows for the estimation of the affinity constant (Ka).

An example of competition is shown in Figure 6: ^{125}I-labeled AL6 was incubated in the

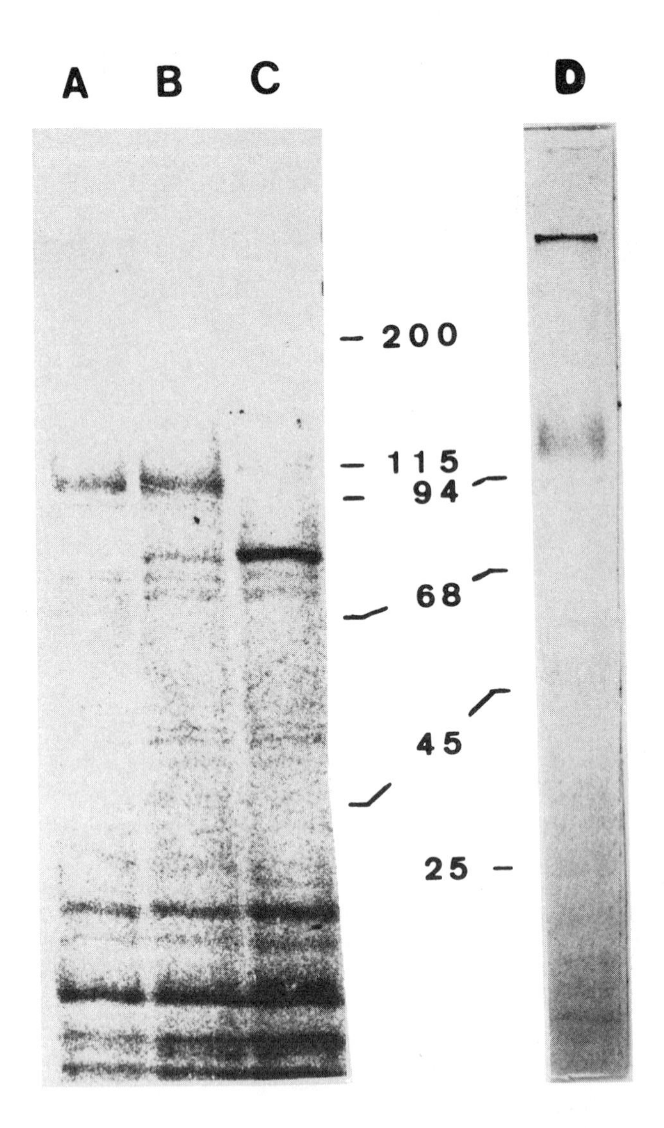

FIGURE 3. Immunoprecipitation of CALLA, labeled by endogenous ^{35}S-methionine incorporation. Lymphoblasts of a patient with cALL were incubated for 24 h in the presence of 0 μg (A), 5 μg/ml (B) or 25 μg/ml (C), of tunicamycin and of ^{35}S-methionine. Granulocytes (D) were cultured for 24 h in the presence of ^{35}S-methionine. Immunoprecipitation was with AL2 followed by SAC coated with rabbit anti-rat Ig antibodies.

presence of Nalm-6 cells and increasing concentrations of different CD9 MAbs. Strongest competition was observed with AL6 itself, 21 D-10, and J2 and a definite but weaker competition with J30. No competition was seen with CLB-Thromb 2, although this MAb is also a CD9 MAb. The affinity constant of AL6 was estimated at 1.3×10^8 M^{-1} by the method of Müller.[26]

Blocking assays of mouse tracer MAbs using rat MAbs (or vice versa) are most easily performed, using rat anti-mouse Ig MAb (or mouse anti-rat Ig MAb) to detect the tracer MAb. No preliminary purification of the tracer MAb is needed.

In the experiment reported in Table 6, different rat MAbs, LO-Tact1, LO-Tact2, LO-Tact5 and LO-CD5-a were tested for their ability to block the binding of the mouse CD 25 MAb, 2A3 (Becton Dickinson). The mouse MAb was added at a subsaturating concentration and detected

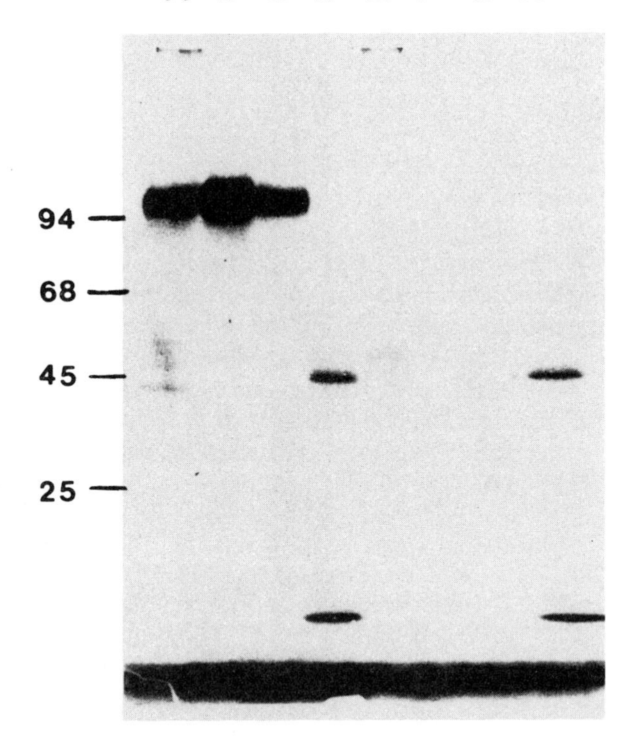

FIGURE 4. Depletion of CALLA by the rat MAb AL2. Cells of a cALL patient were ^{125}I-iodinated and part of the extract was submitted to three cycles of immunoprecipitation using 4 μg of AL2 and 400 μl of SAC coated with rabbit anti-rat Ig antibodies. Fresh (A-D) and depleted (E-H) extracts were then submitted to precipitation using the mouse CD10 MAb J5 (Coulter) (A-E) or the two rat CD10 MAbs, AL2 (B,F) and AL4 (C,G). An anti-HLA-A,-B MAb HL1 was used as a control (D,H).

using the rat MAb, LO-MK-1, which binds to mouse kappa chains (see Chapter 32). Binding of the mouse MAb was totally blocked by LO-Tact1 and LO-Tact2, both directed against the IL2 receptor, while unaffected by LO-Tact5 and LO-CD5-a.

Using the same experimental approach, we demonstrated that the rat MAbs, LO-DR-a, LO-DQ-a, LO-CD5-a, and LO-CD21-a, blocked subsequent binding of the mouse MAbs OKIa (Ortho Diagnostics), Leu-10, Leu-1, (Becton Dickinson), and F74 (LeBien, Third Workshop), respectively.

VI. ANALYSIS OF BIOLOGICAL ACTIVITY OF MAbS

The biological effect of a MAb can be due to the direct interaction of the MAb with the antigen or to the killing of the cell bearing the antigen. Antibody-dependent cell-mediated cytotoxicity and complement-mediated cytolysis depend on the isotype of the MAb and are much more efficient when using rat MAbs than when using mouse MAbs. Chapter 30 deals partially with this topic. Biological effects triggered by the binding of the MAb to the antigen depend primarily on the antigen-binding site of the antibody (specificity and affinity). Assays and results are essentially the same for mouse and rat MAbs.

Inhibition of the binding of the biological ligand of the antigen can be evaluated. LO-Tact1 and LO-Tact2 MAbs were purified by allotype-immunoaffinity chromatography[27] (see Chapter

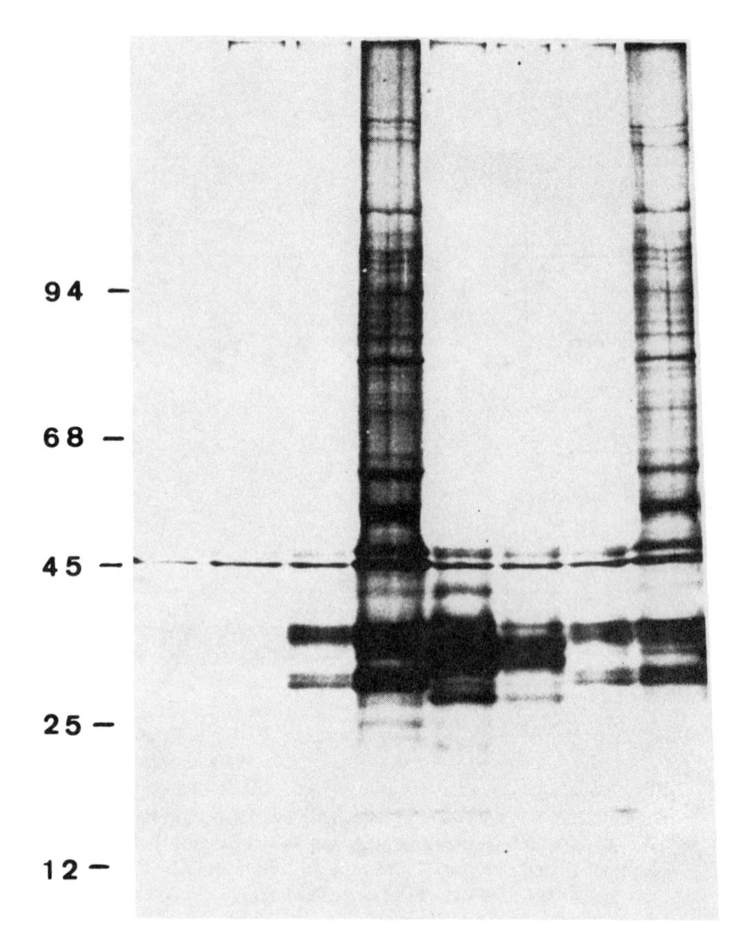

FIGURE 5. Rat MAbs directed against different MHC class II molecules. Raji cells were cultured for 3 h with ^{35}S-methionine and membrane proteins were extracted. Part of the extract equivalent to 1×10^7 cells was depleted by two cycles of immunoprecipitation using LO-D-d and SAC coated with rabbit anti-rat Ig antibodies. Immunoprecipitation of the depleted (A-D) and the fresh (E-H) extracts involved LO-D-d (A,E), LO-D-c (B,F), LO-D-b (C,G), and SAC coated with rabbit anti-rat Ig antibodies or mouse 7-2 MAb (D-H) (Martin; Second International Workshop on Human Leukocyte Differentiation Antigens, 1984) and 50 µl SAC (results of B, Couvreur, unpublished).

15) and compared for their ability to inhibit the binding of ^{125}I-IL2 (Figure 7). LO-Tact2 and to a lesser extent LO-Tact1 inhibit IL2 binding to activated T-lymphocytes. A 50% inhibition of binding was observed at 8.3 n*M* LO-Tact1 or 2.6 n*M* LO-Tact2. The unrelated LO-CD5-a MAb did not interfere with the binding.

Another effect observed with these anti-IL2 receptor MAbs is the inhibition of the mixed lymphocyte culture-induced proliferation (Figure 8). LO-Tact1 and LO-Tact2 brought about an 85% inhibition, even at 10^{-10} *M* concentrations. Only when the amount of added MAb approached the amount of high affinity IL2 receptors in the sample, did the inhibition decrease; we obtained a 50% inhibition with approximately 10,000 MAb-molecules per cell. Although the

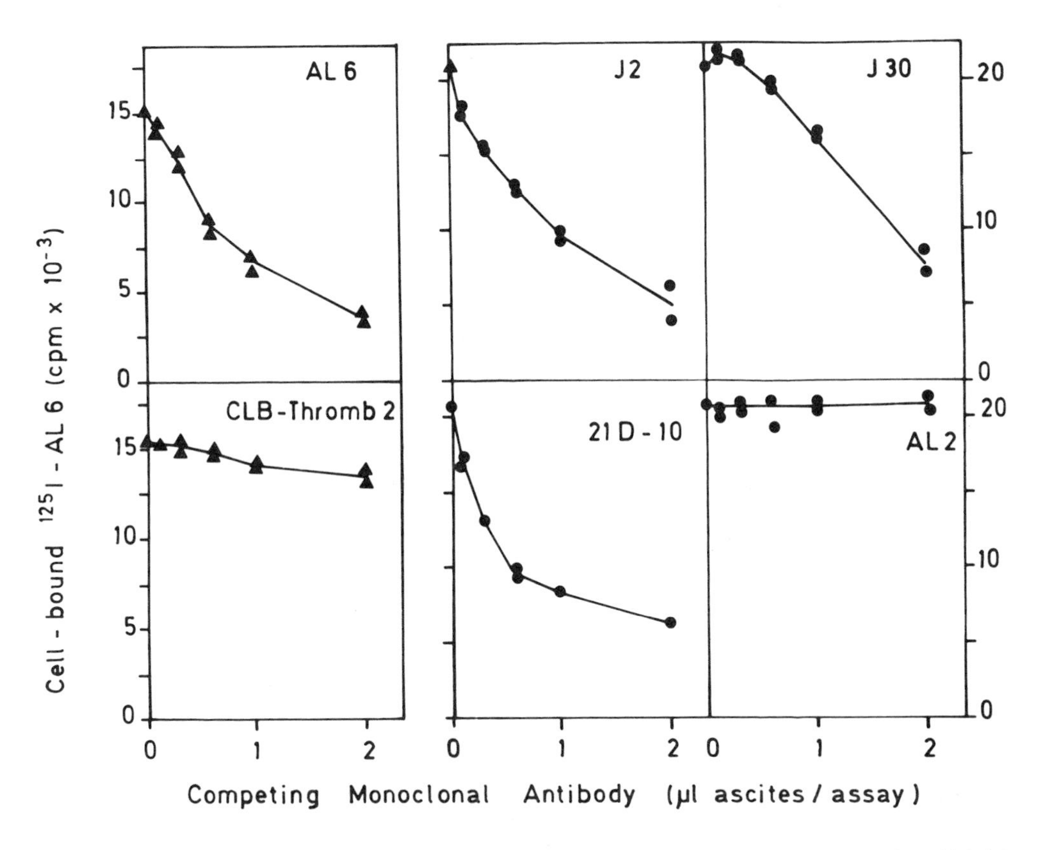

FIGURE 6. Competitive binding of ^{125}I-AL6 and anti-CD9 MAbs. AL6 was purified by electrophoresis and labeled using the Bolton-Hunter reagent (Amersham).[25] CD9 MAbs were obtained from the Second International Workshop on Human Leukocyte Differentiation Antigens: AL6 (Lebacq), J2 and J30 (Ritz), CLB-Thromb2 (van den Borne), and 21D-10 (Garrido). AL2 was a CD10 MAb produced in our laboratory.[2] The competing MAb at different dilutions and 100,000 cpm ^{125}I-AL6 were mixed with Nalm-6 cells. Cell-bound ^{125}I-AL6 was measured after a 2 h incubation at 4°C.

maximal inhibition observed with anti-IL2 receptor MAbs is comparable with the one observed with the anti-MHC class II MAb LO-DR-a, the kinetics of inhibition were different: LO-DR-a had to be present from the initiation of the MLR on, in order to produce its maximal effect. On the contrary, LO-Tact1 added 3 d after initiation of mixed lymphocyte reaction (MLR) still produced a 75% inhibition. This result is consistent with the observation that the alpha-subunit of the IL2 receptor appears on the cell-surface only 4 or 5 d after initiation of the MLR.

VII. CONCLUSION

Rat MAbs directed against human leukocyte differentiation antigens were obtained using the nonsecreting rat immunocytoma fusion cell line IR983F. Specificities were deduced from the reactivity pattern of the MAbs and from the molecular weight of the immunoprecipitated antigen. Very often (but not always) evidence for binding to the same epitope as a reference mouse MAb of the same specificity (same CD) was found.

The facts that rat MAbs can be produced in large amounts and can be easily purified from rat polyclonal antibodies (see Chapter 15), that the rat IgM isotype very efficiently fixes human complement and that the rat IgG2b MAbs activate killer cells (antibody-dependent cell-mediated cytotoxicity) (see Chapter 30), make these MAbs particularly useful clinical reagents.

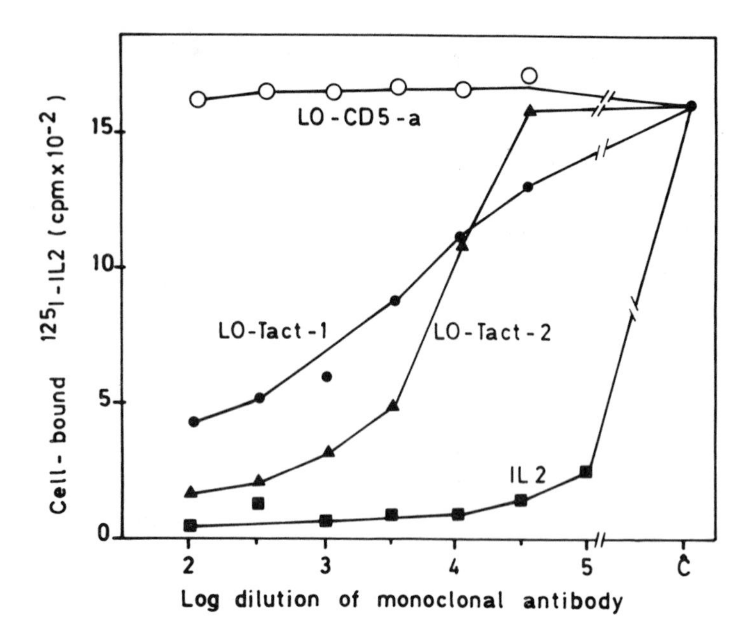

FIGURE 7. Inhibition of IL2-binding by LO-Tact1 and LO-Tact2. MAb at different concentrations or cold IL2 (Amersham), together with 12,500 cpm of (3-^{125}I-iodotyrosyl) interleukin-2 (met°, ala^{125}) (879 Ci/nmol Amersham) were mixed with 6×10^5 PHA-activated lymphocytes in 0.2 ml PBS-5% FCS. After a 2 h incubation, cell-associated ^{125}I-IL2 was measured. The MAbs were purified by allotype-immunoaffinity chromatography.[27] Concentration of LO-Tact1 and LO-Tact2 before dilution was 2.5 mg/ml and of cold IL2 5.5 μM.

VIII. MATERIAL AND METHODS

Equipment and material with references given between brackets are those used in our laboratory. The scientist should adapt the methods to his laboratory facilities.

A. INDIRECT IMMUNOFLUORESCENCE

1. Epifluorescence microscope (Leitz) or cytofluorograph (Epics, Coulter)
2. Bench top centrifuge
3. Test tubes
4. Pipettes

1. Reagents

1. Monoclonal ascites/serum diluted in PBS-5% FCS. Serum of a normal rat is used as a negative control.
2. 0.15 *M* NaCl buffered at pH7.2 with 8 n*M* phosphate known as PBS (Gibco no. 042-04200 *M*)
3. PBS-5% FCS
4. Washing buffer: PBS-2% FCS, 0.2% NaN_3
5. FITC-labeled goat or rabbit anti-rat Ig antibodies: whole Ig or $(Fab')_2$ fragments (RARA-FITC) (Miles, Sigma, Bio Yeda, Tago). Most polyclonal anti-mouse Ig antibody preparations will cross-react with rat Ig, but with variable intensities and must be tested for these properties before use.
6. Formaldehyde 1% in PBS, freshly prepared (if fixing is required).

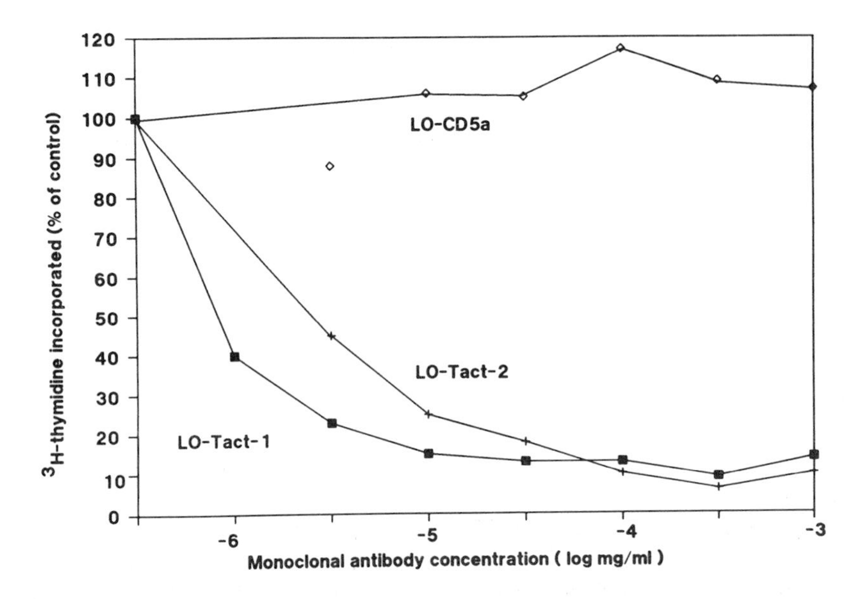

FIGURE 8. Inhibition of mixed lymphocyte culture-induced proliferation by LO-Tact1 and LO-Tact2. Peripheral blood mononuclear cells of normal DR-nonidentical blood donors (10^5 cells/donor/well) were incubated in 0.2 ml RPMI medium containing 10 m*M* Hepes, 20% human AB serum, and different concentrations of purified LO-Tact1, LO-Tact2, or LO-CD5-a. After a 5 d incubation at 37°C with 5% CO_2, 4 µCi ^{3}H-thymidine per well (7 Ci/nmol; Amersham, U.K.) were added and 18 h later, cells were collected on filter paper using a cell harvestor (Skatron, Norway) and incorporated tritium was measured. Control MLR values were 170,000 cpm.

For alternative procedures:

1. Propidium iodide: stock solution: 10 mg propidium iodide is dissolved in 10 ml 0.9% NaCl, 5% EDTA. Before use, dilute 300 times in PBS. Propidium iodide will stain dead cells (also formaldehyde-fixed cells).
2. Glycerol pH 9: to 90 ml glycerol add 10 ml of Tris buffer 1 M pH 9.
3. Wax

2. General Procedure

Dilute MAb-ascites 500 to 2000 times in PBS, 5% FCS. Mix 100 µl diluted MAb with 5×10^5 cells (also in 100 µl), and incubate for 45 min at 4°C or 30 min at room temperature. Wash with 2 ml washing buffer by a 5 min centrifugation at 250 g and aspiration of the supernatant. Resuspend the cell pellet and add a saturating concentration of FITC-labeled anti-rat Ig antibodies, and incubate for 45 min at 4°C (30 min at 20°C). Wash as above. Resuspend in 400 µl PBS (analysis on the same day) or in 400 µl PBS-formaldehyde (analysis within ±10 days), then analyze on a cytofluorograph.

3. Alternative Procedures

For analysis of wet mounts on a microscope, follow the first seven steps as in the general procedure. Add 15 µl of diluted propidium iodide to the cell pellet, transfer the cell suspension to a microscope slide and cover, then read immediately.

For analysis of cell smears on a microscope follow the first seven steps as in the general procedure. Resuspend in 400 to 600 µl washing buffer and use 200 µl of the suspensions for making cytospins. Immediately cover the cell smear with one drop of glycerol pH 9, and a cover slip, and seal by applying melted wax on the edges of the cover slip. The slides are kept for a few days at 4°C or for weeks at –18°C.

B. IODINATION OF CELL SURFACE PROTEINS (LACTOPEROXIDASE METHOD)

1. Material

1. Fume hood for radioactive labeling
2. Disposable tubes with screw caps
3. Bench top centrifuge

2. Reagents

1. PBS-Ca-Mg: PBS containing 0.9 m*M* $CaCl_2$ and 0.5 m*M* $MgCl_2$, pH 7.2
2. PBS containing 5% FCS
3. Lactoperoxidase: (Boehringer no. 107174: ammonium sulfate suspension of enzyme; these enzymes are stable at 4°C for months; just before use, the suspension is centrifuged for 1 min at 15,000 *g* and the precipitate is resuspended in PBS). Make up a 200 U/ml solution in PBS-Ca-Mg.
4. Glucose oxidase (Sigma: G2133): make up a 100 U/ml solution in PBS-Ca-Mg
5. Glucose 25 mg/ml in PBS-Ca-Mg
6. Na ^{125}I carrier-free (Amersham No. IMS30) specific radioactivity (around 15 mCi/µg)

3. Method

For 2×10^7 cells, wash the cells four times in PBS in order to remove all soluble proteins; and resuspend pellet in 200 µl PBS-Ca-Mg. Add 100 µl lactoperoxidase 200 U/ml, and then add 100 µl glucose oxidase 100 U/ml.

Proceed under a fume hood and wear a mask. Add 2 mCi Na ^{125}I and 100 µl glucose. Allow the reaction with occasional mixing for 30 min at room temperature. Add 20 ml of PBS-FCS and allow reaction for 2 to 5 min. Centrifuge at 500 *g* for 5 min, and discard supernate. Wash three more times. Before the last wash it is advisable to transfer the cells into a clean tube since iodine is known to bind to plastic ware.

For labeling of more cells, the amount of enzymes and iodine have to be adapted proportionally.

C. METABOLIC LABELING OF CELLULAR PROTEINS WITH ^{35}S-METHIONINE

1. Reagents

1. EMEM modified medium without methionine (Flow No. 16-222-49 or Gibco 041-1900H)
2. ^{35}S-L-methionine (> 800 Ci/mmol) (Amersham No. SJ235)
3. PBS
4. FCS: it is not necessary to dialyze the serum

2. Method

For 2×10^7 cells, wash once in methionine-free culture medium, to which glutamine, nonessential amino acids, penicillin-streptomycin, and 10% FCS have been added. Resuspend at 0.5×10^7 cells/ml in the same medium and add 0.1 mCi ^{35}S-methionine/ml. Incubate at 37°C in the presence of CO_2 for 3 to 4 h if cell lines in exponential growth are used and for one night if normal cells are used. Harvest cells and wash three times in PBS.

D. EXTERNAL LABELING OF GALACTOSE RESIDUES

1. Reagents

1. Neuraminidase from *Clostridium perfringens* (Sigma N2876); 50 U/ml PBS, pH 7
2. Galactose oxidase from *Dactylium dendroides* (Sigma G3385); 100 U/ml in PBS, pH 7
3. Na B^3H^4 100 mCi at a specific radioactivity >6 Ci/mmol (Amersham)

2. Method

For 5×10^7 cells or 3×10^9 platelets, wash twice in PBS pH 7. Resuspend in 3 ml PBS pH 7 and incubate for 1 h at 37°C with 12.5 U neuraminidase and 5 U galactose oxidase. Wash once in PBS pH 7.4, and resuspend in PBS pH 7.4 and add 20 mCi Na B_3H_4 freshly dissolved in PBS. Allow reaction on ice during 30 min, then wash four times.

E. EXTRACTION OF MEMBRANE PROTEINS

1. Material

1. Microcentrifuge (Heraus)
2. Centrifuge and tubes (Sorvall)
3. Refrigerator or ice

2. Reagents

1. Extraction buffer: 20 m*M* Tris buffer pH 8.2 containing 0.15 *M* NaCl and 0.5% Nonidet P40 (Sigma No. N6507)
2. PMSF (phenyl methane sulfonyl fluoride, Merck No. 7349) 200 m*M* in ethanol.

3. Method

To the labeled cell pellet, add the extraction buffer (1 ml/1×10^7 cells) followed by PMSF (20 µl/ml). Vortex immediately and thoroughly, and keep at 0°C for 5 to 15 min. Centrifuge for 10 min at 3300 *g* or for 1 min at 15,000 *g* (Microfuge), then store the supernate (= cell extract) at –80°C (or –20°C if for a short time).

F. IMMUNOPRECIPITATION OF CELLULAR ANTIGENS

1. Reagents

1. Monoclonal antibodies: ascites or serum containing from 1 to 10 mg MAb/ml
2. Second layer antibodies: affinity-purified rabbit anti-rat Ig antibodies (with Fc fragment);
3. *Staphylococcus aureus* Cowan I strain (SAC): a 10% (w/v) formaldehyde-fixed suspension in phosphate buffer can be obtained from almost every laboratory of microbiology or from Calbiochem under the trade name PANSORBIN (No. 507858). The suspension is stored at –20°C in 1 ml aliquots. Binding capacity of SAC for each batch of second layer is best determined before hand. PANSORBIN binds around 1 mg rabbit Ig/ml of 10% suspension;
4. Wash I: 20 m*M* Tris pH 8.2 containing 0.5 *M* NaCl and 0.5% Nonidet P-40
5. Wash II: 20 m*M* Tris pH 8.2 containing 0.5 *M* NaCl, 0.5% Nonidet P-40 and 0.1% SDS
6. Nonreducing elution buffer: 0.1 *M* Tris pH 6.8 containing 4% SDS, 15% sucrose, and 0.001% bromophenol
7. Reducing elution buffer: same as above, supplemented with 10% 2-mercaptoethanol.

The elution buffers are prepared from stock solutions as follows; 2 ml 0.5 *M* Tris pH 6.8, 4 ml 10% SDS, 0.5 ml 0.2% bromophenol in ethanol, and 2.5 ml 60% sucrose.

The nonreducing elution buffer is stored in 0.9 ml aliquots at –20°C for months or kept at 4°C for weeks. Before use, add 0.1 ml water (nonreducing) or 0.1 ml 2-mercaptoethanol (reducing).

2. Procedure

SAC is washed three times by a 1 min centrifugation at 15,000 *g* and gentle resuspension at 10% w/v with a Pasteur pipette in the wash I solution;

In the preclearing step, add 200 µl of 10% SAC per ml of cell extract (i.e. per 10^7 cellular equivalents). Incubate for 15 to 30 min at 4°C, and centrifuge 1 min at 15,000 *g*. Keep supernate (= cleared cell extract).

For the coating of SAC with second layer (if necessary), add rabbit anti-rat Ig antibodies to a 10% suspension of SAC at a concentration, achieving 80% of saturation. We usually add 70 to 80 µg of purified rabbit antibodies per 100 µl PANSORBIN. Incubate for 30 to 45 min, then wash once by addition of 1 ml wash I and centrifugation for 1 min at 15,000 *g*. Resuspend at 10% (w/v).

In immunoprecipitation for each test, 2.5 × 10^6 cell equivalents (i.e. 0.30 ml cleared cell extract) are used. For IgG1 and IgM MAbs, binding to SAC should be checked first. Only IgM MAbs with a very high affinity are able to precipitate their antigen. For rat MAbs of the IgG2c subclass and for IgG1 and IgM, if they bind to SAC; to 0.30 ml cleared cell extract, add around 10 µg MAb (1 to 5 µl ascites) and incubate for 45 to 60 min at 4°C; then add 50 µl 10% SAC and incubate for 45 to 60 min with continuous end-over-end rotation (or mix every 10 min). For all rat MAbs of the IgG2a, IgG2b isotype, and for IgG1 and IgM, if they do not bind to SAC: to 0.30 ml cell extract, add around 1 µg MAb (0.1 to 1 µl ascites) and incubate for 45 to 60 min at 4°C; then add 100 µl of 10% SAC coated with a second layer and incubate for 45 to 60 min under continuous end-over-end rotation.

For the wash of the immunoprecipitate, centrifuge the mixture of SAC, MAb, and cell extract for 1 min at 15,000 *g*. Remove the supernatant; it can be reused for immunoprecipitation of another antigen. Wash the pellet twice with 1 ml of wash I solution and once with 1 ml of wash II solution. Resuspension can be done by vortexing.

For elution of the antigen, add ±60 µl of elution buffer (reducing or nonreducing) to the last pellet. The volume of the elution buffer should be consistent with the well volumes of the polyacrylamide gel. Resuspend by vigorous vortexing, then boil for 2 to 3 min. After cooling down, centrifuge for 1 min at 15,000 *g*. Load the supernate into an SDS-PAGE well.

The immunoprecipitates can be stored at –20°C (or –80°C), after boiling.

G. SDS-PAGE

1. Reagents

1. Acrylamide (Serva No. 10675)
2. N,N′-methylene-bisacrylamide (Serva No. 29195)
3. ABA (30:0.8) 30 g acrylamide and 0.8 g bisacrylamide in 100 ml H_20 (dark bottle)
4. TEMED = N,N,N′,N′-tetramethylethylenediamine (Serva No. 35925)
5. Tris-HCl 0.5 *M* pH 6.8
6. Tris-HCl 2 *M* pH 8.8
7. SDS (Sigma No. L4509) Sodium Lauryl sulfate 10% in H_2O
8. Sucrose = 60% (w/v) in H_20
9. 2-Mercaptoethanol (Merck No. 805740)
10. Ammonium persulfate (Merck No. 1200): 10% (w/v) in H_20. Prepare freshly every week
11. Sample buffer: see elution buffer (above)
12. Migration buffer: 25 m*M* Tris (3 g/l), 192 mM Glycine (14.5 g/l), and 0.1% SDS. The pH should be around 8.4 (no adjustments necessary if the pH is between 8.2 and 8.6)
13. Staining solution: 250 mg Coomassie blue (Serva No. 17524) dissolved in destaining solution

14. Destaining solution: 25% isopropanol and 10% acetic acid in H_2O or 40% methanol and 10% acetic acid in H_2O.

2. Procedure

The description below is for a 20 ml resolving gel (i.e., 150 mm × 133 mm × 1mm) with a linear gradient from 7 to 15% ABA, and a stacking gel of around 5 ml (i.e. 150 mm × 33 mm × 1 mm).

First assemble the glass plates and spacer strips and stick them together (e.g., with 1% agarose). Prepare the solutions for resolving gel and stacking gel (to be used within 2 or 3 h; keep away from sunlight).

	15%	**7%**	**Stacking**
H_20	0.57 ml	5.74 ml	7.64 ml
Tris pH 8.8	1.8 ml	1.8 ml	0
Tris pH 6.8	0	0	1.2 ml
ABA 30: 0.8	5 ml	2.33 ml	1 ml
Sucrose	2.5 ml	0	0
TEMED	5 µl	5 µl	5 µl
SDS 10%	100 µl	100 µl	100 µl

Start the polymerization in the 7 and 15% solutions by adding 25 µl ammonium persulfate in each flask. The linearly decreasing gradient is brought about in the flask containing the 15% solution. The condition for achieving a linear gradient is that the levels of the two flasks go down at the same rate. We achieve these conditions by means of a peristaltic pump with one tubing carrying the 7% solution into the 15% flask and two tubings carrying the mixture from the 15% flask to the assembled plates. Alternatively, specially designed recipients are commercially available.

As soon as the 20 ml of mixture are poured, carefully add water on top of it. Clean the tubing with water. Wait until polymerization is completed, i.e., 30 min. to 1 h. Remove the water on top of the gel. The top of the gel should be a perfect horizontal line. Rinse once with 5 ml of stacking solution. Then start polymerization of the stacking solution left by addition of 25 µl of ammonium persulfate and pour it on top of the resolving gel, and introduce the comb into the stacking gel. Wait until polymerization is completed (30 to 60 min). Remove the comb and the lower spacer strip and place into the slab gel tank, then fill the upper and lower reservoirs with migration buffer.

With a syringe (Hamilton), apply the sample on top of the resolving gel, i.e., into the well left by the comb and underneath the migration buffer. Start electrophoresis: around 5 h at 130 V (i.e., 10 V/cm) or 16 h at 80 V. Migration through the stacking gel should be under lower voltage (50 V). Turn off power. Separate the plates and carefully transfer the gel into the staining buffer. After a 2 h staining, destain for 2 × 10 min. For drying of the gel, lay the gel on a paper (Whatman No. 3MM) and cover with a plastic sheet. Dry on a gel dryer, i.e., under vacuum and with a thermostatic heat supply (Pharmacia, BioRad,...).

H. AUTORADIOGRAPHY

The following procedure is for detection of ^{35}S and ^{125}I.

In a dark room place the gel into a film pack (Kodak), cover with an X-ray film (Kodak X-Omat film). Close the pack. To ensure a close contact between gel and film, the pack is placed between 2 metal plates, clamped together with clips. After 24 to 48 h at room temperature, develop the film.

I. FLUOROGRAPHY

The following procedure is for the detection of ^3H and ^{14}C.

After staining and destaining (before drying), dehydrate the gel by transfering it into three successive baths of pure dimethylsulfoxide (DMSO) (20 min in each bath). Saturate the gel with 2,5-diphenyloxazole (PPO) by a 2 h incubation in a DMSO bath containing 20% (w/v) PPO for 2 h. For reuse of DMSO and PPO baths, see Bonner and Laskey.[24] Transfer to H_2O (2 or 3 baths of 1 min), then dry. Put into contact with an X-ray film (Kodak X-Omat) for 3 to 7 d at -80°C, then develop.

ACKNOWLEDGMENTS

We thank F. Flemal, M. De Cloedt, F. Nisol, D. Wauters, J. M. Malache, and J. P. Kints for expert technical assistance, F. Bolle for typing the manuscript, and C. Vincenzotto for preparing the figures.

REFERENCES

1. **Lebacq, A. M. and Bazin, H.,** Obtention of rat monoclonal antibodies reactive with human leukaemic lymphoblast, *Bull. Cancer,* 70, 93, 1983.
2. **Lebacq-Verheyden, A. M., Ravoet, A. M., Bazin, H., Sutherland, D. R., Tidman, N., and Greaves, M. F.,** Rat AL2, AL3, AL4, and AL5 monoclonal antibodies bind to the common acute lymphoblastic leukemia antigen (CALLA gp 100), *Int. J. Cancer,* 32, 273, 1983.
3. **Ulmer, A. J. and Flad, H. D.,** Discontinuous density gradient separation of human mononuclear leucocytes using Percoll as gradient medium, *J. Immunol. Methods,* 30, 1, 1979.
4. **Mannoni, P., Janowska-Wieczorek, A., Turner, A. R., McGann, L., and Turc, J. M.,** Monoclonal antibodies against human granulocytes and myeloid differentiation antigens, *Hum. Immunol.,* 5, 309, 1982.
5. **Schneider, U., Schwenk, H. U., and Bornkamm, G.,** Characterization of EBV-genome negative "null" and "T" cell lines derived from children with acute lymphoblastic leukemia and leukemic transformed non-Hodgkin lymphoma, *Int. J. Cancer,* 19, 621, 1977.
6. **Minowada, J., Janossy, G., Greaves, M. F., Tsubota, T., Srivastava, B. I. S., Morikawa, S., and Tatsumi, E.,** Expression of an antigen associated with acute lymphoblastic leukemia in human leukemia-lymphoma cell lines, *J. Natl. Cancer Inst.,* 60, 1269, 1978.
7. **Morikawa, S., Tatsumi, E., Baba, M., Harada, T., and Yasuhira, K.,** Two E-rosette forming lymphoid cell lines, *Int. J. Cancer,* 21, 166, 1978.
8. **Foley, G. E., Lazarus, H., Farber, S., Uzman, B. G., Boone, B. A., and McCarthy, R. E.,** Continuous culture of human lymphoblasts from peripheral blood of a child with acute leukemia, *Cancer,* 18, 522, 1965.
9. **Adams, R. A., Pothier, L., Flowers, A., Lazarus, H., Farber, S., and Foley, G. E.,** The question of stemlines in human acute leukemia. Comparison of cells isolated *in vitro* and *in vivo* from a patient with acute lymphoblastic leukemia, *Exp. Cell Res.,* 62, 5, 1970.
10. **Pulfertaft, R. J.,** Cytology of Burkitt's tumour (African lymphoma), *Lancet,* 238, 1964.
11. **Klein, E., Klein, G., Nadkarni, J. S., Nadkarni, J. J., Wigzell, H., and Clifford, P.,** Surface IgM-kappa specificity on a Burkitt lymphoma cell *in vivo* and in derived culture lines, *Cancer Res.,* 28, 1300, 1968.
12. **Drewinko, B., Mars, W., Minowada, J., Burk, K. H., and Trujillo, J. M.,** ARH-77, an established human IgG-producing myeloma cell line. I. Morphology, B-cell phenotypic marker profile, and expression of Epstein-Barr virus, *Cancer,* 54, 1883, 1984.
13. **Lozzio, C. B. and Lozzio, B. B.,** Human chronic myelogenous leukemia cell-line with positive Philadelphia chromosome, *Blood,* 45, 321, 1975.
14. **Collins, S. J., Gallo, R. C., and Gallagher, R. E.,** Continuous growth and differentiation of human myeloid leukemic cells in suspension culture, *Nature,* 270, 347, 1977.
15. **Sundstron, C. and Nillsson, K.,** Establishment and characterization of a human histiocytic lymphoma cell line (U-937), *Int. J. Cancer,* 17, 565, 1976.
16. **Sutherland, R., Smart, J., Niaudet, P., and Greaves, M. F.,** Acute lymphoblastic leukaemia associated antigen. II. Isolation and partial characterization, *Leuk. Res.,* 2, 115, 1978.
17. **Marchalonis, J. J., Cone, R. E., and Santer, V.,** Enzymic iodination: a probe for accessible surface proteins of normal and neoplastic lymphocytes, *Biochem. J.,* 124, 921, 1971.

18. **Hubbard, A. L. and Cohn, Z. A.,** The enzymatic iodination of the red cell membrane, *J. Cell Biol.*, 55, 390, 1972.
19. **Ravoet, A. M. and Lebacq-Verheyden, A. M.,** Clustering of anti-leukemia and anti-B cell monoclonal antibodies, in *Leukocyte Typing II,* Vol. 2, Reinherz, E. L., Haynes, B. H., Nadler, L. M., and Bernstein, I. D., Eds., Springer Verlag, New York, 1986, 213.
20. **Gahmberg, C. G. and Hakomori, S. I.,** External labeling of cell surface galactose and galactosamine in glycolipid and glycoprotein of human erythrocytes, *J. Biol. Chem.*, 248, 4311, 1973.
21. **Ravoet, A. M., Ninane, J., Seghers, J., Bazin, H., Latinne, D., De Bruyère, M., and Sokal, G.,** Platelet antigens detected by immunoprecipitation with third workshop monoclonal antibodies, in *Leukocyte Typing III,* McMichael et al., Eds., Oxford University Press, New York, 1987, 755.
22. **Schneider, C., Sutherland, R., Newman, R., and Greaves, M.,** Structural features of the cell surface receptor for transferrin that is recognized by the monoclonal antibody OKT9, *J. Biol. Chem.*, 257, 8516, 1982.
23. **Braun, M. P., Martin, P. J., Ledbetter, J. A., and Hansen, J. A.,** Granulocytes and cultured human fibroblasts express common acute lymphoblastic leukemia-associated antigens, *Blood,* 61, 718, 1983.
24. **Bonner, W. M. and Laskey, R. A.,** A film detection method for tritium-labeled proteins and nucleic acids in polyacrylamide gels, *Eur. J. Biochem.*, 46, 83, 1974.
25. **Bolton, A. E., and Hunter, W. M.,** The labeling of proteins to high specific radioactivities by conjugation to a ^{125}I-containing acylating agent, *Biochem. J.*, 133, 529, 1973.
26. **Müller, R.,** Calculation of average antibody affinity in anti-hapten sera from data obtained by competitive radioimmunoassay, *J. Immunol. Methods,* 34, 345, 1980.
27. **Bazin, H., Cormont, F., and De Clercq, L.,** Rat monoclonal antibodies. II. A rapid and efficient method of purification from ascitic fluid or serum, *J. Immunol. Methods,* 71, 9, 1984.

Chapter 20.II

RAT MONOCLONAL ANTIBODIES SPECIFIC FOR HUMAN T LYMPHOCYTES

H. Xia, A. M. Ravoet, D. Latinne, J. Ninane, M. De Bruyère, G. Sokal, and H. Bazin

TABLE OF CONTENTS

I. INTRODUCTION

The development of monoclonal antibodies (MAbs) has offered the opportunity of studying the distribution, structure and function of T lymphocyte membrane antigens. Over the past years a series of molecules present on the T lymphocyte have been defined by mouse MAbs; some have been shown to be associated with T lymphocyte function. For instance, T cell antigen receptor complex (Ti-T3) is recognized by CD3 MAbs (e.g. OKT3) and by WT31;[1,2] the receptor for Interleukin-2 is recognized by CD25 MAbs (e.g., TAC);[3] and the T lymphocyte receptor for sheep erythrocytes (E-receptor), which has recently been proposed as a ligand for LFA-3, is recognized by CD2 MAbs (e.g., OKT11).[4,5] Some antibodies to T-lymphocytes mimic the effect of natural ligands or of antigen stimulation: OKT3 and WT31 activate T lymphocytes through the T cell antigen receptor complex.[6,7] Certain combinations of CD2 MAbs activate T lymphocytes through an alternative pathway.[8] CD4 and CD8 MAbs (e.g., OKT4 and OKT8),[9,10] define distinct subpopulations of T lymphocytes. The antigen defined by CD4 MAbs is mainly expressed on T-helper cells and seems to recognize MHC class II gene products, while antigen defined by CD8 MAbs is mainly expressed on T suppressor/cytotoxic cells and seems to recognize MHC class I antigens.[11,12] We were interested in producing anti-human T lymphocyte MAbs in the rat model because of the difference in immune repertoire between rat and mouse. Moreover, because large quantities of rat MAb are easily produced and purified and because some rat Ig isotypes have been shown to fix human complement and to activate human killer cells, rat MAbs are attractive reagents for clinical therapeutic use. Here, we report five rat MAbs specific for human T lymphocytes.

II. MATERIAL AND METHODS

A. ANIMALS

For LOU/C rats and WISTAR rats see Chapters 5 and 6.

B. CELL LINES

The IR983F (983) nonsecreting rat immunocytoma cells[13] were cultured in DMEM medium (Gibco) supplemented with 15% horse serum, 2 m*M* glutamine, 100 U/ml penicillin, 100 μg/ml streptomycin (PS) and nonessential amino acids (complete DMEM medium). 983 cells were grown at 37°C 5% CO_2.

HPB-ALL, CCRF-CEM, MOLT-4, Jurkat, Raji, Daudi, Nalm-1, Nalm-6, KM3, U937, K562, and HL-60 were cultured at 37° C 5% CO_2 in RPMI-1640 medium containing 10% fetal calf serum, 2 m*M* glutamine and PS.

C. REFERENCE MONOCLONAL ANTIBODIES

OKT3 (CD3), OKT4(CD4), and OKT8 (CD8) were obtained from Ortho Diagnostics; T11 (CD2), T12 (CD6), and B4 (CD19) were purchased from Coulter; Leu3-a (CD4) from Becton-Dickinson; LO-CD5-a (CD5), LO-Tact1 (CD25), MARK-3, LO-MK-1, LO-MG-7, and fluorescein-conjugated rabbit anti-rat Ig antibodies (Fab′) 2 fragments (RARA-FITC) were produced in our laboratories (see Chapters 15, 20.I. and 32).

D. ISOLATION OF LYMPHOCYTES

Peripheral blood mononuclear cells (PBMN) from normal donors were isolated by centrifugation on Ficoll-Hypaque (d: 1,077). T lymphocytes were purified by rosetting with 2-aminoethylisothiouronium-treated sheep red blood cells (SRBC-AET): PBMN at 5×10^6/ml were mixed with 1/10 volume AET-SRBC at about 1.0×10^9/ml and incubated at room temperature for 15 min. After centrifugation at 100 g for 5 min and new incubation at 4°C for 1 to 2 h, rosetted cells (>95% of T lymphocytes) were separated from non-rosetted cells (<5% T lymphocytes) by centrifugation on Ficoll-Hypaque. SRBC were lysed with 0.83% NH_4Cl.

E. IMMUNIZATION

Both T cell line HPB-ALL[14] and purified T lymphocytes were used for immunization. For the first experiment, 2×10^7 HPB-ALL cells were injected intraperitoneally in a LOU/C rat, and 1 month later, 2×10^7 cells and 7 weeks later, 5×10^7 cells were injected i.p. For the second experiment, 5×10^7 purified T lymphocytes were injected twice i.p. into a LOU/C rat at a 3-week interval.

F. PRODUCTION OF RAT-RAT HYBRIDOMAS

Fusion between the splenocytes of an immunized LOU/C rat and rat 983 cells was performed 4 d after the last injection as described in Chapters 6 and 7. From the 7th d after fusion, hybrid clones were observed and they were transferred to 24-well plates. When supernatants in the wells became yellow, screening for specific antibody-secreting hybrid clones was performed by indirect immunofluorescence using immunizing cells as target cells. After a 2nd screening of the whole blood, those hybrid clones which secreted antibodies reacting with lymphocytes, and not with granulocytes and/or monocytes, were selected. The hybrid lines were cloned twice by limiting dilution. The hybridoma cells were either cultured, frozen down in complete DMEM-7.5% dimethylsulfoxide (DMSO) and stored in liquid nitrogen, or injected subcutaneously into LOU rats to induce solid tumors. The tumor cells were frozen down in PBS-10% FCS-7.5% DMSO and stored in liquid nitrogen, or if necessary, tumor cells were injected i.p. to induce ascitic tumor. The sera or ascitic fluid of tumor-bearing rats were collected and stored at –80°C.

G. INDIRECT IMMUNOFLUORESCENCE

5×10^5 cells (mononuclear cells isolated from blood of normal donors or leukemic patients or cell lines) in 100 μl PBS-2% new born calf serum (NBCS)-0.2% NaN_3 or 100 μl whole blood were incubated for 45 min at 4°C with 100 μl antibody-containing supernatant or serum (1/200 dilution). Normal rat serum was used as the control. Red blood cells were lysed in case whole blood was used. Cells were washed once and incubated for 45 min at 4°C with 100 μl of rabbit anti-rat immunoglobulin (Ig) antibody F(ab′)2 fragments labeled with fluorescein isothiocyanate (RARA-FITC), or goat anti-mouse Ig conjugated with FITC (GAM-FITC). Cells were washed once. Immunofluorescence intensity was examined either on a microscope or an EPICS cytofluorograph (Coulter). For double immunofluorescence, cells were labeled with rat MAbs and RARA-FITC as above. Then, cells were incubated with phycoerythrin-conjugated MAbs for 45 min at 4°C. For blocking assays cells were incubated first with rat MAbs, then with a reference mouse MAb at a subsaturating concentration. After washing, cells were incubated for 45 min at 4°C with LO-MK-1-FITC or LO-MG-7-FITC (rat anti-mouse kappa light chain or gamma heavy chain MAb) which do not cross-react with rat Ig.

H. IMMUNOPRECIPITATION AND SDS-PAGE ANALYSIS

The method is described in detail in Chapter 20. I. Briefly, cells were labeled with ^{125}I, using the glucose oxidase/lactoperoxidase method. Membrane proteins were solubilized in 20 m*M* Tris buffer pH 8.2 containing 0.15 *M* NaCl and 0.5% Nonidet P40 (Sigma) and 40 μl 200 m*M* phenyl methane sulfonyl fluoride (PMSF) (Sigma). The cell extract was precleared with 10% *Staphylococcus aureus* Cowan I (SAC). For the immunoprecipitation, 300 μl precleared cell extract (equivalent to 2.5×10^6 cells) was mixed first with 1 μg rat MAb for 60 min at 4° C, then with 100 μl 10% SAC coated with rabbit anti-rat immunoglobulin antibodies for 60 min at 4°C. After washing the immunoprecipitate, the antigens were eluted with 0.1 *M* Tris pH 6.8 containing 4% SDS, 15% sucrose, 0.001% bromophenol, and 10% 2-mercaptoethanol by boiling for 2 to 3 min. The eluate was analyzed by sodium dodecyl sulfate polyacrylamide gel electrophoresis (SDS-PAGE) and autoradiography to determine the molecular weight.

I. FUNCTIONAL STUDIES

For all experiments, cells were cultured at 37°C 5% CO_2 in 96-well plates containing 200 μl

TABLE 1
Reactivity with Normal Hematopoietic Cells

	MAbs				
Cells	LO-CD6-a	LO-CD2-act	LO-Tmat	LO-CD4-a	LO-CD4-b
Blood					
Lymphocytes					
Total	73% ++	70% +	76% ++	54% +++	53% +++
E+	92% ++	90% +	95% ++	68% +++	67% +++
E–	–	–	–	–	–
PHA-PBL	90% ++	96% +++	80% ++	69% +++	74% +++
Monocytes	–	–	–	–	–
Granulocytes	–	–	–	–	–
RBC	–	–	–	–	–
Platelets	–	–	–	–	–
Tonsil cells					
Total	77% ++	63% +	56% +	59% +++	60% +++
E–	–	–	–	–	–
Thymocytes	65% +	86% +	54% +	64% ++	66% ++

RPMI-1640 medium supplemented with 20% human AB serum, 2 m*M* glutamine, 100 U/ml penicillin, 100 μg/ml streptomycin and 25 m*M* Hepes. 10^5 PBMN were cultured with 2 μg/ml phytohemagglutinin (PHA), 20 μg/ml concanavalin A (ConA), 0.5 μg/ml T3 or 10 μg/ml pokeweed mitogen (PWM) or 20 mU/ml tetanus toxoid (TT) in the absence or the presence of rat MAb serum diluted at 1/500. For allogeneic mixed lymphocyte reaction, 10^5 PBMN from two individual donors were cultured together in the presence or absence of rat MAb. ^{3}H-thymidine (3H-TdR, 6.5 Ci/mmol, New England Nuclear) was added to each well (5 μCi/well) 3 d after stimulation with ConA, PHA, PWM, or T3, or 5 d after stimulation with TT or in a mixed lymphocyte reaction (MLR) and incubation was continued overnight. The cells were harvested with a cell harvestor (SKATRON) and ^{3}H-TdR incorporation was measured in a scintillation counter.

III. RESULTS

A. REACTIVITY PATTERNS OF RAT MAb WITH NORMAL CELLS

Fusion of splenocytes of an immunized rat with 983 immunocytoma cells was induced by PEG. Among the hybrid clones secreting antibodies against the immunizing cells, most cross-reacted with granulocytes and/or monocytes and were not analyzed further. Only a few secreted MAbs reacting selectively with lymphocytes. LO-CD6 and LO-CD4-a were obtained after immunization with the T-ALL line HPB-ALL;[14] LO-CD4-b, LO-Tmat, and LO-CD2-act after immunization with purified T lymphocytes.

The five MAbs reported here reacted only with T lymphocytes (Table 1). No reaction was found with monocytes, granulocytes, platelets, red blood cells, or with a B lymphocyte-enriched suspension prepared by depleting T lymphocytes from PBL and tonsil by SRBC rosetting. MAbs LO-CD6-a, LO-Tmat, and LO-CD2-act reacted with about 70% of PBL and more than 90% of SRBC-rosetting lymphocytes. Hence, they are pan-T specific MAbs. MAbs LO-CD4-a and LO-CD4-b reacted strongly with around 55% of PBL and about 70% of T lymphocyte-enriched PBL. So, these two MAbs are specific for a subset of T lymphocytes.

Typical reactivity patterns of our MAbs and of reference pan-T or subset-T specific MAb OKT3, T11, T12, LO-CD5-a, and OKT4 are shown in Figure 1. The profiles of LO-CD6-a and LO-Tmat with lymphocytes in blood and tonsils looked similar to those of reference pan-T specific MAb. The labeling of thymocytes was much weaker than with OKT11. Expression of the LO-Tmat antigen decreases upon activation. On the contrary, the antigen defined by MAb

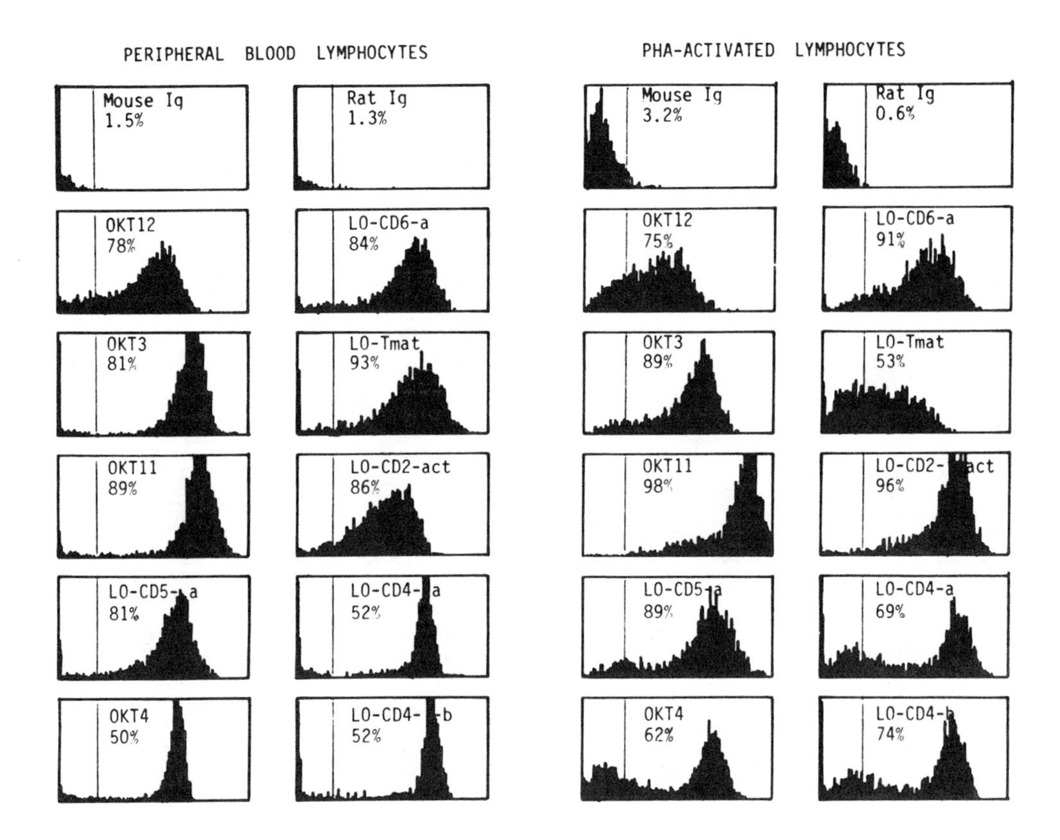

FIGURE 1A. Reactivity pattern of MAb with PBL (left panel) and PHA activated PBL (right panel). X-axis: fluorescence intensity; Y-axis: cell number.

LO-CD2-act was less expressed on normal resting T lymphocytes than after their activation by PHA (Table 1 and Figure 1). This MAb might define an antigen associated with T lymphocyte activation.

Profiles of MAbs LO-CD4-a and LO-CD4-b were identical to those of CD4 MAbs OKT4 and Leu3-a, whatever kind of cell was concerned. They reacted strongly with around 65% of thymocytes.

B. DOUBLE LABELING

In order to confirm that these rat MAbs are T lymphocyte specific and to find out if the subpopulation of T lymphocytes recognized by MAbs LO-CD4-a and LO-CD4-b is indeed the one recognized by OKT4, tonsil cells (Figure 2) or PBMN (Figure 3) were first labeled with these rat MAbs and RARA-FITC, followed by phycoerythrin-conjugated mouse reference MAbs T11, B4 (Figure 2), OKT4, or OKT8 (Figure 3). Among the T11+ cells, we found 95% LO-CD6-a+ cells and 77% LO-Tmat+ cells. Cells that labeled with LO-CD4-a made up about 78% of T11+ lymphocytes. In contrast, the percentage of lymphocytes double-labeled with B4 and our rat MAbs remained low (4 to 6%). This weak positivity could be due to aspecific binding to Fc receptors on B cells. Regarding subsets of peripheral T lymphocytes, more than 99% of T4+ lymphocytes were also labeled with LO-CD4-a or LO-CD4-b and less than 2% of PBL was labeled with OKT4 alone or with the rat MAb alone (Figure 3). Less than 6% of PBL cells were double-labeled with OKT8 and LO-CD4-a or LO-CD4-b. Hence, all these rat MAbs are T lymphocyte specific, except for a possible very low cross-reaction of LO-CD6-a with B lymphocytes and, the subpopulation of T lymphocytes recognized by LO-CD4-a and LO-CD4-b is exactly that recognized by OKT4, not by OKT8 (T4+, T8–).

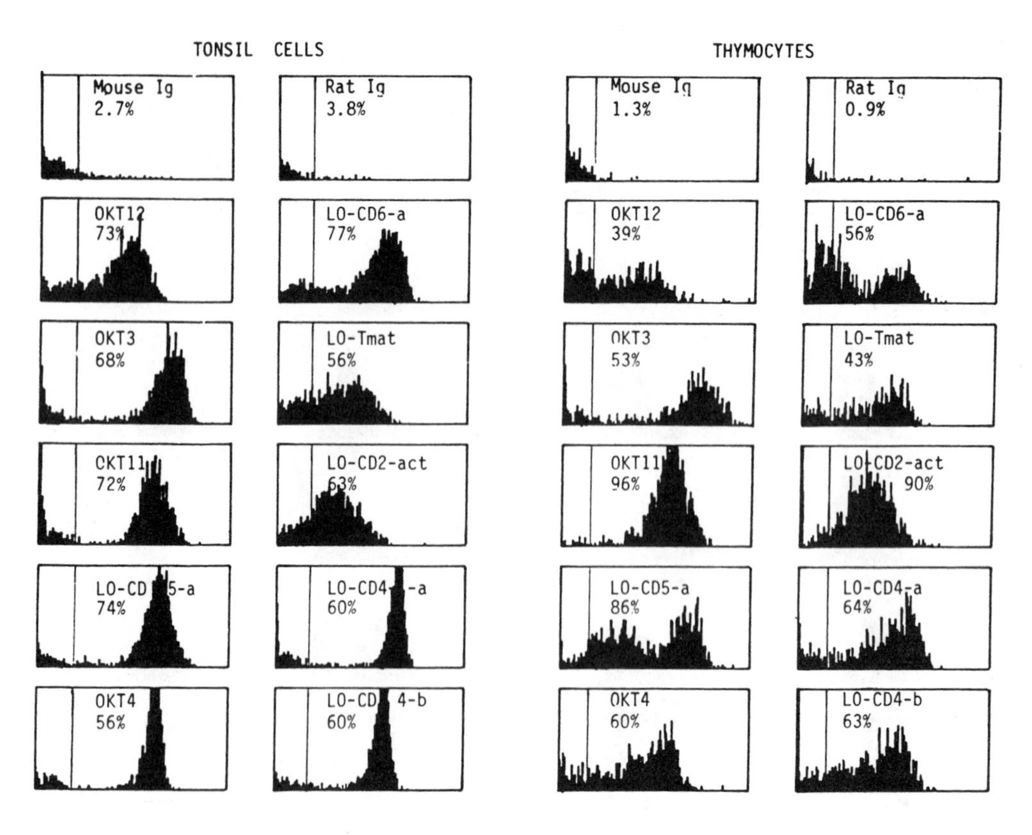

FIGURE 1B. Reactivity pattern of MAbs with tonsil cells (left panel) and thymocytes (right panel). X-axis: fluorescence intensity. Y-axis: cell number.

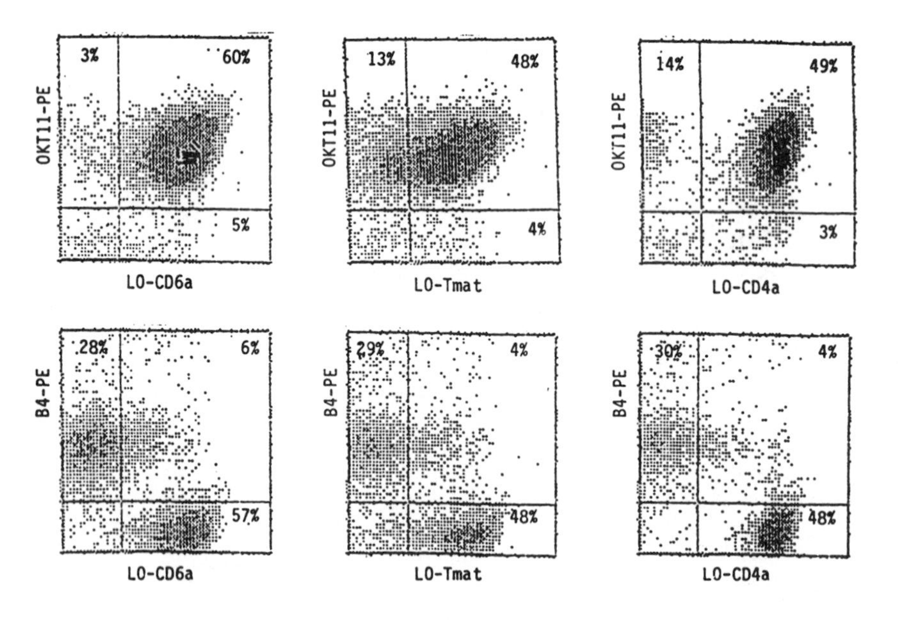

FIGURE 2. Double labeling with pan-T or pan-B MAbs and rat MAbs. Tonsil cells were labeled with MAbs LO-CD6-a, LO-Tmat, LO-CD4-a plus RARA-FITC followed by T11-PE or B4-PE.

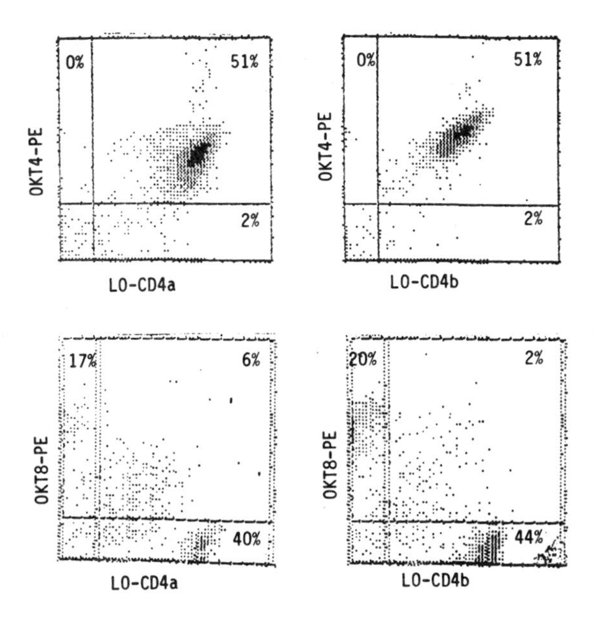

FIGURE 3. Subsets of lymphocytes recognized by LO-CD4-a and LO-CD4-b. PBMN were labeled with rat MAbs LO-CD4-a or LO-CD4-b plus RARA-FITC, followed by OKT4-PE or OKT8-PE.

C. REACTIVITY PATTERNS WITH LEUKEMIC CELLS

In order to detect cross-reactions with immature B or myeloid cells and to delimit at which stage of differentiation the different T antigens appear or disappear, we have studied leukemias, lymphomas (Table 2), and the leukemia-derived cell lines (Tables 3 and 4).

No cross-reaction with non-T cell lines, listed in Table 3, was found, except that LO-CD4-a and LO-CD4-b reacted with U937 just as OKT4 and Leu3-a did. This cell line is known to express the T4 antigen. Non-T non-B lymphoblasts were labeled by none of our MAbs (Table 2). The detection of 2 to 12% positive cells can be assigned to the presence of residual normal lymphocytes. A weak reaction of LO-CD6-a and LO-Tmat with B-CLL and B-lymphoma cells was noticed. Again the 2 to 10% cells brightly labeled with LO-CD2-act, LO-CD4-a, and LO-CD4-b are probably due to residual normal cells.

LO-CD6-a stained 3/4 T cell lines clearly (Table 4), and all T-leukemia samples weakly (Table 2). This pan T specific MAb is different from CD2 and CD5 MAbs, which label the T-ALL (acute lymphoblastic leukemia) cells more brightly. The difference with OKT3 specificity is more subtle, with the only discrepancy observed was with T cell lines. We found no significant difference in reactivity between T12 and LO-CD6-a. LO-Tmat reacted very weakly with 3/6 T-ALL samples and 1/4 cell lines unlike CD2, CD3, CD5 or CD6 MAbs. It might be a new pan-mature-T specific MAb. The LO-CD2-act antigen was present on T-ALL and T-CLL (chronic lymphoblastic leukemia) cells. It was strongly expressed on HPB-ALL, but weakly on MOLT4 and Jurkat, and not on CEM cells (Table 4). LO-CD4-a and LO-CD4-b stained with variable intensity all T-ALL tested, as did OKT4.

D. IMMUNOPRECIPITATION

To identify the molecules defined by these rat MAbs, the antigens were precipitated by MAb from ^{125}I-labeled cell membrane extracts and analyzed by SDS-PAGE. The antigen precipitated by MAb LO-CD6-a migrated as two peptides of 102 kDa and 130 kDa. MAb LO-CD2-act precipitated an antigen with molecular weight of 46 kDa from HPB-ALL cells as well as from PHA-activated lymphocytes (not shown). Up to now, we have not succeeded in precipitating the antigens recognized by MAbs LO-CD4-a, LO-CD4-b, and LO-Tmat.

TABLE 2
Reactivity with Leukemia Cells

	MAbs				
	LO-CD6-a	**LO-Tmat**	**LO-CD2-act**	**LO-CD4-a**	**LO-CD4-b**
T-ALL					
1	38% ±	–	98% ++	94% +++	97% +++
2	43% ±	34% ±	95% +	56% +	64% +
3	22% ±	–	85% +	24% ±	24% ±
4	60% ±	9% ±	93% +	20% ±	19% ±
5	68% ±	–	98% ++	89% ++	89% ++
6	50% ±	20% ±	95% ++	54% ++	61% ++
T-CLL	85% +	33% ±	70% ±	94% ++	93% ++
B-CLL					
1	41% ±	53% ±	10% +	3% ++	5% ++
2	41% ±	–	6% ++	4% ++	4% ++
3	69% ±	36% ±	–	2% ++	3% ++
4	46% ±	–	–	–	–
B-NHL	57% ±	56% ±	8% +	4% +	5% +
cALL					
1	12% +	10% +	9% +	8% ++	7% ++
2	–	–	–	2% +	2% +
3	–	–	–	–	–

TABLE 3
Reactivity with Non-T Cell Lines

	MAbs				
Cells	**LO-CD6-a**	**LO-CD2-act**	**LO-Tmat**	**LO-CD4-a**	**LO-CD4-b**
B lineage					
Nalm-1	–	–	–	–	–
KM3	–	–	–	–	–
Nalm-6	–	–	–	–	–
Raji	–	–	–	–	–
Daudi	–	–	–	–	–
Myeloid lines					
U937	–	–	–	++	++
HL-60	–	–	–	±	±
K562	–	–	–	–	–

E. INHIBITION OF ROSETTE FORMATION

The molecule of 45 to 50 kDa defined by CD2 MAbs bears at least three epitopes. The first two epitopes are present on all T lymphocytes, but only one of them ($T11_1$) functions as an SRBC receptor.[8,15,16] The third epitope is hidden on resting T lymphocytes, but appears fully expressed after activation. MAbs directed against these different epitopes can be distinguished by their inhibition of rosetting with SRBC or AET-SRBC. Normal PBL and PHA-activated PBL were preincubated with LO-CD2-act or OKT11 or control serum and allowed to rosette with SRBC, either AET-treated or untreated (Table 5). OKT11 strongly inhibited rosetting of both PBL and activated PBL with either AET-treated or untreated SRBC. MAb LO-CD2-act blocked the rosetting of normal PBL with untreated SRBC, but had only a slight effect on rosetting of normal PBL with AET-treated SRBC and no effect on rosetting of activated PBL with either AET-treated or untreated SRBC.

TABLE 4
Reactivity with T Cell Lines

MAbs	CEM	HPB-ALL	MOLT-4	JURKAT
OKT3	–	++	–	±
T11	+++	+++	+++	++
T12	+	+	+	–
LO-CD5-a	++	++	++	±
Leu3-a	+++	+++	±	–
OKT4	+++	+++	±	–
LO-CD6-a	+	++	++	–
LO-CD2-act	–	++	+	±
LO-Tmat	±	–	–	–
LO-CD4-a	+++	+++	±	–
LO-CD4-b	+++	+++	±	–

TABLE 5
Inhibition of Rosette Formation

Cell	Percentage of rosetting			
	PBL		PBL-PHA	
Treatment	E+	AET-E+	E+	AET-E+
Mouse IgG	46%	70%	78%	76%
T11	3%	10%	9%	23%
Rat IgG	60%	65%	63%	79%
LO-CD2-act	8%	45%	59%	82%

F. BLOCKING ASSAYS

MAbs LO-CD4-a and LO-CD4-b recognized exactly the same subset of T lymphocytes as OKT4 (Figure 3). Therefore, blocking of binding of mouse CD4 MAbs by our rat CD4 MAbs was investigated. Pre-incubation of purified T lymphocytes with either LO-CD4-a or LO-CD4-b did not prevent OKT4 binding (data not shown), but did prevent MAb Leu3-a binding to the cells as detected by LO-MK-1-FITC (Figure 4 c, d), indicating that LO-CD4-a and LO-CD4-b belong to CD4, and that the epitope recognized by them is the same as that recognized by Leu3-a but is different from that recognized by OKT4.

To determine whether the pan-T specific LO-CD6-a MAb recognizes the same antigenic determinant as mouse pan-T specific MAbs, blocking studies with OKT3 and T12 were performed. No inhibition of OKT3 binding to PBL was found when PBL were pre-incubated with LO-CD6-a (data not shown). However, T12 binding to PBL detected by LO-MK-1-FITC was totally inhibited if PBL were pre-incubated with LO-CD6-a even at high dilution (1/10,000) (Figure 4g). In the same assay, another rat pan-T specific MAb, LO-Tmat, had no inhibitive effect on T12 binding (Figure 4h). These data indicate that MAb LO-CD6-a recognizes the same molecule and even the same epitope as T12 and thus belongs to CD6.

G. FUNCTIONAL ANALYSIS

In order to find out whether the antigens defined by our rat MAbs have any functional significance in immune response, the effect of these MAbs on lymphocyte proliferation induced by lectins, antigens, or OKT3 was studied (Table 6). Pan-T specific MAb LO-CD6-a and LO-Tmat had hardly any effect on lymphocyte proliferation. The effects of the two CD4 MAb LO-CD4-a and LO-CD4-b on lymphocyte proliferation are quantitatively different, LO-CD4-b always having a stronger inhibitive effect than LO-CD4-a. They strongly interfered with the

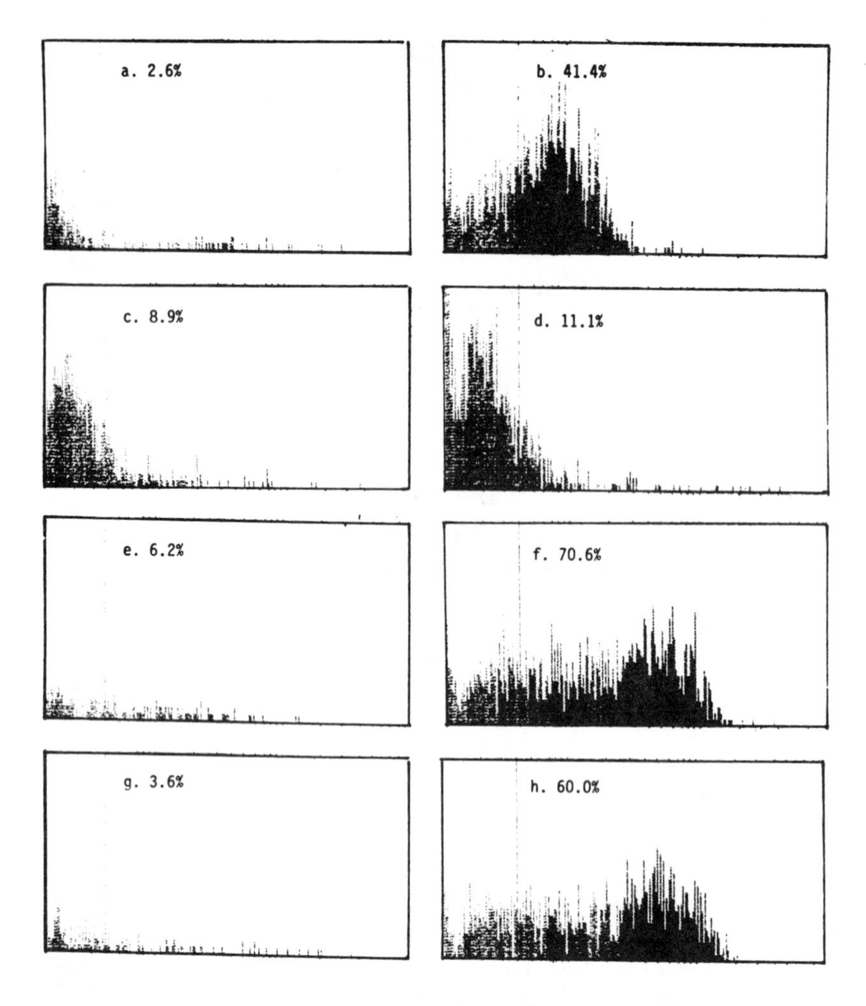

FIGURE 4. A-D: Inhibition of Leu3-a binding. PBL were incubated with PBS (a, b), LO-CD4-a (c), or LO-CD4-b (d) before incubation with normal rat serum (a) or Leu3-a (b, c, d). The binding of Leu3-a to the cells was detected by LO-MK-1-FITC. A-H: Inhibition of T12 binding. PBL were incubated with PBS (e, f), LO-CD6-a (g), or LO-Tmat (h) before incubation with normal rat serum (e) or T12 (f, g, h). The binding of T12 to the cells was detected by LO-MK-1-FITC.

proliferation of lymphocytes induced by the soluble antigen tetanus toxoid (TT) and by cell-bound antigen (MLR): we observed about 70% inhibition by LO-CD4-b and 40 to 60% inhibition by LO-CD4-a. Inhibition of ConA, PWM, and T3-induced proliferation was weak (30 to 40%) by LO-CD4-b, and not significant by LO-CD4-a. None of them had suppressive effect on PHA-induced proliferation.

MAb LO-CD2-act had no effect on PHA-induced proliferation. It lowered the proliferation induced by ConA and PWM (about 50%) and very strongly depressed the antigen-induced proliferation of lymphocytes in the case of TT and MLR (85% inhibition). Moreover, activation of lymphocytes through the antigen receptor complex, induced by OKT3, was also strongly inhibited by LO-CD2-act. We further found that if LO-CD2-act was present during activation of PBL by OKT3, the interleukin-2 (IL-2) receptor expression was reduced from 63 to 36% (data not shown). Furthermore, when LO-CD2-act was added even 3 d after initiation of mixed lymphocyte culture, the inhibition of proliferation remained around 80%, indicating that LO-CD2-act exerts its effect at a late stage in the process of lymphocyte activation.

TABLE 6
Effect of MAbs on Lymphocyte Proliferation Induced by Lectins, OKT3, or Antigens

	Percentage of control		
MAb	**PHA**	**ConA**	**PWM**
Medium	100	100	100
LO-CD6-a	106 ± 8	95 ± 3	99 ± 3
LO-Tmat	108 ± 9	90 ± 10	99 ± 6
LO-CD2-act	101 ± 13	46 ±19	54 ± 5
LO-CD4-a	106 ± 4	88 ± 2	90 ± 6
LO-CD4-b	116 ± 6	59 ± 16	72 ± 1

	Percentage of control		
MAbs	**T3**	**TT**	**MLR**
Medium	100	100	100
LO-CD6-a	80 ± 16	85 ± 2	162 ± 41
LO-Tmat	78 ± 17	112 ± 0	173 ± 92
LO-CD2-act	14 ± 9	13 ± 5	14 ± 1
LO-CD4-a	92 ± 9	43 ± 4	67 ± 13
LO-CD4-b	73 ± 9	32 ± 5	34 ± 9

Note: Results are expressed as means ± standard deviations obtained in two separate experiments.

IV. DISCUSSION

Five rat MAbs specific for human T lymphocytes are reported in this chapter. There was no reaction of these MAbs with peripheral blood non-T cells or with non-T cell lines, except for U937 (Tables 1 and 3). They did not label normal lymphocytes that expressed B lymphocyte specific antigen B4 (Figure 2).

Pan-T specific MAb LO-CD6-a had a reactivity pattern similar to that of OKT3 and T12, but it could be excluded that LO-CD6-a belonged to CD3 because it reacted with CCRF-CEM and Molt4 cells which did not express the antigen recognized by OKT3 (Table 4). It could also be excluded that LO-CD6-a belonged to CD2 and CD5 since LO-CD6-a only reacted with a subset of thymocytes (Table 1 and Figure 1) and reacted very weakly with T-ALL, while CD2 Mab T11 and CD5 MAb LO-CD5-a reacted with around 90% of thymocytes, and T11 strongly reacted with most T-ALL. Several lines of evidence show that LO-CD6-a belongs to CD6: LO-CD6-a and reference CD6 MAb T12 had nearly identical reactivity patterns; and LO-CD6-a defined a molecule with two subunits of 102 and 130 kDa. A molecule recognized by CD6 MAb with similar molecular weight has been reported by Knowles;[17] if LO-CD6-a was allowed to bind to the surface of T lymphocytes, no further binding of CD6 MAb T12 could be detected, indicating that MAb LO-CD6-a binds to the same epitope as T12 or one very close to it.

LO-Tmat reacted with all peripheral blood or tonsil T lymphocytes just as pan-T MAbs did. However, because of its much weaker or absence of reactivity with T-ALL and leukemia-derived T cell lines, it cannot be grouped into CD2, CD3, CD5, or CD6. LO-Tmat might recognize a molecule restricted to resting mature T lymphocytes: (1) we found that LO-Tmat hardly labeled leukemia T cell lines which proliferate rapidly *in vitro* culture (Table 4); (2) LO-Tmat almost did not react with T-ALL (Table 2) which represent immature T lymphocytes at early stages of differentiation; (3) LO-Tmat only reacted with a subset of thymocytes, but no information has been obtained with regard to the stage of differentiation of the labeled thymocytes; and (4) the expression of the molecule defined by LO-Tmat is lowered after activation of T lymphocytes by PHA.

LO-CD2-act defined a 46 kDa molecule present on thymocytes, and resting as well as activated T lymphocytes. This result suggested that the antigen defined by LO-CD2-act might be the E-receptor recognized by CD2 MAbs. Although not identical, the reactivity patterns of LO-CD2-act and OKT11 exhibit similarities. They are both pan-T specific MAbs and their reaction with activated T lymphocytes is stronger than with resting T lymphocytes. Both labeled around 90% of thymocytes and all T-ALL, unlike pan-T specific CD3 and CD6 MAbs. The difference between LO-CD2-act and T11 is that LO-CD2-act always shows weaker reaction with T lymphocytes than T11 and that LO-CD2-act did not react with T cell line CEM while T11 did.

At least three epitopes on the E-receptor molecule have been demonstrated: epitopes 1 and 2 are expressed on resting and activated T lymphocytes and epitope 3 is only expressed on activated T lymphocytes and its expression is rapidly induced by MAbs to epitope 2.[15] Only MAbs to epitope 1 block rosette formation with untreated and AET-treated SRBC. MAb LO-CD2-act blocked rosette formation of resting T lymphocytes with untreated SRBC (slightly with AET-treated SRBC), but had no effect on SRBC rosette formation of activated T lymphocytes. CD2 MAb D66 reported by Boumsell and Bernard[16] has similar properties to LO-CD2-act. Both of them reacted weakly with T lymphocytes and reacted strongly with activated T lymphocytes; both of them inhibited E-rosette formation of PBL with untreated SRBC. The difference is that LO-CD2-act did not block E-rosette formation of activated T lymphocytes while D66 did. Thus the epitope recognized by LO-CD2-act could be different from the epitope recognized by D66.

When function studies were carried out, we found that LO-CD2-act had a strong inhibitory effect on proliferation of lymphocytes induced by soluble antigen TT or cell-bound allogeneic antigen (MLR). It also blocked the activation of lymphocytes through the T cell antigen receptor complex induced by OKT3. This effect of LO-CD2-act is in sharp contrast to that of our CD25 MAb: although CD25 MAb LO-Tact1 and LO-Tact2 totally inhibited MLR, they did not interfere with OKT3-mediated activation. The mechanism by which LO-CD2-act exerts its inhibitory effect on lymphocyte proliferation is not known. On the one hand, LO-CD2-act lowered the expression of IL2 receptor indicating that it interferes with an event before IL2 receptor expression; and on the other hand, LO-CD2-act inhibits proliferation even if it is added 3 d after initiation of MLR which could mean that the CD2 antigen also has a function later on in the proliferation. Clearly, these results need confirmation and comparison with effects of CD25 MAbs and IL2.

MAbs LO-CD4-a and LO-CD4-b had almost identical reactivities. Though the molecular weight of the antigen recognized has not yet been determined, experimental results support the view that these two MAbs belong to CD4. LO-CD4-a and LO-CD4-b reactivity patterns were very similar to those of OKT4 and Leu3-a on all cells or cell lines tested. They reacted with monocytic cell line U937 as OKT4 and Leu3-a did. All T cell markers except T4 are absent from U937 cells. The subset of T lymphocytes recognized by LO-CD4-a and LO-CD4-b is exactly the one defined by OKT4 (Figure 3). This subset of T lymphocytes was T8-. LO-CD4-a and LO-CD4-b blocked the binding of Leu3-a to T lymphocytes but had no effect on OKT4 binding.

We found that LO-CD4-a and LO-CD4-b had an inhibitory effect on TT antigen or allogeneic-antigen induced proliferation of lymphocytes. They had no or a weak inhibitory effect on polyclonal proliferation of lymphocytes induced by mitogens such as PHA and PWM or by OKT3. In contrast to lectin-mediated lymphocyte proliferation, activation of lymphocytes by soluble antigen, or allogeneic antigen is HLA class II restricted. Recently, evidence has emerged, suggesting that T4 antigen recognizes HLA class II antigen.[11,12] Hence the anti-T4 specificities of LO-CD4-a and LO-CD4-b are in agreement with inhibition of proliferation induced by antigen (HLA-class II restricted).

In summary, five human T lymphocyte-specific rat MAbs have been obtained. LO-CD6-a, LO-CD4-a, and LO-CD4-b belong to CD6 and CD4 already defined by mouse MAbs. Our rat LO-CD2-act and LO-Tmat MAbs, however, have specificities rarely or not detected by mouse MAb, indicating a possible difference in immune repertoire between rat and mouse.

ACKNOWLEDGMENTS

The authors thank Dr. V. Deneys for her helpful scientific advice; J. P. Kints, J. M. Malache, F. Nisol, F. Flemal, T. Gervais, A. Cornet, and A. M. Mazzon for their technical help; and F. Bolle for typing the manuscript.

REFERENCES

1. **Spitz, H., Borst, J., Tax, W., Capel, P. J. A., Terhorst, C., and de Vries, J. E.,** Characterization of a monoclonal antibody (WT-31) that recognizes a common epitope on the human T cell receptor for antigen, *J. Immunol.*, 135, 1922, 1985.
2. **Kung, P. C., Goldstein, G., Reinherz, E. L., and Schlossman, S. F.,** Monoclonal antibodies defining distinctive human T cell surface antigens, *Science*, 206, 347, 1979.
3. **Leonard, W. J., Depper, J. M., Uchiyama, T., Smith, K. A., Waldmann, T. A., and Greene, W. C.,** A monoclonal antibody that appears to recognize the receptor for human T-cell growth factor: partial characterization of the receptor, *Nature*, 300, 267, 1982.
4. **Verbi, W., Greaves, M. F., Schneider, C., Koubek, K., Janossy, G., Stein, H., Kung, P., and Goldstein, G.,** Monoclonal antibodies OKT11 and OKT11A have pan-T reactivity and block erythrocyte receptors, *Eur. J. Immunol.*, 12, 81, 1982.
5. **Selvaraj, T., Plunkett, M. L., Dustin, M., Sanders, M. E., Shaw, S., and Springer, T. A.,** The T lymphocyte glycoprotein CD2 binds the cell surface ligand LFA-3, *Nature*, 326, 400, 1987.
6. **van Wauwe, J. P., De Mey, J. R., and Goossens, J. G.,** OKT3: a monoclonal anti-human T lymphocyte antibody with potent mitogenic properties, *J. Immunol.*, 124, 2708, 1980.
7. **Meuer, S. C., Hodgdon, J. C., Hussey, R. E., Protentis, J. P., Schlossman, S. F., and Reinherz, E. L.,** Antigen-like effects of monoclonal antibodies directed at receptors on human T cell clones, *J. Exp. Med.*, 158, 988, 1983.
8. **Meuer, S., Hussey, R., Fabbi, M., Fox, D., Acuto, O., Fitzgerald, K., Hodgdon, J., Protentis, J., Schlossman, S., and Reinherz, E.,** An alternative pathway of T-cell activation: a functional role for the 50 Kd T11 sheep erythrocyte receptor protein, *Cell*, 36, 897, 1984.
9. **Reinherz, E. L., Kung, P. C., Goldstein, G., and Schlossman, S. F.,** Separation of functional subsets of human T cells by a monoclonal antibody, *Proc. Natl. Acad. Sci. U.S.A.*, 76, 4061, 1979.
10. **Reinherz, E. L., Kung, P. C., Goldstein, G., and Schlossman, S. F.,** A monoclonal antibody reactive with the human cytotoxic/suppressor T cell subset previously defined by a heteroantiserum termed TH2, *J. Immunol.*, 124, 1301, 1980.
11. **Meuer, S. C., Schlossman, S. F., and Reinherz, E. L.,** Clonal analysis of human cytotoxic T lymphocytes: T4+ and T8+ effector T cells recognize products of different major histocompatibility complex regions, *Proc. Natl. Acad. Sci. U.S.A.*, 79, 4395, 1982.
12. **Biddison, W. E., Rao, P. E., Talle, M. A., Goldstein, G., and Show, S.,** Possible involvement of the T4 molecule in T cell recognition of class II HLA antigens: evidence from studies of proliferative responses to SB antigens, *J. Immunol.*, 131, 152, 1983.
13. **Bazin, H.,** Production of rat monoclonal antibodies with LOU rat non secreting IR983F myeloma cell line, in *Protides of the Biological Fluids*, Peeters, H., Ed., Pergamon Press, Oxford, 1982, 615.
14. **Morikawa, S., Tatsumi, E., Baba, M., Harada, T., and Yasuhira, K.,** Two E-rosette forming lymphoid cell lines, *Int. J. Cancer*, 21, 166, 1978.
15. **Bernard, A., Brottier, P., Georget, E., Lepage, V., and Boumsell, L.,** The epitopic dissection of the CD2 defined molecule: relationship of the second workshop antibodies in terms of reactivities with leukocytes, rosette blocking properties, induction of positive modulation of the molecule and triggering T cell activation, in *Leukocyte Typing II*, Vol. 1, Reinherz, E.L., Haynes, B.F., Nadler, L.M., and Bernstein, I.D., Eds., Springer-Verlag, New York, 1984, 53.

16. **Bernard, A., Gelin, C., Raynal, B., Pham, D., Gosse, C., and Boumsell, L.,** Phenomenon of human T cells rosetting with sheep erythrocytes analyzed with monoclonal antibodies. "Modulation" of a partially hidden epitope determining the conditions of interaction between T cells and erythrocytes, *J. Exp. Med.*, 155, 1317, 1982.
17. **Knowles, R. W.,** Immunochemical analysis of the T cell-specific antigens, in *Leukocyte Typing II,* Vol. 1, Reinherz, E.L., Haynes, B.F., Nadler, L.M., and Bernstein, I.D., Eds., Springer-Verlag, New York, 1985, 259.

Chapter 21.I

RAT MONOCLONAL ANTIBODIES AGAINST *STREPTOCOCCUS MUTANS* AND*BACTEROIDES GINGIVALIS*

F. Ackermans

TABLE OF CONTENTS

I. ANTIGENS AND IMMUNIZATION SCHEDULES

A. WHOLE BACTERIA OF *STREPTOCOCCUS MUTANS*

Fusion to obtain monoclonal antibodies (MAbs) against the whole bacteria was carried out with popliteal nodes of LOU/C rat.

S. mutans strains OMZ 70 (serotype c) and OMZ 175 (serotype f) obtained from Dr. G. Guggenheim (Zürich) were maintained in brain/heart infusion broth = BHI (Difco Laboratories) and collected in exponential phase culture (O.D. 540 nm = 1). Stimulation of the popliteal node was executed by injection into the footpad.

Day 0: 100 µl of *Bordetella pertussis* 10^{10} cells/ml (Institut Pasteur Production, France) + 150 µl of *S. mutans* (10^9 bacteria/ml) in phosphate buffered saline (PBS); Day 3: 150 µl of *S. mutans* (10^9) bacteria/ml in PBS; Day 6: idem; Day 9: idem; Day 12: fusion of the popliteal node cells of the immunized LOU/C rat and 983 fusion cells.

Another process to obtain specific antibodies against whole bacteria was tried. Day 0: 10^8 bacteria with alumine hydroxide (6 mg/ml) intraperitoneally into a LOU/C rat; Day 14: idem; 6 months later: booster intravenously with 10^8 living *S. mutans*; 4 d later: fusion of the spleen cells of the immunized animal and 983 immunocytoma cells.

B. WHOLE BACTERIA OF *BACTEROIDES GINGIVALIS*

The same immunization schedules were used with success for an anaerobic gram-negative bacteria involved in periodontal disease: *Bacteroides gingivalis* ATCC 33277 (kindly given by Dr. C. Mouton, Quebec, Canada).

Bacteroides gingivalis was maintained in a solid medium containing BHI and blood agar in anaerobic conditions.

C. EXTRACT OF CELL-WALL OF *STREPTOCOCCUS MUTANS* OMZ175

To study the external part of the bacteria, it is sometimes easier to extract a crude fraction of the cell-wall from the microorganism and immunize rats with these extracts.

Crude cell-wall associated antigens were extracted from the bacterial pellet of *S. mutans* OMZ175 with 0.5 *M* phosphate buffer pH 6.0 as described by Schöller et al.[1] = WEA fraction. Protein concentration was determined by the procedure of Lowry et al.[2] with bovine serum albumin as a standard. WEA fraction contained at least 20 proteins with a molecular weight of 30 to 200 kDa.

Several processes were tried to obtain the best immune response in the rats (see Chapter 5).

For our molecules, the best immunization schedule was as follows. Day 0: 50 µg of proteins of WEA fraction of *S. mutans* OMZ175 were injected intraperitoneally in a WISTAR/R rat (because of a lack of LOU rats at that time) with alumine hydroxide (6 mg/ml Amphogel-WYETH); Day 14: 50 µg of WEA fraction + 6 mg/ml alumine hydroxide intraperitoneally; Day 30: tests of the antibody response by Ouchterlony; Day 90, 120, 150, and 180: same as day 30; Day 210: when the concentration of specific antibodies in the rat serum was strongly reduced, 100 µg of WEA fraction were injected in phosphate buffered saline (PBS) intravenously as a booster; Day 214: spleen cells were collected and fused with LOU rat nonsecreting 983 fusion cells (see Chapter 6. I).

II. PARENT FUSION CELL LINE AND FUSION PROCEDURES

The parent myeloma cell line was the rat nonsecreting 983 (see Chapter 4) and spleen cells of the immunized animal were fused in a cell ratio of 5:1. The fusion solution was 50% (wt/wt) polyethylene glycol 4000 (Merck, Darmstadt, West Germany). The fused cells were diluted in complete Dulbecco's Modified Eagle medium (Gibco Laboratories No 041-01965M, Grand Island, W.Y.) containing 5% fetal calf serum and 5% horse serum supplemented with 0.1 m*M* hypoxanthine, 0.4 m*M* aminopterin and 0.015 m*M* thymidin and plated in 24 and 96 wells (see Chapter 6. I.).

III. SCREENING METHODS FOR THE OBTAINMENT OF SPECIFIC MAbs

Antibody production by the hybrid cells was assessed by indirect RIA or by indirect ELISA.

The amount of 50 µg of protein of WEA fraction/ml in carbonate buffer (0.15 *M*, pH 9.6) as coated to each well of specific microtiter plates for RIA or ELISA (see Chapter 9.II).

We used a high concentration of antigens to coat the plates because it is a crude fraction containing several proteins.

Whole bacteria of *S. mutans*, in exponential phase culture (O.D. 540 µm = 1) or *Bacteroides gingivalis*, were washed three times in PBS and were used for coating microtiter plates (50 µl – 10^9 bacteria/ml). These plates were placed for 16 h at 37°C. The ELISA technique was previously described in Chapter 9. II.

IV. PRODUCTION AND PURIFICATION OF RAT MABS

For MAbs obtained after fusion of the spleen cells of rat immunized with WEA fraction of *Streptococcus mutans*, the first passage on animal was performed subcutaneously on a (WISTAR/R × LOU/C)F1 hybrid rat because spleen cells used for the fusion were derived from a WISTAR/R rat and myeloma cells were derived from a LOU/C rat. Later, the teased solid tumor could be injected in irradiated LOU/C rats. Ascitic fluid was collected, tested by agarose electrophoresis (see Chapters 14 and 15). MAbs from IgG isotype were purified on a mouse anti-rat kappa light chain (MARK-3) affinity column (See Chapters 14 and 15). Rat MAbs from IgM isotype (euglobulins) were purified on a chromatography Bio-Gel 5 *M* after precipitation in Tris 1 *M*. IgM proteins were resuspended in phosphate buffered saline (0.05 *M*, pH 7.2) and deposited on a Bio-Gel 5 *M* (see Chapter 15).

V. CHARACTERIZATION AND USE OF BACTERIA MAbs OF *STREPTOCOCCUS MUTANS*

See Chapter 27.

ACKNOWLEDGMENTS

H. Bazin, R. M. Frank, J. P. Klein, J. A. Ogier, T. Bruyère, F. Cormont, C. Genart, F. Nisol, J. M. Malache, and J. P. Kints are gratefully acknowledged for their assistance and advice during this work. This study was supported by the "Fonds de Développement Scientifique" of the University of Louvain, Belgium.

REFERENCES

1. **Schöller, M., Klein, J. P., and Frank, R. M.,** Common antigens of streptococcal and non streptococcal oral bacteria: immunochemical studies of extracellular and cell-wall associated antigens from *Streptococcus sanguis, Streptococcus mutans, Lactobacillus salivarius* and *Actinomyces viscosus, Infect. Immun.*, 31, 52, 1981.
2. **Lowry, O. H., Rosebrough, N.J., Farr, A.L., and Randall, R.J.,** Protein measurement with the folin phenol reagent, *J. Biol. Chem.*, 193, 265, 1951.

Chapter 21.II

A SPECIFIC RAT MONOCLONAL ANTIBODY AGAINST *YERSINIA ENTEROCOLITICA* SEROGROUP 0:8

D. Stynen, A. Goris, L. Meulemans, and E. Briers

TABLE OF CONTENTS

I. INTRODUCTION

Yersinia enterocolitica bacteria are facultatively anaerobic gram negative rods belonging to the family of Enterobacteriaceae.[1] They are medically important organisms, since they can cause a wide variety of infections which are sometimes fatal. The most common clinical forms are acute enteritis (in young children and infants) and appendicitis-like syndrome (in older children). Less frequently and primarily in adults, extra-intestinal inflammations as arthritis and erythema nodosum can be associated with *Y. enterocolitica* infections. Finally, *Y. enterocolitica* disseminated by blood may cause rare infections in different organ systems.[2] *Y. enterocolitica* infections often mimick other diseases, leading to incorrect diagnosis and unnecessary laparatomy in the case of appendicitis-like syndrome. Good diagnostic tools are therefore important.

Y. enterocolitica bacteria possess lipopolysaccharides (LPS) at their surface, which correspond to the O-antigen. This O-antigen shows strain-dependent immunological variations, allowing classification of different strains into serotypes. More than 50 O-serotypes have been described, correlating to some extent with biochemical characteristics and ecological distribution.

Determination of the serotype is important, especially in epidemiological studies. Serotypes 0:3, 0:5, 27, 0:8, and 0:9, the most frequently isolated serotypes in the clinical laboratory, represent the vast majority of pathogenic strains.[3] They can be identified by agglutination of the bacteria with specific polyclonal antisera.

In this study, we demonstrate the feasibility of producing highly specific rat monoclonal antibodies against serotype 0:8, a medically important serotype almost exclusively found in North America. We show that it binds the LPS and we finally evaluate the performance of a latex reagent coated with this monoclonal antibody as a tool for serotyping.

II. MATERIALS AND METHODS

A. BACTERIAL STRAINS

Y. enterocolitica and *Y. pseudotuberculosis* strains were kindly given by Dr. G. Wauters (Université Catholique de Louvain, Brussels). Other bacterial strains were clinical isolates donated by Centraal Laboratorium (Hasselt). *Yersinia* strains were routinely grown on Tryptone Soya Agar (Eco-Bio). Biotyping was carried out according to Ewing.[4] *Y. enterocolitica* Typing Serum (Eco-Bio) was used for serotyping.

B. HYBRIDOMA TECHNOLOGY

A 3-month-old LOU/C rat was immunized by subcutaneous injections of 5×10^7 heat killed *Y. enterocolitica* 0:8 in phosphate buffered saline (pH 7.4, PBS) on days 0, 4, 8, 12, 16, 21, and 25, and intraperitoneal injections on days 39, 55, 115, and 117. Fusion was carried out on day 119, according to the procedure described by Bazin,[5] with a 3:1 ratio of rat spleen cells to IR983F cells. Maintenance of the hybridoma cultures, expansion of selected wells, and cloning by limiting dilution on peritoneal feeder cells was performed as described.[5]

C. *IN VIVO* PRODUCTION AND AFFINITY PURIFICATION OF MONOCLONAL ANTIBODIES

The monoclonal antibody EB-Y8 was produced *in vivo* as described by Bazin.[5] A subcutaneous tumor was induced in a 2-month-old LOU/C rat by injection of 2×10^6 hybridoma cells. The peripheral tissue of this tumor was gently homogenized in a Potter-Elvejhem homogenizer in 10 ml of PBS. Two ml aliquots of the cell suspensions were then injected intraperitoneally in 3-month-old LOU/C.IgK-1b rats. Ascites fluid could be drained after 2 or 3 weeks. Topostasin (Roche) was added (3 U/ml of ascites) and the ascites were allowed to clot overnight at 4°C. The fluid phase was centrifuged (15 min, 1600 *g*) and stored at -20°C in 8 ml aliquots after the addition of merthiolate (0.01%, w/v) and a proteinase inhibitor (phenylmethylsulfonyl fluoride, 0.8 mg/ml).

Monoclonal antibodies from 8 ml aliquots of ascitic fluid were purified by affinity chromatography on a column with 15 ml of an affinity matrix consisting of MARK-3 coupled to Sepharose-4B (Pharmacia), as described by Bazin et al.[6] The immunoglobulin fractions eluted at low pH were collected as 4 ml fractions, immediately neutralized with 1 *M* Tris and dialyzed against the glycine buffered saline (0.1 *M* glycine, 1% (w/v)) NaCl, 0.1% (w/v) sodium azide, pH 8.2) used for coupling to latex beads.

D. ENZYME-LINKED IMMUNOSORBENT ASSAY (ELISA)

Screening for immunoglobulin secreting clones was carried out by a standard ELISA procedure. One hundred microliters of cell culture supernatant were incubated for 2 h at 37°C in microwell plates (Nunc) coated with 100 µl of a *Y. enterocolitica* 0:8 suspension (O.D. 420 nm = 0.260) and blocked with 10% (v/v) normal bovine serum. After three washes, a peroxidase conjugate of MARK-3, obtained from Dr. H. Bazin (Université Catholique de Louvain, Brussels), was added to the wells (1 h at 37°C). Then after four washes, positive reactions were visualised by addition of the peroxidase substrate (3,3′,5,5′,-tetramethylbenzidine) and H_2O_2. PBS with 0.1% (v/v) Tween 20 was used throughout the four washes and dilutions, except for coating and blocking (PBS without Tween 20).

The ELISA protocol for the determination of the specificity of antibodies only differed from that described above by the antigens used for coating. In the determination of the specificity of EB-Y8, 12.5 µg/ml of the purified antibody was used for incubation.

E. DETERMINATION OF ISOTYPES

The isotype of the antibodies was determined by double immunodiffusion.[7] Cell culture supernatants were diffused against a panel of isotype specific polyclonal antisera (Nordic) in a 1% (w/v in PBS) agarose gel.

F. POLYACRYLAMIDE GEL ELECTROPHORESIS AND WESTERN BLOTTING

One inoculation loop full of bacteria was resuspended in 1 ml of a reducing sample buffer containing 1% (w/v) of sodium dodecyl sulfate (SDS), boiled for 5 min and centrifuged. According to Laemmli,[8] 20 µl of supernatant were electrophoretically separated on a 5 to 17.5% gradient gel. Transfer to nitrocellulose paper (0.45 µm pore size, Millipore) was performed on a LKB 2005 Transphor, following the manufacturer's instructions. After the transfer, the blots were pre-incubated for 30 min at 37°C with Tris buffered saline (TBS, pH 7.4, 10 m*M* Tris, 0.15 *M* NaCl) containing 3% (w/v) of defatted powdered milk. Then, monoclonal antibody EB-Y8 was added to a final concentration of 10 µg/ml and incubation continued for another 2 h. After five washes in TBS, the blots were incubated for 1 h in a 1:1000 dilution of a peroxidase-conjugated polyclonal anti-rat immunoglobulin (Nordic) in TBS with 1.5% (w/v) powdered milk. After five more washes in TBS, the binding was visualized by incubation with 1-chloro-4-naphtol and H_2O_2 as the enzyme substrates. A lane with molecular weight markers (Pharmacia) was briefly stained with 0.1% (w/v) amido black in destaining buffer (45% (v/v) methanol, 10% (v/v) acetic acid).

G. LATEX AGGLUTINATION

Affinity purified monoclonal antibody EB-Y8 was covalently coupled to 0.3 µm latex beads (Serva) essentially according to the method of Faure et al.[9] Agglutination was performed by mixing a colony of the bacteria under investigation with one drop of the sensitised latex (0.7% suspension) on a dark glass slide. The result was defined as positive when the agglutination occurred within 1 min of manually rotating the glass slide. Latex beads coated with a monoclonal antibody to an irrelevant antigen never agglutinated *Y. enterocolitica* strains.

TABLE 1
Species and Serogroup Specificity of EB-Y8 as Determined by ELISA[a]

Strain	O.D. 450 nm[b]
Y. enterocolitica 0:8	
A97	0.817 ± 0.020
A99	0.638 ± 0.067
A103	0.626 ± 0.017
A105	0.723 ± 0.096
26/6	0.935 ± 0.058
IP1105B	0.815 ± 0.053
Y. enterocolitica 0:3	
5 strains	< 0.010
Y. enterocolitica 0:5,27	
6 strains	< 0.010
Y. enterocolitica 0:6	
5 strains	< 0.010
Y. enterocolitica 0:9	
6 strains	< 0.010

[a] The following species were negative (O.D. < 0.010) in all three experiments: *Y. pseudotuberculosis* (7 strains), *Serratia marcescens, Klebsiella pneumoniae, Escherichia coli, Haemophilus parainfluenzae, Neisseria meningitidis, Enterobacter cloacae, Shigella sonnei, Pseudomonas aeruginosa, Bacillus subtilis, Proteus vulgaris, Staphylococcus aureus, Streptococcus pneumoniae,* and *Salmonella typhimurium.*

[b] Results are the average of three experiments ± the standard deviation. Results indicated by "<0.010" had O.D. values consistently below 0.010.

III. RESULTS

In the initial screenings, 95 out of 420 wells scored positive in ELISA (O.D. 450 nm > 0.200). About one third of the cultures that remained positive after 3 weeks produced immunoglobulins that bound *Y. enterocolitica* 0:8 but not 0:3. Six of these were selected for cloning by limiting dilution: 2 IgM's, 2 IgG2b's, and 2 IgG2c's. Eventually, one IgG2c producing subclone, EB-Y8, was chosen for *in vivo* antibody production. In our ELISA experiments it had only reacted with *Y. enterocolitica* 0:8, but not with serogroups 0:3, 0:5,27, and 0:9 or *Y. pseudotuberculosis*. In order to investigate the efficiency of the purification, microtiterplates coated with *Y. enterocolitica* 0:8 were incubated with fourfold dilutions of the different column fractions. The highest concentration tested was a 1:100 dilution of each fraction. In the unbound peak, only the leading edge contained some residual ELISA activity: the O.D. obtained with the 1:100 dilution of this fraction, i.e., 0.104, was comparable to the O.D. values obtained with 1:25600 dilutions of intact ascites and fractions in the peak of bound material. This indicates that the 15 ml of affinity matrix efficiently removed the specific immunoglobulins from 8 ml of ascitic fluid. One single purification yielded 28 mg of purified immunoglobulins. ELISA experiments with purified EB-Y8 and plates coated with different strains of *Y. enterocolitica*, representing various serogroups, demonstrate the strict specificity of EB-Y8 for serogroup 0:8 (Table 1). Clearly positive results were obtained with all 6 strains of serogroup 0:8 (O.D. values between 0.600 and 1.000). All other 21 strains, representing serogroups 0:3, 0:5,27, 0:6 and 0:9, were completely negative (O.D. < 0.010). The antibody also failed to bind 14 other bacteria species, including *Y. pseudotuberculosis* (7 strains).

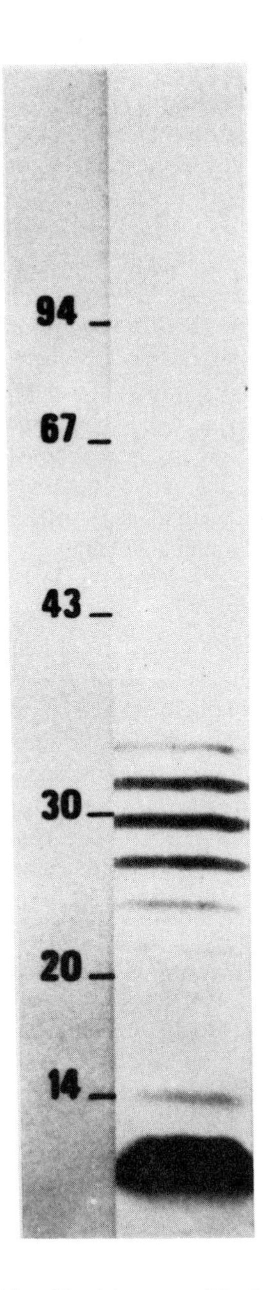

FIGURE 1. Immunoblot of *Yersinia enterocolitica* 0:8 extract with EB-Y8. The electrophoretic mobilities of polypeptide molecular weight markers (in kDa) are indicated for comparison.

Immunoblotting experiments confirmed the specificity of EB-Y8. A ladder-like staining pattern typical for LPS can be observed when extracts of *Y. enterocolitica* 0:8 are probed with EB-Y8 (Figure 1). The heaviest band has an electrophoretic mobility comparable to a polypeptide with a molecular weight of 37 kDa. The other bands then follow at 2 to 3 kDa intervals down to molecular weights below 14 kDa, the smallest marker. The intensities of these bands vary and some can only be observed when the lanes are overloaded (not shown). One very lightly stained band behaves as a heavier macromolecule (56 kDa). This unidentified band could also be found in blots of bacterial extracts from the other serogroups.

Mixing 1 drop of latex beads, sensitized with EB-Y8, on a glass slide with a colony of any of the *Y. enterocolitica* 0:8 strains mentioned in Table 1, resulted in a rapid agglutination of the latex particles and the bacteria. Colonies from *Y. enterocolitica* serogroups 0:3 (5 strains), 0:5,27 (6 strains), 0:6 (5 strains) and 0:9 (5 strains), were not agglutinated. *Y. pseudotuberculosis* (7 strains), *Klebsiella pneumoniae*, *Staphylococcus aureus*, and *S. epidermidis* did not react either.

IV. DISCUSSION

Yersinia enterocolitica serotyping is currently performed by agglutination of bacteria with homemade or commercially available (Eco-Bio) polyclonal rabbit antisera. Although these polyclonal antisera are quite adequate, the establishment of hybridoma cell lines secreting serogroup specific antibodies would have several advantages for the manufacturer. The first aim of this study was to investigate the feasibility of producing a highly specific monoclonal antibody against *Y. enterocolitica* serogroup 0:8. We chose the rat system mainly because of the features of its *in vivo* production system, more particularly the large volumes of ascites in one animal and the possibility of performing a one-step purification of the hybridoma immunoglobulins without contamination by host immunoglobulins.

One fusion sufficed to obtain the specific antibody required. Numerous clones secreted specific antibodies, representing different isotypes. The selected clone, EB-Y8, could be well adapted to *in vivo* production, yielding ascitic fluid with high concentrations of immunoglobulins. The single step affinity purification on a MARK-3-Sepharose 4B proved to be convenient because of its speed, efficiency and simplicity. In ELISA experiments, purified EB-Y8 strongly reacted with all 0:8 strains, while reaction with other strains was identical to blank values (< 0.010). Western blotting experiments confirmed the specificity of EB-Y8 for serogroup 0:8 LPS. Only when applied to blots with cell extracts of this serogroup, it stained several bands, arranged in a pattern, typical for bacterial LPS.[10-12] The heterogeneity of the molecular weights of these LPS is due to the varying numbers of repeating units in the O-antigen region of the LPS molecule.[10] The barely visible band with an electrophoretic mobility comparable to a 56 kDa polypeptide and also present in other serogroups, is presumably not a LPS.

EB-Y8, which first had been shown to be serogroup 0:8 specific in ELISA experiments (Table 1), maintained its reactivity and specificity when linked to latex particles. This latex reagent correctly identified all 0:8 strains and did not give any false-positive results.

More strains, including 0:8 and others, should be tested with the latex reagent in order to further evaluate its diagnostic value. Furthermore, monoclonal antibodies against the other medically important serogroups should be developed. The experiments described above strongly suggest, however, that rat monoclonal antibodies coated onto latex particles could be highly specific tools for serotyping clinical strains of *Y. enterocolitica*.

REFERENCES

1. **Bercovier, H. and Mollaret, H. H.,** *Yersinia*, in *Bergey's Manual of Systematic Bacteriology*, Vol. 1, Krieg, N. R. and Holt, J. G., Eds., Williams & Wilkins, Baltimore, 1984, 498.
2. **Butler, T.,** *Plague and Other Yersinia Infections*, Plenum Press, New York, 1983, chap. 4.
3. **Wauters, G.,** Antigens of *Yersinia enterocolitica, in Yersinia enterocolitica*, Bottone, E.J., Ed., CRC Press, Boca Raton, FL, 1981, 41.
4. **Ewing, W. H.,** *Edwards and Ewing's Identification of Enterobacteriaceae*, 4th ed., Elsevier, New York, 1986, chap. 19.
5. **Bazin, H.,** Rat-rat hybridoma formation and rat monoclonal methods, in *Methods of Hybridoma Formation*, Bartal, A. H. and Hirsault, Y., Eds, Humana Press, Clifton, NJ, 1987, 337.

6. **Bazin, H., Cormont, F., and De Clercq, L.,** Rat monoclonal antibodies. II. A rapid and efficient method of purification from ascitic fluid or serum, *J. Immunol. Methods,* 71, 9, 1984.
7. **Nilsson, L. A.,** Double diffusion-in-gel, *Scand. J. Immunol.,* 17 (Suppl. 10), 57, 1983.
8. **Laemmli, U. K. and Faure, M.,** Maturation of the head of bacteriophage T4, *J. Mol. Biol.,* 80, 575, 1983.
9. **Faure, A., Bladier, D., Fabia, F., and Cornillot, P.,** Quantitative study of tests using latex particles coated with proteins or peptides, in *Protides of the Biological Fluids, Proc. 20th Colloquium,* Peeters, H., Ed., Pergamon Press, Oxford, 1973, 589.
10. **Hitchcock, P. J., Leive, L., Mäkelä, P. H., Rietschel, E. T., Strittmatter, W., and Morrison, D. C.,** Lipopolysaccharide nomenclature - past, present, and future, *J. Bacteriol.,* 166, 699, 1986.
11. **Peterson, A. P., Haug, A., and McGroarty, E. J.,** Physical properties of short- and long-O-antigen-containing fractions of lipopolysaccharides from *Escherichia coli* 0111:B4, *J. Bacteriol.,* 165, 116, 1986.
12. **Preston, M. A. and Penner, J. C.,** Structural and antigenic properties of lipopolysaccharides from serotype reference strains of *Campylobacter jejuni, Infect. Immun.,* 55, 1806, 1987.

Chapter 21.III

PRODUCTION OF RAT HYBRIDOMAS DIRECTED AGAINST *YERSINIA ENTEROCOLITICA*

M. Bodeus, M. P. Sory, J. C. Fang, M. Janssens, N. Delferrière, G. Cornelis, G. Wauters, and G. Burtonboy

Yersiniae are gram-negative bacilli belonging to the family of Enterobacteriaceae. Three species are recognized as human and animal pathogens: *Yersinia pestis*, the agent of plague, *Y. pseudotuberculosis*, and *Y enterocolitica*.[1] The latter can provoke enteritis, mesenteric adenitis, and sometimes septicemia in humans. Such infections may be followed by reactive arthritis or erythema nodosum.[2] The species is heterogeneous and has been subdivided in five biotypes characterized by different metabolic activity patterns.[3] Besides, each biotype includes a number of serotypes with specific antigens,[4] among which there are polysaccharidic antigens referred to as antigens "O", flagellar or "H" antigens, and a few "surface" or "envelope" antigens corresponding, at least in some strains, to a fimbrial structure[5] (Figure 1). The study of these antigens is of interest since only some of the antigenic profiles have so far been characteristic of potentially pathogenic strains.[6] Furthermore, essential virulence determinants of yersiniae are located on plasmids of about 70 kilobases in size.[7,8] The plasmids of the three species are structurally and functionally related.[9-11] Several phenotypic properties, depending on temperature, are specified by these virulence plasmids.[12] One of them is the secretion of large amounts of proteins after incubation at 37°C in a Ca^{++} deficient medium.[13,14] Since these proteins were initially detected in the outer membrane ,they were called YOPs for *Yersinia* Outer Membrane Proteins.[15] As shown by reaction with polyclonal antisera and monoclonal antibodies (MAbs), the YOPs of the different species are immunologically related.[15-17] A number of observations have suggested that these proteins are indeed involved in the pathogenic process of *Yersinia*.[18,19] MAbs could be a powerful means to study their structure and role in pathogenesis. In order to obtain hybridomas secreting MAbs against these antigens, rats have been immunized with *Y. enterocolitica*. For this experiment, the strain chosen was W22703(pGC 565). It is a strain of the O:9 serotype, known to produce YOPs.[14] The antigen was prepared as described in reference 14. The bacteria were first incubated overnight at 28°C in brain-heart infusion (BHI) containing 25 µg/ml kanamycine. With 250 µl of this preculture, we seeded 4 times 10 ml of BHI added with 20 m*M* sodium oxalate, 20 m*M* $MgCl_2$, 20 m*M* glucose, and 25 µg/ml kanamycin. Incubation was performed at 28°C for 2 h, then at 37°C for 4 h. After centrifugation, the cells were washed and resuspended in saline at a density of 3×10^9 bacteria per milliliter according to the optical density at 600 nm.

One milliliter of the suspension was mixed with complete Freund's adjuvant and injected intraperitoneally in a 3-month-old LOU rat. The same amount of *Yersinia* was injected by the same route 20 d later, but without adjuvant. After 3 months, the animal was boosted with a similar injection 3 d before the spleen was removed. Splenic cells were fused with IR983F azaguanine resistant cells,[20] using the polyethylene glycol procedure as described previously.[21] The first hybrid clones were observed to grow at day 9 and new colonies appeared during the next two weeks, to reach a total of 172 clones. The supernatants of the growing cells were tested by agglutination of *Yersinia*. Five specimens were found to agglutinate serotype 0:9 and not 0:3. Besides, the culture medium of each clone was analyzed by immunoblot analysis against the YOPs produced by the strain used for immunization of the rat[22,23] (Figure 2). Eighteen other hybridomas were shown to secrete antibody which recognizes one of the peptides made by the virulent strain as summarized in Table 1. Six of these hybridomas have already been cloned in soft agar and production of these MAbs was obtained by injecting cells intraperitoneally into

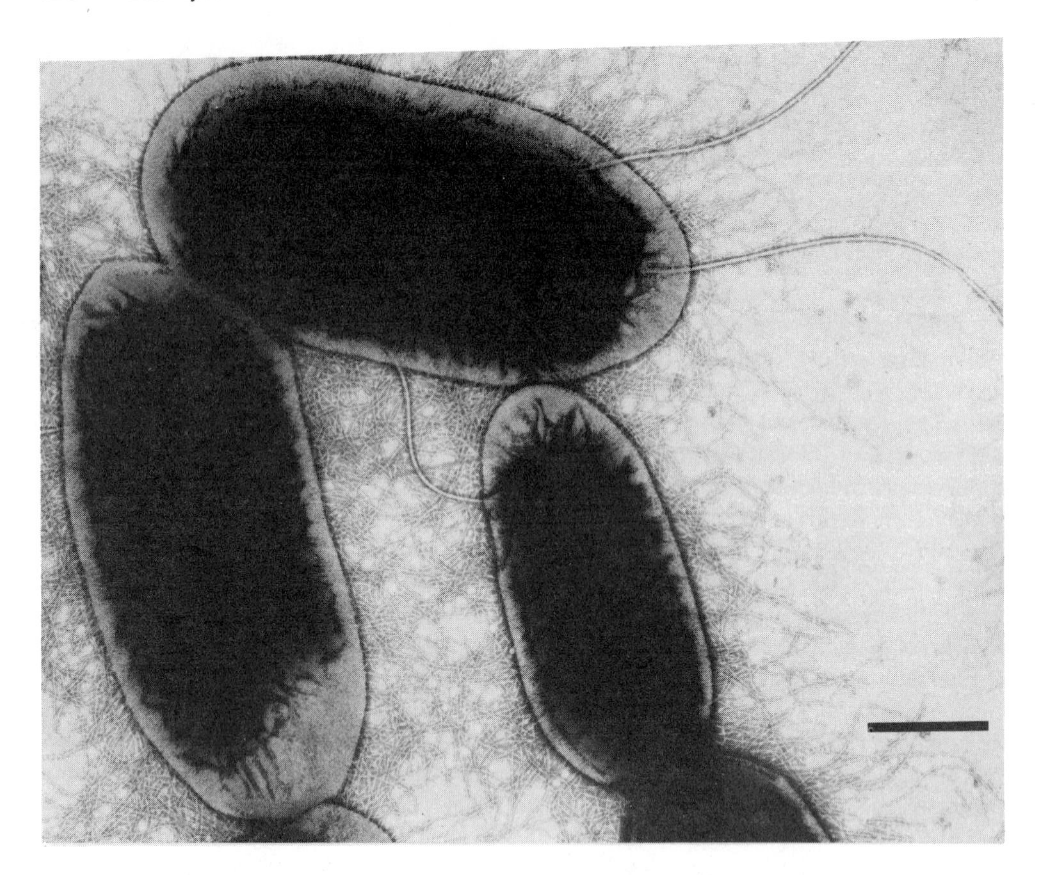

FIGURE 1. An example of *Yersinia enterocolitica* as observed in the electron microscope by negative staining. This 0:1 strain possesses pili and flagellae.

LOU rats and collecting the ascitic fluid as discussed thoroughly in Chapter 14. Two other hybridomas have been produced *in vitro*.[24] These MAbs are now being further characterized. The anti YOPs antibodies are being analyzed in the laboratory of G. Cornelis. The epitopes involved in the agglutination process are being studied by G. Wauters. This work is still in progress.

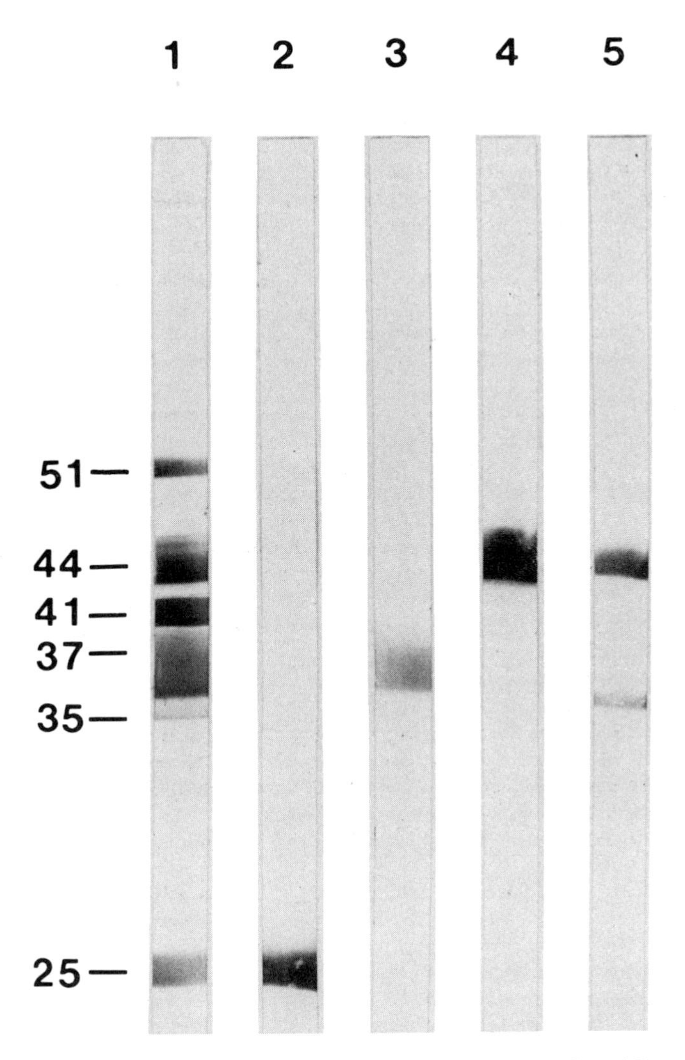

FIGURE 2. Immunoblot analysis of MAbs directed against YOPs of *Y. enterocolitica* W22703(pGC 565). Lane 1: control showing YOPs as recognized by a rabbit serum, diluted 1:100, raised against formolized *Y. enterocolitica* incubated at 37°C. Lane 2 to 5: MAbs 6G1, 13A4, 9B7, and 6H1. Molecular mass of YOPs are indicated in kDa.

TABLE 1
Hybridomas Secreting Antibodies Against YOPs

YOPs M.W.(kDa)	Hybrid colonies	Hybridomas cloned in agar
51	—	—
48	1	ND
44	5	1
41	—	—
37	7	2
35	1	1
25	4	2[a]

[a] One of these antibodies recognizes both YOP25 and YOP44.

REFERENCES

1. **Bercovier, H. and Mollaret, H.,** Genus XIV, *Yersinia,* in *Bergey's Manual of Systematic Bacteriology,* Vol. I, Holt, J. G., Ed., Williams & Wilkins, Baltimore, 1984, 498.
2. **Wormser, G. P. and Keusch, G. T.,** *Yersinia enterocolitica,* clinical observations, in *Yersinia Enterocolitica,* Bottone, E., Ed., CRC Press, Boca Raton, FL, 1981, 83.
3. **Bercovier, H., Brenners, D. J., Ursing, J., Steigerwalt, A. G., Fanning, G. R., Alonso, J. M., Carter, G. P., and Mollaret, H. H.,** Characterization of *Yersinia enterocolitica* sensu stricto, *Curr. Microbiol.,* 4, 201, 1980.
4. **Wauters, G.,** Antigens of *Yersinia enterocolitica,* in *Yersinia Enterocolitica,* Bottone, E., Ed., CRC Press, Boca Raton, FL, 1981, 41.
5. **Aleksic, R., Rohde, R., Müller, G., and Wohlers, B.,** Examination of the envelope antigen K1, in *Yersinia enterocolitica* which was identified as fimbriae, *Zentralbl. Bakteriol. Parasitenkd. Infektionskr Hyg. Abt. I. Orig. A,* 234, 513, 1976.
6. **Wauters, G.,** Correlation between ecology, biochemical behaviour and antigenic properties of *Yersinia enterocolitica,* in *Contributions to Microbiology and Immunology* Vol. 2, *Yersinia Pasteurella* and *Francisella,* Winblad, S., Ed., S. Karger, Basel, Switzerland, 1973, 38.
7. **Gemski, P., Lazere, J. R., and Casey, T.,** Plasmid associated with pathogenicity and calcium dependency of *Yersinia enterocolitica, Infect. Immun.,* 27, 682 1980.
8. **Zink, D. L., Feeley, J. C., Wells, G., Vanderzant, C., Vickery, J. C., Roofs, W. D., and O'Donovan, G. A.,** Plasmid-mediated tissue invasiveness in *Yersinia enterocolitica, Nature,* 283, 224, 1980.
9. **Heesemann, J., Keller, C., Morawa, R., Schmidt, N., Siemens, H. J., andLaufs, R.,** Plasmids of human strains of *Yersinia enterocolitica,* molecular relatedness and possible importance for pathogenesis, *J. Infect. Dis.,* 147, 107, 1983.
10. **Portnoy, D. A. and Falkow, S.,** Virulence associated plasmids from *Yersinia enterocolitica* and *Yersinia pestis, J. Bacteriol.,* 148, 877, 1981.
11. **Biot, T. and Cornelis, G.,** The replication, partition and YOPs regulation of the pYV plasmids are highly conserved in *Y. enterocolitica* and *Y. pseudotuberculosis, J. Gen. Microbiol.,* 138, 1988, in press.
12. **Cornelis, G., Laroche, Y., Balligand, G., Sory, M. P., andWauters, G.,** *Yersinia enterocolitica,* a primary model for bacterial invasiveness, *Rev. Infect. Dis.,* 9, 64, 1987.
13. **Heesemann, J., Algermissen, B., and Laufs, R.,** Genetically manipulated virulence of *Yersinia enterocolitica, Infect. Immun.,* 46, 105, 1984.
14. **Cornelis, G., Sory, M. P., Laroche, Y., and Derclaye, I.,** Genetic analysis of the plasmid region controlling virulence in *Yersinia enterocolitica* 0:9 by mini Mu insertion and lac gene fusions, *Microb. Pathogenesis,* 1, 349, 1986.
15. **Bölin, I., Portnoy, D. A., and Wolf-Watz, H.,** Expression of the temperature inducible outer membrane proteins of yersiniae, *Infect. Immun.,* 48, 234, 1985.
16. **Martinez, R. J.,** Plasmid-mediated and temperature-regulated surface properties of *Yersinia enterocolitica, Infect. Immun.,* 41, 921, 1983.
17. **Heesemann, J., Kalthoff, H., and Koch, F.,** Monoclonal antibodies directed against plasmid-encoded released proteins of enteropathogenic *Yersinia, FEMS Microbiol. Lett.,* 36, 15, 1986.
18. **Lian, C. J. and Pai, C. H.,** Inhibition of human neutrophil chemiluminescence by plasmid-mediated outer membrane proteins of *Yersinia enterocolitica, Infect. Immun.,* 49, 145, 1985.
19. **Straley, S. C. and Bowmer, W. S.,** Virulence genes regulated at the transcriptional level by Ca_2+ in *Yersinia pestis* include structural genes for outer membrane proteins, *Infect. Immun.,* 51, 445, 1986.
20. **Bazin, H.,** Production of rat monoclonal antibodies with LOU rat non secreting IR983F myeloma cell line, in *Protides of the Biological Fluids,* Peeters, H., Ed., Pergamon Press, Oxford, 1982, 615.
21. **Burtonboy, G., Bazin, H., and Delferrière, N.,** Rat hybridoma antibodies against Canine Parvovirus, *Arch. Virol.,* 71, 291, 1982.
22. **Towbin, H., Staehelin, T., and Gordon, J.,** Electrophoretic transfer of proteins from polyacrylamide gels to nitrocellulose sheets procedure and some applications, *Proc. Natl. Acad. Sci. U.S.A.,* 76, 4350, 1979.
23. **Surleraux, M., Bodeus, M., and Burtonboy, G.,** Study of canine parvovirus polypeptides by immunoblot analysis, *Arch. Virol.,* 95, 271, 1987.
24. **Bodeus, M., Burtonboy, G., and Bazin, H.,** Rat monoclonal antibodies. IV. Easy method for *in vitro* production, *J. Immunol. Methods,* 79, 1, 1985.

Chapter 22

RAT MONOCLONAL ANTIBODIES AGAINST VIRAL ANTIGENS

G. Burtonboy, M. Bodeus, N. Delferrière, M. Surleraux, A. Rovayo, V. Vercruysse, and J. C. Fang

TABLE OF CONTENTS

I. INTRODUCTION

In the field of virology, the many advantages and possibilities of monoclonal antibodies (MAbs) are now well documented. Owing to the hybridoma technology, it is indeed possible to produce in large amounts antibody specific for given epitopes. For the diagnosis of viral diseases, it means the supply of specific and reproducible tools which are suitable not only to detect the presence of a virus, but also to differentiate from each other strains with closely related antigenic make-up. In more basic research, the approach provides new ways to characterize, quantify, and purify virions or viral peptides; besides it allows the selection of viruses mutated at a given antigenic determinant and opens new perspectives for those who in virology analyze the relationship between structure and function. Since the pioneering work of Koprowski et al.,[1] MAbs have been widely used in the study of virus variants. Reviews on the subject are to be found in the literature.[2-6]

Until recently, most of the work in this domain has been based on the mouse hybridoma technology as it was originally described by Köhler and Milstein.[7] An interesting alternative has been made available by the development of the rat system where hybridomas are obtained by fusing LOU rat myeloma cells with LOU rat spleen cells.[8,9] In this method the secreting clones can be grafted into histocompatible animals, the production being further facilitated by the fact that a single rat yields up to 50 ml of ascitic fluid versus a maximum of 10 ml in a mouse. Part of the hard work is therefore taken out of the procedure (see Chapter 2). As far as purification is concerned, it is made easier by means of immunoadsorbants using mouse MAbs specific for rat immunoglobulin (Ig) class, subclass, or light chains; and there is even a very elegant technique founded on allotypic differences (see Chapter 3).

Moreover, the rat system may be advisable because, for some antigens, the immunological response is apparently better in rats than in mice.[10] Production of rat MAbs has been reported by other investigators either in the field of animal virology[11] or against plant viruses.[12,13] In our laboratory, this approach has been used to develop a repertoire of MAbs directed against canine parvovirus structural proteins and to obtain hybridomas secreting antibodies which recognize various polypeptides coded by the Human Immunodeficiency Virus (HIV).

These two examples will be discussed in more detail in this chapter.

II. RAT MONOCLONAL ANTIBODIES AGAINST CANINE PARVOVIRUS ANTIGENS

A. THE CANINE PARVOVIRUS

In 1978 an apparently new disease was observed among dogs. Curiously enough, similar observations were made almost simultaneously in the U.S., Canada, South Africa, Australia, and in European countries, where severe epidemics of canine hemorrhagic enteritis were reported to occur in many kennels , the mortality rate reaching up to 100% in some cases.[14-16] Autopsy of the animals revealed distinctive intestinal lesions with the disappearance of villi and necrosis or cystic dilatations of the crypts in the duodenum and jejunum.[17] Clinical and pathological features were thoroughly described,[18] taking into account the various forms of the disease and drawing attention to the leukopenia, which is part of a syndrome quite similar to what is known in cats and minks, as the feline panleukopenia or the mink enteritis.[19]

This canine hemorrhagic enteritis has been shown to be associated with a viral infection.[20] Large numbers of viral particles are present in feces of symptomatic dogs. In the electron microscope, the virions examined by negative staining display a profile suggestive of an icosahedral shape with a maximum diameter of around 25 nm (Figure 1).[21] The virus can be grown in tissue culture of feline or canine kidney cells, where it produces nuclear inclusions.[22] It agglutinates pig, cat, horse, and rhesus red cells at 4°C and acid pH. Hemagglutination is inhibited by the serum of dogs recovering from hemorrhagic enteritis or by Igs of animals

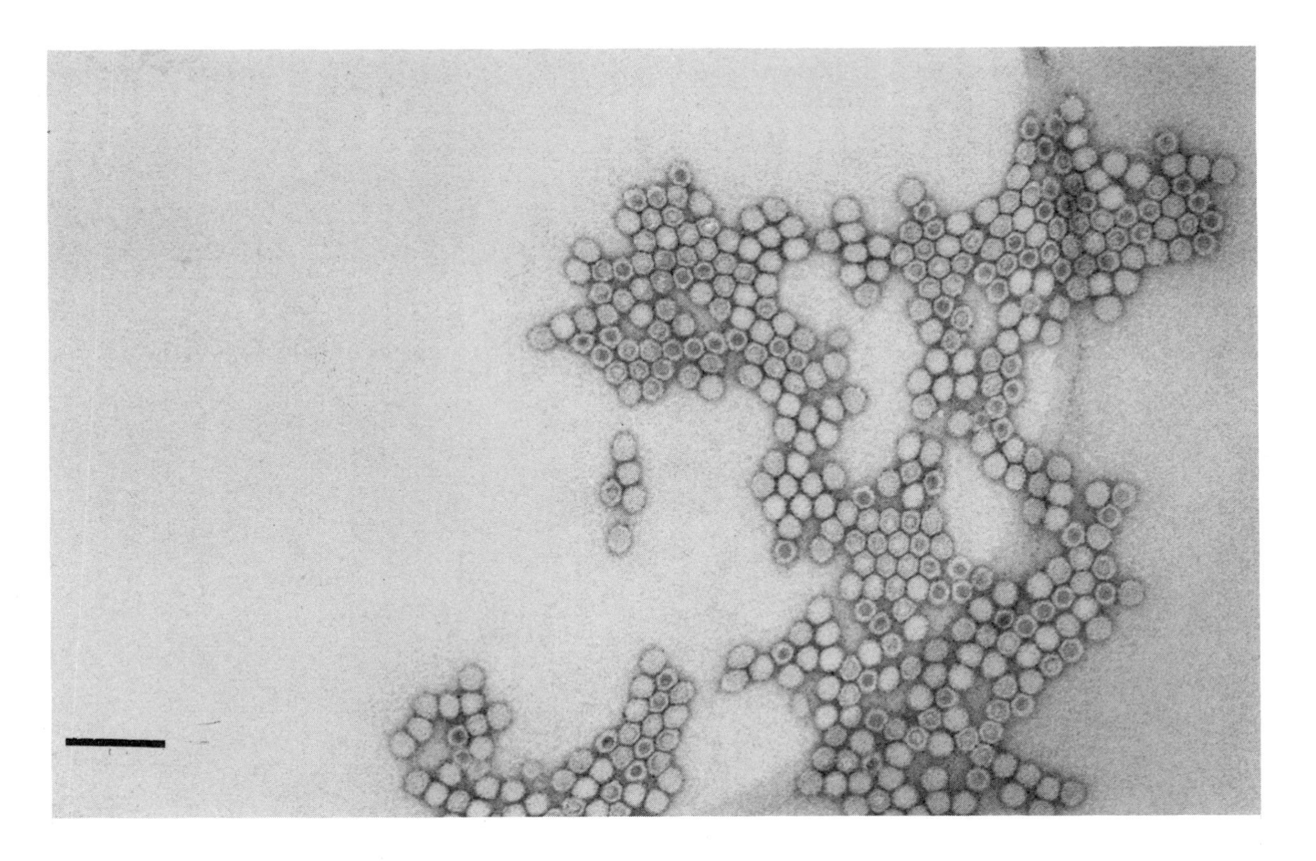

FIGURE 1. Image of viral particles of canine parvovirus. The bar represents 0.1 μm.

TABLE 1
Anti-CPV IHU Titer After Immunization

Serum of animal	Mice (BALB/c)			Rat (LOU/C)		
	1	2	3	1	2	3
Day 0	4	4	<4	4	<4	8
Day 15	16	64	32	100	2000	2000

immunized against the virus.[23] As expected from the morphological features, this agent has been thoroughly characterized as an autonomous parvovirus.[24]

The genome is a single stranded DNA,[25] and the three main structural polypeptides share a common C terminal sequence.[26-29] The virus is now referred to as the canine parvovirus or in short CPV.[30]

There is a quite close relationship between this dog virus and the viruses associated with a very similar syndrome in either cats or minks. The serum of an animal immunized against CPV reacts with the feline panleucopenia (FPLV) and the mink enteritis virus (MEV). In fact, it precipitates the viral antigens in immunodiffusion; it detects infected cells by immunofluorescence and also binds to viral particles as shown by immunoelectron microscopy; it inhibits hemagglutination and neutralizes infectivity for the three viruses, whereas a serum sample taken before immunization does not recognize any of them. Vice versa, an antiserum against FPLV or MEV reacts with the canine virus.[31]

Clearly, these parvoviruses belong to the same serotype. However, there are important differences among them. As far as hemagglutination is concerned, optimal conditions and range of susceptible erythrocytes are not the same. Immunodiffusion can even reveal that the cross reactivity is incomplete.[32] Besides, the canine virus is not virulent for cats or minks and the virus which causes disease in feline does not affect dogs.

In order to investigate this phenomenon, it was interesting to produce a large panel of MAbs directed against the new parvovirus isolated from dogs with hemorrhagic enteritis.

B. PRODUCTION OF MAbs AGAINST CPV

The animals to be immunized were injected intraperitoneally with 0.5 ml of a suspension containing 10^5 HA units/ml of the canine parvovirus. The virus isolated in our laboratory from the feces of a dog with hemorrhagic enteritis and referred to as CPV/B was grown in NLFK cells. The cells were frozen and thawed three times, 10 d after infection. Viral antigen was purified from the clarified supernatant by adsorption on rhesus erythrocytes at 4°C, pH 6 followed by elution at 37°C, pH 7 in 2% of the original volume.

As a preliminary experiment, three rats, 6-month-old LOU/C males, were inoculated with this material and their serum was analyzed 15 d later. The antibody level, which was below 8 hemagglutination inhibition units (IHU) before the injection, already reached 100 or more IHU 2 weeks afterwards. No seroconversion was observed in mock infected controls maintained in the same cage. A comparative trial had been carried out in BALB/c mice, but with the same immunization schedule, the antibody response was apparently lower than in the rats (Table 1).[10]

The production of hybridomas was based on the procedure recommended by Fazekas and Scheidegger.[33] A rat immunized intraperitoneally with twice 5×10^4 HAU was boosted with a similar inoculum after a delay of up to 6 months. The animal was bled 5 days later, the spleen was removed and splenic cells were gently dispersed in MEM, buffered with HEPES. These cells were washed and 3×10^7 of them were mixed with 5×10^6 azaguanine resistant rat myeloma cells, namely 983 cells.[9] After centrifugation, the cell pellet resuspended in 1 ml of a 50% solution of polyethylene glycol (PEG 4000 Merck or PEG 1000 BDH) was slowly diluted in a selective medium containing aminopterine, hypoxanthine, and thymidine (Chapter 2). The cell suspension was then distributed in 14 microplates with 96 wells in which a feeder layer of rat peritoneal cells had been seeded for 2 d. The culture was incubated at 37°C in a 5% CO_2 atmosphere. Twice a week the supernatant was removed and fresh medium added. At the fourth change, aminopterin was omitted and a medium rich in thymidine and hypoxanthine (HT Medium Gibco) was used instead of the selective solution.

The hybrid clones growing in the microplates were transferred to 24-well macroplates (COSTAR). Selection of the colonies was based on the presence in the culture supernatant of specific antibody as demonstrated by immunofluorescence (Figure 2).

The method has been described elsewhere.[10] Briefly, cells infected by the virus used for immunization were fixed in acetone and covered with the medium to be tested. Antibodies bound to this antigen were then revealed by means of an anti-rat Ig antiserum labeled with fluorescein. Many noninfected cells were still observed in the culture monolayer and provided the test with an internal control, so that antibodies against cellular nonviral antigen were not selected.

Out of 7 fusions, 57 hybridomas were obtained, each of them derived from a different heterokaryon and producing a MAb against one of the epitopes of the canine parvovirus antigens. They are referred to as RH, for rat hybridoma, plus a serial number.

The hybrid cells secreting a specific antibody were cloned twice in soft agar. Large amounts of MAb were obtained either *in vitro* by growing the cells in a roller bottle,[34] or *in vivo* where ascitis was generated in histocompatible rats by intraperitoneal injection (see Chapter 3).

Class and subclass of the Igs were characterized thanks to the specific antisera of H. Bazin.[25] Interestingly enough, a large majority of the monoclonal Igs were found to be made of IgG2b heavy chains. As shown in Table 2, the distribution of subclasses was apparently not modified by the lapse of time between the first injection and the fusion. As expected, most of the Ig had kappa light chains except two which had lambda chains.

Concentration of the MAbs in the culture medium was estimated according to the method described by Mancini.[36] It varied from one hybridoma to the other. After 3 d incubation at 37°C, a flask seeded with 10^5 cells/ml can contain between 5 and 50 μg/ml of the MAbs.[34]

C. SPECIFIC ACTIVITY OF ANTI-CPV ANTIBODIES

Antibody activity can be analyzed in a variety of ways. But the principles of the various tests which are available for this purpose can basically differ from one another and a MAb will be recognized in a given technique only if its binding to the corresponding epitope modifies the result of the test. From this point of view, it was interesting to compare the results obtained when measuring antibody activity by different approaches. Since it had been assumed that immunofluorescence offered the advantage of detecting antibodies directed against almost any viral antigen, this method was chosen as the screening procedure for the selection of hybrid clones,[10] and accordingly, all the rat hybridomas described here of course gave a positive reaction when analyzed by immunofluorescence.

Another approach, immunoelectron microscopy, allows the recognition of the antibodies which agglutinate the viral particles,[37] and are supposed to be directed against antigenic determinants located at the outer surface of the virion. Out of 57 anti -CPV MAbs, 45 were found to display such a property (Figure 3). Some of these antibodies were analyzed in immunodiffusion according to Ouchterlony and opposite viral particles purified in a CsCl density gradient generated a precipitation line.

Hemagglutination inhibition is another means to measure anti CPV antibody activity.[23] The titer is expressed as the reciprocal of the highest dilution which still inhibits the agglutination of 5×10^5 rhesus erythrocytes by 4 times the minimal virus amount required for hemagglutination.

Of the monoclonals, 38 were found to display a titer of more than 8 IHU. Of course, the concentration of the Ig had to be taken into account and the specific activity, that is, the titer per

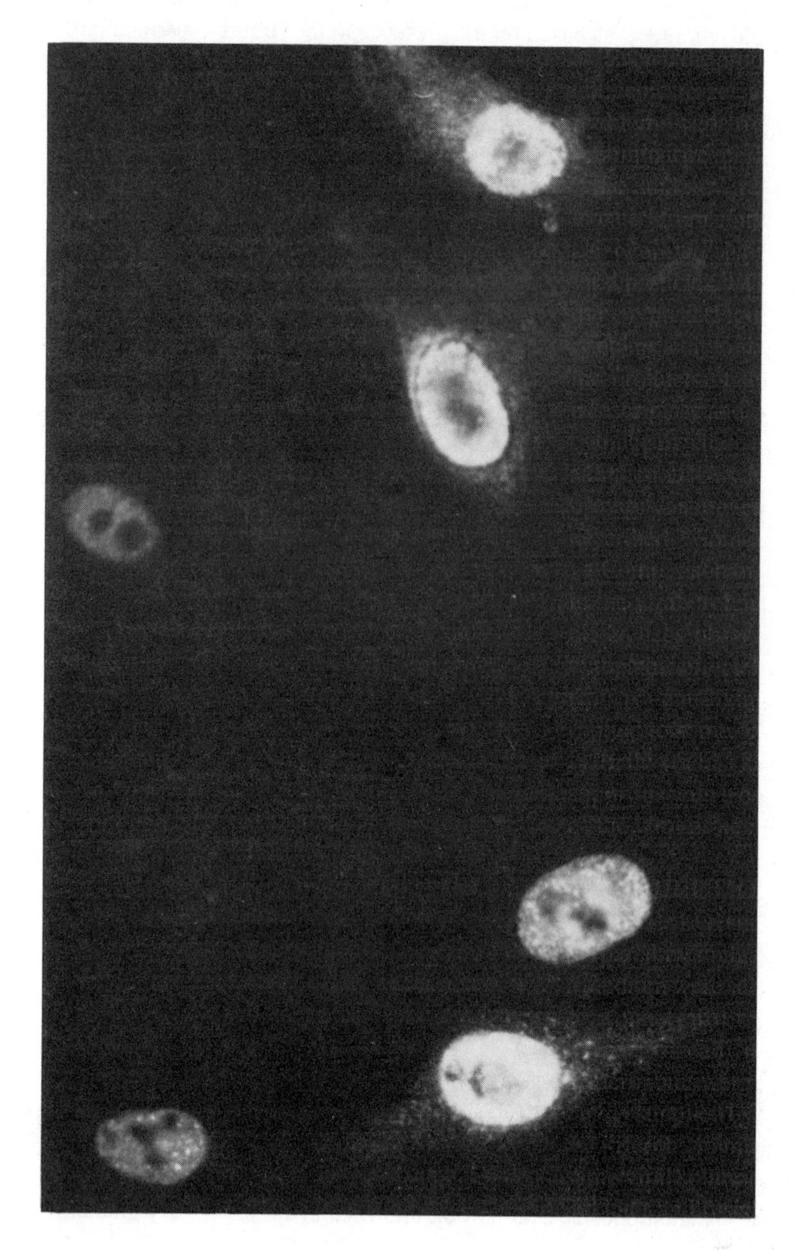

FIGURE 2. Indirect immunofluorescence of cells infected by CPV. The viral antigen present in nuclei is recognized by MAb RH 1.

TABLE 2
Isotypes of Immunoglobulins Produced by Hybridomas Against Canine Parvovirus

Fusion at day	Positive clones	Class or subclass of Igs				
		IgG1	IgG2a	IgG2b	IgG2c	IgM
25	18	2	1	11	—	4
150	10	—	—	9	—	1
180	10	—	1	8	—	1
201	10	—	1	9	—	—

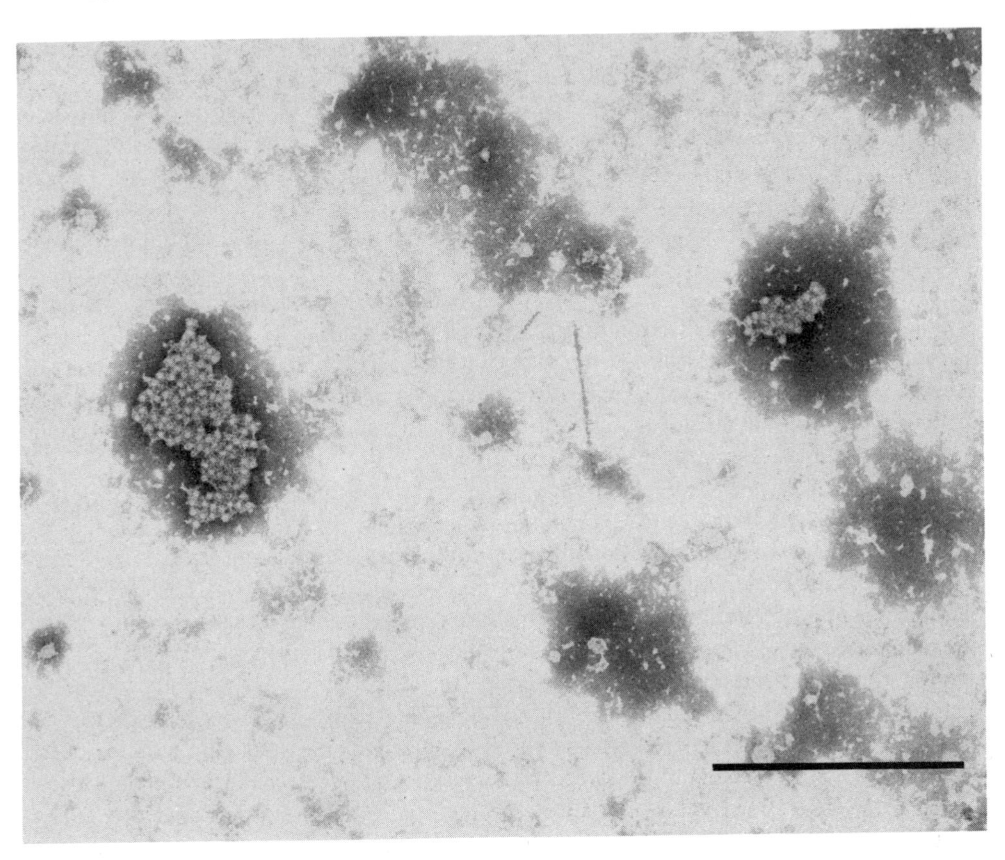

FIGURE 3. Immunoelectron microscopy: the viral particles of CPV are agglutinated by a specific antibody. The bar represents 1 μm.

microgram, was calculated for each. This value varied from one to the other even if the Igs belonged to the same subclass.

A plaque reduction assay was developed in order to estimate the neutralizing activity. The technique is similar to what has been described by Siegl for the Feline Panleukopenia virus.[38] A culture of NLFK cells was infected when at 25% confluence, with a very low concentration of the Canine Virus and further incubated under agarose for 10 d. The monolayer was then stained with neutral red and foci of infected cells appeared as clear plaques. Several dilutions of each antibody solution were mixed with an inoculum of 100 PFU/ml. A reduction of at least 80% was considered as an indication of neutralizing activity. It was quantitated as the reciprocal of the highest dilution with activity. Although it was clear that such values were indeed a function

TABLE 3
Specific Activity Against CPV of Rat Monoclonal Antibodies (RH)

Number of RH	Reaction tested by IF	IEM	IH	SN
12	+	-	-	-
38	+	+	+	+
4	+	+	-	-
5	+	+	-	+

Note: 57 rat monoclonals were tested.

of the viral preparation used in the analysis, since all the MAbs had been tested against the same sample of virus, the results could be compared among each other. Table 3 summarizes these data and shows that 43 antibodies were found to neutralize CPV infectivity. Most, but not all of them, inhibited the viral hemagglutination. The discrepancy observed in five cases was confirmed in repeated experiments. It may be worthwhile to notice that the two biological effects are not absolutely linked, suggesting that they involve at least slightly different structures. On the other side, although a few MAbs analyzed in this study bound to the outer surface of the nucleocapsid as exemplified by immunoelectron microscopy, they did not modify the infectivity. Further investigations demonstrated that if the virions were coated with this type of antibody,the addition of rabbit anti-rat Ig antiserum had the effect to dramatically drop the infectivity. Since this antiserum alone has no neutralizing activity, it can be assumed that the phenomenon is quite similar to what has been reported for enteroviruses.[39]

Indeed, it is likely that viral particles are aggregated by an antibody only if they are present at a concentration high enough to favor cross-linking. On the contrary, the assay of infectivity requires working with a very small number of virions. The neutralization effect is then due to the mere influence of the antibody binding on the protein which recognizes the host cell receptor. The hypothesis generally admitted is that some strain has to be applied on the antigen-antibody complex. According to the epitope the antibody alone will or not exert such an effect.

Immunoblotting analysis has been performed to localize, on the viral polypeptides, the epitopes corresponding to the rat MAbs . This work is reported in detail by Surleraux et al.[40] The polypeptides of the Canine Parvovirus had been characterized previously by sodium dodecyl sulfate (SDS) and polyacrylamide gel electrophoresis (PAGE) combined with immunoadsorption.[41] Three structural proteins were described with molecular weight estimated at 85,000, 70,000 Da for the most abundant, and 67,000 Da. Western blot analysis was made according to the method of Towbin et al.,[42] and the attachment of anti-CPV MAb revealed by means of a mouse anti-rat kappa chain MAb (MARK-1) (see Chapter 1) labeled with peroxidase. This reagent was kindly supplied by H. Bazin. With this procedure, 47 of the rat hybridoma antibodies were tested against the polypeptides of CPV particles. As expected, a large number of these MAbs, in fact 31, did not recognize any polypeptide on the nitrocellulose sheet. It is likely that owing to denaturing conditions, and in the presence of beta mercaptoethanol, some conformational epitopes were lost. Low reactivity of MAbs in immunoblot analysis has already been reported by others.[43]

Of the 47 RH, 16 gave a positive result. In the first group, 12 monoclonals had in common the fact that they detected the three structural polypeptides of CPV in proportions corresponding to their molar ratio in the viral particle. A second pattern was observed with three antibodies recognizing only VP85. However, RH 4, which reacted with this polypeptide, also revealed an additional band at an apparent molecular weight of 75,000 Da. Finally the monoclonal RH 16 was found to bind to VP70, but not to VP85 or VP67. If one takes into account the data available on the Parvovirus genome and proteins,[44,45] these observations allow us to map the epitopes on the CPV polypeptides. According to the nucleotide sequence of CPV, the righthand part of the genome contains only one open reading frame that is transcribed as the 3′ part of the various messenger RNAs and the three structural polypeptides have the same amino acid sequence at the C terminal side.[28] So that an epitope which is found on these three polypeptides is likely to be located on the common C terminal domain. On the contrary, the antigenic determinant recognized by a MAb specific for VP85 is probably part of the N terminus region unique to this polypeptide. The monoclonal RH 16 reacts exclusively with VP70. Now, the smallest of the structural proteins VP67 is supposed to derive from VP70 by proteolytic processing and such a cleavage is mimicked by trypsin treatment of the viral particles. The antibody does not recognize VP70 after proteolytic cleavage to VP67. Therefore it can be assumed that the corresponding antigenic structure is on the part of the polypeptide which is removed by the trypsin treatment. The enzyme sensitive site has been mapped, at least for another parvovirus, at 20 amino acids from the N terminus.[46]

Moreover, by knowing the biological activity of the monoclonals, it is sometimes possible to localize the epitopes either on the surface or inside the capsid. Indeed any antibody which agglutinates the viral particles, neutralizes the infectivity, and inhibits the hemagglutination, must recognize an epitope exposed on the surface of the virion. Since epitopes at the C as well as at the N terminals of VP85 correspond to agglutinating antibodies, it is good evidence that both ends of the polypeptide are exposed at the surface of the capsid. The region in between displaying a highly hydrophobic sequence,[27] it can be assumed that it is buried in the capsomere arrangement and the shape of the protein should involve some kind of a hairpin structure.[40]

D. ANALYSIS OF VIRAL VARIANTS

Further work was done with the rat MAbs directed against the Canine Parvovirus, using them as a tool to study antigenic variations among viral strains. The goal of the program was indeed to analyze the differences between this virus associated with the epidemics of hemorrhagic enteritis observed in kennels, and the closely related viruses referred to as FPLV and MEV.

To analyze this relationship, 54 rat MAbs were tested by immunofluorescence on cells infected by either the mink, the cat, or the dog virus. It was found that although most hybridomas secrete an antibody which recognizes the three parvoviruses equally well, one MAb did detect an antigen in cells inoculated with the canine and the feline but not the mink agent, furthermore 7 MAbs were found to be specific to the dog enteritis virus and did not react with FPLV nor MEV. However, it must be emphasized that if these antigenic modifications can be used as markers, they are not sufficient to explain the biological differences observed among the three viruses. Good evidence for that arose from the study of a non-hemagglutinating variant of CPV.[11] To optimize the production of viral particles, a strain of Parvovirus isolated from a dog with hemorrhagic enteritis was cultured on feline cells for a number of passages. In the course of these experiments it was noticed that in the supernatant of the cultures, the viral titer measured by hemagglutination was found to be lower and lower and eventually became undetectable although cytopathic effect was obvious and the number of particles estimated by electron microscopy was very high. A non-hemagglutinating virus was cloned and characterized thoroughly. It was compared to a low passage of the original isolate. The two viruses were tested by immunofluorescence with 52 of the rat MAbs and no differences were seen. This result was confirmed by C. Parrish using a panel of mouse monoclonals anti-CPV and rat MAbs against FPV in an enzyme linked immunosorbent assay. By this approach the hemagglutinating and the non-hemagglutinating viruses were antigenically similar. In fact DNA sequence analysis of both strains revealed that they were almost identical except for point mutations at base 2243 and 2797 introducing two amino acid changes in the capsid proteins, respectively, Arg to Lys and Val to Leu. Even if the role of these apparently minor modifications has not been defined, at the molecular level it should explain the disappearance of the erythrocyte binding site.

The important point here is that the loss of a biological activity does not necessarily correlate with an easily perceived antigenic shift.

On the other side, mutants were selected by cloning CPV in the presence of various neutralizing antibodies. The variants which appeared at a frequence of about 10^{-7} and were not recognized any more by the monoclonal used for the selection, still displayed biological activities similar to those of the original virus: hemagglutination conditions and red cell repertoire, host range and cytopathic effect were apparently the same. These neutralization mutants were further analyzed in order to better understand the linkage of epitopes and the antigenic make-up of the viral capsid.

The appearance of such resistant viruses is very important to keep in mind if one considers the possibility of protecting animals by injection of a neutralizing MAb.

E. A MULTIPOTENT MATERIAL

The rat MAbs directed against the Canine Parvovirus have been used for various purposes and proved to be very helpful tools.[20]

To diagnose the viral infection, they appear as quite interesting reagents to detect antigen by immunofluorescence or immunoperoxidase, in autopsy material as well as in tissue culture, especially because the indirect technique can be made highly specific and efficient by means of anti-rat Ig light or heavy chain mouse MAbs.[35] Radioimmunoassays, enzyme linked immunoassays, were developed in our laboratory using some of the rat anti-CPV monoclonals. Besides, a latex test was devised: polysterene beads coated with RH 1 were agglutinated in the presence of the viral antigen. The system can be used as a slide test and provides us with a simple way to detect the presence of a virus in field studies. On the other side, a quantitative approach is obtained by counting the non-agglutinated beads in a particle counter. The number of isolated beads is inversely related to the amount of antigen following a sigmoid curve. Antibodies added to the antigen will prevent the reaction and thus their presence can be estimated by measuring this inhibition. The procedure is as sensitive as ELISA and radioimmunoassay detecting antigen or antibodies at a concentration as low as 150 ng/ml, and it does not use radioactive nor mutagenic material and is homogeneous in the sense that no phase separation and no washing step are needed. Furthermore, this latex agglutination test is easily automated, which increases precision, speed, and facility.[47]

The rat monoclonals against CPV also appear as an invaluable tool to purify viral particles and polypeptides by affinity chromatography.

A cascade of immunoadsorbents is currently used in our laboratory. In the first step, Igs obtained from rat serum by ammonium sulfate precipitation and purification on DEAE cellulose are covalently linked to AcA 22 activated with glutaraldehyde. On this column a mouse MAb directed against rat Ig light or heavy chains is easily purified from culture medium as well as from ascitic fluid. In its turn, the mouse monoclonal is then immobilized on a solid phase and Ig secreted by a rat hybridoma can be adsorbed on and eluted from such a column (see Chapter 15). A third immunoadsorbant is prepared with this rat antibody and used to purify the viral antigen bearing the corresponding epitope (see Chapter 15).

Each step of the procedure is carefully monitored by measuring the concentration of the material before and after adsorption, during the washing step and elution. The epitope bearing proteins which are not bound when passing through the column can be reused for a second passage. The percentage lost in the procedure is negligible and the recovery sheet shows a yield of up to 95%.

This approach was highly rewarding. Important amounts of viral polypeptides were obtained in a reasonably pure state and different antigens were separated from each other. This raises interesting questions, for instance, a viral polypeptide VP75 is purified by RH 1, a monoclonal that does not recognize it, suggesting that we are dealing with a fourth structural protein.[40]

Indeed, in the course of these experiments, the fact that in the rat system large volumes of antibody rich ascitic fluid were easy to supply for each hybridoma, was certainly a definite advantage.

III. RAT MONOCLONAL ANTIBODIES AGAINST HUMAN IMMUNODEFICIENCY VIRUS

A. THE HUMAN IMMUNODEFICIENCY VIRUS (HIV)

Even if the production of hybridomas relies on quite general principles, it is also true that each antigenic target gives the investigator difficulties of its own. In the field of virology, the problems vary from one virus to the other. As a matter of example, we report here some preliminary experiments made to obtain rat MAbs directed against the retrovirus associated with AIDS and now referred to as the human immunodeficiency virus, or HIV.[48] This agent was isolated from contaminated individuals by culturing lymph node or peripheral blood mononuclear cells.[49,50] Nowadays permanently infected cell lines are available, which keep producing viral particles.[51] The virions get shape by budding at the plasma membrane, and their aspect is quite typical (Figure 4). Inside the lipid membrane coated with glycoproteins, the nucleocapsid contains a diploid genome made of two identical plus strand RNAs. The structural proteins are the products of three genes ; GAG codes for the capsid polypeptides, POL for the reverse transcriptase and endonuclease, and ENV for the glycoproteins. Viral replication begins in the cytoplasm where the genetic information of the viral genome is copied in a sequence of double stranded DNA which eventually becomes part of the host-cell chromosome. Analysis of this provirus has revealed additional open reading frames and non-structural proteins have been described.[52] Some of them were shown to play a role in the control of the genome expression.[53,54]

Comparison of various isolates attracted attention to the fact that the amino acid sequences to the viral proteins vary largely among strains.[55] That has to be put in parallel with the many differences observed if one considers the biological properties such as host-cell range and cytopathic effect. Furthermore, another immunodeficiency virus called HIV 2 has been isolated from patients living in West Africa. The antigenic relationship between the two serotypes seems limited to some of the structural polypeptides.[56]

In this field a large panel of MAbs directed against these viruses would be very helpful and supply us with a powerful tool not only for diagnostic purpose, but also to analyze the structure of the viral proteins and better understand the molecular basis of their functions.

B. PRODUCTION OF RAT MAbs AGAINST HIV

Female LOU rats were immunized by intraperitoneal injection of concentrated viral particles prepared by high speed centrifugation of antigen rich culture supernatant. Eight liters of such a medium were submitted to 20,000 RPM for 4 h to obtain a pellet resuspended in a final volume of 2 ml. Half of this material was mixed with complete Freund Adjuvant and inoculated intraperitoneally into a 3- month-old LOU rat which was given the rest of the antigen at day 15. Experiments were made in parallel with two different strains of HIV: LAV cultured in CEM cells kindly supplied by F. Barré and J.C.Chermann,[49] and ARV grown in HUT cells given by Levy.[50]

After the immunization process, the rats were kept for 6 months and then boosted with the same viral solution which had been used to immunize them. Their splenic cells were removed 3 d later and fused with 983 cells in the presence of PEG (see Chapter 6. I). The method was similar to what had been successful for the production of monoclonal antibodies against Canine Parvovirus.[10]

Hybrid clones were found to appear from the 2nd to the 5th week. For the detection and follow-up of the colonies, the microplates were examined every 2 d by means of a mirror allowing us to look at the bottom of the wells ,so that it was easy to monitor a number of plates.

When the cells had grown enough to fill half of the cupule surface, the clones were transferred to 24-well macroplates. There, as soon as the growth was obvious, the supernatant was taken and

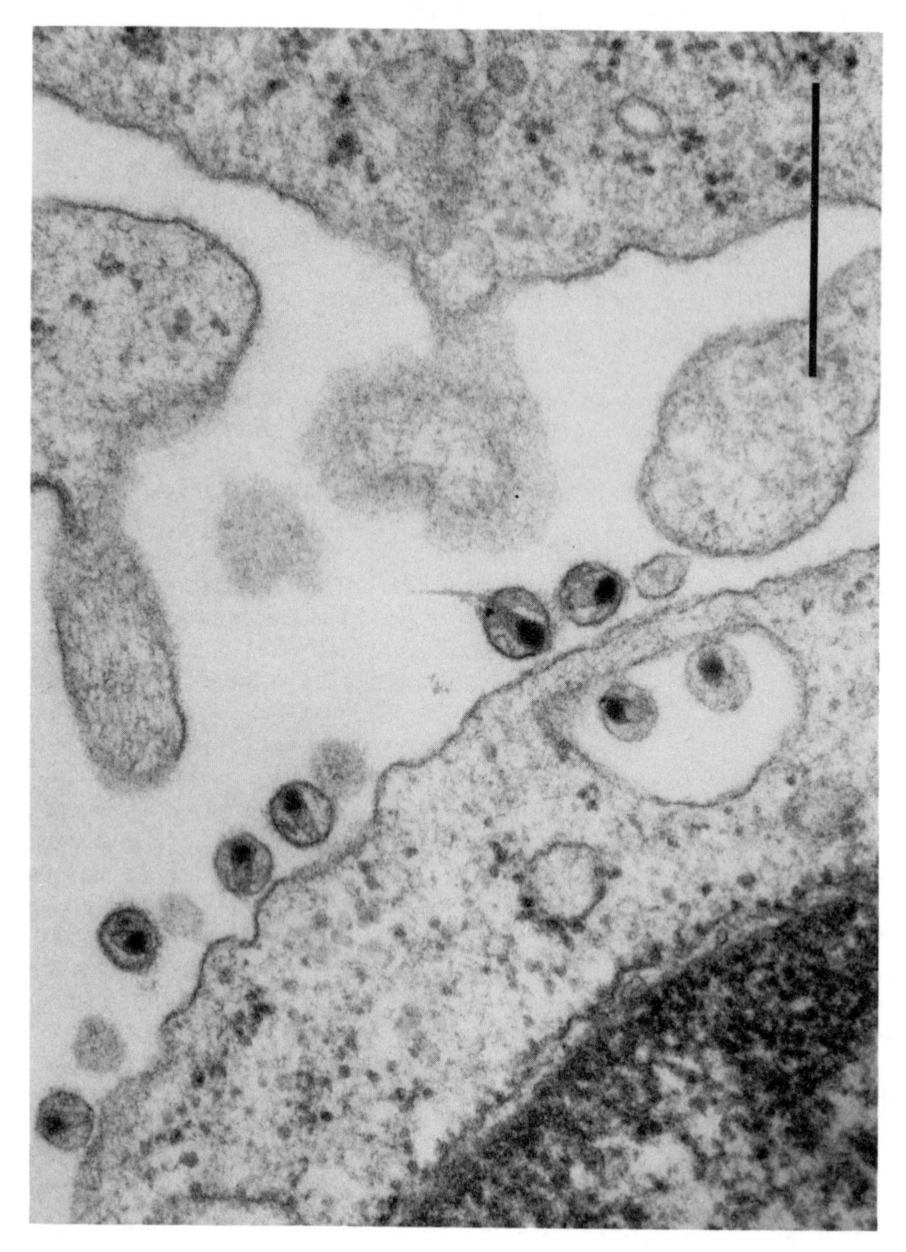

FIGURE 4. Image of HIV particles. The bar represents 0.5 µm.

tested for the presence of a specific antibody by indirect immunofluorescence on cells infected by the virus chosen for the immunization procedure. Uninfected cells were used as the control. Although this technique was very convenient for the screening of the hybridomas, it has possible drawbacks. One of them is that the method could not be sensitive enough to detect the presence of some MAbs directed against the glycoproteins. On the other hand, a monoclonal which recognizes infected and not uninfected cells could be directed against non-viral antigen expressed in the presence of a HIV transactivating protein.[55] Therefore it is necessary to analyze the specificity of these antibodies by other approaches.

From Western Blot analysis, it was possible to obtain more information. Four monoclonals recognize viral polypeptides: LO-HIV1-1 and LO-HIV1-2 bind to the major protein P25 and the precursors Pr40 and 55. The monoclonal LO-HIV1-3 recognizes P13 and the same precursors as expected since the two peptides are coded by the same gene GAG, the product of which being subsequently processed by proteolytic cleavage in three different pieces.[57] A polypeptide with a molecular weight around 34 and supposed to be endonuclease is detected by LO-HIV1-4 and 5.

Thus these antibodies allow the identification of some structural proteins and provide a way of purifying them and their polycistronic precursors, which means the amino acid sequences coded by the corresponding gene.

As far as the variability of isolates is concerned, some rat MAbs are directed against viral epitopes which seem to be a specific strain: LO-HIV1-4 and 5 recognize an epitope of ARV but not present on LAV nor HTLV3; on the contrary LO-HIV1-3 recognizes LAV and HTLV 3 but not the californian strain ARV.

In these preliminary experiments made to obtain rat MAbs against HIV, the yield was extremely poor. From one fusion where 126 hybrid clones were growing, only 4 produced specific antibodies. That is likely to be due to the immunization process. Here the main problem seems to be the very low quantity of antigenic material produced in the cell culture and eventually inoculated into the animal.

To improve it, several approaches have been considered. It is now possible to obtain a sufficient amount of recombinant proteins for immunization. Some rats have already been immunized with this type of antigen and are waiting for the fusion experiments. Besides, the rat monoclonals already obtained against the HIV antigens have been produced in large quantities and the purified immunoglobulins covalently linked to a solid phase in order to obtain immunoaffinity column and purify the various polypeptides. A small amount of antigen, around 10 ng/l, can be recovered from the supernatant of the tissue culture.[59] However, from cell pellet in good conditions, it is sometimes possible to obtain 5 times more. It is thus possible to concentrate enough material to immunize animals. If the monoclonal taken for the immunoadsorption also recognizes precursor polypeptides, this procedure supplies us with some kind of a feed back amplifying system and the possibility to produce hybridomas against the various proteins coded by a given messenger. This work is still in progress.

IV. RAT MONOCLONAL ANTIBODIES TO PLANT PATHOGENS

See Chapter 29.

ACKNOWLEDGMENTS

We wish to thank H. Bazin for his help in acquiring the art of making the hybridomas, F. Nisol, J.M. Malache, and J.P. Kints for their friendly and skillful assistance, as well as F. Bolle and P. Lambrecht for typing the manuscript.

REFERENCES

1. **Koprowski, H., Gerhard, W., and Croce, C.,** Production of antibodies against influenza virus by somatic cell hybrids between mouse myeloma and primed spleen cells, *Proc. Natl. Acad. Sci. U.S.A.*, 74, 2985, 1977.
2. **Yewdell, J. W. and Gerhard, W.,** Antigenic characterization of viruses by monoclonal antibodies, *Annu. Rev. Microbiol.*, 35, 185, 1981.
3. **Oxford, J.,** The use of monoclonal antibodies in virology, *J. Hyg. Cambridge*, 88, 361, 1982.
4. **Yolken, R. H.,** Use of monoclonal antibodies for viral diagnosis, in *Current Topics in Microbiology and Immunology*, Bachman, P., Ed., Springer-Verlag, New York, 104, 177, 1983.
5. **Carter, M. J. and Ter Meulen, V.,** The application of monoclonal antibodies in the study of viruses, *Adv. Virus Res.*, 29, 95, 1984.
6. **Mc Cullough, K. C.,** Monoclonal antibodies implications for virology, *Arch. Virol.*, 87, 1, 1985.
7. **Köhler, G. and Milstein, C.,** Continuous cultures of fused cells secreting antibody of predefined specificity, *Nature*, 256, 495, 1975.
8. **Galfré, G. and Milstein, C.,** Preparation of monoclonal antibodies, strategies and procedures, *Methods Enzymol.*, 73, 1, 1981.
9. **Bazin, H.,** Production of rat monoclonal antibodies with the LOU rat non secreting IR983F myeloma cell line, in *Protides of the Biological Fluids*, Peeters, H., Ed., Pergamon Press, Oxford, 1982, 615.
10. **Burtonboy, G., Bazin, H., and Delferriére, N.,** Rat hybridoma antibodies against canine parvovirus, *Arch. Virol.*, 71, 291, 1982.
11. **Parrish, C. R., Burtonboy, G., and Carmichael, L. E.,** Characterization of non hemagglutinating mutant of canine parvovirus, *Virology*, 163, 230,1988.
12. **Koenig, R. and Torrance, L.,** Antigenic analysis of potato virus X by means of monoclonal antibodies, *J. Gen. Virol.*, 67, 2145, 1986.
13. **Torrance, L., Pead, M. T., Larkins, A., and Butcher, G.,** Characterization of monoclonal antibodies to a U.K. isolate of Barley yellow Dwarf virus, *J. Gen. Virol.*, 67, 549, 1986.
14. **Eugster, A. K. and Nairn, C.,** Diarrhea in puppies parvovirus like particles demonstrated in their feces, *Southwest. Vet.*, 30, 59, 1977.
15. **Appel, M., Meunier, P., Pollock, R., Greiser, H., and Carmichael, L.,** Canine Viral Enteritis, *Canine Practice Medecine*, 7, 22, 1980.
16. **Carmichael, I. F. and Binn, L. N.,** New canine infections, in *Advances in Veterinary Science*, Academic Press, New York, 1981, 1.
17. **Coignoul, F. and Dewaele, A.,** Canine haemorrhagic enteritis, pathology of a syndrome, *Ann. Med. Vet.*, 123, 47, 1979.
18. **Pollock, R. V. and Carmichael, L.,** Canine viral enteritis recent developments, *Mod. Vet. Pract.*, 60, 375, 1979.
19. **Johnson, R. H.,** Feline panleukopenia virus *in vitro*, Comparison of strains with a mink enteritis virus, *J. Small Anim. Pract.*, 8, 319, 1967.
20. **Burtonboy, G.,** Identification d'un nouveau parvovirus, Ph.D. thesis, Université Catholique de Louvain, Faculté de Médecine, Bruxelles,
21. **Burtonboy, G., Coignoul, F., Delferriére, N., and Pastoret, P. P.,** Canine haemorrhagic enteritis, detection of viral particles by electron microscopy, *Arch. Virol.*, 61, 1, 1979.
22. **Appel, M., Scott, F., and Carmichael, L.,** Isolation and immunisation studies of a canine parvovirus like virus form dogs with haemorrhagic enteritis, *Vet. Rec.*, 105, 156, 1979.
23. **Carmichael, L. E., Joubert, J. C., and Pollock, R. V. H.,** Hemagglutination by canine parvovirus, serological studies and diagnostic applications, *Am. J. Vet. Res.*, 41, 784, 1980.
24. **Siegl, G., Bates, R., Berns, K., Carter, B., Kelly, D., Kurstak, E., and Tattersall, P.,** Characteristics and taxonomy of parvoviridae, *Intervirology*, 23, 61, 1985.
25. **Mc Master, G., Tratschin, J. D., and Siegl, G.,** Comparison of canine parvovirus with mink enteritis virus by restriction site mapping, *J. Virol.*, 38, 368, 1981.
26. **Paradiso, P. R., Rhode S. L., III, Singer, II,** Canine parvovirus, a biochemical and ultrastructural characterization, *J. Gen. Virol.*, 62, 113, 1982.
27. **Rhode, S. L., III,** Nucleotide sequence of the coat protein gene of canine parvovirus, *J. Virol.*, 54, 630, 1985.
28. **Reed, P., Jones, E. V., and Miller, T. J.,** Nucleotide sequence and genome organization of canine parvovirus, *J. Vet. Inv.*, 396, 1987.
29. **Johnson, B.,** Parvovirus proteins, in *The Parvoviruses*, Plenum Press, New York, 1984, 259.
30. **Siegl, G.,** Canine parvovirus, origin and significance of a new pathogen, in *The Parvoviruses*, Plenum Press, New York, 1984.
31. **Tratschin, J. D., Mc Master, G. K., Kronauer, G., and Siegl, G.,** Canine parvovirus, relationship to wild type and vaccine strains of feline panleukopenia virus and mink enteritis virus, *J. Gen. Virol.*, 61, 33, 1982.

32. **Flower, R., Wilcox, G., and Robinson, W.,** Antigenic differences between canine parvovirus and feline panleukopenia virus, *Vet. Rec.*, 107, 254, 1980.
33. **Fazekas De St. Groth, S., and Scheidegger, D.,** Production of monoclonal antibodies, strategy and tactics, *J. Immunol. Meth.*, 35, 1, 1980.
34. **Bodeus, M., Burtonboy, G., and Bazin, H.,** Rat monoclonal antibodies. IV. Easy method for *in vitro* production, *J. Immunol. Meth.*, 79, 1, 1985.
35. **Bazin, H.,** Rat-Rat hybridoma formation and rat monoclonal antibodies, in *Methods of Hybridoma Formation*, Bartal, A. and Hirshaut, Y., Eds., Humana Press, Clifton, NJ, 1987, 337.
36. **Mancini, G., Carbonara, A., and Heremans, J. F.,** Immunochemical quantitation of antigens by single radial immunodiffusion, *Immunochemistry*, 2, 235, 1965.
37. **Almeida, J. D. and Waterson, A. P.,** The morphology of virus antibody interaction, *Advances in Virus Research*, Vol. 12, Smith, K. M. and Lauffer, M., Eds., Bang F.A.P., Academic Press, New York, 15, 307, 1969.
38. **Siegl, G. and Kronauer, G.,** A plaque assay for feline panleukopenia virus, *J. Gen. Virol.*, 46, 211, 1980.
39. **Emini, E., Jameson, B., and Winner, E.,** Antigenic structure of poliovirus, in *Immunochemistry of Viruses., The Bases for Serodiagnosis and Vaccine*, Van Regenmortel, M. and Neurath, A. Eds., Elsevier, New York, 1985, 281.
40. **Surleraux, M., Bodeus, M., and Burtonboy, G.,** Study of canine parvovirus polypeptides by immunoblot analysis., *Arch. Virol.*, 95, 271, 1987.
41. **Surleraux, M. and Burtonboy, G.,** Structural polypeptides of a canine parvovirus study by immunoadsorption and sequential analysis of infected cells, *Arch. Virol.*, 82, 233, 1984.
42. **Towbin, H., Staehelin, T., and Gordon, J.,** Electrophoretic transfer of proteins from polyacrylamide gels to nitrocellulose sheets: procedure and some applications., *Proc. Natl. Acad. Sci. U.S.A.*, 76, 4350, 1979.
43. **Irons, L., Ashworth, L., and Wilton-Smith, P.,** Heterogeneity of the filamentous haemagglutinin of bordetella pertussis studied with monoclonal antibodies., *J. Gen. Microbiol.*, 129, 2769, 1983.
44. **Tattersall, P. and Shatkin, A.,** Sequence homology between the structural polypeptides of minute virus of mice., *J. Mol. Biol.*, 111, 375, 1977.
45. **Jongeneel, C. V., Sahli, R., McMaster, G. K., and Hirt, B.,** A precise map of splice junctions in the mRNAs of MVM., *J. Virol.*, 59, 564, 1986.
46. **Paradiso, P. R., Williams, K. R., and Costantino, R. L.,** Mapping of the amino terminus of the H1 parvovirus major capsid protein., *J. Virol.*, 52, 77, 1984.
47. **Bodeus, M., Cambiaso, C., Surleraux, M., and Burtonboy, G.,** A latex agglutination test for the detection of canine parvovirus and corresponding antibodies, *J. Virol. Methods*, 19, 1, 1988.
48. **Marx, J. L.,** AIDS virus has new name, *Science*, 232, 699, 1986.
49. **Barré-Sinoussi, F., Chermann, J. C., Rey, F., Nugeyre, M. T., Chamaret, S., Gruest, J., Dauguet, C., Axler-Blin, C., Brun-Vezinet, F., Rouzioux, C., Rozenbaum, W., and Montagnier, L.,** Isolation of a T lymphotropic retrovirus from a patient at risk for acquired immune deficiency syndrome (AIDS)., *Science*, 220, 868, 1983.
50. **Levy, J. A. and Shimabukuro, J.,** Recovery of AIDS associated retrovirus from patients with AIDS or AIDS related conditions and from clinically healthy individuals, *J. Infect. Dis.*, 152, 734, 1985.
51. **Popovic, M., Sarngadharan, M., Read, E., and Gallo, R.,** Detection isolation and continuous production of cytopathic retroviruses (HTLVIII) from patients with AIDS and pre-AIDS., *Science*, 224, 497, 1984.
52. **Wain-Hobson, S., Sonigo, P., Danos, O., Cole, S., and Alizon, M.,** Nucleotide sequence of the AIDS virus LAV., *Cell*, 40, 9, 1985.
53. **Arya, S. K., Guo, C., Josephs, S. F., and Wong-Staal, F.,** Transactivation gene of human T. Lymphotropic virus type III (HTLV III), *Science*, 229, 69, 1985.
54. **Wong-Staal, F., Chanda, P., and Ghrayeb, J.,** Human immunodeficiency virus the eight gene, *AIDS Res. Hum. Retrov.*, 3, 33, 1987.
55. **Robey, W., Nara, P. L., Poore, C., Popovic, M., Mc.Lane, M., Barin, F., Essex, M., and Fischnger, P.,** Rapid assessment of relationship among HIV isolates by oligopeptides analyses of external envelope glycoproteins., *AIDS Res. Hum. Retrov.*, 3, 401, 1987.
56. **Clavel, F., Guetard, D., and Brun-Vezinet, F.,** Isolation of a new human retrovirus from west African patients with AIDS., *Science*, 233, 343, 1986.
57. **Dimarzo Veronese, F., Rahman, R., Copeland, T., Oroszlan, S.,Gallo, R., and Sarngadharan, M.,** Immunological and chemical analysis of P6, the carboxy terminal fragment of HIV P15., *AIDS Res. Hum. Retrov.*, 3, 253, 1987.
58. **Pyle, S., Bess, J., Robey, W., Fishinger, P., Gilden, R., and Arthur, L.,** Purification of 120,000 Dalton envelope glycoprotein from culture fluids of HIV-infected H9 cells, *AIDS Res. Hum. Retrov.*, 3, 387, 1987.

Chapter 23

CONTRIBUTION OF RAT MONOCLONAL ANTIBODIES TO THE STUDY OF SCHISTOSOMA MANSONI INFECTIONS

J. M. Grzych, C. Verwaerde, M. Capron, and A. Capron

TABLE OF CONTENTS

I. INTRODUCTION

The introduction of hybridoma technology in parasitology has greatly contributed to our understanding of parasitic diseases. If a large majority of the observations published since 1979 has been performed with mouse monoclonal antibodies, several studies have particularly evolved with the construction of rat × rat hybridomas. We will further illustrate this topic with some of the original observations performed in *Schistosoma mansoni* infection with particular emphasis on the characterization of putative protective antigens, the analysis of effector and regulatory mechanisms, and the monoclonal anti-idiotypic approach of this infection. But before coming to the heart of the subject we would like to focus on some of the essential characteristics of *S. mansoni* infection that could also justify the choice of the rat model.

The immune response elicited against *S. mansoni* is essentially directed against the larval stage of the parasite, the schistosomula, and leads to the destruction of the larvae through several effector mechanisms. These mechanisms involve specific anaphylactic antibodies, IgG or IgE class in man, IgG2a and IgE isotypes in rat, and different effector cell populations like macrophages, eosinophils, and platelets.[1]

The close relation established between such ADCC mechanisms and the acquisition of a partial or total protective immunity in the corresponding infected host together with the demonstration that similar ADCC mechanisms were involved in rat and human schistosomiasis,[2] makes the target antigens of these mechanisms potential candidates for the human immunoprophylaxis. In order to investigate further the major role of IgG2a and IgE antibodies in the acquired immunity of rat against *S. mansoni,* we have applied hybridoma technology.

In a previous paper, we reported preliminary results on the isolation of mouse/rat hybrid cell supernatants containing IgG2a or IgE antibodies able to induce cytotoxicity against *S. mansoni* schistosomula *in vitro* in the presence of different effector cell populations.[3]

But the relative instability of such hybrids together with the limited use of heterologous system (mouse/rat hybridomas) led us to develop a rat-rat system of hybridization and to study the *in vitro* and *in vivo* biological functions of these rat monoclonal anti *S. mansoni* antibodies.

II. MONOCLONAL APPROACH OF EFFECTOR MECHANISMS

Among the different rat monoclonal antibodies produced in our fusion experiments, two antibodies deserve particular attention in respect to their biological properties, namely the IPL Sm1 IgG2a rat antibody[4] selected from the fusion of IR983F rat myeloma cell line[5] and B lymphocytes collected from *S. mansoni* infected rat, and the B48-14 antibody[6] of IgE isotype isolated in rat fusion experiment involving spleen cells of LOU rat immunized with a glycoprotein-enriched fraction of *S. mansoni* incubation products.

The *in vitro* approach of the effector functions of IPL Sm1 monoclonal antibody carried on supernatants or ascitic fluid showed that this antibody elicited, in the presence of normal rat eosinophils, a strong heat-stable cytotoxic activity comparable to those currently observed in the case of infected rat sera (Figure 1A). Further investigations concerning the biological properties of IPL Sm1 antibody indicated the involvement of mast cells in the killing of schistosomula. It was suggested that this antibody perfectly reproduce the activity of polyclonal antibodies present in the sera of *S. mansoni* 4-week-infected rats.[7]

The biological relevance of the *in vitro* finding has been confirmed *in vivo* in passive transfer experiments. The injection of IPL Sm1 monoclonal antibody in native LOU rat 4 h previously infected with *S. mansoni* cercariae significantly reduced the parasitic burden of these animals (50 to 60%) (Figure 1B).

A similar approach applied to the B48-14 antibody of IgE isotype also confirmed the effector functions of anaphylactic antibodies. Native B48-14 antibody tested in the different cell-mediated cytotoxicity mechanisms previously reported, revealed its strong capacity to kill

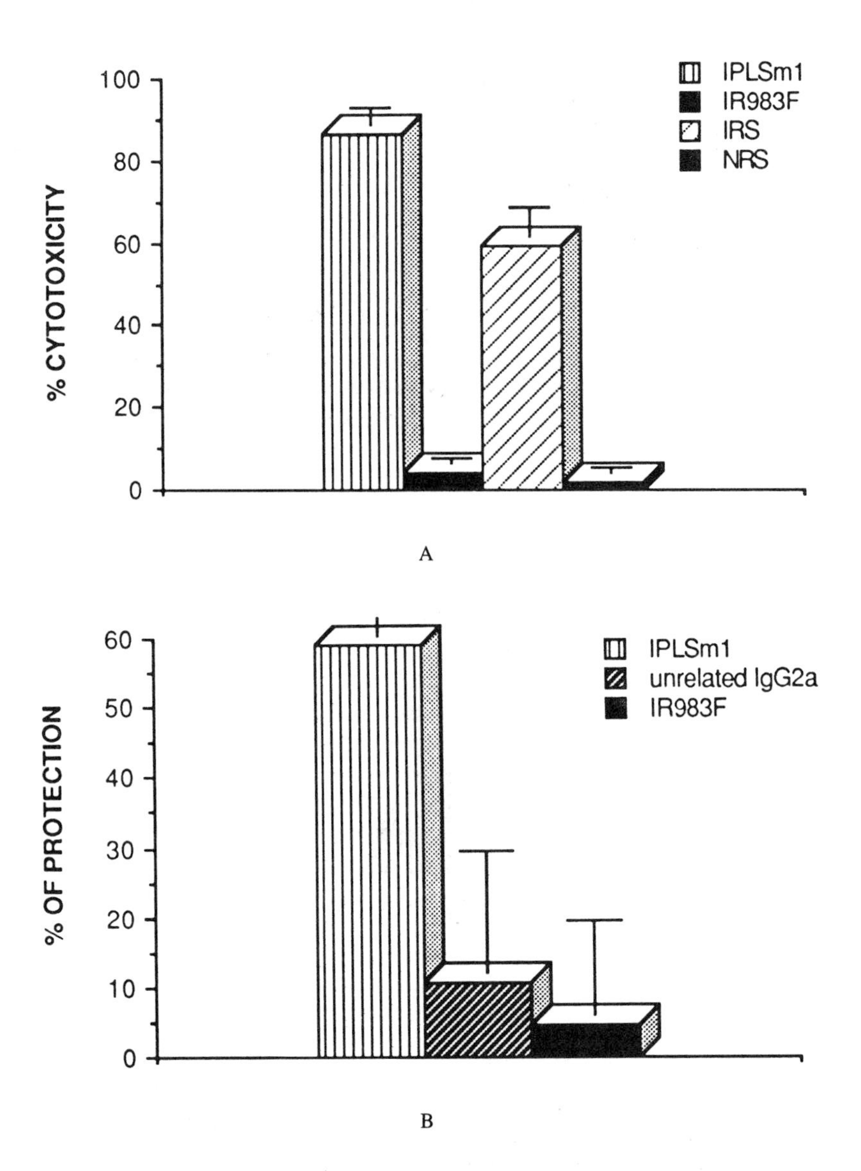

FIGURE 1. *In vitro* and *in vivo* effector functions of IPL Sm1 (IgG2a) monoclonal antibody. (A) The percentage of cytotoxicity was measured after 48 h incubation with effector cells (LOU rat peritoneal cells enriched in eosinophils) (means of two duplicate experiments ± SD). Ascitic fluid from IPL Sm1 (10^{-3} final dilution), normal rat sera (NRS) 4 week-infected rat sera (IRS), and ascitic fluid from IR983F myeloma (10^{-3} final dilution) were heat-inactivated 2 h at 56°C. (B) Ascitic fluid (2 ml) obtained from IPL Sm1 hybridoma, C1-19 hybridoma (unrelated IgG2a monoclonal antibody), or IR983F myeloma were injected i.v. to rats infected 4 h previously with *S. mansoni* cercariae. The parasitic burden was evaluated 3 weeks later by liver perfusion. The number of worms in rats injected with IPL Sm1 was compared with number of worms in rats injected with 2 ml of physiological saline.

schistosomula in the presence of normal rat macrophages, eosinophils, or platelets in a dose-dependent manner comparable to the levels of mortality observed in infected rat sera (Figure 2A). As in the case of IgG2a monoclonal antibody, B48-14 IgE monoclonal antibody passively transferred to normal recipient rats is able to achieve a significant level of protection ranged between 40 and 60% towards a challenge infection with *S. mansoni* cercariae (Figure 2B).

This first set of experimental data once more confirmed and emphasized the essential role

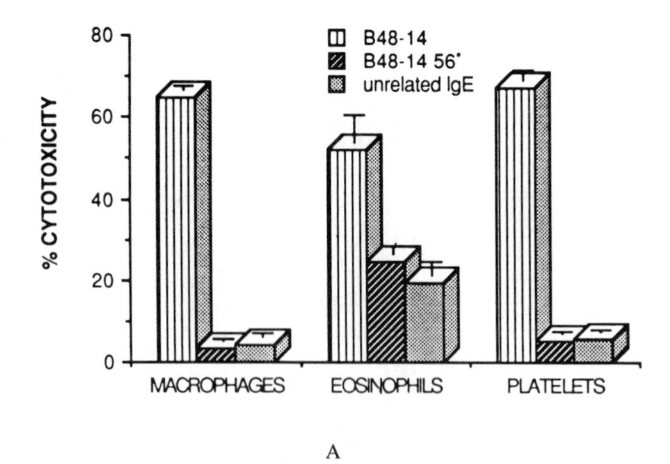

A

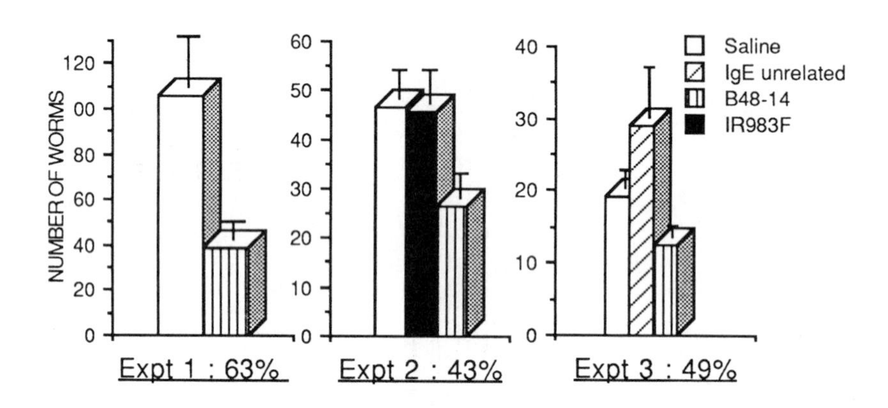

B

FIGURE 2. Biological properties of B48-14 IgE monoclonal antibody (A) Schistosomula were incubated with effector cells and with native or heated (4 h at 56°C) B48-14 IgE or unrelated monoclonal IgE. The percentages of cytotoxicity were expressed after deduction of the value obtained without effector cells. (B) Sera (0.5 ml) of rats bearing B48-14 hybridoma IR983F myeloma, or unrelated IgE hybridoma were injected i.p. into LOU rats infected with 800 cercariae (groups of 5 rats were used for each test). Parasitic burdens were evaluated 3 weeks later by liver perfusion. The number of worms in rats injected with B48-14 IgE was compared with the number of worms collected from rats injected with 0.5 ml of physiological saline.

played by anaphylactic antibodies in the induction of antibody-dependent cell-mediated cytotoxicity and brought a clear cut argument of the close correlation between such *in vitro* mechanisms and the *in vivo* expression of a protective immunity.

III. CHARACTERIZATION OF *SCHISTOSOMA MANSONI* ANTIGENS

In light of these observations, IPL Sm1 and B48-14 rat monoclonal antibodies were further applied in the identification of potentially protective antigens.

The use of IPL Sm1 IgG2a antibody has allowed the identification from the *Schistosoma mansoni* schistosomula membrane of a 38 kDa glycoprotein antigen (GP 38).[8] This antigen is present on the cercariae, but is no longer detectable at the lung stage of the infection. GP 38 antigen was also shown to react with the polyclonal antibodies present in various infected hosts including mice, rats, monkeys, and humans.[9] Additional studies revealed that 97% of a group of 120 Brazilian patients with *S. mansoni* infection produced antibodies against circulating antibodies against this molecule, suggesting that GP38 corresponds to a potent immunogen in human. More interestingly, in human the antibody response to GP 38 shows a close parallelism with the prevalence and the intensity of the infection, this indicated that antibody response to the 38 kDa could be considered as an important marker of *S. mansoni* infection.[10]

More recently, using B48-14 IgE monoclonal antibody, a 26 kDa molecule was characterized among the products released *in vitro* by the schistosomula (SRP, schistosomula released products). These results must be related to previous work performed on SRP where Auriault et al.[11] described an allergen recognized by polyclonal IgE antibodies in infected rat sera and having the same molecular weight.

Moreover, the immunization of rats with the SRP led to the production of IgE antibodies preferentially directed against two molecules of 22 and 26 kDa. These IgE antibodies exhibit a marked cytotoxicity *in vitro* against schistosomula and partly supported the protection conferred to normal rat against experimental schistosomiasis by passive transfer of anti-SRP antiserum.[12]

IV. REGULATORY MECHANISMS

In the case of rat IgE and IgG2a monoclonal antibodies, our results bring clear cut evidence in the role played by anaphylatic antibodies in immunity to schistosomiasis and allowed us to establish a close relation between several *S. mansoni* antigens and their capacity to elicit a protective humoral response. The production of other monoclonal antibody isotypes has allowed for some surprising observations.

A rat anti-*S. mansoni* monoclonal antibody of IgG2c isotype (IPL Sm3) was obtained from the fusion of IR983F cells and spleen cells from *S. mansoni* infected rat.[13] This antibody which does not exhibit any killing activity for schistosomula recognizes the 38 kDa antigen previously defined as the target antigen of the IPL Sm1 protective antibody. However, it was able to inhibit in a dose-dependent manner the cytotoxic effect of IPL Sm1 and eosinophils (Figure 3A).

This inhibitory effect observed *in vitro* was confirmed *in vivo* by the demonstration that IgG2c monoclonal antibody could strongly decrease the protective role of IgG2a monoclonal antibody (Figure 3B). The inhibition could also be achieved with F(ab′)2 fragments of IgG2c monoclonal antibody which indicates a possible blocking action on the recognition of the schistosomulum surface by IgG2a antibody, hypotheses also supported by the demonstration in immunoprecipitation experiments that the IPL Sm3 monoclonal antibody inhibits the fixation of IPL Sm1 antibody to its target antigen GP38.

The data obtained by applying different monoclonal antibodies raised the question of the existence of such regulatory mechanisms during the time course of *S. mansoni* infection in rat and human.

In rat, we could show that the inhibitory effect of IgG2c antibodies was not restricted to monoclonal antibodies. The IgG2c depletion of immune rat sera by protein A Sepharose absorption led to increase the eosinophil-mediated cytotoxicity *in vitro*. The *in vivo* relevance was confirmed by the kinetic study of the cytophilic IgG antibodies bound to eosinophils from infected rats according to their status of immunity to reinfection: the presence of IgG2a corresponding to the period of immunity, whereas the detection of IgG2c antibodies was linked to the decrease of immunity to reinfection.[14]

In human infection, recent evidence has been obtained showing the existence of IgM blocking antibodies which were able to bind to the GP38 antigen.[15] These observations lead to the concept that whereas potential markers of immunity in man defined by protective monoclo-

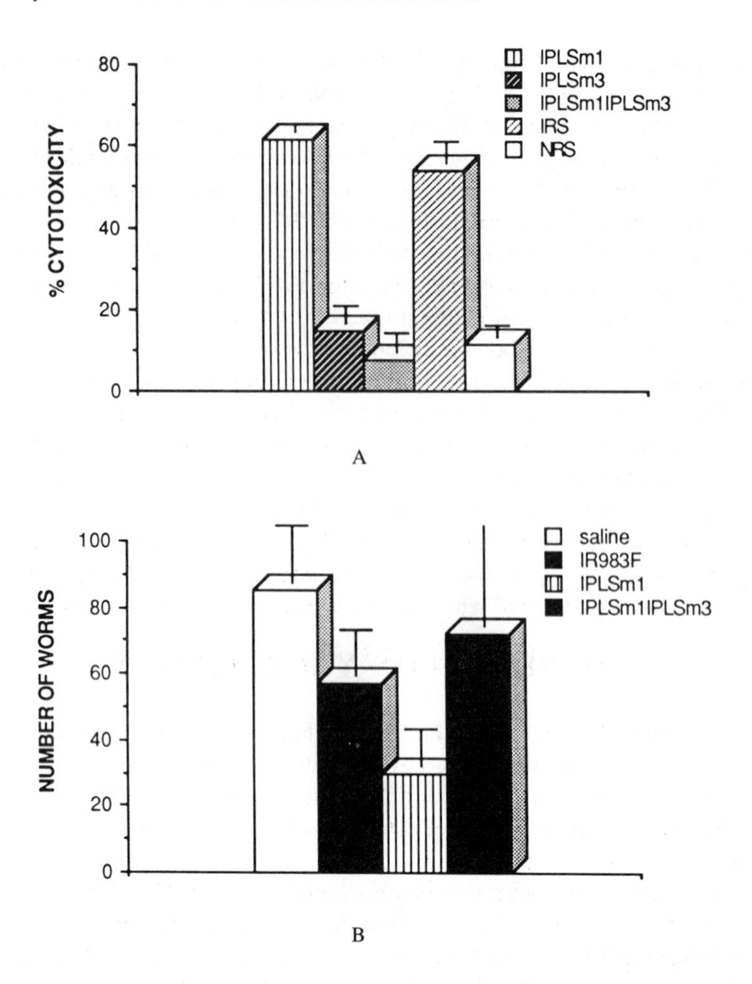

FIGURE 3. *In vitro* and *in vivo* blocking activities of IPL Sm3 (IgG2c) monoclonal antibody. (A) The percentage of cytotoxicity was measured after 48 h incubation with effector cells (LOU rat peritoneal cells enriched in eosinophils) (mean of two duplicate experiments ± SD). Ascitic fluids and sera (1/32 final dilution) were heat-inactivated for 2 h at 56°C (B) Two milliliters of ascitic fluid (either IgG2a-producing IPL Sm1, IgG2c-producing IPL Sm3, or IR983F control) or two milliliters of saline were injected i.v. in each rat (groups of 5 rats were used for each test). The number of worms was estimated by liver perfusion 21 d after the challenge infection with 1000 cercariae.

nal antibodies, might be present in many individuals, and this appears to be the case with GP38, the expression of immunity might be controlled by the presence or the absence of blocking antibody response.

The observations reported here with the two rat monoclonal antibodies IPL Sm1 and IPL Sm3 appear to have two major consequences. From a biological point of view, the demonstration that the same molecule could elicit both protective and blocking antibody responses represents a new way for the parasite to escape the immune response that it elicits. Secondly the identification of antagonist antibodies in schistosomiasis brings new insight to the existence of regulatory mechanisms which might affect the expression of protective immunity.

V. PRODUCTION OF ANTI-IDIOTYPIC ANTIBODIES

Although from these studies, GP38 antigen represents theoretically a good candidate for a

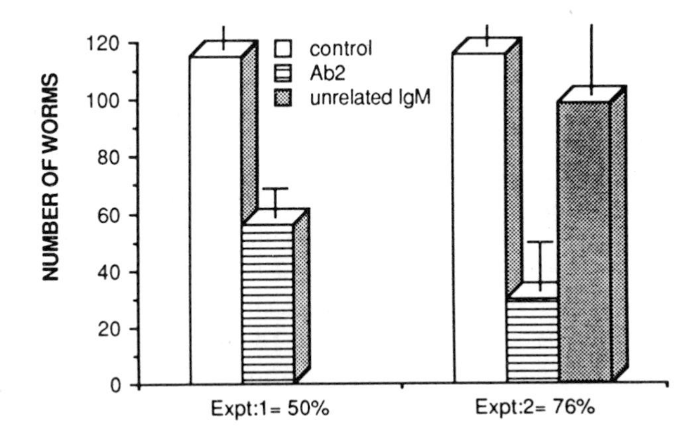

FIGURE 4. Protective effect of anti-idiotype immunization. LOU rats immunized with JM8-36 antibody (Ab2) were infected with 800 *S. mansoni* cercariae. Parasite burdens were measured 3 weeks later by liver perfusion. The number of worms obtained from Ab2 immunized rats was compared to those obtained in the control groups, i.e., LOU rats injected with physiological saline or with IgM purified from normal rat sera.

potential vaccine against schistosomiasis, the glycanic nature of the epitope recognized by the IPL Sm1 antibody together with its capacity to induce blocking antibodies limits its application in immunoprophylaxis of *S. mansoni* infections. Therefore on the basis of Jerne's[16] network theory, we considered as an alternative approach, the immunization with anti-idiotype antibodies.

Monoclonal antibodies to IPL Sm1 antibody were obtained in a homologous system of hybridization by the fusion of IR983F rat myeloma cell line and spleen cells from LOU rat immunized with IPL Sm1 antibody. From the 200 hybrid cell supernatants obtained from two successive fusion experiments, 29 supernatants induced a significant level of inhibition (70%) in the fixation of IPL Sm1 antibody to its target 38 kDa antigen, and suggested that the antibodies produced were able to fix an epitope of IPL Sm1 which is closed to a part of its combining site, indicating therefore that the Ab2 were paratope-induced antibodies and that they could bear the internal image of the original epitope.

Immunization experiments carried out with the particular JM8-36 Ab2 antibody allow to reproduce several parameters of *S. mansoni* infection. Successive injections of purified IgM JM8-36 anti-idiotype antibody elicited anti-*S. mansoni* antibodies (Ab3) capable of inhibiting the binding of IPL Sm1 antibody to its target antigen indicating a close specificity for the same molecule. These Ab3 antibodies could also be detected by indirect immunofluorescence reactions on the surface of schistosomula and were strongly cytotoxic for schistosomula in the presence of rat eosinophils. Further inhibition or depletion experiments revealed that Ab3 antibodies were of the IgG2a subclass, similar to Ab1 (IPL Sm1). The *in vivo* relevance of this mechanism was demonstrated by showing that Ab3 conferred significant levels of protection by passive transfer, whereas rats immunized by Ab2 demonstrated a marked protection (50 to 76%) toward a challenge infection[17] (Figure 4).

These results clearly demonstrate that immunization with an anti-idiotype monoclonal rat antibody could reproduce several parameters of the acquired immunity observed in the rat schistosomiasis, suggesting the potential use of anti-idiotype antibodies as an alternative approach to immunization against pathogens.

The recent demonstration of cross-reactive idiotopes between rat and human models together with the fact that similar effector mechanisms were described in rat and human infections, suggests that one can favorably consider the use of anti-idiotype antibodies in human to elicit anti-*S. mansoni* antibody response.

VI. CONCLUSIONS

From this brief review, it now seems possible to draw the first conclusions concerning the application of rat-rat hybridomas in the study of *S. mansoni* infection.

Our knowledge has considerably moved forward with the development of rat monoclonal antibodies. The restricted specificity of such material has allowed us to establish a close correlation between two different *S. mansoni* major immunogens and their capacity to elicit a strong cytotoxic humoral response in which anaphylactic antibodies play an essential role.

The fine dissection of the extremely complex mechanisms observed during the course of the infection is at the origin of the demonstration in rat and human of some regulatory mechanisms that could considerably limit the efficiency of protective antibody response and might have deep consequences in the selection of protective molecules available for immunoprophylaxis.

Aside from their applications in the analysis of the immune response, monoclonal antibodies could also represent an elegant alternative in the production of vaccines, by the construction of anti-idiotype antibodies bearing the internal image of carbohydrate epitope of glycoconjugated major immunogens which cannot be produced by DNA recombinant methodology.

By revealing the diversity of the regulatory mechanisms in parasite infections, these data emphasize once again the necessity of a careful dissection of immunity before attempting any active immunization, and amply justify the monoclonal approach of parasite infection.

REFERENCES

1. **Capron, A. and Dessaint, J. P.,** Effector and regulatory mechanisms in immunity to schistosomes: a heuristic view, *Rev. Immunol.*, 3, 455, 1985.
2. **Capron, M. and Capron, A.,** Rat, mice and men models for immune effector mechanisms against schistosomiasis, *Parasitol. Today*, 2, 69, 1986.
3. **Verwaerde, C., Grzych, J. M., Bazin, H., Capron, M., and Capron, A.,** Production d'anticorps anti-*Schistosoma mansoni*. Etude préliminaire de leurs activités biologiques, *C.R. Acad. Sci.*, 289, 725, 1979.
4. **Grzych, J. M., Capron, M., Bazin, H., and Capron, A.,** *In vitro* and *in vivo* effector functions of rat IgG2a monoclonal anti-*S. mansoni* antibodies, *J. Immunol.*, 129, 2739, 1982.
5. **Bazin, H., Grzych, J. M., Verwaerde, C., and Capron, A.,** A LOU rat non-secreting myeloma cell line suitable for the production of rat-rat hybridomas, *Ann. Immunol.*, 131D, 359, 1980.
6. **Verwaerde, C., Joseph, M., Capron, M., Pierce, R. J., Damonneville, M., Velge, F., Auriault, C., and Capron, A.,** Functional properties of rat monoclonal IgE antibody specific for *Schistosoma mansoni, J. Immunol.*, 138, 4441, 1987.
7. **Capron, M., Capron, A., Torpier, G., Bazin, H., Bout, D., and Joseph, M.,** Eosinophil-dependent cytotoxicity in rat schistosomiasis. Involvement of IgG2a antibody and role of mast cells, *Eur. J. Immunol.*, 8, 1270, 1978.
8. **Dissous, C., Grzych, J. M., and Capron, A.,** *Schistosoma mansoni* surface antigen defined by a rat protective monoclonal IgG2a, *J. Immunol.*, 129, 2232, 1982.
9. **Dissous, C. and Capron, A.,** Isolation of surface antigens from *Schistosoma mansoni* schistosomula, in *Protides of the Biological Fluids*, Peeters, H., Ed., Pergamon Press, Oxford, 1982, 179.
10. **Dissous, C., Prata, A., and Capron, A.,** Human antibody response to *Schistosoma mansoni* surface antigens defined by protective monoclonal antibodies, *J. Infect. Dis.*, 149, 227, 1984.
11. **Auriault, C., Damonneville, M., Verwaerde, C., Pierce, R. J., Joseph, M., Capron, M., and Capron, A.,** Rat IgE directed against schistosomula-released products is cytotoxic for *Schistosoma mansoni* schistosomula *in vitro, Eur. J. Immunol.*, 14, 132, 1984.
12. **Damonneville, M., Aurieault, C., Verwaerde, C., Delanoye, A., Pierce, R., and Capron, A.,** Protection against experimental *Schistosoma mansoni* schistosomiasis achieved by immunization with schistosomula-released products antigens (SR-A): role of IgE antibodies, *Clin. Exp. Immunol.*, 65, 244, 1986.
13. **Grzych, J. M., Capron, M., Dissout, C., and Capron, A.,** Blocking activity of rat monoclonal antibodies in experimental schistosomiasis, *J. Immunol.*, 133, 998, 1984a.

14. **Khalife, J., Capron, M., Grzych, J. M., Bazin, H., and Capron, A.,** Fc receptors on rat eosinophils: isotypic regulation of cell activation, *J. Immunol.,* 135, 2780, 1985.
15. **Khalife, J., Capron, M., Capron, A., Grzych, J. M., Butterworth, A. E., Dunne, D. W., and Ouma, J. H.,** Immunity in human Schistosomiasis mansoni. Regulation of protective immune mechanisms by IgM blocking antibodies, *J. Exp. Med.,* 164, 1626, 1986.
16. **Jerne, N. K.,** Towards a network theory of the immune system, *Ann. Immunol. Inst. Pasteur,* 125C, 373, 1974.
17. **Grzych, J. M., Capron, M., Lambert, P. H., Dissous, C., Torres, S., and Capron, A.,** An anti-idiotope vaccine against schistosomiasis, *Nature,* 316, 74, 1985.

Part V
Specific Uses and Properties of Rat Monoclonal Antibodies

Chapter 24.I

MOUSE IMMUNOGLOBULIN ISOTYPING

T. Delaunay and P. Manouvriez

TABLE OF CONTENTS

I. IMMUNODOT ASSAY

The immunodot assay, originally described by McDougal et al.,[1] and adapted in our laboratory, allows the determination of the heavy chain and the light chain classes of mouse immunoglobulins (Igs). It is based on the principle of a "sandwich immunoassay".

The main feature of our assay is the use of rat monoclonal antibodies (MAbs) anti-mouse heavy and light chain isotype. These MAbs are dotted in sequence on strips of nitrocellulose filter paper (5 × 80 mm, 0.22 μm porosity Schleicher and Schuell). After saturation, the strips are incubated with the mouse hybridoma supernatants and the mouse Igs, bound to the specific anti-mouse isotype rat MAbs, are detected with a peroxidase-conjugated goat polyclonal antibody directed against mouse Igs. The procedure for the test is the same as that used for the determination of rat Ig isotypes described in Chapter 11. The nature and the concentrations of the specific anti-mouse Ig isotype rat MAbs are given in Table 1.

Figure 1 shows the results of immunodot isotyping of hybridoma culture supernatants containing mouse MAbs. The control dots and the dots made with the rat MAbs which did not bind specifically for the tested Ig isotype appear almost colorless, whereas the dots binding a mouse Ig are darkly stained and easily distinguished. Non-specific immunodot reactions obtained with polyclonal antibodies[1] do not exist in our test owing to the use of MAbs. Figure 1 again shows that a positive test can be carried out with only 50 μl of mouse hybridoma cell culture supernatant. A rat MAb directed against mouse lambda chain is still not available. Lambda light chain Igs can be suspected when a positive result is obtained with an anti-heavy chain dot and a negative result is obtained with a rat MAb anti-mouse kappa chain such as LO-MK-1.

II. ELISA

The ELISA determination of the mouse Ig isotypes is performed in the same way as the determination of the rat Ig isotypes described in Chapter 11. The antibodies used are the same as those used in the immunodot assay and their concentrations are given in Table 1.

Figure 2 shows an ELISA test made with culture supernatants obtained from hybridomas secreting the different Ig isotypes.

III. CONCLUSIONS

These immunodot and ELISA isotyping assays using rat MAbs anti-mouse heavy and light chain Igs have three qualities:

1. Quickness: only 3 h is necessary for the isotype determination of a MAb. This allows a rapid selection of hybridoma cells secreting a given isotype.
2. Specificity: the reactions obtained with MAbs are unambiguous, unlike those sometimes obtained with polyclonal antibodies.
3. Sensitivity: this assay can detect dual or polyclonal antibody secretion and isotype switching in cell cultures.

TABLE 1
Use of Rat Monoclonal Antibodies for the Determination of the Mouse Immunoglobulin Isotypes

Rat monoclonal antibodies		Specificity for mouse Ig isotype	Avidity M^{-1}	Used concentration in µg/ml	
Name	Class			Immunodot	ELISA
LO-MM-9	IgG2a kappa	IgM	7.4×10^8	25	1
LO-MG1-2	IgG1 kappa	IgG1	9.3×10^8	50	10
LO-MG2a-2	IgG1 kappa	IgG2a	6.8×10^9	50	10
LO-MG2b-2	IgG1 kappa	IgG2b	1.1×10^{10}	50	10
LO-MG3-7	IgM kappa	IgG3	2.4×10^{10}	50	10
LO-MK-1	IgG2a kappa	Ig kappa	2.7×10^9	50	10

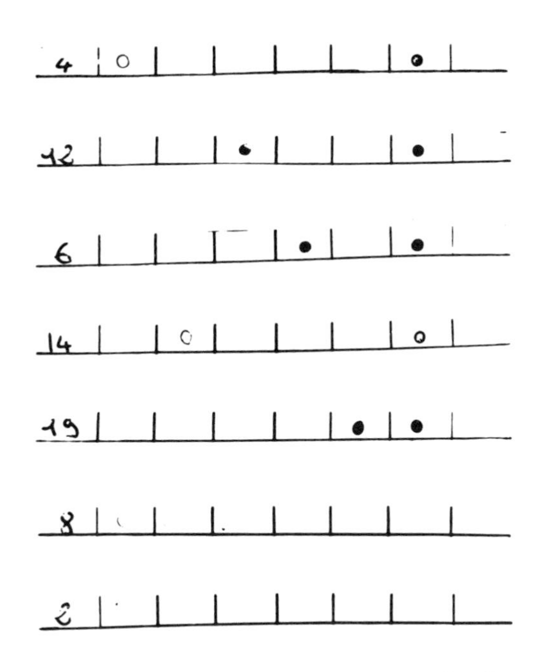

FIGURE 1. Immunodot determination of mouse Ig isotype in culture supernatant. Nitrocellulose strips were dotted with the following rat MAbs anti-mouse Ig isotype: LO-MM-9, anti-µ; LO-MG1-2, anti-gamma1; LO-MG2a-2, anti-gamma2a; LO-MG2b-2, anti-gamma2b; LO-MG3-7, anti-gamma3; and LO-MK-1, anti-kappa. The strips were incubated with 40-fold diluted culture supernatant and the specifically adsorbed mouse Igs were revealed with peroxidase labeled goat anti-mouse Ig. From left to right, the strips detected the following isotypes in the culture supernatants; (2) negative control, PAIO culture supernatant; (8) IgM non-kappa; (19) IgG3-kappa; (14) IgG1-kappa; (6) IgG2b-kappa; (12) IgG2a-kappa; and (4) IgM-kappa

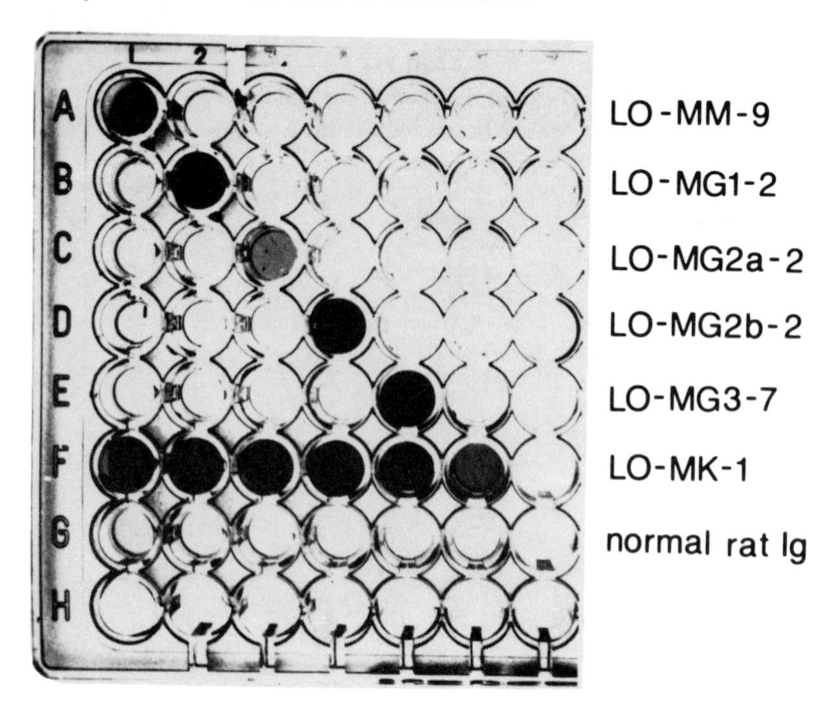

FIGURE 2. Determination of mouse Ig isotype by a sandwich immunoassay. Each row was coated with the same rat MAb anti-mouse Ig isotype than for the immunodot assay (Figure 1). A row was also coated with normal mouse Ig to detect the presence of rheumatoid factors which can give false positive tests. The specifically adsorbed mouse Igs were revealed with peroxidase-labeled goat anti-mouse Ig. Each column was incubated with 16-fold diluted culture supernatant. The following isotypes were detected; (1) IgM-kappa; (2) IgG1-kappa; (3) IgG2a-kappa; (4) IgG2b-kappa; (5) IgG3-kappa; (6) IgK unknown heavy chain (it was LB4 an IgE); (7) negative control, PAIO culture supernatant.

REFERENCE

1. **McDougal, J. S., Browning, S. W., Kennedy, S., and Moore, D. D.,** Immunodot assay for determining the isotype and light chain type of murine monoclonal antibodies in unconcentrated hybridoma culture supernatants, *J. Immunol. Methods,* 63, 281, 1983.

Chapter 24.II

HUMORAL IMMUNE SUPPRESSION OF MICE BY MONOCLONAL RAT ANTI-MOUSE M-HEAVY CHAIN ANTIBODIES

A. Cerny

TABLE OF CONTENTS

I. INTRODUCTION

Neonatally initiated continuous treatment of mice with heterologous anti-IgM antibodies leads to a state of B cell and antibody deficiency. Such mice were initially generated to study B cell ontogeny.[1-4] They have also been used to study the contribution of B cells and antibodies in various experimental situations. In this paper, anti-IgM induced humoral immune suppression is reviewed with special emphasis on the use of monoclonal rat anti-mouse IgM antibodies.

II. HISTORICAL BACKGROUND

To study pathways of the ontogeny of immunoglobulin synthesis in the chicken, Kincade et al.[5] treated chicken embryos *in ovo* with anti-IgM antibodies. This treatment caused a transient elimination of IgM producing cells. Combining injections with bursectomy at hatching resulted in virtually complete and permanent eradication of B cells and their products. Susequently, the same group elaborated data on the effect of neonatally initiated treatment of germ free mice with goat anti-mouse μ antibodies on various B cell parameters including histology and immunofluorescence analysis of lymphoid tissues, serum immunoglobulin levels and synthetic rates of spleen fragments, and response to antigen.[1] In 1972, Manning and Jutila used rabbit anti-μ for the treatment of conventionally housed mice including nude mice.[6] These workers showed that neonatally initiated anti-μ treatment was suppressive, whereas if started later in young adult mice it was not. Furthermore suppression seemed to be T cell independent, being operative in nude mice as well. Since then many groups have worked with this model, and several reviews have been published.[7-10]

The criteria used to demonstrate successful humoral immune suppression used by most groups are listed in Table 1.

III. METHODOLOGY

Although the mechanism of anti-IgM mediated B cell depletion is not completely understood, the target for suppression is the newly formed surface IgM positive B cell.[11,12] This cell type can be detected at low frequency before birth in the 16 to 17 d fetal liver, but is mainly produced after birth.[13] During further differentiation, B cells become insensitive to the suppressive effects of anti-IgM both *in vitro* and *in vivo*.[14,15]

Anti-μ treatment can be started either prenatally, neonatally, or later in adult mice. Prenatal treatment is achieved by continuously injecting μ-suppressed females during pregnancy. Heterologous anti-IgM antibodies are transmitted from mother to young during its intrauterine development and postnatally via the mother's milk.[16] We have shown, based on an observation by Manning et al.,[17] that this way, complete B cell and antibody suppression resulting in agammaglobulinemia can be induced.[18] Neonatal initiation of anti-IgM treatment is most commonly used. It may however be delayed up to 6 d after birth.[10] It is important to realize, that neonatally suppressed mice have high residual mouse immunoglobulin levels mainly of the IgG1 subclass and are therefore only partially suppressed with respect to humoral immunity.[19]

Treatment of adult mice with anti-IgM has a variety of effects on the immune response and leads to B cell suppression only in combination with other immunosuppressive treatments such as X-irradiation or Cyclophosphamide.[8,20] Suppression of adult mice circumvents the need for time and reagent consuming continuous treatment from birth until the age at which the experiment is performed but has on the other hand several disadvantages: (1.) the additionally required immunosuppression may influence cell compartments other than B cells; (2.) although B cell deficiency can be achieved, serum immunoglobulin levels are not reduced except for IgM; and (3.) injection of anti-μ into adult mice induces immune complex formation with mouse serum IgM, which in turn may alter the immune status of such mice.[21]

TABLE 1
Criteria Used to Demonstrate B Cell Depletion

1. Absence of surface IgM positive cells in lymphoid tissues.
2. Reduction of serum immunoglobulin levels.
3. Histomorphology of depleted B cell dependent areas in lymphoid tissues.
4. Detection of free anti-IgM in the sera of treated mice.
5. Absence of plaque-forming cells.
6. Suppression of mitogen (LPS) induced proliferative response.
7. Lack of antigen-specific antibody response.

IV. ANTI-IgM REAGENTS

The production of well defined anti-IgM reagents in sufficient amounts is obviously of major importance for the generation of B cell-deficient mice. The proper choice of monoclonal versus polyclonal antibodies, species origin, degree of purity, specificity, dosage, and choice of appropriate control reagents have to be considered.

A. PROBLEMS ASSOCIATED WITH POLYCLONAL REAGENTS

The most commonly used reagents are polyclonal goat or rabbit derived anti-IgM sera, IgG enriched fractions thereof, or affinity purified antibodies. Their properties depend on several parameters. The choice of species used for the generation of anti-IgM antibodies seems to be important. Drewes et al.[22] compared standardized anti-mouse μ antisera produced in rabbit, goat, chicken, and turkey and found that avian sera were not suppressive and that goat derived anti-μ only suppressed the direct anti-SRBC plaque forming cell (PFC) response, whereas rabbit anti-μ suppressed both direct as well as indirect PFC responses. In addition, goat anti-μ was found to be at least 8 times less potent than rabbit anti-μ when compared on the basis of radial immunodiffusion titers.

For immunization, the antibody titers and isotype pattern depend on the immunization protocol which defines the antigen dose, route of immunization, and adjuvant use. Most workers work with the BALB/c myeloma protein MOPC 104E which induces an anti-μ heavy chain response and an anti-lambda light chain response.

In processing and purification additional factors may influence the *in vivo* potency of a particular antibody preparation, e.g., the methods of purification and standardization.

We have estimated the doses reported in the literature depending on the degree of purity and standardization. They range between 1 and 5 mg of pure antibody per week for maintenance of suppression. In addition the dose of anti-IgM needed may depend on the mouse strain used. We have recently shown that the autoimmune prone mouse strain (NZW×BXSB)F1 requires a higher dose of anti-IgM for humoral immune suppression than non-autoimmune mouse strains.[23] These points should emphasize the need for thorough *in vivo* titration of a particular antibody preparation. In our hands affinity purified rabbit anti-μ antibodies were still suppressive at a dose of 30 μg injected 3 to 4 times per weeks.[24]

B. MONOCLONAL REAGENTS

In addition to anti-IgM specific monoclonal antibodies anti-IgD and anti-Ia specific monoclonal antibodies, inducing depletion of certain B cell subsets have successfully been used.[25-27] Also due to the efforts involved to produce large amounts of antibodies required, monoclonal reagents have not been as commonly used as polyclonal reagents for *in vivo* B cell depletion. A procedure to produce and purify large amounts of pure rat monoclonal antibodies has been recently described.[28]

Three advantages of monoclonal over polyclonal anti-IgM reagents are:

TABLE 2
Characterization of Rat Anti-Mouse μ Monoclonal Antibodies

Name	Rat isotype	Heavy chain specificity	Ref.
LO-MM-5	IgG1, kappa	μ	18, 30
LO-MM-8	IgG1, kappa	μ	30
LO-MM-9	IgG2a, kappa	μ	30
b-7-6	IgG1, kappa	μ (C2 domain)	24, 29
C-2-23	IgG2a, kappa	μ (C4 domain)	24, 29

1. Monospecificity: monoclonal antibodies are monospecific with respect to the antigen they recognize. Polyclonal reagents, even if affinity purified, may react both with heavy and light chain common and variable region determinants.
2. Homogeneity: monoclonal reagents are homogeneneous in terms of the isotype they express whereas polyclonal preparations are made up of a mixture of different isotypes to which various biological functions are linked.
3. Reproducibility: monoclonal reagents are widely distributed and the procedures to produce and purify them are standard in most laboratories; as discussed earlier this is not the case for the production of antisera. Results obtained with a particular monoclonal reagent can therefore be more easily reproduced and generalized.

We have so far used five monoclonal rat anti-mouse μ antibodies; b-7-6, C-2-23, LO-MM-5, LO-MM-8, and LO-MM-9.[29,30] Their properties are shown in Table 2. Our experience can be summarized as follows:

1. All monoclonal antibodies tested were both inducing and maintaining B-cell suppression *in vivo* according to the criteria given in Table 1.[31]
2. The b-7-6 monoclonal antibody used *in vitro* induces in soluble form DNA synthesis in small resting splenic B cells.[29] Administered to neonatal mice however, b-7-6 suppressed the B cell development.[24]
3. The amount of antibody required for successful B cell suppression was in a comparable range to the one found for polyclonal rabbit derived antibodies.[24]
4. These reagents may be used either alone or in combination with each other possibly displaying a synergistic effect. We have successfully used a mixture of equal amounts of b-7-6 and C-2-23 or LO-MM-5 and LO-MM-8.[24,32]
5. LO-MM-5 has been shown to pass from mother to offspring, and thus can be used for prenatal treatment in order to generate both B cell deficient and agammaglobulinaemic mice.[18]

V. CONCLUSION

Since 1970 anti-IgM treated B cell depleted mice have proven to be a useful tool to study the role of humoral versus cell-mediated immune functions. In this article dealing with methodological aspects of this animal model, we advocate the use of pure, well characterized monoclonal anti-μ reagents which due to monospecificity and homogeneity permit a high degree of reproducibility.

ACKNOWLEDGMENTS

The author thanks R.M. Zinkernagel and C. Heusser for reviewing the manuscript.

REFERENCES

1. **Lawton, A. R., III, Asofsky, R., Hylton, M. B., and Cooper, M. D.,** Suppression of immunoglobulin class synthesis in mice. I. Effects of treatment with antibody to μ-chain, *J. Exp. Med.,* 135, 277, 1972.
2. **Manning, D. D.,** Recovery from anti-Ig induced immunosuppression: implications for a model of Ig-secreting cell development, *J. Immunol.,* 113, 455, 1974.
3. **Manning, D. D. and Jutila, J. W.,** Immunosuppression of mice injected with heterologous anti-immunoglobulin heavy chain antisera, *J. Exp. Med.,* 135, 1316, 1972.
4. **Murgita, R. A., Mattioli, C. A., and Tomasi, T. A., Jr.,** Production of a runting syndrome with selective IgA deficiency in mice by the administration of anti-heavy chain antisera, *J. Exp. Med.,* 138, 209, 1973.
5. **Kincade, P. W., Lawton, A. R., Bockman, D. E., and Cooper, M. D.,** Suppression of immunoglobulin G synthesis as a result of antibody-mediated suppression of immunoglobulin M synthesis in chickens, *Proc. Nat. Acad. Sci. U.S.A.,* 67, 1918, 1970.
6. **Manning, D. D. and Jutila, J. W.,** Immunosuppression of congenitally athymic (nude) mice with heterologous anti-immunoglobulin heavy-chain sera, *Cell. Immunol.,* 14, 453, 1974.
7. **Manning, D. D.,** Heavy chain isotype suppression: a review of the immunosuppressive effects of heterologous anti-Ig heavy chain antisera, *J. Reticuloendothel. Soc.,* 18, 63, 1975.
8. **Gordon, J.,** The B lymphocyte deprived mouse as a tool in immunobiology, *J. Immunol. Methods,* 25, 227, 1979.
9. **Jacobson, R. H.,** Immunodeficiency models in characterization of immune responses to parasites-an overview, *Vet. Parasitol,* 10, 141, 1982.
10. **Lawton, A. R., III and Cooper, M. D.,** Modification of B lymphocyte differentiation by anti-immunoglobuoins, in *Contemporary Topics in Immunobiology,* Vol. 3, Cooper, M.D. and Warner, N.L., Eds., Plenum Press, New York, 1974, chap. 8.
11. **Aden, D. P., Manning, D. D., and Reed, N. D.,** Exclusion of cooperating T cells as targets for heterologous anti-μ serum, *Cell. Immunol.,* 14, 307, 1974.
12. **Nelson, S. J. and Manning, D. D.,** Fc dependence of anti-μ antibody-induced isotype suppression in mice, *Cell. Immunol.,* 62, 1, 1981.
13. **Owen, J. J. T., Wright, D. E., Habu, S., Raff, M. C. and Cooper, M. D.,** Studies on the generation of B lymphocytes in fetal liver and bone marrow, *J. Immunol.,* 118, 2067, 1977.
14. **Sidman, C. L. and Unanue, E. R.,** Receptor-mediated inactivation of early B lymphocytes, *Nature,* 257, 149, 1975.
15. **Raff, M. C., Owen, J. J. T., Cooper, M. D., Lawton A. R., III, Megson, M., and Gathings, W. E.,** Differences in susceptibility of mature and immature mouse B lymphocytes to anti-immunoglobulin-induced immunoglobulin suppression in vitro, *J. Exp. Med.,* 142, 1052, 1975.
16. **Brambell, F. W. R.,** The transmission of passive immunity from mother to young, *Frontiers of Biology,* Vol. 18, Neuberger, A., Tatum, E. L., and Holborow, E. J., Eds., North-Holland Publishing, Amsterdam, 1970.
17. **Manning, D. D.,** Complete humoral immunosuppression of mice by rabbit anti-μ antibodies passing the murine placenta, *J. Immunol.,* 118, 1109, 1977.
18. **Cerny, A., Heusser, C. H., Sutter, S., Huegin, A. W., Bazin, H., Hengartner, H., and Zinkernagel R. M.,** Generation of agammaglobulinaemic mice by prenatal and postnatal exposure to polyclonal or monoclonal anti-IgM antibodies, *Scand. J. Immunol.,* 24, 437, 1986.
19. **Friend, P. S., Theofilopoulos, A. N., Fidler, J. M., and Dixon, F. J.,** B-cell suppression by anti-IgM antibody: humoral and cellular analyses, *Cell. Immunol.,* 61, 404, 1981.
20. **Mizutani, T., Kimura, H., Wagi, K., and Okayama, T.,** Suppression of IgM-producing cells by administration of anti-IgM serum in vivo before sheep red blood cells, *Chem. Pharm. Bull.,* 30, 1880, 1982.
21. **Nagase, F., Ponzio, N. M., Waltenbaugh, C., Katz, I., and Thorbecke, G. J.,** Nonspecific immune modulating effects of ascites fluid and hyperimmune sera in vivo, *Cell. Immunol.,* 90, 92, 1985.
22. **Drewes, M. A. and Manning, D. D.,** Effect of antiserum-producing species on isotype suppression efficacy in mice, *J. Immunol.,* 124, 2016, 1980.

23. **Cerny, A., Starobinski, M., Hügin, A. W., Sutter, S., Zinkernagel, R. M., and Izui, S.,** Treatment with high doses of anti-IgM prevents, but with lower doses accelerates autoimmune disease in (NZWxBXSB)F1 hybrid mice, *J. Immunol.,* 138, 4222, 1987.
24. **Cerny, A., Hügin, A. W., Sutter, S., Heusser, C. H., Bos, N., Izui, S., Hengartner, H., and Zinkernagel, R. M.,** Suppression of B cell development and antibody responses in mice with polyclonal rabbit and monoclonal rat anti-IgM antibodies. I. Characterization of the suppressed state, *Expl. Cell. Biol.,* 53, 301, 1985.
25. **Gause-Pfreundschuh, A., Kappen, C., Dildrop, R., and Rajewsky, K.,** Anti-μ suppression by monoclonal antibody, Poster presented at the 6th Int. Congress of Immunology, July 6-11, 1986, Toronto, Canada, Session 1, 11, 36.
26. **Xue, B., Hirano, T., Pernis, B., Ovary, Z., and Thorbecke, G. J.,** Physiology of IgD. III. Effect of treatment with anti-IgD from birth on the magnitude and isotype distribution of the immune response in the spleen, *Eur. J. Immunol.,* 14, 81, 1984.
27. **Fultz, M. J., Scher, I., Finkelman, F. D., Kinkade, P., and Mond, J. J.,** Neonatal suppression with anti-Ia antibody. I. Suppression of murine B-lymphocyte development, *J. Immunol.,* 129, 992, 1982.
28. **Bazin, H., Cormont, F., and De Clercq, L.,** Rat monoclonal antibodies. II. Rapid and efficient method of purification from ascitic fluid or serum, *J. Immunol. Methods,* 71, 9, 1984.
29. **Julius, M. H., Heusser, C. H., and Hartmann, K. U.,** Induction of resting B cells to DNA synthesis by soluble monoclonal anti-immunoglobulin, *Eur. J. Immunol.,* 14, 753, 1984.
30. **Cormont, F., Manouvriez, P., De Clercq, L., and Bazin, H.,** The use of rat monoclonal antibodies to characterize, quantify and purify polyclonal or monoclonal mouse IgM, *Methods Enzymol.,* 121, 622, 1986.
31. **Cerny, A., Zinkernagel, R. M., and Bazin, H.,** Results for LO-MM-8 and LO-MM-9, unpublished data, 1985.
32. **Cerny, A., Zinkernagel, R. M., and Bazin, H.,** Results for LO-MM-5 and LO-MM-8, unpublished data, 1985.

Chapter 24.III

RAT MONOCLONAL ANTIBODIES AGAINST MOUSE IgE

M. V. Chavez, P. Manouvriez, and H. Bazin

TABLE OF CONTENTS

I. INTRODUCTION

Rat monoclonal antibodies (MAbs) against mouse IgE were obtained by fusing spleen cells of a LOU/C rat (immunized with a mouse IgE MAb IgEL-b4) with the 983 fusion cell line.[1] The nomenclature of hybridomas used hereafter is LO-ME for LOUVAIN rat MAb (LO), against mouse (M), IgE isotype (E), and these abbreviations are followed by a number.

LO-ME can be used for quantitative and qualitative determinations of mouse IgE and studies on mouse immunological system when IgE is involved.

II. QUANTITATIVE AND QUALITATIVE DETERMINATIONS

Concentrations of mouse IgE from, e.g., ascitic fluid, culture supernatants, or serum can be determined by a non-competitive EIA; a simple sandwich ELISA test with two LO-ME directed against different mouse IgE epitopes. One LO-ME (e.g., LO-ME-2) is coated on the EIA, i.e., microplates; it will recognize a first antigenic determinant on the mouse IgE. A second antigenic determinant will be recognized by another LO-ME (e.g., LO-ME-3); the latter MAb must be peroxidase-labeled:

1. 100 μl of 5 μg/ml of LO-ME (e.g., LO-ME-2) are coated at 4°C overnight.
2. Wells are washed 3 times with 0.15 *M* phosphate buffered saline (PBS) pH 7.2 0.1% Tween-20 (Sigma No. 1379) and saturated for 2 h at room temperature with PBS-5% gelatin (Merck No. 4078).
3. 3 washings are carried out with PBS-0.1% Tween-20.
4. 50 to 100 μl of dilutions of the sera which must be tested are incubated in each well for 1 to 2 h at room temperature.
5. Wells are washed again with PBS-0.1% Tween-20 and incubated for 1 to 2 h at room temperature with 100 μl at 1 μg/ml of a peroxidase-labeled LO-ME (e.g., LO-ME-3) different from that employed for the coating (both LO-ME-2 and LO-ME-3 do not recognize the same mouse IgE epitopes).
6. Wells are extensively washed and revealed with OPD, as described in the protocol of ELISA test (see Chapter 9). With this simple sandwich ELISA nanograms of mouse IgE can be determined (Figures 1 A,B).

Detection of mouse IgE may also be done using a peroxidase-labeled rat MAb against mouse kappa light chain (LO-MK-1) as a second layer, but the sensitivity of this assay is lower than those obtained with peroxidase-labeled LO-ME.

III. *IN VIVO* STUDIES

A. CELLS

1. Direct Fluorescence Technique

As an example of these rat MAbs anti-mouse IgE, we shall briefly describe our present studies performed on them. Rats infected with some parasites show a high percentage of cells carrying IgE on their surface membrane,[2] the origin of which is highly interesting to know. These IgE molecules may have an intrinsic or cytophilic nature which must be identified. Cytophilic rat IgE can be displaced by incubating the cell in a high concentration of mouse IgE. By this artifice, rat lymphocytes bearing membrane receptors for IgE (rat or mouse) can be detected, using rat MAb anti-mouse IgE. The procedure is the following:

1. 1 to 2 × 10^6 rat mesenteric lymph node cells (MLN) are centrifuged 5 min at 200 *g*.
2. The pellet is incubated for 1 h at room temperature with 100 μl of ascitic fluid containing about 1 mg/ml of mouse IgE.

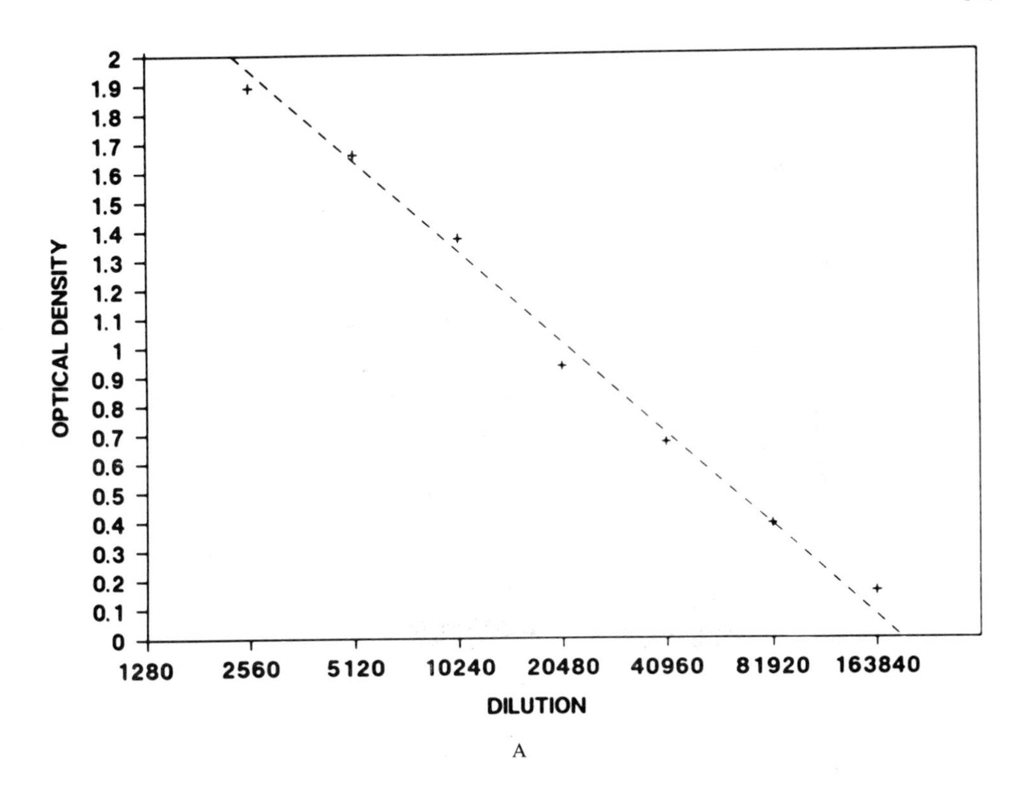

A

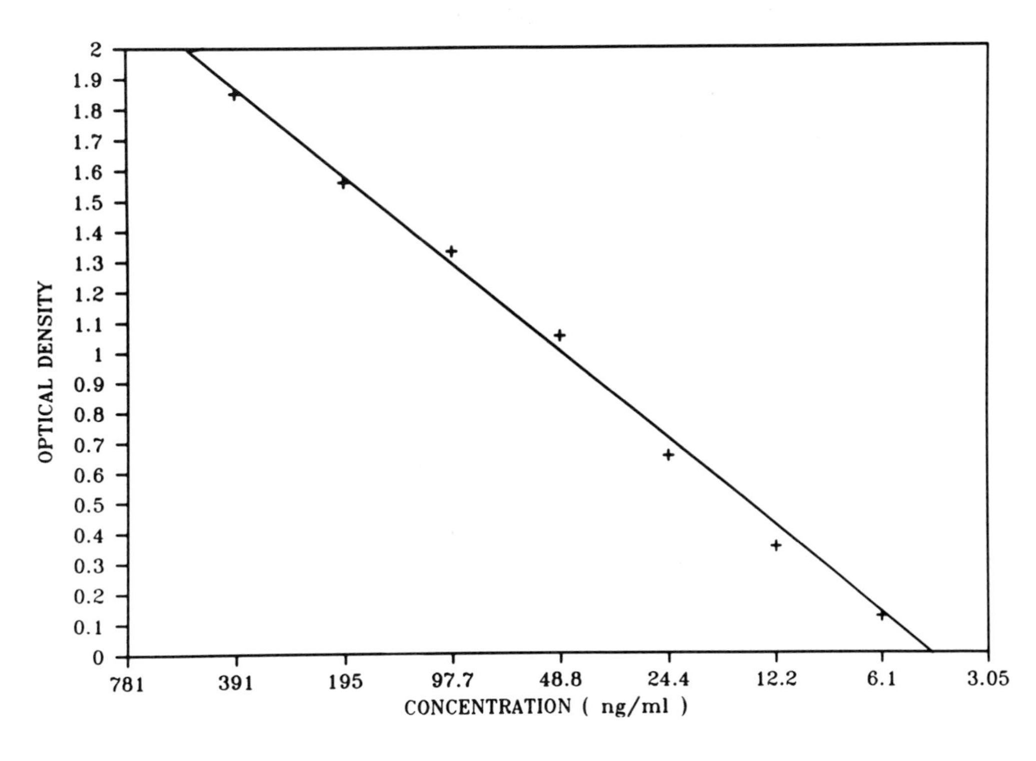

B

FIGURE 1. (A) EIA microplates were coated with 5 μg of LO-ME-2. After plastic saturation, ascitic fluid containing ± 1 mg/ml of mouse IgE (Figure A) or (B) purified mouse IgE were incubated for 2 h at room temperature. Microplates were washed and 0.1 mg of peroxidase-labeled LO-ME-3 was added on each well for 1 h at room temperature.

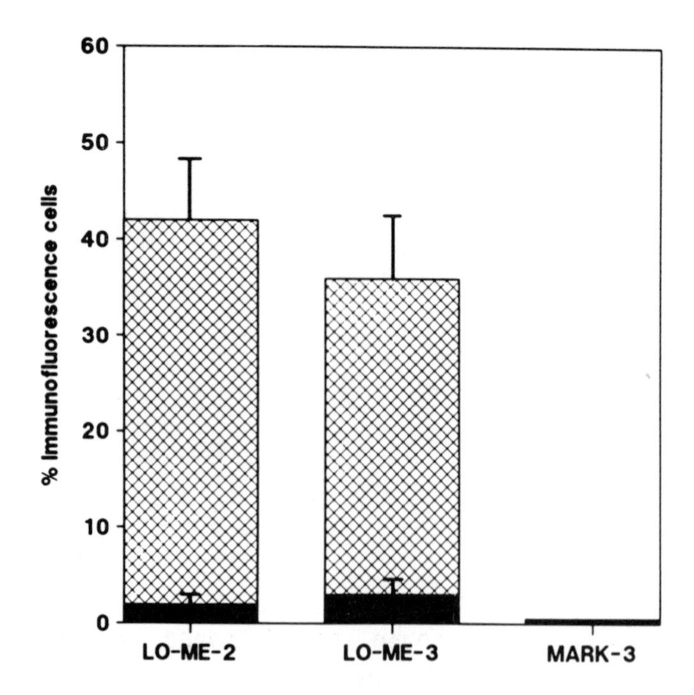

FIGURE 2. Immunofluorescence: ❑ Mesenteric lymph node (MLN) cells of infected BALB/c mice; ■ MLN cells of pathogen-free BALB/c mice. MLN cells were first incubated with different purified MAbs (LO-ME-2 and LO-ME-3) and then subjected to a second incubation with MARK-3-FITC.

3. The cells are washed twice with 0.15 *M* phosphate buffered saline (PBS) 0.1% azide 5% complement inactivated fetal calf serum.
4. Cells are incubated with 50 µl of fluorescein-labeled LO-ME-2 or LO-ME-3 for 30 min at 4°C in presence of complement inactivated rat serum (FCS).
5. Wash twice with 0.15 *M* PBS, 0.1% azide, and 5% complement inactivated FCS.
6. Cells are observed with a fluorescence microscope.

2. Indirect Technique

We have also studied sIgE by indirect immunofluorescence. A suspension of MLN cells of mice infected 14 d before with 400 *Nippostrongylus brasiliensis* (N. b.) larvae was made in 0.15 *M* PBS pH 7.2, 0.1% azide supplemented with 5% complement inactivated FCS. All cell washings were carried out with the same buffer. 50 µl containing 1×10^6 cells were incubated for 30 min at 4°C with 50 µl of different purified MAbs. Cells were washed twice and incubated again with a mouse anti kappa light chain MAb (MARK-3) labeled with fluorescein isothyocyanate. Specific pathogen-free mice were taken as controls because they were expected to have little sIgE B lymphocytes (Figure 2).

B. TISSUE

LO-ME labeled with fluorescein isothyocyanate allows us to detect IgE-secreting plasmocytes of MLN sections of infected BALB/c mice (Figure 3). The number of fluorescent cells decreased when MLN sections were incubated with an excess of unlabeled MAb compared with FITC labeled MAb showing that the labeling was specific against IgE-secreting plasmocytes. No fluorescent cells were seen in MLN sections of pathogen-free BALB/c mice. The procedure is the following:

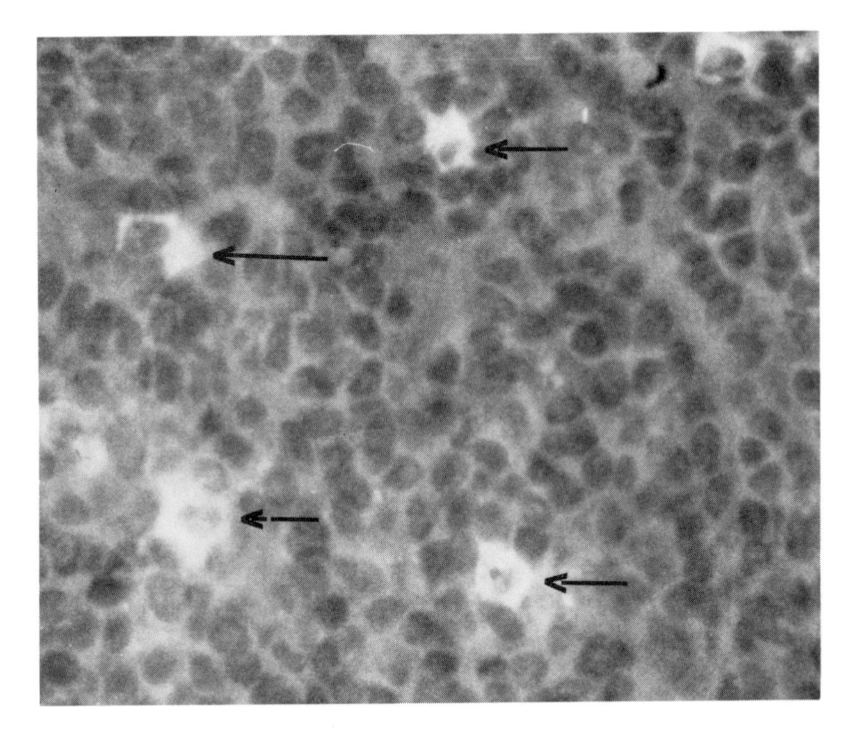

FIGURE 3. MLN sections of a N.b. infected BALB/c mouse. Note the IgE containing plasma cells.

1. 5 μm MLN sections were fixed in ice methanol for 5 min.
2. They were washed in PBS for 10 min.
3. 50 μl of LO-ME-2-FITC at 100 μg/ml were put on the sections for 30 min at room temperature and then washed with PBS.
4. Stained sections were mounted in glycerol/PBS at a concentration of 9:1 in order to observe the immunofluorescence.

REFERENCES

1. **Bazin, H.,** Production of rat monoclonal antibodies with the LOU rat non secreting IR983F myeloma cell line, in *Protides of the Biological Fluids,* Peeters, H., Ed., Pergamon Press, Oxford, 1982, 615.
2. **Manouvriez, P. and Bazin, H.,** *In vivo* kinetics and nature of rat IgE-bearing lymphocytes after IgE stimulation, *J. Immunology,* 133, 3274, 1981.

Chapter 24.IV

PURIFICATION OF MOUSE MONOCLONAL ANTIBODIES BY IMMUNOAFFINITY CHROMATOGRAPHY

H. Bazin, F. Nisol, and J. M. Malache

TABLE OF CONTENTS

I. INTRODUCTION

Many reviews have been published on the purification of mouse polyclonal or myeloma immunoglobulins (Igs)[1-5] or mouse monoclonal antibodies (MAbs).[6] Most of the points developed in Chapter 15 (rat Ig or MAb purification) can also be applied for mouse Ig purification.

They generally consider conventional procedures such as ammonium sulfate precipitation, ion exchange chromatography, gel filtration, and sometimes zone electrophoresis. Modern techniques include the use of a fast protein liquid chromatography apparatus (Pharmacia; MAPS preparative system, BioRad) which is useful and efficient, but expensive to buy and operate and not always necessary. Affinity chromatography on Protein A or Protein G is also very often presented as a universal method of purification. All these techniques are valuable, although they all deal with physicochemical properties which are owned not only by the MAb molecules which must be purified, but also to some extent by many other Ig molecules which are coming from the host (ascitic fluid-serum) or from the media (supernatant).

Some publications have presented the immunoaffinity chromatography as a method of purification of mouse MAb which can be performed on the insolubilized antigen corresponding to the antibody sites of the MAb, at least when its avidity is suitable for adsorption/elution and the antigen is available in large quantities. The technique is particularly convenient when using a hapten such as the dinitrophenol, which can be used in competition in the column to elute the MAb.

This small chapter is devoted to the use of rat MAb anti-mouse Ig as a technique of mouse-MAb purification. This method is based on properties of the MAb to be purified, more or less restricted to a limited population of molecules to which the MAb belongs.

Immunoaffinity purification using polyclonal antibodies anti-mouse Ig is generally difficult to perform because of the great range of avidities which characterize the antibodies produced by all polyclonal immune responses. Moreover, the fact that the antigen (in the present case, a mouse MAb) to be purified, can be bound to the insoluble matrix-rat MAbs by one antibody combining site or more, means that the difficulties to release it, will also vary according to this parameter. Finally, the number of antigenic epitopes which can be recognized, can vary according to the class of the MAb to be purified (IgM versus IgG, for example). All in all, these various difficulties lead us to consider this technology as unadapted to purified Igs and particularly MAbs of any species.

The use of rat MAb anti-mouse Ig is a very efficient technology which is easy to handle and requires very limited equipment.

II. IMMUNOAFFINITY COLUMNS OF RAT MONOCLONAL ANTIBODY ANTI-MOUSE Ig ISOTYPES

A. COUPLING THE RAT MAb ANTI-MOUSE Ig TO AN INSOLUBLE MATRIX

See Chapter 15.

B. CAPACITY OF MOUSE MAb BINDING PER MILLIGRAM OF RAT MAb

This capacity depends on three main parameters: the rat MAb itself, the density of antibody combining sites per gram of gel, and the previous use of the immunoaffinity columns which are denatured by each run. In Table 1, there are some examples of binding capacities of mouse Ig by rat MAb. Clearly, there are differences in binding capacities between the various rat MAbs. However, they seem to be directly linked not so much to the avidities of the MAb (see Chapter 19.I.), but rather more to the experimental conditions of the experiments. The quantities of MAb bound per milliliter of Sepharose-4B are also an important parameter. Preliminary experiments seem to indicate the existence of an optimal ratio of milligrams of MAb to milliliters of gel, so

TABLE 1
Ranges of Binding Capacities of Mouse Ig by Various Rat MAbs, in the Conditions Indicated in Section II.A

MAb coupled to Sepharose-4B (7.5 mg of MAb per gram of gel)	Mouse Ig considered[a]	Range capacity (mg) of mouse Ig bound to 1 mg of rat MAb
LO-MK-2	polyclonal, monoclonal PB1	0.45 to 0.50
LO-MM-9	monoclonal PB1, XH3	0.20 to 0.90
LO-MG1-2	monoclonal MARD-3, 2-E5,B8407	0.45 to 1.00
LO-MG1-13	monoclonal 2-E5,B8407	0.50 to 0.60
LO-MG2a-2	monoclonal A6202	0.55
LO-MG2a-3	monoclonal MOPC-173, A6202	0.40 to 0.55
LO-MG2b-1	monoclonal 307-E9,C1907	0.20 to 0.50
LO-MG2b-2	monoclonal 307-E9,C1907	0.20 to 0.40

[a] Kindly, in part, provided by J. Van Snick, University of Louvain, Experimental Medicine Unit, Brussels, Belgium.

TABLE 2
Capacities of Mouse IgG1 Binding by LO-MG1-2 Rat MAb Coupled to Sepharose-4B

MAb	Quantities of MAb coupled to 1 ml ml of gel	Capacities per ml of gel	Capacities per mg of MAb
LO-MG1-2	7.5 mg	3.9 mg	0.48 mg
LO-MG1-2	3.0 mg	1.7 mg	0.59 mg

Table 2 shows capacities of mouse IgG1 binding by rat-MAb LO-MG1-2 slightly higher at a ratio of 3 mg than at 7.5 mg/ml of gel. Quite clearly, further investigation is necessary before we know the optimal conditions of use of immunoaffinity chromatography of mouse (or rat) MAb purification.

C. CONDITIONS OF ELUTION

The conditions of elution of mouse Ig from immunoaffinity columns made with various rat MAbs are different. Table 3 gives some examples. The LO-MG2b-1 MAb has an interesting property which releases its antigen with only a limited increase of ionic strength, even at a neutral pH. This gives a purification method which is not denaturing. Most other MAbs release their antigen at a pH between 3.8 and 2.8 which is weakly denaturing when maintained for a short period of time.

III. PURIFICATION OF MOUSE MAb BY RAT MAb COUPLED TO AN INSOLUBLE MATRIX

A. PURIFICATION OF MOUSE MAb FROM ASCITIC FLUID OR SERUM

1. Preparation of Serum

See Chapter 15.

TABLE 3
Indicative Conditions of Elution of Mouse Ig from Rat MAb Immunoaffinity Columns

MAb-Sepharose-4B	PBS + 2.5 *M* NaCl, pH 7.2	Glycine-HCl buffer 0.1 *M* at pH 4.5	3.8	2.8
LO-MK-2	0	0	×	××
LO-MM-9	0	0	××	—
LO-MG1-2	0	0	×	××
LO-MG1-15	0	0	××	—
LO-MG2a-2	0	×	××	—
LO-MG2a-3	0	0	××	—
LO-MG2a-7	0	0	×	××
LO-MG2b-1	××	—	—	—
LO-MG2b-2	0	0	××	—
LO-MG-7	0	×	××	—
LO-MG-11	0	×	××	—

Note: 0: no elution; x: partial elution; xx: complete elution.

TABLE 4
Various Possibilities to Purify Mouse Monoclonal Antibodies from Ascitic Fluid or Serum by Affinity Chromatography

Rat monoclonal antibodies coupled to Sepharose 4B	Degree of purity	Approximative contamination	Comments
Anti-kappa type	MAb + host polyclonal Ig of the kappa type	95% of the host polyclonal Ig	Interesting technique if followed by a DEAE chromatography to remove most polyclonal Igs of the host
Anti-lambda type	MAb + host polyclonal Ig of the lambda type	5% of the host polyclonal Ig	Good technique when it can be applied
Anti-isotype	MAb + host polyclonal Ig of the same Ig	Variable from nearly none (IgE class) to about 20% of the host Ig (IgG1 or IgG2a isotypes)	Purity greatly dependent on the serum Ig concentration of the host. Most of the polyclonal Ig from the host can generally be eliminated by DEAE chromatography

2. Technique of Immunoaffinity Purification

In Table 4, there is a brief description of the various possibilities and their individual advantages and disadvantages.

Step 1: "n" ml of serum or ascitic fluid from BALB/c mice bearing (or not) a histocompatible hybridoma (or plasmacytoma) is applied at a rate of about 2 ml/min, at room temperature to a column of Sepharose-4B (Pharmacia, Sweden) on which "y" mg of rat MAb anti-mouse Ig has been immobilized.

Step 2: The column is washed with 10 to 12 "n" ml of phosphate buffered saline (PBS), then with 10 "n" ml of PBS containing 2.5 *M* NaCl, and then again with about 10 "n" ml of PBS.

TABLE 5
Examples of Purification of Mouse Polyclonal Ig by Immunoaffinity Chromatography Using Sepharose-4B as Insoluble Matrix and Rat MAb Anti-Mouse Heavy or Light Chains of Ig

Rat MAb coupled to Sepharose-4B	Number of ml of ascitic fluid applied on the column	Number of g of Sepharose-4B	Quantity of MAb (mg) coupled to the matrix	Purified polyclonal mouse Ig
LO-MK-2	15	2	35	12 mg of polyclonal Ig carrying kappa light chains
LO-MM-9	45	1.5	30	3.7 mg of polyclonal IgM
LO-MG1-2	30	1	25	16 mg of polyclonal IgGl
LO-MG2a-3	25	1	25	9.1 mg of polyclonal IgG2a
LO-MG2b-2	25	2	50	7.5 mg of polyclonal IgG2b

Note: These data do not represent the maximum capacities of the various affinity columns, but just examples chosen at random in the laboratory books.

TABLE 6
Example of Purifications of Mouse MAb or Myeloma Proteins of Various Isotypes by Immunoaffinity Chromatography Using Sepharose-4B as Insoluble Matrix and Rat MAb Anti-Mouse Heavy or Light Chains of Ig

Rat MAb coupled to Sepharose-4B	Number of ml of normal fluid containing mouse MAb	Number of g of Sepharose-4B	Quantity of MAb (mg) coupled to the matrix	Purified MAb + polyclonal Ig
LO-MK-2	7.5 ml of MOPC 17 (IgG2a plasmacytoma)	2	35	12.4 mg of MOPC 173 + polyclonal kappa Ig
LO-MM-9	5 ml of PB-1 (IgM MAb)	1.5	30	7.8 mg of PB-1 + polyclonal kappa Ig
LO-MM-9	3 ml of MARE-IR2[a] at 17 mg/ml	1	40	42 mg of MARE-IR2[a]+ polyclonal IgM (percentage of specific MAb recovery: 82%)
LO-MG1-2	10 ml of MARD-3 (IgG1 MAb)	1	25	20 mg of MARD-1 + polyclonal IgGl
LO-MG2a-3	5 ml of MOPC 173 (IgG2a plasmacytoma)	1	25	17.2 mg of MOPC 173 + polyclonal IgG2a
LO-MG2b-2	7 ml of C1907B3 (IgG2b MAb)	2	50	21.9 mg of C1907B3 + polyclonal IgG2b

Note: These data do not represent the maximum capacities of the columns, but are just examples taken at random in the laboratory books.

[a] Titration of the MARE-IR2 itself by its binding on IR2 rat myeloma immunoglobulins.

Step 3: Mouse Ig is then eluted by decreasing the pH of the column with a glycin-HCl 0.1 *M* + 0.15 *M* NaCl buffer at a pH, depending on the rat MAb, between 2.8 and 4.5 or even 7.2 with just an increased NaCl molarity (see Table 3). The purified MAb is neutralized as soon as possible when necessary. Tables 5 and 6 give some examples of mouse Ig purifications.

TABLE 7
Example of Use of Immunoaffinity Columns Made of Rat MAbs Anti-Mouse Immunoglobulins Isotypes

Antibody coupled to Sepharose-4B	Number of runs already done	Still in use
LO-MM-9	54	yes
LO-MG1-2	29	no[a]
LO-MG1-2	36	yes
LO-MG2a-3	49	yes
LO-MG2b-2	31	yes
LO-MK-2	85	yes

[a] Prematurely denaturated by "dirty" ascitic fluid.

TABLE 8
Examples of Immunopurification of Mouse MABs from *In Vitro* Supernatants Concentrated Four Times by Ultrafiltration

Isotypes	Culture medium	Concentration of mouse MAb in the supernatants[a] (mg/ml) (tested by ELISA)	Percentage of recovery after a preconcentration followed by the purification and a final concentration at about 5 mg/ml
IgM	Medium 1[b]	0.044	44.4
	Medium 2	0.029	62.5
IgG3	Medium 1	0.039	57.0
	Medium 2	0.049	83.0

[a] Mouse MAbs prepared by P. Nabet and M. Maugras (Unité INSERM U284, Vandoeuvre-les-Nancy, France).
[b] Medium 1 was supplemented in fetal calf serum, and medium 2 with lactoserum.

3. Longevity of the Immunoaffinity Columns Using Rat MAbs Anti-Mouse Immunoglobulin Isotype

Table 7 gives examples of use of the rat MAbs anti-mouse Ig isotypes. The stabilities of the various rat MAbs seem to be more or less equivalent to those of mouse MAbs.

B. PURIFICATION OF MOUSE MAb FROM *IN VITRO* SUPERNATANT

When the same immunoaffinity columns as those given in Section III. A. are used, it is possible to purify mouse MAbs from *in vitro* supernatant. This technique allows purification of mouse MAbs from calf or horse Igs added to the culture media with fetal calf or horse sera. Table 8 gives examples of purification for two different culture media and for two mouse MAbs, one being an IgM and the other an IgG3, which are both difficult to purify by conventional techniques. IgM cannot bind to Protein A and IgG3 can be easily denatured at the low ionic power by DEAE chromatography. The results show a recovery coefficient of 44 to 83%, which is interesting in view of the concentration of MAb in the supernatant, even after a concentration step which is always a major source of protein loss.

IV. CONCLUSIONS

Purification of mouse MAbs by immunoaffinity chromatography using rat MAb is perfectly possible either from ascitic-fluid, sera or from *in vitro* culture supernatants. The recovery percentage is interesting and perfectly comparable to other methods. The purity is good and is not particularly impaired by immunoglobulin brought by the horse or fetal calf serum added to the culture medium. Economically speaking, the immunoaffinity method of mouse MAb purification is not expensive and can certainly compete with most the other techniques which have not produced an identical purity and are not so quick.

REFERENCES

1. **Fahey, J. L., Wunderlich, J., and Mishell, R.,** The immunoglobulins of mice. I. Four major classes of immunoglobulins: 7Sgamma2-, 7Sgamma1-, gamma1A (Beta2A) and 18Sgamma1M-globulins, *J. Exp. Med.,* 120, 223, 1964.
2. **Fahey, J. L., Wunderlich, J., and Mishell, R.,** The immunoglobulins of mice. II. Two subclasses of mouse 7Sgamma2-globulins: gamma2a and gamma2b-globulins, *J. Exp. Med.,* 120, 243, 1964.
3. **Bazin, H. and Malet, F.,** Metabolism of different immunoglobulin classes in irradiated mice. I. Catabolism, *Immunology,* 17, 345, 1969.
4. **Potter, M.,** Immunoglobulin producing tumors and myeloma proteins of mice, *Phys. Rev.,* 52, 631, 1972.
5. **Smith-Gill, S. J., Finkelman, F. D., and Potter, M.,** Plasmacytomas and murine immunoglobulins, *Methods Enzymol.,* 116, 121, 1985.
6. **Goding, J. W.,** *Monclonal Antibodies: Principles and Practice,* Academic Press, London, 1983.

Chapter 24.V

SCREENING OF MOUSE MONOCLONAL ANTIBODIES

F. Ackermans, C. Vincenzotto, and H. Bazin

When the hybridomas are growing, a few days after the fusion, the choice of an appropriate screening assay is essential to quickly and easily identify the positive cultures. The use of rat monoclonal antibodies (MAbs) directed against heavy or light chains of mouse immunoglobulins (Igs) is one of the best and easiest methods to screen mouse fusion. We can use Enzyme Linked Immuno Sorbent Assay (ELISA), radioimmunoassay, or immunofluorescence techniques (Figure 1).

The antigen which is used for immunization is coated on the bottom of the specific microtiter plates at a predetermined concentration as described at Chapter 9. Any remaining protein-binding sites of the plastic are saturated by a large excess of irrelevant protein, which is usually bovine serum albumin (BSA) or dried skim-milk (3%). The supernatant of hybridoma cultures is added for 2 h at 37°C and any unbound material is washed out. Rat MAbs against mouse kappa light chain (LO-MK-1 and LO-MK-2) directly labeled with peroxidase or alcaline phosphatase (see Chapter 19) for ELISA or with I^{125} for RIA (see Chapter 19) are added for 2 h at 37°C. The rest of the screening techniques are the same as described in Chapter 9. Owing to the use of MAbs directly labeled with the enzyme I^{125} or fluorescein, the level of the background is very low and the screening very specific.

It is also possible to select the isotype(s) of the mouse MAb that must be selected, for example kappa, lambda, mu, total gamma, or gamma1, gamma2a, gamma2b, gamma3, alpha, or even epsilon or delta. The screening can be made in the opposite direction: the plates were coated with rat MAbs against the selected isotype and revealed with labeled antigen.

An example is given here with a fusion to obtain anti-idiotype MAbs. Rat MAbs coupled to cellulose were injected twice at 14 d intervals on a BALB/c mouse (3-months-old). Fusion with PAIO myeloma cells was performed 4 months later after an i.v. injection 4 d before fusion. For screening the production of specific antibodies, we used an ELISA technique. Rat MAbs utilized for immunization were coated on NUNC microtiter plates. Supernatant of mouse hybridomas was added (2 h at 37°C) and after washing, LO-MK-1-peroxidase (1/500) was platted out into the microtiter plates. On the positive hybridomas, other immunological tests were carried out to determine whether the mouse MAbs were anti-idiotypes and whether they had the internal image of the first antigen.

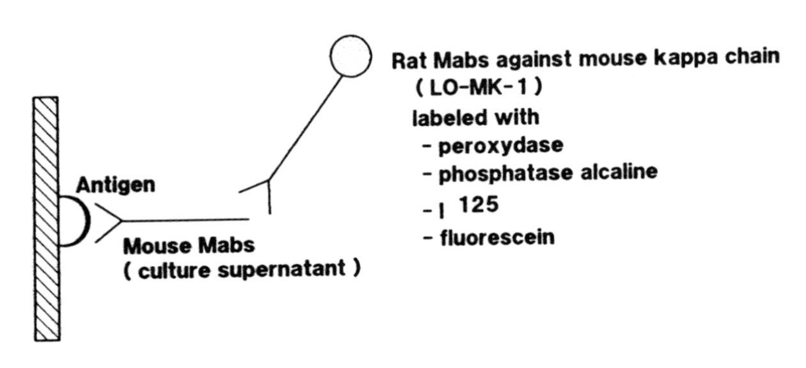

FIGURE 1. Diagram of the principles of the screening of mouse MAbs for a given antigen.

Chapter 24.VI

USE OF RAT MONOCLONAL ANTIBODIES IN IMMUNOHISTOCHEMISTRY INVESTIGATIONS

B. Platteau and H. Bazin

TABLE OF CONTENTS

I. INTRODUCTION

Rat monoclonal antibodies (MAbs) may be used as immunospecific staining reagents in direct and indirect peroxidase methods. Immunoperoxidase procedures allow detection, visualization, and localization of antigens in tissues of many species, particularly of rats, mice, or humans.

The direct peroxidase method is the simplest way to localize a certain antigen by using a MAb specifically directed against it (Figure 1). The specific antibody is linked to an enzyme-like peroxidase and applied to the tissue, it will react with the antigen. To localize the reaction, we add 3,3′-diaminobenzidine-tetra-hydrochloride (DAB) which will procedure a brown color. The direct immunoperoxidase method can be performed very quickly with low non-specific reactions if there is a specific conjugated MAb for every antigen to be localized.

The indirect peroxidase method on frozen sections is a two-step procedure (Figure 2). The tissue is first incubated with an unconjugated antibody specific for the antigen. A second peroxidase conjugated antibody capable of binding the primary antibody is then applied. The reaction is localized by adding DAB.

The indirect peroxidase method takes approximately twice as long to complete as the direct method, but one conjugated secondary antibody can be used for a variety of primary antibodies made in the same animal species.

II. MATERIALS AND METHOD

A. ANTIBODIES

Rat MAbs highly specific for rat, mouse, or human immunoglobulins (Igs) are used unconjugated or chemically linked to peroxidase by the periodate method described by Nakane and Kawai.[1] Each antibody is optimally diluted in 0.05 *M* Tris HCl buffered saline at pH 7.6 (TBS) with 0.2% bovine serum albumine (BSA).

B. IMMUNOPEROXIDASE STAINING METHOD

As many surface antigens are destroyed by fixation, frozen sections are used mostly for direct and indirect staining methods. However, some antigens are well preserved after fixation, for example, in Carnoy's fluid and remains recognizable by the antibody as well as in frozen sections.

1. Frozen Tissue

Fresh slices of tissues (generally spleen or lymph nodes) are impregnated in tissue-Tek II (embedding medium for frozen tissue specimens, Division Miles Laboratories, Naperville, Illinois, U.S.), snap frozen in liquid nitrogen, and stored at –70°C.

The frozen tissue block is mounted to the metal chuck with fresh tissue-Tek II and cut with a cryo microtome (Bright 5030, Bright Instrument Company Ltd., England) at –20°C.

Sections of 4-μm thickness are picked up on formol-gelatin coated slides, thoroughly dried at room temperature, fixed in acetone at 4°C for 20 min, and air-dried again for at least 30 min. Before beginning the first incubation with the antibody, sections are washed in TBS. All incubations are carried out at room temperature in a humidity chamber.

When using the direct immunoperoxidase technique, the slides are incubated for 60 min with optimally diluted peroxidase conjugated MAb. By the indirect immunoperoxidase technique, the sections are incubated with the first antibody for 45 min, thoroughly washed in TBS and incubated for 45 min with the second peroxidase-coupled antibody solution.

After the incubation with the peroxidase conjugated antibody, the slides are washed in TBS for 15 min (3 changes). Freshly prepared 0.05% 3,3′-diaminobenzidine is applied on the section

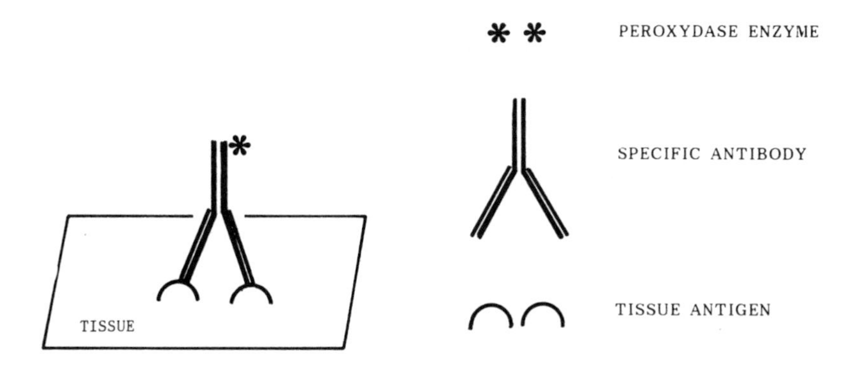

FIGURE 1. Direct peroxidase method.

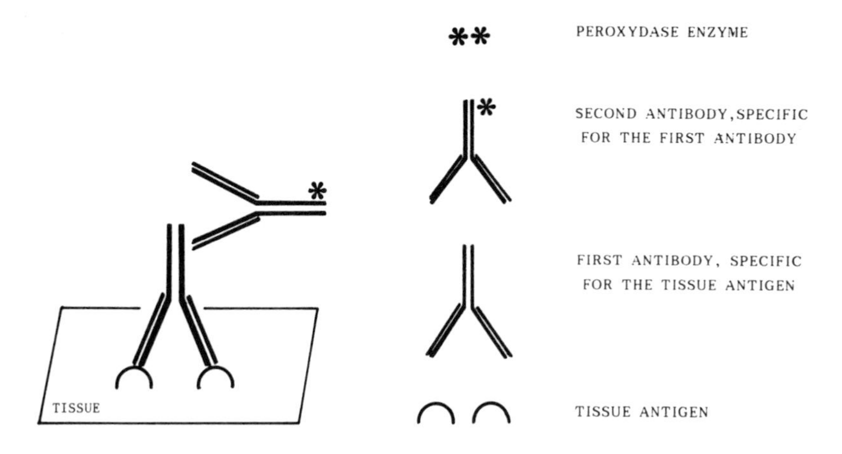

FIGURE 2. Indirect peroxidase method.

to give a colored end product. The slides are then rinsed in water, counterstained in Mayer's haemalun for 2 min, followed by bluing in tap water, dehydration in ethanol, clearing in xylene, and mounting in "EUKITT" (O. Kindler, Freiburg) medium.

2. Fixed Tissue

Small blocks of freshly removed tissue are placed in Carnoy's fluid and fixed for 18 h at 4°C, then dehydrated in ethanol for 4 h at 4°C, cleared in xylene for 18 h at 4°C, and impregnated in paraplast tissue embedding medium (Monoject Scientific Inc., Ireland) for 2 h at 56°C (2 changes) before embedding. Fixed sections are cut 3 to 4 μm thick and collected on thoroughly cleaned microsocope slides. The slides are deparaffinized in fresh xylene and rehydrated through ethanol to TBS at pH 7.6. The indirect immunoperoxidase staining procedure is then applied. The sections are subjected to the following steps:

1. Blocking of endogenous staining in 1% H_2O_2 in methanol for 20 min at room temperature
2. Several changes (stirred) of Tris-HCl buffered saline pH 7.6
3. Applying the first antibody at appropriate dilution for an overnight incubation (18 to 20 h) in a humidity chamber at 4°C
4. Washing in TBS (3 changes) at room temperature
5. Second peroxidase coupled MAb optimally diluted for 45 min at room temperature
6. Tris-HCl-buffered saline (stirred) for 15 min (3 changes)

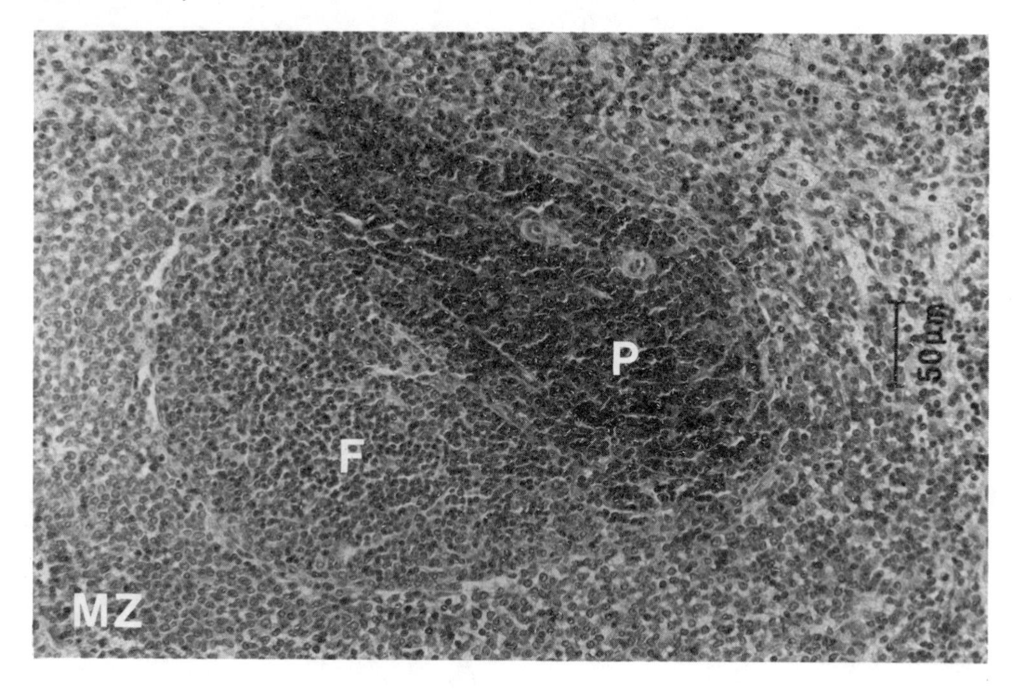

FIGURE 3. Normal rat spleen, Carnoy fixed. Paraffin section stained with the W3/13 MAb and detected, using LO-MK-1 peroxidase-conjugated second antibody showing T cells accumulated in the PALS (P) around the central arteriol and scattered in the follicle (F) and marginal zone (MZ).

7. Freshly made 0.05% 3,3′-diamino-benzidine tetrahydrochloride (DAB) peroxide solution (5 mg DAB in 10 ml Tris-HCl saline buffer plus 50 µl of 3% hydrogen peroxide)
8. Rinse in water, counterstain in Mayer's haemalun for 2 min, followed by bluing, dehydration, cleaning, and mounting in "EUKITT" medium.

III. MONOCLONAL ANTIBODIES COMMONLY USED FOR CHARACTERIZATION AND LOCALIZATION OF LYMPHOID CELLS

A. IN RAT LYMPHOID TISSUES

1. Detection of T Cells

For this purpose, the indirect peroxidase staining is used. Frozen or paraffin sections are incubated with mouse MAbs against T cell antigens; W3/13,[2] W3/25,[2] OX8,[3] and OX19,[4],which were kindly provided by Dr. A. Williams.

As a second antibody, a peroxidase-conjugated rat MAb (LO-MK-1) against the mouse kappa light chain (Figure 3) is used.

2. Detection of B Cells

When mouse MAbs against rat IgK-1b (Table 1) are used, the direct or indirect peroxidase staining can be applied only to frozen tissue sections; staining on paraffin sections has not succeeded until now. For indirect peroxidase staining, LO-MK-1 peroxidase conjugate is used as second antibody (Figure 4).

When rat MAb against rat kappa b light chain (LO-RK-1b-1) is used, only the direct staining on frozen sections gives good results (see Figure 1 in Chapter 25.II.).

On paraffin sections, IgD and IgM positive B cells can be demonstrated by using rabbit anti-IgM or -IgD specific antibodies as first antibody and as second antibody, a rat MAb against rabbit IgG conjugated to peroxidase (Figure 5).

TABLE 1
Mouse MAbs Against Rat Igs

Mouse monoclonal antibody to	Hybridoma
Rat IgM	MARM-4
Rat IgE	MARE-1
Rat IgA	MARA-2
Rat IgD	MARD-3
Rat IgG1	MARG1-1
Rat IgG2a	MARG2a-8
Rat IgG2b	MARG2b-8
Rat IgG2c	MARG2c-2
Rat kappa light chain	MARK-1
Rat lambda light chain	MARL-5
Rat kappa-1a allotype	MARK-3

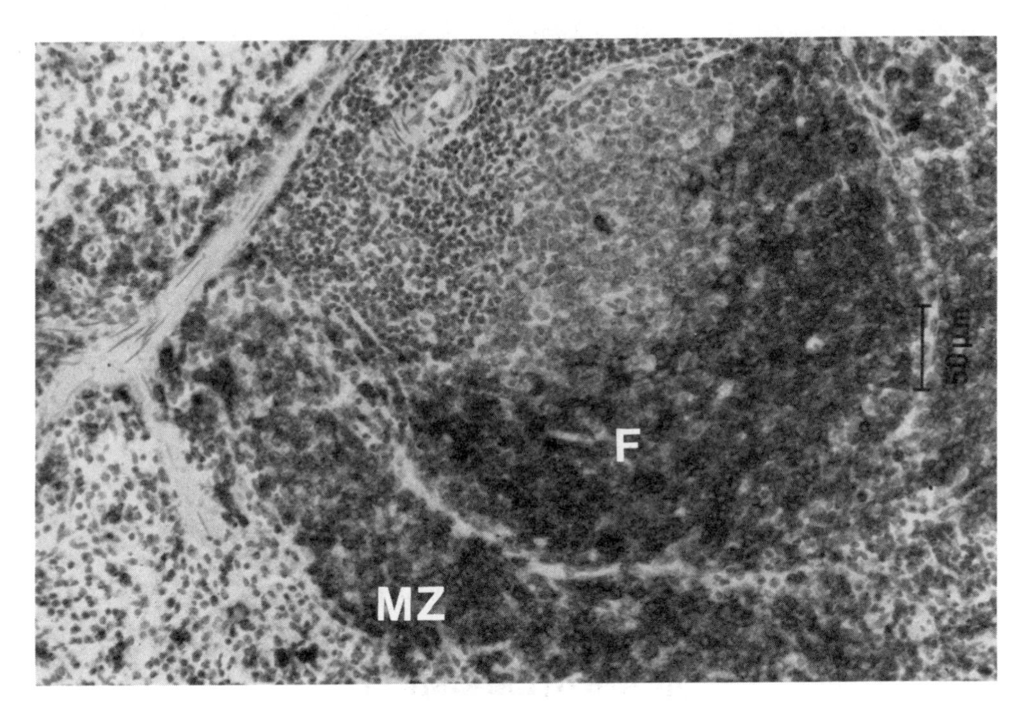

FIGURE 4. Normal rat spleen. Cryostat section direct stained for IgM with the MARM-4 peroxidase conjugate shows B cells with ring-like membrane staining in the follicle (F) and marginal zone (MZ); few cells show cytoplasmic IgM.

B. IN MOUSE LYMPHOID TISSUES

Rat MAb against mouse IgM (LO-MM-5) and mouse kappa light chain (LO-MK-1, LO-MK-2), when conjugated to peroxidase can be used on cryo-frozen tissue sections of the various lymphoid organs (spleen, lymph nodes, Peyer's patches) to determine the distribution of B cells with surface markers and cells with cytoplasmic Igs.

C. IN HUMAN TISSUES

Rat MAb against mouse kappa light chain (LO-MK-1) can be used as second peroxidase-

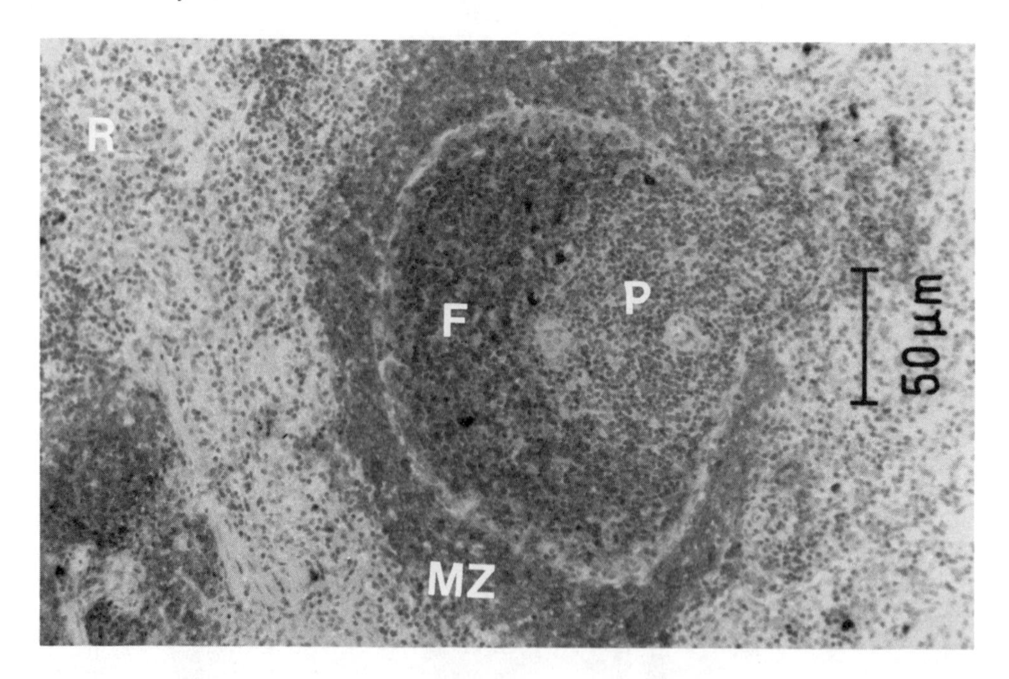

FIGURE 5. Normal rat spleen, Carnoy fixed. Paraffin section stained for IgM with a rabbit serum anti-IgM as first antibody and a rat anti-rabbit IgG MAb, peroxidase conjugate as second antibody. The follicle (F) and marginal zone (MZ) show positive B cells for surface IgM. In the red pulp (R) and the outer periarteriolar lymphocytic sheath (P) cells with cytoplasmic staining can be recognized.

conjugated antibody to detect mouse MAb to human antigens. An example is illustrated in Figure 6. The staining reaction is achieved in a 4-µm thick paraffin section of a human lymph node biopsy from a patient with Hodgkin's disease after fixation in Bouin's fixative. The section is incubated overnight with Leu-M1 (Becton Dickinson), a mouse MAb detecting CD15, normally present mainly on granulocytes and on Reed-Sternberg cells in Hodgkin's disease. As a second antibody, the LO-MK-1 peroxidase conjugate is used.

IV. CONCLUSION

The use of LO-MK-1 peroxidase conjugate possesses the advantage of employing a single universal labeling reagent as a second antibody for most, if not all the different mouse specific MAbs against human or rat antigens.

ACKNOWLEDGMENTS

We thank H. Noel, J. Rahier, P. Camby, and L. Wendrickx from the Department of Pathology, School of Medicine, University of Louvain, in Brussels, for their contribution to part of this work.

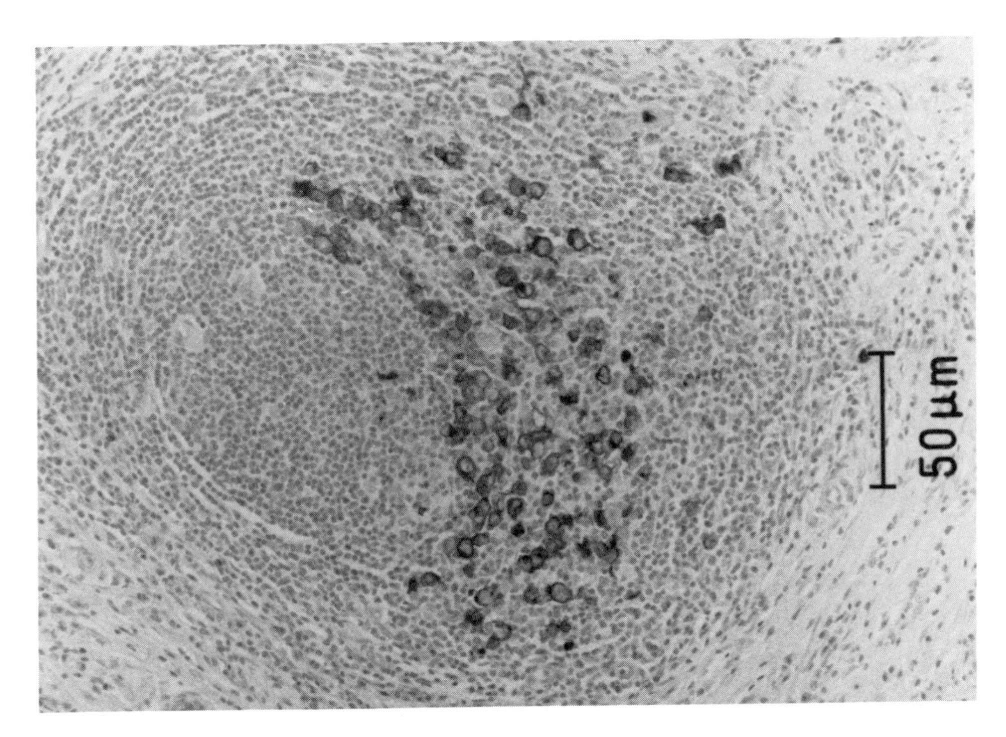

FIGURE 6. Human lymph node, Bouin fixed. Paraffin section stained with Leu-M1 and detected, using LO-MK-1 peroxidase-conjugated second antibody, showing abnormal cells of the Reed-Sternberg cell type in the periphery of the lymphoid nodule.

REFERENCES

1. **Nakane, P. K. and Kawaoi, A.,** Peroxidase-labeled antibody a new method of conjugation, *J. Histochem. Cytochem.,* 22, 1084, 1974.
2. **Williams, A. F., Galfré, G., and Milstein, C.,** Analysis of cell surfaces by xenogenic myeloma-hybrid antibodies: differentiation antigens of rat lymphocytes, *Cell,* 12, 663, 1977.
3. **Brideau, R. J., Carter, P. B., McMaster, W. R., Mason, D. W., and Williams, A. F.,** Two subsets of rat T lymphocytes defined with monoclonal antibodies, *Eur. J. Immunol.,* 10, 609, 1980.
4. **Dallman, M. J., Mason, D. W., and Webb, M.,** Role of thymus-derived and thymus-independent cells in murine skin allograft rejection, *Eur. J. Immunol.,* 12, 511, 1982.

Chapter 24.VII

USE OF RAT-RAT MONOCLONAL ANTIBODIES LABELED WITH FLUOROCHROME

P. Manouvriez and A. Cornet

I. INDIRECT FLUORESCENCE LABELING WITH MAbs

Mouse MAbs are now widely used for the immunodetection of cellular antigens. They are mostly revealed by indirect fluorescence using fluorochrome-labeled polyclonal goat or rabbit antibodies. Good quality heterosera recognizing all Ig classes and subclasses are difficult to produce (see Chapter 10). Differences in quality between batches may also be encountered. These problems should be avoided by the use of MAbs as second layer reagents. Some attempts have been made with fluorescein-labeled (FITC) rat-rat MAbs anti-mouse Igs. Figure 1 shows the compared results obtained with a FITC-labeled commercial goat anti-mouse antiserum and FITC-labeled rat anti-mouse Ig, as observed by analysis on an EPICS flow cytofluorometer (Coulter). A mixture of different MAbs will in the future give an excellent and reproducible labeled second layer reagent to detect any mouse MAb.

II. DIRECT FLUORESCENCE LABELING WITH MAbs

Most mouse MAbs are generally used in ascitic fluid or culture supernatant. They are then revealed by indirect fluorescence with a labeled-polyclonal anti-mouse Ig.

One of the reasons for the use of the indirect labeling technique is that mouse MAbs are expensive to purify in great quantities before conjugation with a fluorochrome. Rat MAbs are easy to purify in one step as described in Chapter 15 and can thus be easily labeled with FITC or another fluorochrome. The main risk with direct immunofluorescence is to have a lower signal and thus a poor discrimination between labeled- and unlabeled-cells. With most tested rat MAbs the fluorescence signal was strong enough for fluorescence microscopic or flow cytofluorometric application . Figure 2 shows an example of human HLA class II DRa positive cells detection using an EPICS flow cytofluorometer (Coulter) for the cell analysis.

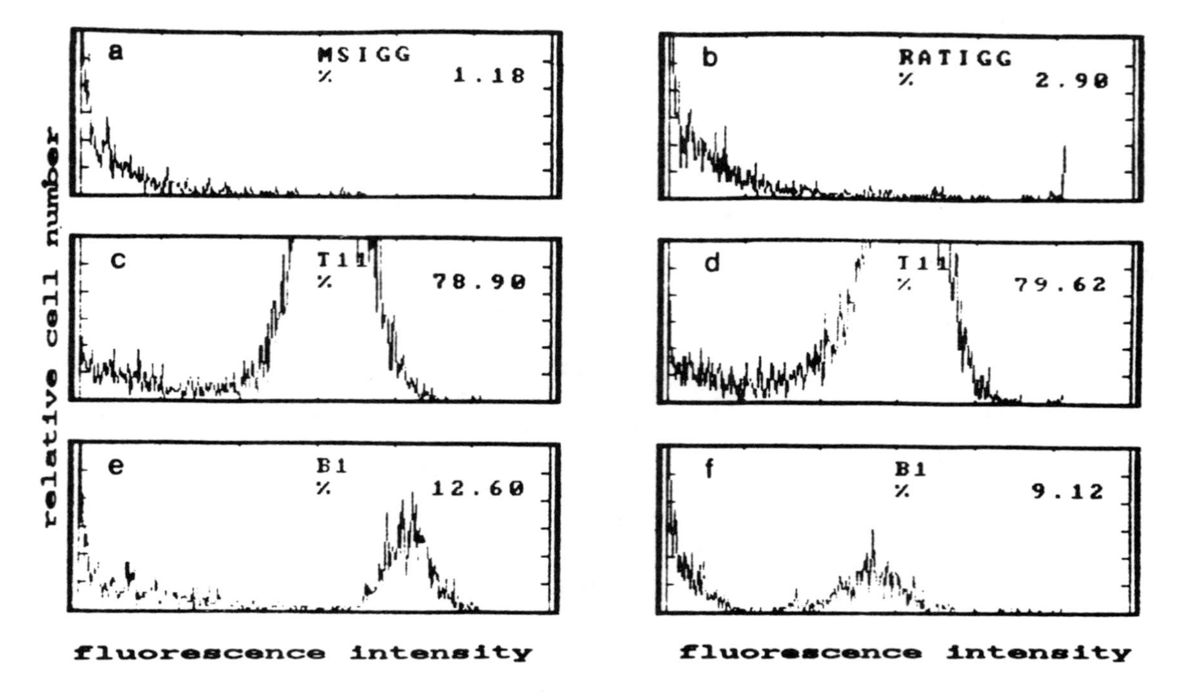

FIGURE 1. Indirect immunofluorescence labeling of human peripheral blood cells, using mouse MAbs. Comparison between the use of FITC-labeled polyclonal second layer antibodies and FITC-labeled rat MAbs anti-mouse Ig. Controls: (a) Normal mouse IgG plus FITC-labeled goat anti-mouse IgG (GAM-FITC); (b) Normal mouse IgG plus FITC-labeled normal rat Ig. Labeling of CD2 positive T lymphocytes with OKT11 mouse MAb (Coulter), using (c) GAM-FITC, and (d) LO-MG1-2-FITC (FITC-labeled LOU rat MAb anti-mouse IgG1). Labeling of CD20 positive B lymphocytes with B1 mouse MAb (Coulter), using (e) GAM-FITC, and (f) LO-MG2a-2-FITC (FITC-labeled LOU rat MAb anti-mouse IgG2a).

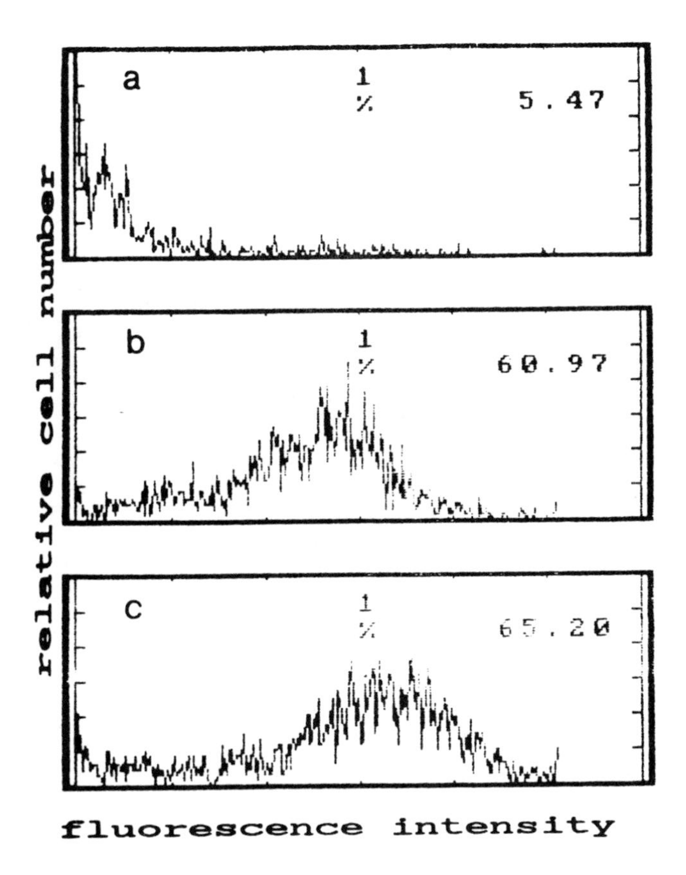

FIGURE 2. Labeling of HLA class II DRa positive cells. Comparison between indirect labeling with mouse MAb and direct labeling with rat MAb. (a) Control normal mouse Ig plus FITC-labeled goat anti-mouse Ig (GAM-FITC). (b) Indirect labeling with OKIa (Ortho-clone) and GAM-FITC. (c) Direct labeling with FITC-labeled LO-DRa (rat MAb anti-human HLA class II DRa). This rat MAb inhibits the binding of OKIa.

Chapter 25.I

USE OF RAT MONOCLONAL ANTI-RAT ALLOTYPE ANTIBODIES IN IMMUNOHISTOPEROXIDASE TECHNIQUE AND ELISA

B. Platteau and H. Bazin

TABLE OF CONTENTS

I. INTRODUCTION

The obtainment of a mouse anti-rat IgK-1a monoclonal antibody (MAb) (MARK-3)[1] and anti-rat IgK1-b rat MAbs (LO-RK-1b-1) (see Chapter 19. II) can contribute greatly to the study of the survival and localization of cells of B lymphocyte lineage in lymphoid organs (see Chapter 1). For that purpose, B lymphocytes or their stem cells of a kappa allotype were transferred into sublethally irradiated rat recipients of the other kappa allotype (manuscript in preparation). The same techniques could be used for immunoglobulin (Ig) molecules or specific antibodies of the various Ig (sub)classes.

II. PURIFICATION OF RAT MONOCLONAL ANTIBODIES FROM ASCITIC FLUID, SERUM, OR CULTURE SUPERNATANT

See Chapter 15.

III. MATERIALS

A. ANIMALS AND CELL TRANSFER

The LOU/C strain of rats bearing the IgK1-a allotype was used as recipient rats and they received an X-irradiation dose of 5.5 Gy. LOU/C.IgK1b (OKA) rats are from a congenic strain bearing the IgK-1b (OKA) allotype on a LOU/C background and have been derived at our laboratory.[2] Spleen cells from LOU/C.IgK-1b (OKA) rats (donor) were injected intravenously into LOU/C rats 4 h after irradiation.

B. ANTISERA

Staining for Ig-light chain carrying the IgK-1a or the IgK-1b allotypes in rat lymphoid tissue can be easily performed by using peroxidase- conjugated anti-rat IgK-1a mouse MAb (MARK-3) or anti-rat IgK-1b rat MAb (LO-RK-1b-1).

IV. DETECTION OF SPECIFIC ANTIBODIES BY ANTI-KAPPA ALLOTYPE MAb

A. THE DIRECT IMMUNOPEROXIDASE METHOD

In order to detect the cells from the donor or from the host, the direct immunoperoxidase method is applied on frozen sections as it has been described in Chapter 24. VI. Figure 1 illustrates the use of the peroxidase-conjugated anti-rat IgK-1b rat MAb (LO-RK-1b-1) in order to visualize the survival of IgK-1b spleen donor cells in a 6-Gy-irradiated LOU/C.IgK-1a recipient rat 5 d after transfer.

B. THE ELISA METHOD

This method has been applied to the detection of IgK-1a and IgK-1b serum Ig levels. The indirect enzyme-linked immunosorbent assay used here involves a four-step procedure. As an example, we are giving our method for the IgM detection.

In our method microtiter plates are coated with 100 µl of purified anti-rat mouse MAb (MARM-4, 10 µg/ml in 0.1 *M* carbonate buffer pH 9.6). After incubation overnight at 4°C the plates are washed on phosphate buffered saline pH 7.2 (PBS) and saturated with 200 µl of PBS containing 10% normal goat serum (NGS). After washing 3 times with PBS, 100 µl of the test serum samples, containing IgK-1a and/or IgK-1b IgM, are incubated for 1 h at 37°C. The plates are washed 3 times and 100 µl of anti-IgK-1a (MARK-3) or anti-IgK-1b (LO-RK-1b) allotypic MAb conjugated to peroxidase is applied to determine IgK-1a and IgK-1b IgM serum concentrations, respectively. After an incubation of 1 h at 37°C, the plates are washed and the

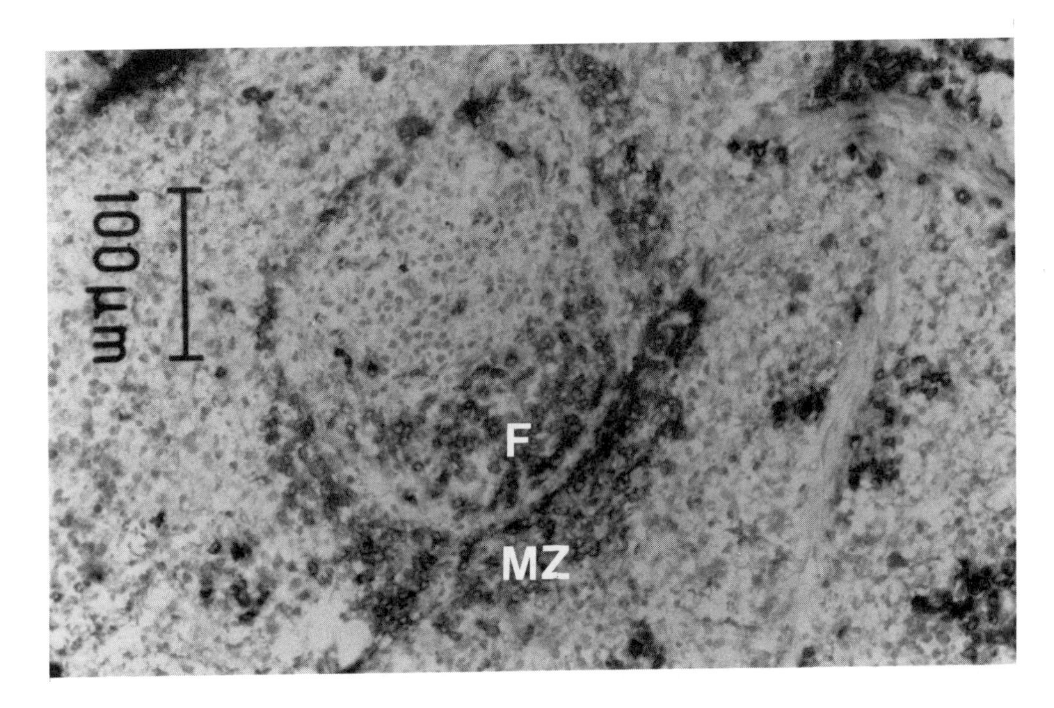

FIGURE 1. Direct immunoperoxidase staining on a frozen section of IgK-1b cells in the spleen taken from a LOU/C·IgK-1a rat, 5 d after a 6-Gy irradiation followed by a transfer of 1×10^8 LOU/C.IgK-1b spleen cells. F = follicle (recirculating B cells); MZ = marginal zone (sessile B cells).

substrate solution containing 5 µg O-phenylene diamine in 10 ml citrate phosphate buffer 0.1 *M* pH 5.5 and 0.06% ureumperoxide is added. The enzyme concentration is measured by determining the rate of color development, using a photometric reading of the absorbance at 492 nm. The IgM serum level for the IgK-1a and IgK-1b light chains is calculated by comparison to a reference serum from LOU/C.IgK-1a rats and LOU/C.IgK-1b (OKA) rats, respectively, with known IgM concentration and tested in several dilutions on the same plate as the test samples.

In Figure 2, there is an example of such a spleen cell transfer from LOU/C.IgK-1a into an irradiated LOU/C.IgK-1b showing the appearance of IgM produced by the transferred cells and correlatively, the disappearance of the host IgM production.

Evidently, the same technique could be used for the other rat Ig classes or subclasses, using mouse MAbs anti-rat Ig classes and subclasses such as those developed in our laboratory.

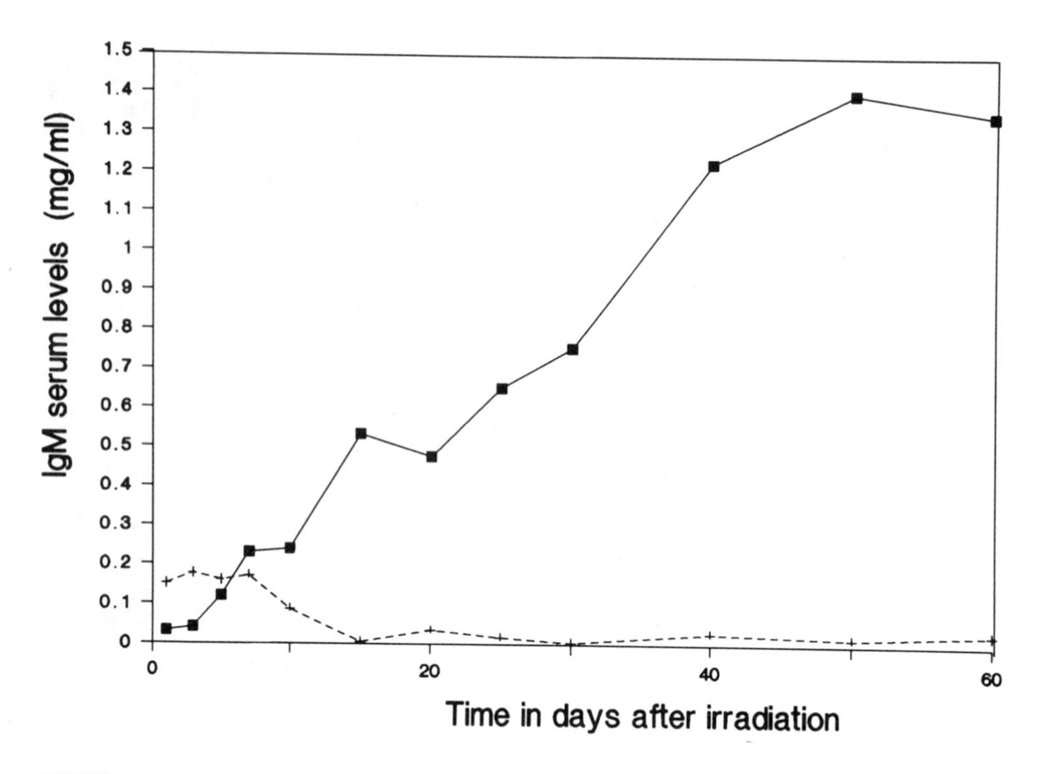

FIGURE 2. IgK-1a and IgK-1b IgM serum concentrations produced by transfers of 1.5×10^7 IgK-1a (donor) spleen cells in 5.5 Gy X-irradiated LOU/C.IgK-1b (host) rats. ■—■—■ shows the IgK-1a IgM serum level detected by the MARK-3 P.O. antibody. +—+—+ shows the IgK-1b (host) IgM serum level detected by the LO-RK-1b antibody in the same serum sample.

REFERENCES

1. **Bazin, H., Cormont, F., and De Clercq, L.,** Rat monoclonal antibodies. II. Rapid and efficient method of purification from ascitic fluid or serum, *J. Immunol. Methods,* 71, 9, 1984.
2. **Beckers, A., Querinjean, P., and Bazin, H.,** Allotypes of rat immunoglobulins. II. Distribution of the allotypes of kappa and alpha chain loci in different inbred strains of rat, *Immunochemistry,* 11, 605, 1974.

Chapter 25.II

USE OF ANTI-RAT KAPPA LIGHT CHAIN ALLOTYPE MONOCLONAL ANTIBODIES FOR THE QUANTIFICATION OF RAT MONOCLONAL ANTIBODIES IN ASCITIC FLUID OR SERUM

P. Manouvriez

TABLE OF CONTENTS

I. INTRODUCTION

The monoclonal antibody (MAb) content of a serum or ascitic fluid can be quantified by a specific antigen-antibody reaction. However, each MAb needs its own standard and if the corresponding antigen is very difficult to produce, it will be impossible to use it for this purpose. In the rat species, the existence of two kappa light chain allotypes can be used for the titration of MAbs. LOU/C rats bear the IgK-1a allotype. Hybridomas obtained by fusion of LOU/C lymphoblasts will thus secrete IgK-1a immunoglobulins (Igs). They can be produced in congenic rats of the LOU/C.IgK-1b(OKA) strain[1] which are perfectly histocompatible with LOU/C rats (Platteau and Bazin, unpublished results). Bazin et al.[2] developed two mouse MAbs anti-rat IgK-1a allotype (MARK-2 and MARK-3) which can be used for the specific and quantitative detection of IgK-1a rat Igs in a serum or ascitic fluid also containing IgK-1b Igs. MARK-2 and MARK-3 are not perfectly specific for the IgK-1a allotype; however, their affinity for IgK-1b is some 250 times lower. In the tests we developed, this specificity problem was circumvented and results are very reproducible as they were controlled with reconstituted ascites, i.e., MAb-free IgK-1b ascites to which a known quantity of IgK-1a MAb was added.

II. IgK-1a BEARING RAT Ig QUANTIFICATION BY A DIRECT SANDWICH ASSAY

A. REAGENTS AND MATERIAL

See Chapter 10 ELISA determination of rat Ig isotypes. For this test the following antibodies are used: MARK-2 or MARK-3 at 10 μg/ml in 0.1 *M* borate buffer, pH 9.5, for the coating of the plates, horseradish peroxidase-labeled MARK-3 (MARK-3-PO) at 0.5 μg/ml in PBS + 0.1% Tween 20, and for the standard, IgK-1b ascites or serum containing a known concentration of IgG or IgM with IgK-1a light chain.

B. PROCEDURE

For the details, see Chapter 10 ELISA determination of rat Ig isotypes. The plates are activated by overnight incubation with 100 μl MARK-2 in borate buffer. After saturation of the remaining binding sites on the plastic, the wells are incubated for 1 h at 37°C with different dilutions of the sera to test and of the standard starting with a 1/125 dilution. The specifically bound IgK-1a Igs are thereafter revealed with MARK-3-PO.

C. RESULTS

Typical curves obtained with IgG2b, IgE, and IgM are shown in Figure 1. The regression lines obtained by this method have a relatively low slope. This together with the low sensitivity of the test due to the competition of the solid phase adsorbed MARK-2 with MARK-3-PO for binding on the same epitope, permits the quantification of IgK-1a MAb in a wide range of concentrations, with only one 1/125 dilution.

III. IgK-1a BEARING RAT Ig QUANTIFICATION BY IMMUNOCOMPETITIVE BINDING ASSAY

This immunocompetitive binding assay has been adapted from Springer,[3] who developed it for a radioimmunoassay. The principle is to perform an antigen-antibody reaction with a constant concentration of labeled antibody and serial dilutions of unlabeled antigen. The amount of labeled antibody remaining free is thereafter revealed by adsorption on insolubilized antigen. In our test, the antigen is IgK-1a and the antibody is MARK-3-PO.

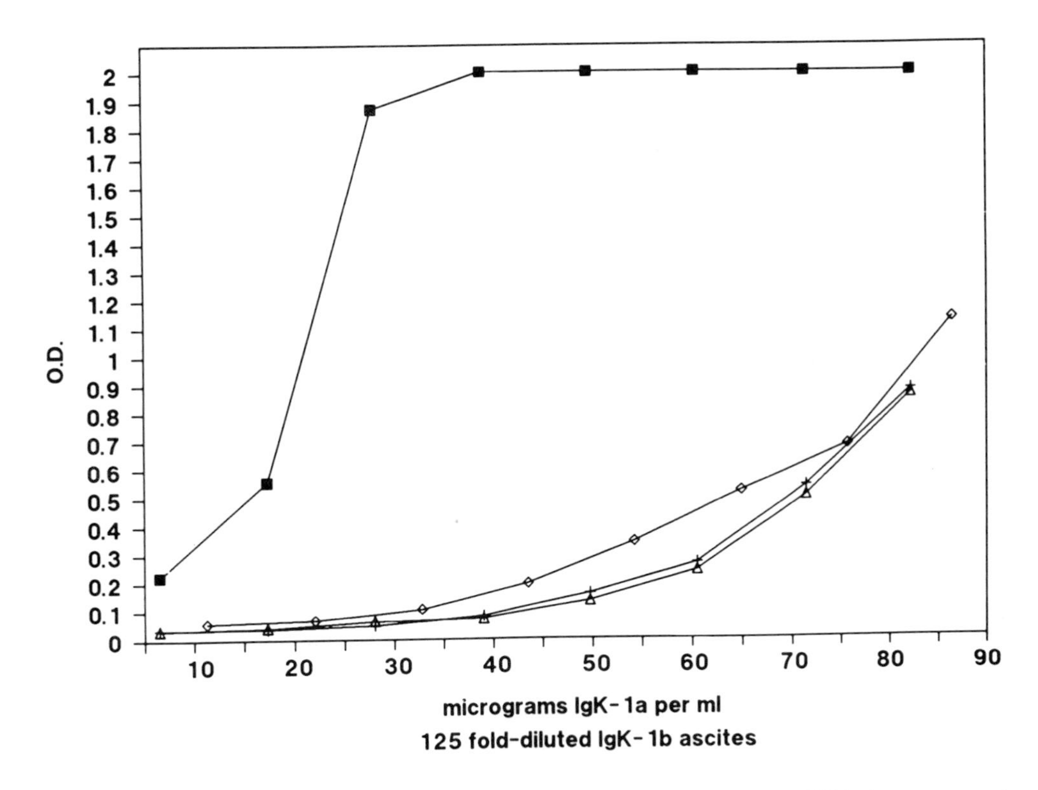

FIGURE 1. Quantification of IgK-1a MAbs in IgK-1b ascites by a direct sandwich assay. ELISA plates were coated with MARK-2 (mouse MAb anti-rat IgK-1a allotype). Standard IgK-1a containing ascitic fluids were made by the addition of purified IgK-1a allotype rat IgM, LO-CD24-3 (■); IgG1, LO-HA1 (Δ); IgG2b, LO-CD10-1/AL2 (◊); or IgE, IR2 (+) in LOU/C.IgK-1b ascites. Ascites were diluted 125 times before incubation in the coated plate. Bound IgK-1a MAbs were revealed with peroxidase labeled MARL-3. The concentrations given in the graph are that of diluted ascitic fluids.

A. REAGENTS AND MATERIAL

1. 96-Well EIA plate coated with 100 μl per well of 5 μg/ml rat IgG-IgK-1a and the remaining free binding sites saturated with an unrelated protein
2. 5-ml plastic tubes
3. 0.15 *M* PBS pH 7.2 + 0.1% Tween 20
4. Unlabeled or PO-labeled MARK-3 at 2.5 μg/ml
5. If unlabeled MARK-3 is used, PO-labeled polyclonal anti-mouse Ig antibodies, checked for non-crossreactivity with rat Igs (in general an appropriate adsorption is required for such a quantity)
6. O-Phenylenediamine dihydrochloride at 0.4 mg/ml in citrate 0.1 *M* buffer, pH 5 + 0.1 μl/ml 35% hydrogen peroxide

For the details of buffer preparation, see Chapter 10: ELISA determination of rat immunoglobulin isotypes.

B. PROCEDURE

Starting at 1/125, twofold dilutions in PBS, 0.1 % Tween 20, of the sera to titrate and of the standard are mixed with an equal volume of MARK-3 at a concentration of 2.5 μg/ml. After 1 h incubation at room temperature, 100 μl of these solutions are transferred in microtitration wells

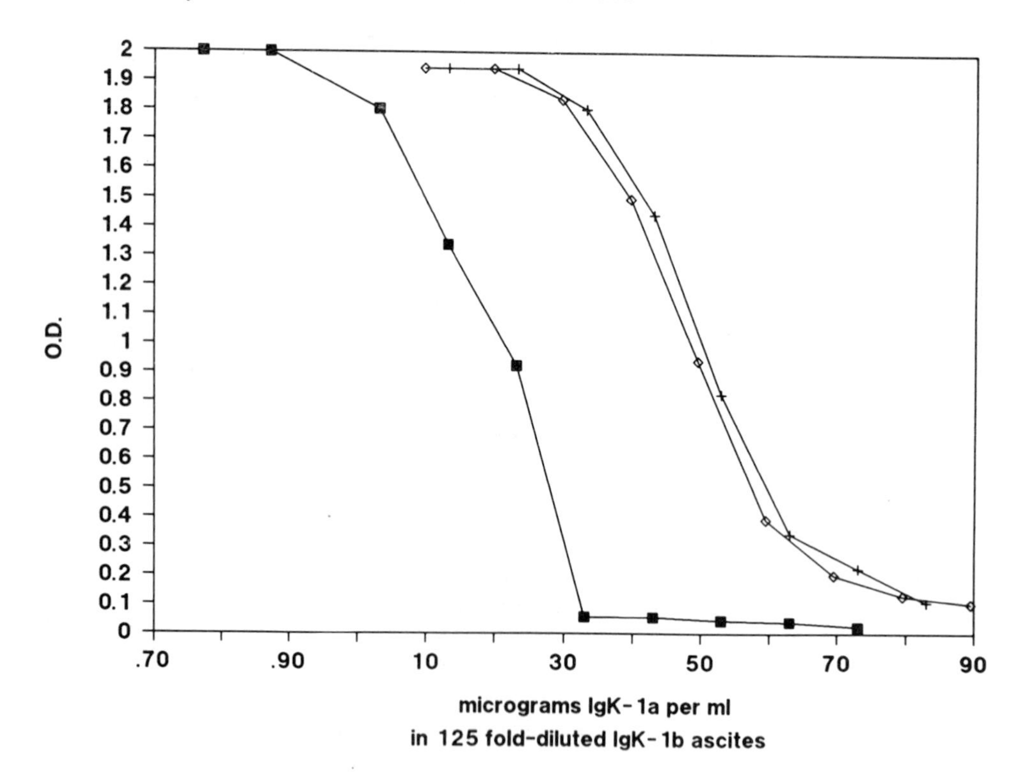

FIGURE 2. Quantification of IgK-1a MAbs in IgK-1b ascites by an immunocompetitive binding assay. ELISA plates were coated with antigen. Standard IgK-1a containing ascitic fluids were made by addition of purified IgK-1a allotype rat IgM (■); IgG1 (+), and IgE (◊). Ascitic fluid was diluted 125-fold and preincubated with peroxidase labeled MARK-3 (mouse MAb anti-rat IgK-1a allotype). The free-remaining labeled MARK-3 was thereafter revealed by adsorption on the antigen-coated ELISA plate. The concentrations given in the graph are that of diluted IgK-1a containing ascitic fluids.

coated with rat IgG allotype IgK-1a. After an additional hour of incubation at 37°C and three washes with PBS, 0.1% Tween, the MARK-3 bound to the plate is revealed with polyclonal rabbit or goat (not rat) anti-mouse IgG labeled with peroxidase. If MARK-3-PO has been used, this last step is omitted.

C. RESULTS

Previous tests have shown that the sensitivity of this immunocompetitive binding assay is not significantly different if MARK-3-PO is used instead of indirect labeling with polyclonal labeled anti-mouse Ig antibodies. In all our tests we use MARK-3-PO, which saves antibodies and reduces the time needed to perform the test. Figure 2 shows the results obtained by immunocompetitive binding assay with IgK-1b ascites from the host containing IgG2b, IgE, and IgM carrying IgK-1a light chains from the hybridomas. They were the same samples as those used for the sandwich test shown in Figure 1. The steeper slopes of the regression lines obtained by the immunocompetitive binding assay allowed a very accurate quantification; however, different dilutions of the sera to test are necessary to be sure to have at least one dilution in the good range of concentrations.

IV. CONCLUSIONS

Using both these methods, the sandwich assay and the immunocompetitive binding assay, monomeric Igs bearing light chains of the IgK-1a allotype can be quantified with only one standard curve. IgM-IgK-1a Igs are pentamers, they need a standard of the same IgM isotype to allow a non-overestimated quantification. The sandwich assay is easier to perform but its accuracy is lower than that of the immunocompetitive binding assay. The former one is used for routine tests, the latter is used when a very accurate quantification is needed.

These techniques could be adapted to quantify IgK-1b allotype Igs contained in IgK-1a ascites using LO-RK1b-1 or LO-RK1b-2 MAbs (rat IgG allotype IgK-1a, anti-rat IgK-1b allotype).

REFERENCES

1. **Beckers, A., Querinjean, P., and Bazin, H.,** Allotypes of rat immunoglobulins. II. Distribution of the allotypes of kappa and alpha chain loci in different inbred strains of rat, *Immunochemistry,* 11, 605, 1974.
2. **Bazin, H., Cormont, F., and De Clercq, L.,** Rat monoclonal antibodies. II. Rapid and efficient method of purification from ascitic fluid or serum, *J. Immunol. Methods,* 71, 9, 1984.
3. **Springer, T. A.,** Quantification of hybridoma immunoglobulins and selection of light-chain variants, *Methods Enzymol.,* 92, 147, 1983.

Chapter 26

EXTRACORPOREAL REMOVAL OF ARTERIOGENIC PARTICLES BY USE OF RAT MONOCLONAL ANTIBODY COUPLED TO SEPHAROSE AS SPECIFIC IMMUNOSORBENT

T. Delaunay, H. Bazin, M. Koffigan, A. Tacquet, and J. C. Fruchart

TABLE OF CONTENTS

I. INTRODUCTION

Familial hypercholesterolemia (FH) is an inherited metabolic disorder charaterized by an increased level of plasma apolipoprotein B containing lipoprotein, which is attributed to a genetic deficiency in the specific surface receptor of LDL.[1,2] Patients suffering from this disease are unable to clear LDL by LDL-receptor-mediated endocytosis and so they have 5 times or more the normal concentration of LDL in their plasma. Homozygous patients develop fatal premature atherosclerotic vascular lesions in the first decade of their lives and heterozygotes develop these lesions in the third decade.[3]

Since 1975, plasmapheresis at 2-week intervals has been used in the successful management of the plasma cholesterol level of homozygous patients;[4] other researchers have developed techniques for lowering LDL, using sorbents as heparin agarose beads.[5,6] Recently, another research group[7] has proposed a technique using anti-LDL polyclonal antibodies coupled to Sepharose as immunosorbent (LDL-pheresis). After some treatments, a visible reduction in clinical manifestations of atherosclerosis was observed and it seems that cholesterol pools of the body were depleted.[8,9]

To our knowledge, no LDL-pheresis experiments with low density lipoprotein immunoadsorption on columns containing mouse or rat monoclonal monospecific antibody to human apolipoprotein have been performed.

In this report, we describe preliminary experiments with a new material which absorbs ApoB containing lipoprotein. Our *in vitro* and *in vivo* experiments enabled a selective and efficient LDL-pheresis system that uses the new material mentioned above to remove ApoB containing lipoprotein from patients with familial hypercholesterolemia.

The rat model which allows the mass production of monoclonal antibody (MAb) (above 80 mg per animal), circumvents the problems of supply and cost of the technique of LDL-pheresis.

II. MATERIAL

The rat MAbs described here are the same as those characterized in Chapter 19.

A. *IN VITRO* AND *IN VIVO* EPURATION

1. *In Vitro* Assay

Rat MAbs (25 mg) were coupled with 1 g of CNBr-activated Sepharose-4B (7 mg of monoclonal antibody per millilitre swollen gel). Six small-sized immunosorbent columns were prepared for the preliminary studies. Normal and hypercholesterolemic plasmas and lipoprotein preparations were passed through each column and the retained fractions (lipids and apoproteins) were then determined by a specific assay.

2. *In Vivo* Mass Production of LO-HApoB-4

From the six rat MAbs tested, only one, LO-HApoB-4, was selected for the *in vivo* mass production and for the preparation of a large-sized immunosorbent column (see Section III).

Three hundred rats were required to produce 10 g of LO-HApoB-4 and 10 l of ascitic fluid were collected (approximately 1 mg of LO-H ApoB4 per millilitre of ascitic fluid). LO-HApoB-4 rat monoclonal antibody was purified with 1 g of MARK-3-Sepharose-4B immunosorbent according to a "batch" technique, and 250 ml of prepared ascitic fluid was incubated under constant agitation with 1 g of MARK-3 coupled with CNBr-activated Sepharose-4B, then 1 h later the normal ascitic proteins and IgK-1b immunoglobulins were eliminated by washing on a fritted-glass funnel with PBS and PBS-NaCl 2.5 *M*. The proteins retained were then eluted with a glycine HCl 0.1 *M*, NaCl 0.15 *M*, pH 2.8 buffer, and collected in a neutralizing buffer.

From 1 to 1.5 g of MAb per day could be purified. After concentration and dialysis, 25 mg of purified LO-HApoB-4 were coupled with 1 g of CNBr-activated Sepharose-4B. Two LDL

TABLE 1
Removal Efficiency of Monoclonal and Polyclonal Immunosorbents

Immunosorbents	Retained fractions (% of the initial samples)							Samples (μg/ml)
	B1	B2	B3	B4	B5	B6	B7	
Apolipoproteins								plasma
AI	0	0	0	0	0	0	4	610
AII	0	0	0	0	0	0	5	160
B	90	89	94	99	91	53	91	1350
CIII	68	52	45	30	32	33	49	43
E	40	38	39	24	26	11	30	52
Apolipoproteins								LDL2
AI	0	0	0	0	0	0	0	18
AII	0	0	0	0	0	0	0	1, 32
B	64	60	57	66	40	32	47	2083
CIII	31	30	28	30	29	17	36	19
E	21	29	21	20	25	10	25	28
Apolipoproteins								HDL
AI	3	2	1	0	1	2	5	2100
AII	3	0	0	0	0	0	2.7	425
B	26	25	0	0	0	0	0	114
CIII	1	0	0	0	0	0	0	54
E	1	1	0	0	0	0	0	40

immunosorbent columns (800 ml each) containing 5 g of LO-HApoB-4 were prepared and used for *in vivo* extracorporeal LDL-removal (LDL-pheresis).

B. LDL-PHERESIS PROCEDURE

Two silanised glass columns with sintered glass filters were each packed with 800 ml anti-LDL Sepharose and stored at 4°C in 0.02% sodium azide in sterile saline and saturated with chloroform. Prior to each LDL-pheresis procedure, the preservative solution was washed from each column with sterile saline (3 to 4 l). Arterial blood was pumped through a plasma separator, and the resulting plasma flow was directed through one anti-LDL Sepharose column at a flow rate of approximately 30 ml/mn, allowing immunoabsorption of 3 to 5 l plasma for each 3 $^1/_2$ to 4 h treatment.

Coagulation was prevented by using an infusion of heparin with anticoagulant-citrate-dextrose formula B or anticoagulant-citrate-dextrose formula A without heparin. The low density lipoprotein-free plasma stream exiting the column was mixed with the cell-rich stream from the cell separating centrifuge and pumped back to the patient. Regeneration of the column was effected, using sequential 1-l washes of 0.15 *M* sodium chloride, 0.2 *M* glycine, HCl buffer pH 3.0, phosphate buffered saline pH 7.4, and 0.15 *M* sodium chloride. Samples were taken before and after the treatment and all lipid measurements were performed by conventional enzymatic analysis with commercial kits.

III. RESULTS

A. *IN VITRO* REMOVAL

The capacities of the six monoclonal immunosorbents and of a polyclonal immunosorbent (B7) for binding the apolipoprotein B were compared on the basis of the amount of apolipoprotein B retained per milligram of antibody per single passage. The specificity was also determined. As shown in Tables 1 and 2, the capacities of five monoclonal immunosorbents (except LO-HApoB-6) were quite similar to that of the polyclonal immunosorbent. The

TABLE 2
Maximal Removal Capacity of Mono- and Poly-Immunosorbents

Immunosorbent	B1	B2	B3	B4	B5	B6	B7
mg of removal ApoB per mg of coupled antibody	0.54	0.53	0.47	0.57	0.35	0.14	0.45

TABLE 3
Clinical Data and Results of a Familial Hypercholesterolemic Heterozygote Patient

	Plasma concentration (g/l)		Removed		8 d after the treatment (g/l)
	Steady state	After treatment (5 l plasma)	Total (g)	%[a]	
ApoAI	0.84	0.59	1.25	29.7	0.657
ApoB	3.6	0.255	16.72	92.9	2.41
ApoCIII	0.110	0.512	—	>	0.047
ApoE	0.110	0.027	0.41	75	0.096
Total chol.[b]	3.123	0.546	12.88	82.5	3.41
HDL chol.[b]	0.105	0.068	0.185	35	0.115
Tg[b]	2.192	0.624	7.84	71.5	1.88
P1[b]	2.984	0.90	10.4	69.8	3.18

[a] Removed component as a percentage of total component before treatment.
[b] Chol.: cholesterol; Tg: Triglycerids; Pl: Phospholipids.

maximal removal capacity was 0.5 mg of apolipoprotein B per milligram of MAb for LO-HApoB-1, -2, -3, -4, 0.35 mg/ml for LO-HApoB-5, and 0.14 mg/ml for LO-HApoB-6.

From these results, it was clear that LO-HApoB-4 offered the best significant immunosorbent capacity and recovery over the other monoclonal antibodies. Therefore, it was decided to produce this monoclonal antibody in large quantity in order to prepare LDL-pheresis columns.

B. *IN VIVO* REMOVAL

LDL-pheresis was carried out according to the method described by Stoffel et al.[10] with one significant difference: a specific rat MAb bound to Sepharose-4B was used as immunosorbent.

Two patients with familial hypercholesterolemia, one homozygote, and one heterozygote, were treated by immunoadsorption in the extracorporeal system described above and are still being treated by LDL-pheresis (Prof. A. Tacquet, Centre Régional d'Hémodialyse, Hôpital A Calmette, Lille, France). No undesirable side effects or changes in clinical, chemical, or other parameters have yet been observed.

The pre- and post-treatment plasma apolipoproteins and lipid concentrations in each experimental treatment are presented in Tables 3 and 4. As was expected, the immunoadsorption column effectively removed apolipoprotein B containing lipoproteins.

As shown in Table 3 (heterozygote patient), 92.9% of ApoB was removed from the plasma and more exactly (see Table 4, homozygote patient) 65% of ApoB, 62% LpB-cIII, and 83.8% LpB-E (LpB-cIII and LpB-E: ApoB containing lipoproteins associated with ApocIII and ApoE). This reduction in ApoB plasma level led to the lowering of total cholesterol, Lp-B cholesterol, and LDL cholesterol concentrations ranging from 82 to 40%. The effectiveness of ApoB removal appeared to be different in the two treated patients. In the homozygote patient, who presented a very high level of total cholesterol (7.23 g/l), only 65.5% of LDL was removed and 41% of the total cholesterol was eliminated. But, as in the heterozygote patient, approximately the same quantity of total cholesterol was retained by the immunosorbent (15.05 g homozygote patient and 12.88 g heterozygote patient).

TABLE 4
Clinical Data and Results of a Familial Hypercholesterolemic Homozygote Patient

	Plasma concentration (g/l)		Removed	
	Steady state	After treatment (5 l plasma)	Total (g)	%
ApoAI	1.21	1.02	0.95	15
ApoAII	0.19	0.19	0	—
LpAII-AI	0.605	0.43	0.875	29.0
ApoB	2.87	0.99	9.4	65.5
ApoCIII	0.032	0.02	0.06	37.5
LpCIII-B	0.267	0.102	0.82	62.2
ApoE	0.114	0.044	0.35	61.4
LpE-B	0.52	0.084	2.18	83.8
Total chol.	7.23	4.22	15.05	41.6
Total Tg	0.77	0.31	2.3	59.0
Total Pl	4.64	2.99	8.25	35.0
HDL chol.	0.7	0.56	0.7	20.0
HDL Tg	0.31	0.21	0.5	32.2
HDL Pl	1.08	0.91	0.85	15.7
LpA chol.	0.44	0.36	0.4	18.1
LpA Pl	1.19	1.1	0.45	7.1
LpB chol.	6.79	3.86	14.6	43.0
LpB Pl	3.45	1.89	7.8	45.2
LDL chol.	4.89	2.3	12.9	52.9
LDL Pl	2.71	1.19	7.6	56.0

Figures 1 and 2 present the time-course data for changes in concentrations of total cholesterol, LDL cholesterol, LpB cholesterol, and ApoB protein during a single LDL-removal treatment in homozygote and heterozygote patients. The total cholesterol level decreased in the same manner as the ApoB protein concentration or LpB cholesterol and LDL cholesterol. Approximately 2.5 g of ApoB containing lipoproteins per gram of rat monoclonal antibody were retained and 3 g of cholesterol per gram of rat MAb was removed from hyperlipidemic plasma (from results not shown here, the effectiveness of 5 g of rat MAb linked to CNBr-activated Sepharose-4B was equivalent to 11 g of goat polyclonal antibody linked to CNBr-activated Sepharose-4B).

The plasma levels of total triglycerids and phospholipids were also reduced since LDL and VLDL particles are associated not only with cholesterol, but also with phospholipids and triglycerids. Parts of ApoE, ApocIII, ApoAI, and ApoAII were also eliminated, because heterogenic ApoB containing particles bound by the MAb contained other apolipoproteins such as ApoE, ApocIII, etc.

The cholesterol concentration rebound that happened 8 or 10 d after the treatment (see Table 3) as described by other researchers,[9,10] resulted most probably from a blood LDL replenishment by the liver cholesterol or by the tissular cholesterol.

IV. COMMENTS

This experiment, sponsored by the Commission of the European Communities' Bioengineering Program and developed in collaboration with SERLIA, Prof. J. C. Fruchard (Lille), Behring Diagnostic, France, Dr. P. J. Volle (chemical and clinical charaterization of rat monoclonal antibodies), Prof. W. S. Reisen (Berne) (preliminary extracorporeal removal studies), and with Prof. A. Tacquet (Lille) (routine use of immunosorbent column), has led to the production and the charaterization of the first rat MAbs directed against human apolipoprotein B. The mass

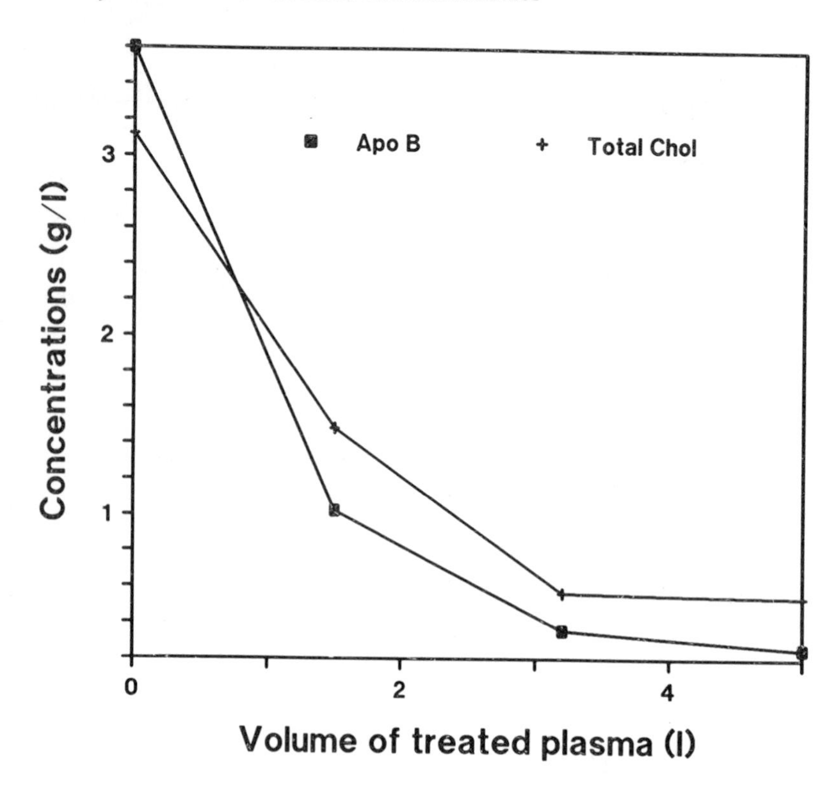

FIGURE 1. Time-course of cholesterol and apolipoprotein B changes during extracorporeal removal of LDL in heterozygote patient (+—+, total cholesterol; ■—■ apolipoprotein B). Plasma samples were taken from the plasma line before entering the immunoadsorption column.

production of the MAb (LO-HApoB-4), thanks to the rat model, has allowed preparation of the first monoclonal immunosorbent column used in LDL-pheresis. Our results show that LDL-pheresis, when performed with mono-immunosorbent reduces plasma cholesterol level to a normal range in two patients suffering from the most severe form of familial hypercholesterolemia. The impact of LDL-pheresis on the long term clinical course of patients with familial hypercholesterolemia requires further study, and we should like the routine use of LDL-pheresis which achieves clinical improvement and perhaps regression of disease being taken over by an industrial partner.

ACKNOWLEDGMENTS

This work was sponsored by the Commission of the European Communities in the framework of the Biomolecular Engineering Program. We are grateful to W.S. Reisen (Berne) for his valuable suggestions at various stages of this study and would also like to acknowledge our thanks to J.P. Kints and F. Nisol. F. Bolle is acknowledged for her help in preparing this manuscript.

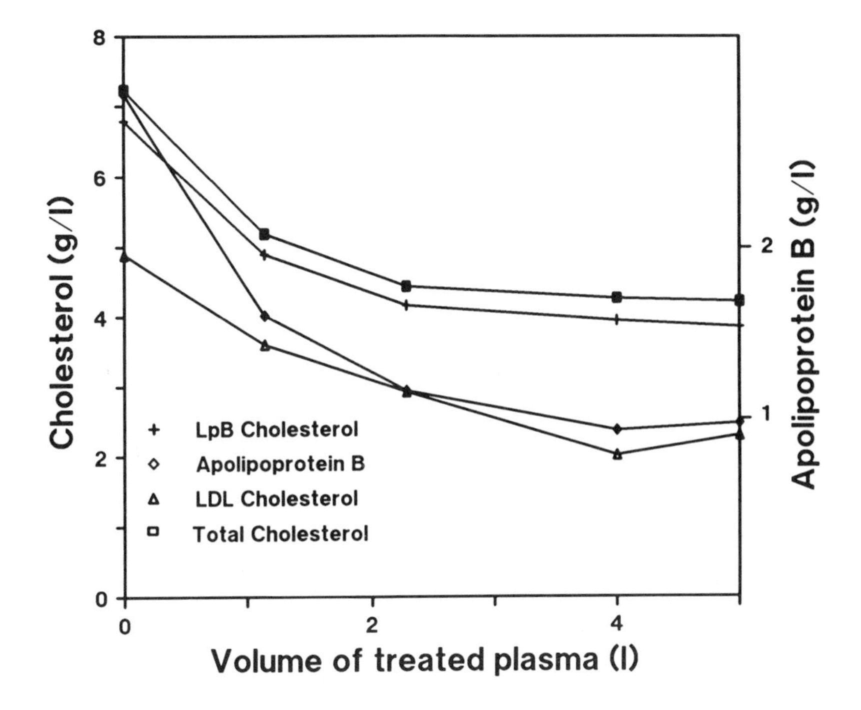

FIGURE 2. Time-course of cholesterol and apolipoprotein B changes during extracorporeal removal of LDL in homozygote patient (□—□, total cholesterol; +—+, LpB-cholesterol; Δ—Δ, LDL-cholesterol; and ◊—◊, apolipoprotein B).

REFERENCES

1. **Golstein, J. L. and Brown, M. S.,** Familial hypercholesterolemia, in *The Metabolic Basis of Inherited Diseases,* Stanbury, J. B., Wijngaarden, J. B., Frederickson, D. S., Goldstein, J. L., and Brown, M. S., Eds., McGraw-Hill, New York, 1983, 622.
2. **Langer, T., Strober, W., and Levy, R. I.,** The metabolism of low density lipoprotein in familial type II hyperlipoproteinemia, *Lancet,* 2, 1147, 1974.
3. **Stone, N. J., Levy, R. I., and Fredrickson, D. S.,** Coronary artery disease on 116 kindred with familial type II hyperlipoproteinemia, *Circulation,* 49, 457, 1974.
4. **Thompson, G. R., Lowenthal, R., and Myant, N. B.,** Plasma exchange in the management of homozygous familial hypercholesterolemia, *Lancet,* 1, 1208, 1975.
5. **Lupien, P. J., Moorjani, S., and Awad, J.,** A new approach to the management of familial hypercholesterolemia; removal of plasma cholesterol based on the principle of affinity chromatography, *Lancet, 1,* 1261, 1975.
6. **Graisely, B., Cloarec, M., and Salmon, S.,** Extracorporeal plasma therapy for homozygous familial hypercholesterolemia, *Lancet,* 2, 1147, 1980.
7. **Stoffel, W. and Demant, T.,** Selective removal of apolipoprotein B containing lipoproteins from blood plasma, *Proc. Natl. Acad. Sci. U.S.A.,* 78, 611, 1981.
8. **Parker, T. S., Gordon, B. R., Saal, S. D., Rubin, A. L., and Amrens, E. M.,** Plasma high density lipoprotein is increased in man when low density lipoprotein (LDL) is lowered by LDL-pheresis, *Proc. Natl. Acad. Sci. U.S.A.,* 83, 777, 1986.
9. **Saal, S. D., et al.,** Removal of low density lipoprotein in patients by extracorporeal immunoadsorption, *Am. J. Med.,* 80, 583, 1986.
10. **Stoffel, W., Greve, V., and Borberg H.,** Application of specific extracorporeal removal of low density lipoprotein in familial hypercholesterolemia, *Lancet,* 2, 1005, 1981.

Chapter 27

ANTIBACTERIA ANTIGENS, ESPECIALLY *STREPTOCOCCUS MUTANS*

F. Ackermans

TABLE OF CONTENTS

I. INTRODUCTION AND PURPOSE OF THESE MONOCLONAL ANTIBODIES

Streptococcus mutans is the major bacteria included in the plurietiological factors of caries. The preliminary step of caries formation is the adsorption and fixation of different specific bacteria to the enamel of the tooth surface.

Actually, it seems that *S. mutans* binding to hydroxylapatite of enamel is a two-step reaction involving initial attachment of cells followed by co-aggregation with other bacterial species.[1] Adherence of *S. mutans* seems to be mediated by salivary glycoproteins, which are selectively adsorbed onto the hydroxylapatite of tooth enamel and form "the acquired pellicle".

Recent studies have suggested that ionic and hydrophobic interactions may be important binding mechanisms which hold the bacteria to the pellicle;[2] another mechanism implicated in bacterial adherence has been identified for *S. sanguis,* consisting of a lectin-like interaction between streptococcal adhesins and salivary glycoproteins.[3]

The components of *S. mutans* cell wall, which seem to play an important role in bacterial adherence or agglutination, are composed of two major classes of macromolecules, i.e., serotype antigens and wall- associated proteins.[3]

Studies of the proteins which bind more or less strongly to the streptococcal cell wall and which interact with salivary glycoproteins on the tooth surface started not so long ago. Several of these proteins were purified;[6-9] some of them were characterized as glucosyltransferase,[10] glucan-binding proteins,[8] or adhesin,[11] and some of them were cloned.[12]

Mouse monoclonal antibodies (MAbs) have been successfully used for bacterial serotyping and characterization of antigen. They have been used to analyze lipopolysaccharides of several gram negative[13] or to characterize the cell surface proteins of various bacterial strains.[14,15]

First, we made rat MAbs to facilitate the protein typing of *S. mutans* serotypes. Since MAbs react with a single antigenic determinant, before the immunization of animals we did not extensively purify the streptococcal wall-extracted antigens.

Later, we used one selected rat MAb to purify and characterize a protein (adhesin 74K-SR) from the crude fraction of the cell wall of *S. mutans* serotype f by MAb immunoaffinity chromatography.

To extensively study this protein, rat MAbs were produced against purified 74K-SR and used to make an epitope analysis of this protein and were later used to clone this protein in *Escherichia coli*.

Using the whole bacteria of *Streptococcus mutans* as immunogens, rat MAbs were made to detect the most immunogenic constituant of the cell surface of this microorganism and to obtain MAbs against specific serotype carbohydrates to study the immunological relationship between the eight serotypes of *S. mutans*.

The hybridoma lines which reacted with *S. mutans* or antigens of *S. mutans* were called LO-SM-1, LO-SM-2, LO-SM-3, LO-SM-4, LO-SM-5, etc., and LO for LOUVAIN rat and SM for *S. mutans*.

II. RAT MONOCLONAL ANTIBODIES AGAINST THE CELL WALL OF *STREPTOCOCCUS MUTANS*

In the first experiments in which we used the crude cell wall of *S. mutans* (see Chapter 21) serotype f OMZ 175 (called WEA fraction) for immunization, four cell lines producing MAbs directed against a component of WEA *S. mutans* OMZ 175 were established.[16]

All four MAbs reacted only with two antigens of WEA from *S. mutans* by Western blotting, immunoprecipitation techniques, ELISA, and competitive ELISA.

Western blot analysis of WEA showed that the four MAbs recognized two related cell-wall associated proteins with apparent molecular weight of 125 kDa and 74 kDa.

Immunoprecipitation of whole cells with the MAb confirmed the surface localization of the two antigens.

ELISA and competitive ELISA were used to analyze the distribution of the epitopes on seven *S. mutans* serotypes. All *S. mutans* serotypes were found to express the recognized epitopes; however, different reactivity patterns could be distinguished among the various strains tested.

LO-SM-2, a rat MAb obtained from this fusion, specific for an adhesion of the WEA fraction of *S. mutans* serotype f[16,17] was used to prepare a MAb immunoaffinity chromatography to purify the *S. mutans* protein. Pure adhesine (= SR) was obtained from the crude WEA fraction (in PBS + 1% Triton X-100 + 0.05% SDS) with a single chromatographic step. The active SR could be eluted from the column in a highly purified form with 0.2 *M*-glycine / HCl pH 2.8. The final yield was about 32% in terms of binding activity. Characterization of the adhesine was accomplished by classical methods of crossed-immunoelectrophoresis, SDS, or 4 to 30% native gradient-polyacrylamide gel electrophoresis, Western blotting, etc. These proteins had a MW of 74 kDa.

III. RAT MONOCLONAL ANTIBODIES OBTAINED FROM THE PURIFIED 74K-SR

Five cell lines secreting MAb against adhesine 74 kDa were obtained from the same schedule as previously described for the WEA fraction (see Chapter 21). These antibodies were used to study the antigenic structure of these molecules and the relation between this and saliva binding sites.[17,18] Three partially different epitopes could be delineated on protein 74 kDa by using unlabeled and alkaline phosphatase-labeled MAbs in competitive ELISA. Two of them were involved in the binding of salivary glycoproteins to *S. mutans* cells, as demonstrated by the inhibition of saliva binding to *S. mutans* with these two MAbs. These results confirmed that saliva binding was epitope specific.[19]

IV. CONCLUSION

We used rat MAbs for a number of specific applications; determination of the most antigenic components in a crude fraction, purification by immunoaffinity chromatography of a specific component from a mixture of bacterial protein, and the study of the antigenic and biological structure of a cell-wall protein of *S. mutans*.

ACKNOWLEDGMENTS

H. Bazin, R. M. Frank, J. P. Klein, J. A. Ogier, T. Bruyère, F. Cormont, C. Genart, F. Nisol, J. M. Malache, and J. P. Kints are gratefully acknowledged for their assistance and advice during the preparation of this work.

REFERENCES

1. **Clark, W. B. and Gibson, R. J.,** Influence of salivary components and extracellular polysaccharide synthesis from sucrose on the attachment of *Streptococcus mutans* 6715 to hydroxyapatite surfaces, *Infect. Immun.*, 18, 514, 1977.
2. **Gibbons, R. J. and Etherden, I.,** Comparative hydrophobicities of oral bacteria and their adherence to salivary pellicles, *Infect. Immun.*, 41, 1190, 1983.
3. **Ericson, T. and Rundegren, S.,** Characterization of a salivary agglutinin reacting with a serotype c strain of *Streptococcus mutans*, *Eur. J. Biochemic*, 133, 225, 1983.

4. **Hamada, S. and Slade, H. D.,** Biology, immunology and cariogenicity of *Streptococcus mutans, Microbiol. Rev.,* 44, 331, 1980.
5. **Ogier, J. A., Klein, J. P., Sommer, P., and Frank, R. M.,** Identification and preliminary characterization of saliva-interacting surface antigens of *Streptococcus mutans* by immunoblotting, ligand blotting and immunoprecipitation, *Infect. Immun.,* 45, 107, 1984.
6. **Forester, H. N., Hunter, N., and Knox, K. W.,** Characteristics of high molecular weight extracellular protein of *Streptococcus mutans, J. Gen. Microbiol.,* 129, 2779, 1985.
7. **Russell, M. W., Zanders, E. D., Bergmeier, L. A., and Lehner, T.,** Affinity purification and characterization of protease susceptible antigen I of *Streptococcus mutans, Infect. Immun.,* 29, 999, 1980.
8. **Russell, R. R. B.,** Glucan-binding proteins of *Streptococcus mutans* serotype c, *J. Gen. Microbiol.,* 112, 197, 1979.
9. **Schöller, M., Klein, J. P., Sommer, P., and Frank, R. M.,** Common antigens of Streptococcal and non streptococcal oral bacteria: characterization of wall associated protein and comparison with extracellular protein antigen, *Infect. Immun.,* 40, 1186, 1983.
10. **Hamada, S.,** Role of glucosyltransferase and glucan in bacterial aggregation and adherence to smooth surfaces, in *Special Supplement to Chemical Series, Glucosyltransferase Glucans, Sucrose and Dental Caries,* Doyle, R. S. and Ciardi, J. E., Eds., ERL Press, Washington, 1983, 31.
11. **Ackermans, F., Klein, J. P., Ogier, J., Bazin, H., Cormont, F., and Frank, R. M.,** Purification and characterization of a saliva-interacting cell-wall protein from *Streptococcus mutans* serotype f using monoclonal antibody immunoaffinity chromatography, *Biochem. J.,* 228, 211, 1985.
12. **Holt, R. G., Abiko, Y., Saito, S., Smorawinska, M., Hansen, J. B., and Curtiss, R.,** *Streptococcus mutans* genes that code for extracellular proteins in *Escherichia coli* K-12, *Infect. Immun.,* 38, 147, 1982.
13. **Apicella, M. A., Bennet, K. M., Hermerath, C. A., and Roberts, D. E.,** Monoclonal antibody analysis of lipopolysaccharide from *Neisseria gonorrhea* and *Neisseria meningitiotis, Infect. Immun.,* 34, 751, 1981.
14. **Cisar, J. O., Barshumian, E. L., Curl, S. H., Vatter, A. E., Sanberg, A. N., and Siracawian, R. P.,** Detection and localization of a lectin on *Actinomyces viscosus* t1 4V by monoclonal antibodies, *J. Immunol.,* 127, 1318, 1981.
15. **Robertson, S. M., Frisch, C. F., Gulig, P. A., Kettman, J. R., Johnstow, K. M., and Hansen, E. J.,** Monoclonal antibodies directed against a cell-surface exposed outer membrane protein of *Haemophilus influenzae* type b, *Infect. Immun.,* 36, 80, 1982.
16. **Ackermans, F., Klein, J. P., Cormont, F., Bazin, H., Ogier, J. A., Frank, R. M., and Vreven, J.,** Antibody specificity and antigen characterization of rat monoclonal antibodies against *Streptococcus mutans* cell wall associated protein antigens, *Infect. Immun.,* 49, 344, 1985.
17. **Ogier, J. A., Bruyère, T., Ackermans, F., Klein, J. P., and Frank, R. M.,** Specific inhibition of *Streptococcus mutans* interactions with saliva components by monoclonal antibodies binding to different epitopes on the 74K cell wall saliva receptor (74K-SR), *FEMS Microbiol. Lett.,* 30, 233, 1985.
18. **Ogier, J. A., Klein, J. P., Schöller, M., and Frank, R. M.,** Studies of *Streptococcus mutans* interactions with saliva glycoproteins by an enzyme linked immunosorbent assay, *J. Microbiol. Meth.,* 5, 157, 1986.
19. **Bruyère, T., Ackermans, F., Klein, J. P., Pillat, M., and Frank, R. M.,** Monoclonal antibodies against *Streptococcus mutans* 74KD wall saliva receptor (74K-SR): correlation between antigenic structure and saliva binding sites, *FEMS,* 50, 211, 1988.

Chapter 28

MERCURY-INDUCED AUTOIMMUNITY PRODUCTION OF MONOCLONAL AUTOANTIBODIES

J. C. Guery, F. Hirsch, B. Bellon, and P. Druet

TABLE OF CONTENTS

I. INTRODUCTION

Several drugs or toxic agents are known to induce autoimmune manifestations in human or experimental situations. The mechanism of these drug-induced disorders is however badly understood. During the last few years we have developed an experimental model in the rat using $HgCl_2$ as an inducing agent. Among the strains tested, the Brown-Norway (BN) strain appeared to be the most susceptible. This strain bears the RT1n haplotype at the major histocompatibility complex. Lewis rats (RT11) are resistant. We showed that $HgCl_2$-injected BN rats develop an autoimmune glomerulonephritis.[1] The autoimmune glomerulonephritis observed in this model is biphasic. During the 2nd and 3rd week, anti-glomerular basement membrane (GBM) antibodies are produced which are responsible for the linear pattern of fixation observed by immunofluorescence. From the end of the 1st month a granular pattern of fixation of the fluoresceinated anti-rat IgG antiserum is observed. The origin of these deposits is still unknown. They could be due to the deposition of circulating immune complexes, or alternatively, to the binding of free-circulating antibodies to an irregularly distributed antigen in the glomerular capillary wall as recently described in Heymann's nephritis. Several antibodies to self (DNA) and non-self antigens (TNP, sheep red blood cells) are also produced. This suggested that $HgCl_2$ triggered polyclonally BN rat B cells. This hypothesis was still strengthened since we observed that mercury-injected BN rats exhibited a hyperimmunoglobulinemia mainly affecting the total serum IgE level. Other experiments have shown that $HgCl_2$-induced polyclonal activation of B cells required the presence of T cells both *in vitro* and *in vivo* and that autoreactive T cells were probably at play. It is also important to note that this autoimmune disease is spontaneously autoregulated. Autoimmune abnormalities appear from day 8, reach a plateau during the 2nd and 3rd week, and then decline although $HgCl_2$ injections are pursued.[1]

This model, which can be considered a model of drug-induced lupus syndrome, therefore offered the possibility to obtain monoclonal autoantibodies by fusing spleen cells from mercury-injected rats with rat myeloma cell lines.[2-4]

II. METHODS

A. EXPERIMENTAL PROCEDURE

Initially obtained from the CSEAL (Orléans-La Source, France), 8- to 10-week-old BN rats were then bred in our own animal house. LOU/M rats were provided by Dr. H. Bazin (Brussels, Belgium). (LOU/M × BN)F1 hybrids were produced in our laboratory. Male or female BN rats were subcutaneously injected three times a week with 100 μg of $HgCl_2$ per 100 g body weight as previously described.

B. HYBRIDIZATION

The non secreting myeloma cell line (IR983F) kindly provided by Dr. H. Bazin was cultured as previously described. Spleen cells were removed from $HgCl_2$-injected BN rats during the induction phase of the disease (between day 11 and 14). Otherwise, rats were not intentionally immunized. After gentle teasing, spleen cells were mixed with exponentially growing IR983F cells in a 5:1 ratio and fused in the presence of polyethylene glycol. The fused cells were 96- or 24-well plates at 1×10^6 living cells per milliliter of HAT medium on a feeder layer of peritoneal cells from normal rats. Supernatants were screened 17 to 25 d after fusion. Selected hybridomas were then cloned by limiting dilution.

C. SCREENING FOR PRODUCTION OF ANTIBODIES

Supernatants of growing cultures were tested for the production of kidney fixing antibodies by indirect immunofluorescence on normal rat kidney cryostat sections. As previously described a linear pattern of fixation along the glomerular capillary wall suggested the presence of anti-GBM antibodies. Anti-GBM activity was further studied using collagenase-digested rat GBM as an antigen in an ELISA. Kidney-fixing antibodies that recognize other non GBM renal structures by indirect immunofluorescence were also selected.

The presence of anti-nuclear antibody activity was looked for by indirect immunofluorescence using normal rat liver cryostat sections as a target. The specificity of anti-nuclear antibodies was then investigated using the Farr and the Chritidia luciliae assays.

Because one of the hallmarks of mercury-induced autoimmunity was an increased production of serum IgE, all the supernatants were tested for their IgE content using an ELISA previously described.[5] Otherwise, heavy chain isotype of the selected MAbs was determined by Ouchterlony's technique with specific antisera.

Finally several supernatants were also screened for natural antibody-like activity against a panel of self and non-self antigens.[4]

D. GRAFTING OF HYBRIDOMAS

Positive clones were grafted subcutaneously and then intraperitoneally on (LOU/M × BN)F1 hybrids for ascites production. Some clones did not grow when grafted on F1 hybrids. They were first grafted on Rnu/Rnu rats and then on F1 hybrids. It was also observed that such clones grew much more easily when first injected intravenously into F1 hybrids. In that situation, hybridomas developed mainly in the recipient's liver. Liver cells were then collected, injected intraperitoneally into F1 hybrids, and for unclear reasons, often grew quite easily.[6]

E. PURIFICATION OF MONOCLONAL ANTIBODIES

IgG1 MAbs were purified from ascitic fluid precipitated at 40% ammonium sulfate saturation, by anion exchange chromatography using TSK DEAE 5 PW preparative column (LKB, Bromma, Sweden) on a Beckman high performance liquid chromatography binary gradient system.[7] If necessary, a gel filtration chromatography using a Sepharose S200 column (Pharamcia Fine Chemicals, Uppsala, Sweden) was performed. IgM antibodies were purified from ascitic fluid precipitated at 50% ammonium sulfate saturation, by gel filtration chromatography on a Sepharose S200 column (Pharmacia).

Preparations were assessed for purity by SDS-PAGE.

III. RESULTS AND DISCUSSION

Several fusions were performed, but supernatants from all the clones obtained were tested for their Ig content in only three of them. In these experiments, 407 wells were sown and 269 Ig-producing clones were obtained. By contrast, 288 wells were sown from two fusions performed using spleen cells from normal BN rats. In that case, only nine clones were obtained. This confirms that $HgCl_2$ induces polyclonal activation of B cells.

We will now describe the antibody specificity of some of the clones that we obtained following the fusions described above.[2-4]

A. ANTI-GBM ANTIBODIES

Five clones (four IgG1 and one IgM) fixed along the GBM in a linear pattern. They also recognized arteriolar walls, and to a much lesser extent, tubular basement membrane (TBM) and Bowmann's capsule. Four of them also bound to human GBM. However none of them reacted with collagenase-digested GBM in an ELISA. The fine specificity of the GBM determinant has therefore not yet been characterized.

Failure of the MAbs to react with collagenase-digested GBM could be due to the absence of the determinant recognized in the GBM preparation used. Interestingly, all the MAbs when injected intravenously into a normal recipient were found deposited in a typical linear pattern along the GBM. Moreover, one of them could induce weak and transient proteinuria suggesting that it could be of a pathogenic significance. More recently, anti-idiotypic antibodies were raised against these anti-GBM antibodies. It has been observed that four of the five anti-GBM MAbs shared a common idiotypic determinant that was also found on anti-GBM IgG eluted from the kidneys of diseased rats.

These anti-GBM monoclonal antibodies are, to the best of our knowledge, the only ones that were produced and since they also recognize human GBM, they are potentially of great interest to characterize the GBM epitopes recognized and their idiotypes in anti-GBM mediated glomerulonephritis.

B. NON ANTI-GBM, KIDNEY FIXING ANTIBODIES

One clone produced a MAb (IgG1) which recognized only the TBM of cortical tubules in BN rats but not in LEW rats. The antigen is therefore an alloantigen only expressed in BN rats. It has been reported that (LEW × BN)F1 hybrids develop an anti-TBM mediated nephritis when immunized with BN-TBM. Anti-TBM mediated nephritis due to the production of alloantibodies has also been reported to occur in human following kidney transplantation, with a TBM alloantigen only expressed on the kidney of the donor probably being at play. This anti-TBM MAb may be of great interest in this respect.

Six other MAbs recognized glomerular structures which are not yet defined. One (IgM) probably recognized visceral glomerular epithelial cells. Two other IgM MAbs recognized mesangial structures. One (IgE) MAb fixed on the wall of arterioles and to a lesser extent on the mesangium. The last two (IgM) MAbs stained endothelial and/or mesangial cells. These antibodies may be of great interest to characterize better the second phase of the glomerular disease. It is indeed possible that one or more of these MAbs are specific for an important antigen.

C. CLONES RECOGNIZING CYTOPLASMIC ANTIGENS

Three supernatants (two IgM and one IgG2b) reacted by indirect immunofluorescence with renal cytoplasmic antigens present in tubular cells. One of them brightly stained the cytoplasm of distal tubular cells and to a lesser extent that of proximal tubular cells. It did not stain liver cells. This pattern corresponds to that described for anti-mitochondrial (anti-M2) antibodies.

D. ANTI-NUCLEAR ANTIBODIES

Three clones produced IgM antibodies that reacted by indirect immunofluorescence with liver nuclei. The pattern of fixation was either homogeneous or peripheral. These antibodies were negative with the Chritidia Luciliae assay, but they had an anti-SS DNA activity in the Farr assay.

E. NATURAL AUTOANTIBODIES

Using ELISA, 59 supernatants of the hybridomas obtained were also tested for antibody activity against a panel of autologous and heterologous antigens (actin, tubulin, myosin, myoglobulin, ds-DNA, peroxidase, ovalbumin, bovine serum albumin), as well as against several haptens such as TNP. Six clones (four IgM and two IgE) were found to react with at least one antigen of the panel, two (one IgM and one IgE) reacted only with TNP, one (one IgM) reacted mainly with peroxidase, while the three last clones (two IgM and one IgE) recognized several antigens. The latter MAbs were found to be polyspecific antibodies and behave exactly like mouse natural autoantibodies. In addition, these polyspecific natural autoantibodies shared a common idiotype with mouse natural antibodies.

These results indicate that mercury-induced polyclonal activation of B cells also affects in this model clones producing natural antibodies. Whether these antibodies play or not, a role in the pathogenesis of the disease is not known.

F. IgE-PRODUCING CLONES

All the supernatants obtained were also tested for their IgE content; 110 of them were found to contain detectable (at least 0.01 μg/ml) amounts of this isotype. Ten of the supernatants were cloned and an antibody activity was looked for using the assays described above, in 37 of the supernatants containing IgE. As previously described, only three supernatants were found to react with at least one of the antigen tested. Interestingly only one of them stained glomerular structures, although very weakly. These results confirm that $HgCl_2$ affects mainly IgE-producing cells and that their antibody specificity is still unknown.

IV. CONCLUSION

Monoclonal antibody technology applied to this rat model of autoimmune disease allowed several conclusions. It has been confirmed that mercury induces polyclonal activation of B cells which mainly affects the IgE-producing cells. It is possible to obtain autoantibodies against numerous autoantigens and monoclonal IgE. These MAbs will be in the future of great interest to understand better the pathogenesis of this autoimmune disease.

Moreover, using MAb technology in this model could have other potential interests. Numerous other autoantibodies, such as anti-thyroglobulin, anti-collagen II antibodies, and rheumatoid factor (unpublished), have recently been described in this model. Therefore it would be theoretically possible to produce MAbs against these and probably other autoantigens. Furthermore, it has been shown that mercury is able to potentiate the immune response towards an exogenous antigen such as ovalbumin. This model could therefore allow to facilitate the obtention of MAbs, including IgE MAb, specific for an exogenous antigen.

REFERENCES

1. **Pelletier, L., Hirsch, F., Rossert, J., Druet, E., and Druet, P.,** Experimental mercury-induced glomerulonephritis, *Springer Semin. Immunopathol.*, 9, 359, 1987.
2. **Hirsch, F., Druet, E., Venderville, B., Cormont, F., Bazin, H., and Druet, P.,** Production of monoclonal anti-glomerular basement membrane antibodies during autoimmune glomerulonephritis, *Clin. Immunol. Immunopathol.*, 33, 425, 1984.
3. **Hirsch, F., Kuhn, J., Ventura, M., Vial, M. C., Fournie, G., and Druet, P.,** Autoimmunity induced by $HgCl_2$ in Brown-Norway rats, *J. Immunol.*, 136, 3273, 1986.
4. **Lymberi, P., Hirsch, F., Kuhn, J., Ternync, T., Druet, P., and Avrameas, S.,** Autoimmunity induced by $HgCl_2$ in Brown-Norway rats. II. Monoclonal antibodies sharing specificities and idiotypes with mouse natural monoclonal antibodies, *J. Immunol.*, 136, 3277, 1986.
5. **Sapin, C., Hirsch, F., Delaporte, J. P., Bazin, H., and Druet, P.,** Polyclonal IgE increase after $HgCl_2$ injections in BN and LEW rats: a genetic analysis, *Immunogenetics*, 20, 227, 1984.
6. **Hirsch, F., Vendeville, B., De Clercq, L., Bazin, H., and Druet, P.,** Rat monoclonal antibodies. III. A simple method for facilitation of hybridoma cell growth in vivo, *J. Immunol. Methods*, 78, 103, 1985.
7. **Buchiel, S. W., Billman, J. R., and Alber, T. R.,** Rapid and efficient purification of mouse monoclonal antibodies from ascites fluid using high performance liquid chromatography, *J. Immunol. Methods*, 69, 33, 1984.

Chapter 29

RAT MONOCLONAL ANTIBODIES TO PLANT PATHOGENS

J. Hutschemackers, H. Bazin, and M. Verhoyen

TABLE OF CONTENTS

I. IMMUNOLOGICAL METHODS IN PHYTOPATHOLOGY

Since the discovery of antigenic properties of bacteria and viruses, serological methods have been used in phytopathology to detect and identify the plant pathogens. From the early qualitative diagnostic methods of precipitating the virus in the clarified sap of an infected plant and of agglutinating the bacterial isolate, vast progress has been made in improving the sensitivity of the methods for plant pathogen detection.

The introduction of enzyme-linked immunosorbent assay (ELISA) in plant virology[1,2] has allowed detection of plant viruses in nanogram quantities. Because of its suitability, ELISA has become the major routine method for checking plant material for the presence of plant viruses.[3-5] ELISA has also been extended for identification and detection of plant pathogenic bacteria,[6-8] mycoplasma-like organisms,[9] fungi,[10] as well as for the detection of plant growth regulators.[11-13]

In plant pathology, the first mouse monoclonal antibodies (MAbs) to tobacco mosaic virus[14] were obtained in 1981. Since then, rapid progress has been made in plant virology and now mouse MAbs to more than 40 plant virus species are available.

Hybridoma technology has also allowed the production of mouse MAbs to plant pathogenic bacteria (*Corynebacterium sepedonicum,*[15] *Xanthomonas campestris* pv. *campestris,*[16] *Pseudomonas syringae* pv. *savastanoi,*[17] mycoplasma-like organisms (aster yellow causal agent,[18] corn stunt spiroplasma, and *Spiroplasma citri*[19,20]), fungi (*Fusarium* spp,[21] *Phytophthora cinnamomi*[22]) and their toxins (aflatoxin B1 of *Aspergillus* spp,[23] toxic glycopeptide of *Ophiostoma ulmi*[24]).

In this chapter the feasibility of producing rat MAbs to plant pathogens and using them as reagents for diagnostic and research purposes are demonstrated through two examples; a phytopathogenic bacterium and a phytovirus.

II. RAT MABS TO A PLANT BACTERIUM: *ERWINIA AMYLOVORA*

A. INTRODUCTION

Erwinia amylovora is the causal bacterium of a plant disease; the fireblight of the Rosaceae species.

The disease affects fruit trees (pear, apple, and quince) and ornamental trees (hawthorns, *Cotoneaster, Pyracantha,* rowan trees, and loquat trees). It is a peril not only for the orchards and the ornamental nurseries, but also for the environment. For example, between 1978 and 1981 approximately 20,000 pear trees were lost in Somerset, U.K.[25] and a lot of hawthorns were destroyed. Fireblight has in fact been known since 1780, when it first appeared in the U.S. Today, there is still no effective treatment against it. Moreover, since its first outbreak in the U.K. in 1957, this disease has continued its spread through Europe (Denmark in 1968, Germany in 1971, France and Belgium in 1972, The Netherlands, Ireland, Norway, Sweden and Poland) and through Mediterranean regions (Cyprus, Egypt, and Israel in 1985).

Confronting this menace, governments started organizing controls to prevent the spread of the disease. Research programs were undertaken on three important aspects: the pathogenesis and genetics of the bacterium, the disease epidemiology, and the treatments against the disease.[26]

In the field of epidemiology, immunological methods especially immunofluorescence were used to detect bacterial populations from apparently healthy host plants.[27]

All these methods, including immunofluorescence,[28-30] agglutination,[31] immunodiffusion,[32] and ELISA,[33] used antisera as reagents.

But we produced rat MAbs in an attempt at improving immunological detection and identification of *Erwinia amylovora* with more specific reagents.

B. ANTIGENS AND IMMUNIZATION SCHEDULES

Whole bacterial cells of *E. amylovora*, strain B87, were used as immunogen. This strain was isolated from an infected hawthorn in Belgium in 1981.

The bacteria were maintained on medium containing yeast extract (5 g l-1), bacto-peptone (5 g l-1), glucose (10 g l-1) and bacto-agar (20 g l-1), and harvested in PBS 24 h after inoculation.

Fusions to obtain MAbs were performed with spleen or popliteal lymph node cells from immunized LOU/C rats.

For fusions made with popliteal lymph nodes, the best immunization schedule was as follows:

- Day 1: 100 µl containing 10^8 *Bordetella pertussis* (Perthydral, Institut Pasteur Production, France) and 10^8 *Erwinia amylovora* B87 were injected in each hind footpad of LOU/C rat
- Days 4, 7, 10: same injection as on day 1 but without *Bordetella pertussis*
- Day 13: popliteal lymph nodes cells were fused with 983 fusion cells (see Chapter 4).

For fusions made with spleen, the best schedule was:

- Day 1: 1 ml containing 3×10^9 *B. pertussis* and 3×10^7 *Erwinia amylovora* B87 was injected intraperitoneally in LOU/C rat
- Days 8, 15, 22, and 30: same injection as on day 1
- Days 151, 152, and 153: same injection as on day 150
- Day 154: spleen cells were harvested for fusion with 983 fusion cells (see Chapter 4).

C. PARENT MYELOMA LINE AND FUSION PROCEDURES

The partner myeloma cell line was the rat non-secreting 983 fusion cell line (see Chapter 4).

Fusions of spleen or popliteal lymph node cells with myeloma cells at a cell ratio of 5:1 were made with polyethylene glycol 4000 as a fusing agent (see Chapter 6). Fused cells from each fusion experiment were suspended in HAT selective medium and seeded directly into 96-well plates containing outbred WISTAR rat peritoneal cells as feeder layer (see Chapter 7).

D. SCREENING METHODS TO SELECT MAbs

An indirect ELISA and an indirect immunofluorescence test were used to screen hybridomas.

For ELISA, 100 µl of whole bacteria (10^7 bacteria ml^{-1}) was dried onto a plate. Nonspecific adsorption of reagents onto the plate was prevented with 1% bovine serum albumin in PBS or 1% milk powder in PBS, and then 50 µl of hybridoma culture supernatants were added into the plate. Reaction between bacteria and rat antibodies was monitored with mouse MAb anti-rat kappa light chain, MARK-134, labeled to alkaline phosphatase and with p-nitrophenyl phosphate substrate. The level of substrate conversion was quantified by measuring optical density at 405 nm.

For an immunofluorescence test, whole bacteria (10^7 bacteria ml^{-1}) were dried on slides. Hybridoma culture supernatants were incubated upon bacteria and the presence of rat antibodies was revealed with mouse MAb, MARK-134, conjugated to fluorescein isothiocyanate at 30 µg protein ml^{-1}.

E. CLONING, PRODUCTION, AND PURIFICATION OF MAbs

Positive cell cultures were cloned by limiting dilution (see Chapter 8).

The immunoglobulin class and subclass of the MAbs were determined by immunodiffusion precipitation with *in vitro* culture supernatants of hybridomas and rat immunoglobulin subclass specific antisera (see Chapter 10). MAbs were produced *in vivo* by injecting hybridoma cells intraperitoneally in LOU/C.IgK-1b (OKA) rats (see Chapter 12). MAbs were purified from ascitic fluids by affinity chromatography on sepharose 4B (Pharmacia, Belgium) linked to

TABLE 1
Specificity by ELISA of LO-Ea-10, LO-Ea-15, and LO-Ea-18 Monoclonal Antibodies

		cOD		
Species	**n**	**LO-Ea-10**	**LO-Ea-15**	**LO-Ea-18**
Erwinia amylovora	93	≥1.26	≥1.11	≥0.97
Erwinia spp. (except *E. amylovora*)	21	≤0.11	≤0.14	≤0.11
Xanthomonas spp.	14	≤0.09	≤0.04	≤0.09
Pseudomonas spp.	27	≤0.12	≤0.07	≤0.12
Enterobacter spp.	7	≤0.06	≤0.07	≤0.09

Note: These tests were carried out with a purified MAb and with pure bacterial cultures. The results are expressed as corrected optical density (cOD) obtained by the difference between the OD of the complete reaction (antigen-MAb-conjugate-substrate) and the OD of the control reaction (MAb-conjugate-substrate). For *E. amylovora,* the minimal cOD, which is obtained after 1 h of enzymatic reaction is given. For the other species, the maximal cOD, obtained after 1 h of enzymatic reaction is given. n = number of strains.

From Hutschemackers, J., Verhoyen, M., and Bazin, H., *Bulletin OEPP/EPPO Bulletin,* 17, 211, 1987. With permission.

mouse MAb anti-rat kappa 1a light chain, MARK-3 (see Chapter 14). The selected hybridoma lines and their MAbs were named LO-Ea-x referring to LOUVAIN, *E. amylovora,* file number.

F. SELECTION OF HYBRIDOMA LINES SECRETING MAbs SPECIFIC TO *ERWINIA AMYLOVORA*

To be used as reagents for the immunological identification and detection of *E. amylovora,* the MAbs must react specifically with *E. amylovora.* To determine their specificity, purified MAbs were tested by the screening ELISA method against different pure strains of bacterial species. The ELISA results were confirmed by the screening immunofluorescence method.

Three rat × rat hybridoma cell lines named LO-Ea-10, LO-Ea-15, and LO-Ea-18 were selected because they produced MAbs specific to *E. amylovora.*[35] These MAbs reacted with 93 strains of *E. amylovora* from seven geographic areas, but did not react with 21 strains of other *Erwinia* species (*E. mallotivora, E. quercina, E. herbicola, E. milletiae, E. carotovora, E. chrysanthemi,* and *E. uredovora*). No cross-reaction was observed with 14 strains of *Xanthomonas,* 7 strains of *Enterobacter agglomerans,* and 27 strains of *Pseudomonas* (see Table 1).

The LO-Ea-10 hybridoma was obtained from fusion with popliteal ganglionary and 983 fusion cells. The LO-Ea-15 and LO-Ea-18 hybridomas resulted from the fusion between splenocytes and 983 fusion cells.

G. IDENTIFICATION OF *ERWINIA AMYLOVORA* IN BACTERIAL ISOLATES FROM INFECTED PLANTS

In classic identification methods, bacteria are isolated from infected plants by inoculation and cloning on suitable media; the cultures thus obtained are then identified by biochemical or immunological methods. In this case, after isolation and growth of bacteria, diagnosis of *E. amylovora* could be performed by means of immunological methods using as reagents the LO-Ea-10, LO-Ea-15, or LO-Ea-18 MAbs specific to this bacterial species.

The immunological methods which were used included the two screening methods (indirect immunofluorescence and indirect ELISA) and different forms of ELISA. As the LO-Ea-18 MAb had been labeled with biotin,[36] a direct and a sandwich ELISA were carried out. Moreover, the

TABLE 2
Detection of *Erwinia Amylovora* in Infected Plant Samples with LO-Ea-18 Monoclonal Antibody

Samples	IF	ELISA
Pear "Durondeau" leaves	+	1.19
Pear "Doyenné de Comice" leaves	+	1.27
Pear "Conférence" leaves	+	0.53
Healthy pear leaves	–	0.07

Note: After soaking infected plant samples overnight in PBS, the washing buffers were centrifuged. The pellets were tested by indirect immunofluorescence (IF) using as reagents LO-Ea-18 MAb and MARK-1 labeled to fluorescein. The washing buffer supernatants were used to inoculate suitable media. After 48 h of incubation, supernatants of liquid culture media were tested by ELISA using as reagents LO-Ea-18 MAb and MARK-1 labeled to alkaline phosphatase. The ELISA results were obtained after 4 h of enzymatic reaction and expressed in cOD as described in Table 1.

indirect ELISA using as reagents biotin-labeled MARK-1 mouse MAb and streptavidin-alkaline phosphatase conjugate could be used. With these methods, the detection limit of bacterial concentration was 10^5 to 10^6 bacteria per milliliter. It was equal to the limit obtained with rabbit antisera.[37] The LO-Ea-10, LO-Ea-15, and LO-Ea-18 MAbs could not be used in agglutination tests because their precipitating properties have not been observed.

H. DETECTION OF *ERWINIA AMYLOVORA* ON PLANT SAMPLES

The isolation of the causal organism generally requires time-consuming procedures. Moreover if, for some bacterial species, good selective media allow relatively easy isolation from diseased plants, many plant pathogenic bacteria are difficult to isolate and then the process of isolation is long and tedious.

To detect *E. amylovora,* its isolation however is not necessary. Indeed, as shown in column IF of Table 2, the LO-Ea-18 MAb could detect *E. amylovora* directly from plant material surface. In this case, fireblighted plant samples were washed overnight in PBS and the washing buffers thus obtained were centrifuged. The pellet suspended in a smaller volume of PBS was used as antigen and tested with LO-Ea-18 MAb by the indirect immunofluorescence screening method. This method can be used when the bacterial concentration is higher than 10^6 bacteria per milliliter, which is the smallest bacterial concentration that could be detected by this method.

If the bacterial concentration is smaller than 10^6 bacteria per milliliter, bacterial multiplication without isolation is required. Indeed, Laroche et al.[33] have observed that the anti- *E. amylovora* rabbit antisera reacted by ELISA with supernatants of liquid culture media seeded with *E. amylovora*. To control the application of this method with rat MAb, liquid culture media were inoculated with different bacterial species. After 20 h of incubation, culture media were harvested and centrifuged; their supernatants were tested by indirect ELISA with LO-Ea-18 MAb. This MAb was able to detect what Laroche et al.[33] call *E. amylovora* "metabolites". Indeed, it reacted with supernatants of liquid culture media inoculated with *E. amylovora* strains and did not react with supernatants of liquid culture media inoculated with other bacterial species (see Table 3).

This method was applied to fireblighted plant samples. Plant samples were washed overnight in PBS. Suitable liquid media were inoculated with washing buffers. After 48 h of incubation, supernatants of liquid culture media were tested. The results presented in column ELISA of Table 2 show that this method could be used to detect *E. amylovora* from plant samples, by

TABLE 3
ELISA for Detection of *Erwinia Amylovora* in Supernatants of Liquid Culture Media Inoculated with Bacterial Species

Species	n	cOD
Erwinia amylovora	91	>0.69
Erwinia spp. (except *E. amylovora*)	20	<0.16
Enterobacter spp.	7	<0.05
Xanthomonas spp.	14	<0.09
Agrobacterium spp.	6	<0.07
Corynebacterium spp.	7	<0.14
Pseudomonas spp.	34	<0.12

Note: ELISA was carried out as described by Laroche et al.[33] with LO-Ea-18 MAb. Results are presented as in Table 1. n = number of strains.

means of LO-Ea-18 MAb. It should be noticed that the growth of a given bacterial species can be inhibited by the presence of antagonistic bacterial species or bacteriophages. For antagonistic bacteria, Laroche et al.[33] have observed that *Erwinia herbicola, Pseudomonas fluorescens,* and *P. syringae,* three species commonly found on plant material, did not interfere in this test with the detection of *Erwinia amylovora.*

III. RAT MAbs TO A PLANT VIRUS: POTATO VIRUS X

A. INTRODUCTION

Potato virus X (PVX), a potexvirus, essentially attacks the Solanaceae species. It causes mild mosaic of potato, mosaic and slight stunting of tomato, and mottle or necrotic ring spotting of tobacco. Worldwidely distributed in potato-growing areas, it provokes yield depressions of over 10%, varying to virus strain and potato cultivar. Moreover, multiple infection with other potato viruses, such as potato virus A and potato virus Y, is much more damaging than a single infection.[38]

Because presently there is no effective treatment against plant viruses, a practical solution is the selection of free-plant virus seed tubers. This selection consists of detecting the plant viruses in seed samples and eliminating those that are infected. Among detection methods, the serological ones are the most common.

Rat MAbs to PVX were produced and their utilization in detection methods was studied.

B. ANTIGENS AND IMMUNIZATION SCHEDULE

PVX, strain MM, isolated in the CAR (Center of Agronomical Research, Gembloux, Belgium) was purified from infected *Nicotiana tabacum* cv "Xanthi" leaves by the butanol/chloroform method and injected in LOU/C rats.

The immunization schedule consisted of three weekly intraperitoneal injections of 200 μg of virus emulsified in complete adjuvant (first injection) or incomplete adjuvant of Freund (next two injections). Four days before fusion at day 211, 200 μg of virus were injected intravenously.

C. PARENT MYELOMA LINE AND FUSION PROCEDURE

Fusion between spleen cells and 983 fusion cells at a cell ratio of 5:1 was performed with polyethylene glycol 4000 as the fusing agent (see Chapter 6). Fused cells were suspended in HAT selective medium and seeded directly into 96-well plates containing outbred WISTAR rat peritoneal cells as feeder layer (see Chapter 7).

TABLE4
Reactivity in HADAS-ELISA of LO-VX-4, LO-VX-6, LO-VX-9, and LO-VX-10 MAbs to PVX Strains

Antigen		cOD			
Isolate	**Source**	**LO-VX-4**	**LO-VX-6**	**LO-VX-9**	**LO-VX-10**
PVX strain					
MM	CAR	>1.79	>1.80	1.28	>1.70
PV117	ATCC	>1.79	>1.80	1.25	>1.70
804	de Bokx	>1.81	1.77	0.91	1.68
805	de Bokx	1.70	1.68	1.49	1.71
806	de Bokx	>1.81	>1.78	0.80	1.40
807	de Bokx	>1.81	>1.78	1.52	1.68
903	de Bokx	>1.81	>1.78	1.46	1.68
904	de Bokx	1.42	1.38	1.52	1.71
905	de Bokx	1.26	1.15	>1.77	1.69
906	de Bokx	>1.79	>1.80	1.17	1.66
907	de Bokx	1.49	1.47	1.59	>1.73
982	de Bokx	>1.79	>1.80	1.29	>1.70
Healthy plant		0.00	0.01	0.00	0.00

Note: The results after 1 h of enzymatic reaction were expressed as cOD obtained by the difference between the OD of the complete reaction (IgY-plant sap-MAb-conjugate-substrate) and the OD of control reaction (IgY-MAb-conjugate-substrate).

D. SCREENING METHOD TO SELECT MAbs

A heterologous antiglobulin double antibody sandwich form of ELISA (HADAS-ELISA) was used as a screening method. This assay was carried out as described in Snacken and Verhoyen.[39] Plates were coated with purified anti-PVX yolk globulins and incubated for 4 h at 35°C. After washing, 100 μl of purified PVX diluted at 200 ng/ml in PBS + 0.05% Tween 20 was added to each well and incubated overnight at 4°C. Then 100 μl of hybridoma culture supernatants were incubated for 2 h at 35°C. Mouse MAb anti-rat kappa light chain, MARK-134, labeled with alkaline phosphatase (100 μl/well) was incubated for 4 h at 35°C. The level of p-nitrophenylphosphate conversion was measured at optical density at 405 nm.

E. CLONING, PRODUCTION, AND PURIFICATION OF MAbs

Cloning, production, and purification of MAbs were performed in the same way as they were for *E. amylovora*. From this fusion, 15 hybridomas indexed LO-VX-1 to LO-VX-15 were selected, but to date only 4 (LO-VX-4, LO-VX-6, LO-VX-9, and LO-VX-10) have been studied.

F. REACTIVITY OF ANTI-PVX RAT MAbs

The LO-VX-4, LO-VX-6, LO-VX-9, and LO-VX-10 MAbs purified from ascitic fluids were tested by ELISA against 12 strains of PVX and 5 other potexviruses.

As shown in Table 4, the 4 MAbs reacted by HADAS-ELISA with the 12 tested PVX strains. The HADAS-ELISA was performed as described above except that the antigen consisted of either PVX infected sap or healthy control sap (dilution 1/40 in PBS) and the conjugate was incubated for 1 h instead of 4 h.

The reactivity of these MAbs against the other potexviruses was determined by indirect ELISA, in which infected or healthy sap diluted 40 times in 0.05 *M* sodium carbonate buffer pH 9.6 were directly added into plates.

In indirect ELISA (see Table 5), the LO-VX-9 and LO-VX-10 MAbs reacted with the narcissus and the white clover mosaic virus strains. For the four MAbs, weak cross-reactions were observed with cassava common mosaic virus.

TABLE 5
Reactivity in Indirect ELISA of LO-VX-4, LO-VX-6, LO-VX-9, and LO-VX-10 MAbs to Other Potexviruses

	cOD			
Antigen	**LO-VX-4**	**LO-VX-6**	**LO-VX-9**	**LO-VX-10**
Cactus virus X				
Infected plant sap	0.00	0.01	0.03	0.01
Healthy plant sap	0.00	0.00	0.00	0.00
Cassava common mosaic virus				
Infected plant sap	0.07	0.12	0.14	0.19
Healthy plant sap	0.00	0.01	0.04	0.05
Narcissus mosaic virus				
Infected plant sap	0.01	0.11	0.43	0.48
Healthy plant sap	0.03	0.09	0.08	0.20
Tulip virus X				
Infected plant sap	0.00	0.01	0.00	0.00
Healthy plant sap	0.00	0.01	0.00	0.01
White clover mosaic virus				
Infected plant sap	0.02	0.04	0.18	1.52
Healthy plant sap	0.02	0.06	0.00	0.00
Potato virus X 805				
Infected plant sap	1.26	0.93	1.05	0.81
Healthy plant sap	0.04	0.01	0.07	0.00

Note: The cOD obtained after 4 h of enzymatic reaction was calculated by the difference between the OD of the complete reaction (plant sap-MAb-conjugate-sbustrate) and the OD of the control reaction (MAb-conjugate-substrate).

G. UTILIZATION OF RAT MAbs TO PVX

An immunological method using only anti-PVX rat MAbs was carried out for detection of PVX. The LO-VX-6 MAb was chosen as the model.

A double antibody sandwich form of ELISA (DAS-ELISA) was carried out as follows: plates were coated with LO-VX-6 rat MAb, and after incubation of plant sap, antigens were detected with the LO-VX-6 MAb labeled with alkaline phosphatase and p-nitrophenylphosphate. The temperature and incubation time were the same as in HADAS-ELISA.

As shown in Table 6, an immunological method using only anti-PVX rat MAbs can be carried out. The results presented in this table confirmed the cross-reaction between LO-VX-6 MAb and cassava common mosaic virus strain.

The reactivity between rat MAbs and viruses can also be observed by immunosorbent electron microscopy followed by a decoration technique. For these assays, LO-VX-6 MAb was coated on an electron microscopy grid. After 20 min of incubation, a purified PVX was allowed to react with the coating MAb for 20 min. The virus was then decorated with LO-VX-6 MAb for another 20 min. After staining with 1% uranyl acetate, the grid was observed with electron microscope (Philips EM301). The result is presented in Figure 1 and shows the reaction of the LO-VX-6 MAb and PVX.

As can be seen the immunosorbent electron microscopy followed by decoration technique can be used for the studies of relationships between viral strains or species and the studies of surface antigens.

Torrance et al.[40] obtained hybridomas from fusions between spleen cells of rats immunized against PVX and one of two rat myeloma lines, Y3.Ag1.2.3 and YB2/3.0Ag20. These rat MAbs against PVX were used to compare the serotypes of PVX with the resistance groups[40] and to study the antigenic determinants of viral strains.[41]

TABLE 6
Detection of PVX with DAS-ELISA

Antigen	cOD LO-VX-6
Cactus virus X (n = 1)	
Infected plant sap	0.05
Healthy plant sap	0.02
Cassava common mosaic virus (n = 1)	
Infected plant sap	0.83
Healthy plant sap	0.00
Tulip virus X (n = 1)	
Infected plant sap	0.00
Healthy plant sap	0.00
White clover mosaic virus (n = 1)	
Infected plant sap	0.00
Healthy plant sap	0.00
Potato virus X (n = 11)	
Infected plant sap	>1.70
Healthy plant sap	0.00

Note: The cOD obtained after 1 h of enzymatic reaction were calculated by difference between the OD of the complete reaction (LO-VX-6-plant sap-enzyme-labeled-LO-VX-6-substrate) and the OD of the control reaction (LO-VX-6-enzyme-labeled-LO-VX-6-substrate). For PVX strains, the minimal cOD is listed. n = number of strains.

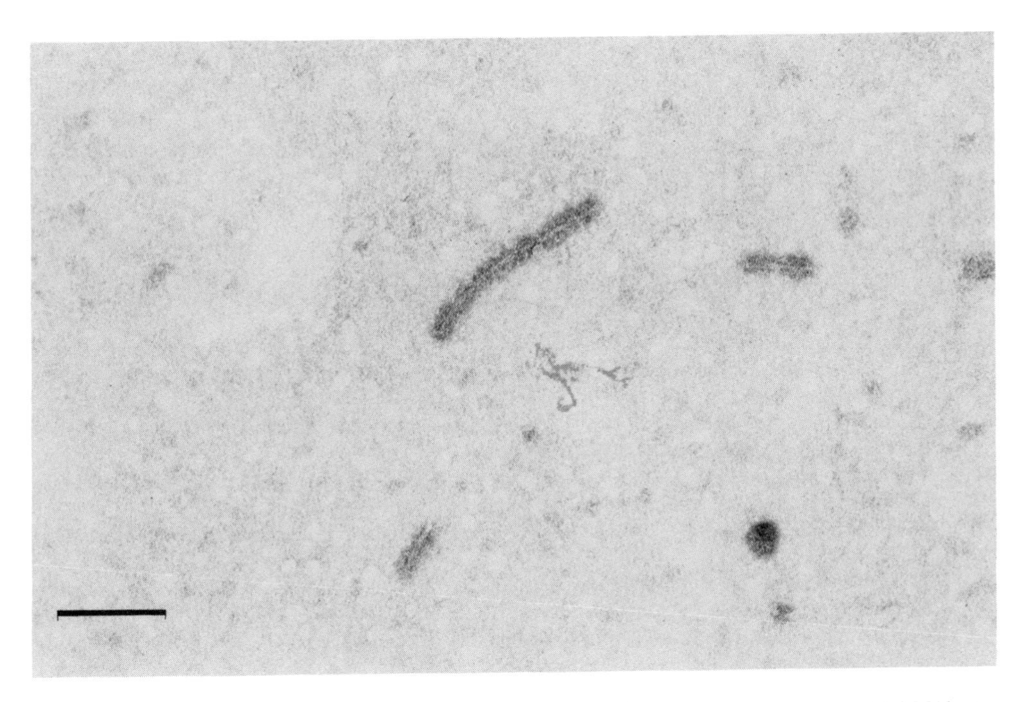

FIGURE 1. Reaction between PVX and LO-VX-6 MAb. The reaction between PVX and LO-VX-6 MAb was revealed by immunosorbent electron microscopy followed by decoration technique. The bar represents 200 nm.

IV. CONCLUSIONS

The production of rat MAbs to plant pathogens has been demonstrated in this chapter. These MAbs, like those specific to *E. amylovora,* could replace polyclonal antisera in immunological diagnostic methods. If MAbs are produced for diagnostic purposes, their specificity must be studied with care in order to select hybridoma lines producing MAbs specific to a given species. Indeed, for the diagnosis of a given species, the MAb reagent must recognize all its strains, but may not cross-react with other species, especially with those that are related. When specificity for a given species is detected, a mixture of MAbs may be used for two reasons. The first reason is to amplify the immunological response. In this case, the two (or more) MAbs of the mixture must be specific to the given species but directed to two (or more) different epitopes. The second reason is to obtain a reagent specific to all strains of the given species. If MAb 1 recognizes only some serogroups of the given species and MAb 2 recognizes the others, then the resulting mixture of the two MAbs will recognize all serogroups of the species.

An immunological method using as a reagent the chosen MAb or mixture must then be tested and applied to detect the phytopathogen in naturally infected plants.

With their high degree of specificity, MAbs are precious analytical tools for research purposes. Examples of such applications exist but are not very numerous yet because of the recent utilization of hybridoma technology in phytopathology. The serotypes of potato virus X[40] and of barley yellow dwarf virus[41] were studied by means of rat MAbs. As shown by Koenig and Torrance,[42] rat MAbs can also be used to study the antigenic determinants of viral strains.

As shown above, with immunosorbent electron microscopy followed by the decoration technique, rat MAbs can be used to study the relationships between viral strains and surface antigens.

ACKNOWLEDGMENTS

This research was sponsored by the "Institut pour l'Encouragement de la Recherche Scientifique dans l'Industrie et l'Agriculture" (I.R.S.I.A), Brussels, Belgium. The authors want to thank A. Cluyse, C. Leal Lopez, C. Negel, and D. Wauters for their technical assistance.

REFERENCES

1. **Voller, A., Bartlett, A., Bidwell, D. E., Clark, M. F., and Adams, A. N.,** The detection of viruses by enzyme-linked immunosorbent assay (ELISA), *J. Gen. Virol.,* 33, 165, 1976.
2. **Clark, M. F. and Adams, A. N.,** Characteristics of the microplate method of enzyme-linked immunosorbent assay for the detection of plant viruses, *J. Gen. Virol.,* 34, 475, 1977.
3. **Van Regenmortel, M. H. V. and Burckard, J.,** Detection of a wide spectrum of tobacco mosaic virus strains by indirect enzyme-linked immunosorbent assays (ELISA), *Virology,* 106, 327, 1980.
4. **Edwards, M. L. and Cooper, J. I.,** Plant virus detection using a new form of indirect ELISA, *J. Virol. Meth.,* 11, 309, 1985.
5. **Zrein, M., Burckard, J., and Van Regenmortel, M. H. V.,** Use of the biotin-avidin system for detecting a broad range of serologically related plant viruses by ELISA, *J. Virol. Meth.,* 13, 121, 1986.
6. **Kishinevsky, B. and Bar-Joseph, M.,** Rhizobium strain identification in *Arachis hypogaea* nodules by enzyme-linked immunosorbent assay (ELISA), *Can. J. Microbiol.,* 24, 1537, 1978.
7. **Stevens, W. A. and Tsiantos, J.,** The use of ELISA for the detection of *Corynebacterium michiganense* in tomatoes, *Microbios Lett.,* 10, 29, 1979.
8. **Laroche, M. and Verhoyen, M.,** Adaptation et application du test ELISA, méthode indirecte, à la détection d'*Erwinia amylovora* (Burrill) Winslow et al., *Parasitica,* 40, 197, 1984.

9. **Boudon-Padieu, E., Schwartz, Y., Larrue, J., and Caudwell, A.,** ELISA and immunoblotting detection of grapevine flavescence dorée ML0-induced antigens in individual vector leafhoppers, *Bulletin OEPP/EPPO Bulletin,* 17, 305, 1987.
10. **Amouzou-Alladaye, E., Dunez, J., and Clerjeau, M.,** Immunoenzymatic techniques for the detection of *Phytophthora fragariae* in strawberry: antiserum specificity, comparison of sandwich ELISA and indirect ELISA, *Bulletin OEPP/EPPO Bulletin,* 17, 307, 1987.
11. **Weiler, E. W., Jourdan, P. S., and Conrad, W.,** Levels of indole-3-acetic acid in intact and decapitated coleoptiles as determined by a specific and highly sensitive solid-phase enzyme immunoassay, *Planta,* 153, 561, 1981.
12. **Daie, J. and Wyse, R.,** Adaptation of the enzyme-linked immunosorbent assay (ELISA) to the quantitative analysis of abscisic acid, *Anal. Biochem.,* 119, 365, 1982.
13. **Atzorn, R. and Weiler, E. W.,** The immunoassay of gibberellins. II. Quantitation of GA3, GA4 and GA7 by ultra-sensitive solid-phase enzyme immunoassays, *Planta,* 159, 7, 1983.
14. **Dietzgen, A. G. and Sander, E.,** Monoclonal antibodies against a plant virus, *Arch. Virol.,* 74, 197, 1982.
15. **De Boer, S. H. and Wieczorek, A.,** Production of monoclonal antibodies to *Corynebacterium sepedonicum, Phytopathology,* 74, 1431, 1984.
16. **Alvarez, A. M., Benedict, A. A., and Mizumoto, C. Y.,** Identification of xanthomonads and grouping of strains of *Xanthomonas campestris* pv *campestris* with MAbs, *Phytopathology,* 75, 722, 1985.
17. **Casano, F. J., Hung, S. Y., and Wells, J. M.,** Differentiation of some pathovars of *Pseudomonas syringae* with monoclonal antibodies, *Bulletin OEPP/EPPO Bulletin,* 17, 173, 1987.
18. **Lin, C. P. and Chen, T. A.,** Comparison of monoclonal antibodies and polyclonal antibodies in detection of the aster yellows mycoplasmalike organism, *Phytopathology,* 76, 45, 1986.
19. **Jordan, R., Konai, M., Lee, I. M., and Davis, R. E.,** Production and characterization of monoclonal antibodies to *Spiroplasma citri* and corn stunt spiroplasma, *Phytopathology,* 75, 1351, 1985.
20. **Lin, C. P. and Chen, T. A.,** Production of monoclonal antibodies against *Spiroplasma citri, Phytopathology,* 75, 848, 1985.
21. **Iannelli, D., Capparelli, R., Marziano, F., Scala, F., and Novallo, C.,** Production of hybridomas secreting monoclonal antibodies to the genus *Fusarium, Mycotaxon.,* 18, 523, 1983.
22. **Hardham, A. R., Suzaki, E., and Perkin, J. L.,** The detection of monoclonal antibodies specific for surface components on zoospores and cysts of *Phytophthora cinnamomi, Exp. Mycol.,* 9, 254, 1985.
23. **Candlish, A. A. G., Stimson, W. H., and Smith, J. E.,** A monoclonal antibody to aflatoxin B1: detection of the mycotoxin by enzyme immunoassay, *Lett. Appl. Microbiol.,* 1, 57, 1985.
24. **Benhamou, N., Ouellette, G. B., Lafontaine, J. G., and Joly, J. R.,** Use of monoclonal antibodies to detect a phytotoxic glycopeptide produced by *Ophiostoma ulmi,* the Dutch elm disease pathogen, *Can. J. Bot.,* 63, 1177, 1985.
25. **Stead, D. E.,** A brief review of the status of fireblight in U.K., *Bulletin OEPP/EPPO Bulletin,* 17, 219, 1987.
26. **Paulin, J. P.,** Réalisations récentes de la recherche sur le feu bactérien, *Bulletin OEPP/EPPO Bulletin,* 17, 177, 1987.
27. **Van Vaerenbergh, J., Crepel, C., and Vereecke, M.,** Monitoring fireblight for official phytosanitary legislation in Belgium, *Bulletin OEPP/EPPO Bulletin,* 17, 195, 1987.
28. **Thomson, S. V. and Schroth, M. N.,** The use of immunofluorescent staining for rapid detection of epiphytic *Erwinia amylovora* on healthy pear blossom, *Proc. Am. Phytopath. Soc.,* 3, 321, 1976.
29. **Paulin, J. P.,** Overwintering of *Erwinia amylovora:* sources of inoculum in spring, *Acta Hortic.,* 117, 49, 1981.
30. **Laroche, M. and Verhoyen, M.,** Détection d'*Erwinia amylovora* directement sur du matériel végétal par la technique d'immunofluorescence, *Med. Fac. Landbouww. Rijksuniv. Gent,* 48, 647, 1983.
31. **Laroche, M. and Verhoyen, M.,** Identification sérologique des phytobactéries en Belgique, *Med. Fac. Landbouww. Rijksuniv. Gent,* 46, 757, 1981.
32. **Laroche, M. and Verhoyen, M.,** Identification sérologique d'*Erwinia amylovora* par immunodiffusion, *Med. Fac. Landbouww. Rijksuniv. Gent,* 47, 1083, 1982.
33. **Laroche, M., Givron, C., and Verhoyen, M.,** Utilisation de la méthode ELISA pour identifier *Erwinia amylovora* par l'intermédiaire de ses métabolites, *Bulletin OEPP/EPPO Bulletin,* 17, 205, 1987.
34. **Bazin, H., Xhurdebise, L. M., Burtonboy, G., Lebacq, A. M., De Clercq, L., and Cormont, F.,** Rat monoclonal antibodies. I. Rapid purification from *in vitro* culture supernatants, *J. Immunol. Methods,* 66, 261, 1984.
35. **Hutschemackers, J., Verhoyen, M., and Bazin, H.,** Production d'anticorps monoclonaux de rat spécifiques à *Erwinia amylovora, Bulletin OEPP/EPPO Bulletin,* 17, 211, 1987.
36. **Bayer, E. A., Wilchek, M., and Skutelsky, E.,** Affinity cytochemistry: The localization of lectin and antibody receptors on erythrocytes via avidin-biotin complex, *FEBS Lett.,* 68, 240, 1976.
37. **Hutschemackers, J., Bazin, H., and Verhoyen, M.,** Seuil de détection d'*Erwinia amylovora* au moyen d'anticorps monoclonaux, *Med. Fac. Landbouww, Rijksuniv. Gent,* 52, 1065, 1987.

38. **Beemster, A. B. R. and de Bokx, J. A.,** Survey of properties and symptoms, in *Viruses of Potatoes and Seed-Potato Production,* de Bokx, J. A. and van der Want, J. P. H., Eds., Pudoc, Wageningen, 1987, chap. 6.
39. **Snacken, F. and Verhoyen, M.,** Utilisation des anticorps isolés de jaune d'oeufs de poule pour détecter le potato virus X par le test ELISA multicouche, *Parasitica,* 42, 61, 1986.
40. **Torrance, L., Pead, M. T., Larkins, A. P., and Butcher, G. W.,** Characterization of monoclonal antibodies to a U.K. isolate of barley yellow dwarf virus, *J. Gen. Virol.,* 67, 549, 1986.
41. **Torrance, L., Larkins, A. P., and Butcher, G. W.,** Characterization of monoclonal antibodies against potato virus X and comparison of serotypes with resistance groups, *J. Gen. Virol.,* 67, 57, 1986.
42. **Koenig, R. and Torrance, L.,** Antigenic analysis of potato virus X by means of monoclonal antibodies, *J. Gen. Virol.,* 67, 2145, 1986.

Chapter 30.I

RAT MONOCLONAL ANTIBODIES FOR COMPLEMENT-MEDIATED AND CELL-MEDIATED CYTOLYSIS

M. Sekhavat, A. M. Ravoet, A. Neirynck, and G. Sokal

TABLE OF CONTENTS

I. INTRODUCTION

Rat MAbs are especially attractive for therapeutic use when *in vitro* or *in vivo* elimination of target cells is considered.

Autologous bone marrow transplantation for hematopoietic reconstitution of cancer patients after intensive chemotherapy (radiotherapy) is used when the bone marrow graft is free of cancer cells. In cases where residual leukemic cells or infiltrating solid tumor cells are detected or suspected in the bone marrow, these malignant cells can be killed selectively by *in vitro* treatment of the graft with MAbs and complement,[1,2] Optimalization of purging conditions, using a blend of MAbs rather than a single MAb and multiple cycles of treatment, led to a higher than 4-log reduction in cancer cell contamination.[3,4] Alternatively, the cancer cells can be coated with specific MAbs and removed, using immunomagnetic beads and magnets.[5,6] Until now, the benefits of very efficient purging have been masked by the insufficient conditioning of the patient, and relapses have been frequent.

Regarding allogeneic bone marrow transplantation, *in vitro* removal of immunocompetent T cells aims at the reduction of incidence and severity of graft vs. host disease (GVHD). This purging has mostly been performed by treatment with MAbs and complement[7-12] soybean agglutination, rosetting with sheep erythrocytes,[13] or counterflow contrifugation.[14] T cell depletion has been shown to reduce or even eliminate the severe GVHD,[7] but does not necessarily result in prolonged disease-free survival as both late graft failure (graft rejection rate) and relapse rate are increased.[9,11] These drawbacks of purging have been observed when using pan-T cell specific cocktails (CD4 + CD5 + CD8; CD2 + CD5 + CD7; CD2 + CD3 + CD4 + CD5 + CD6 + CD8 + Tp44) or a less specific MAb (CAMPATH-1).[8] Therefore, additional conditioning of the patient[9,15] or *in vivo* treatment of the patient with MAbs have been attempted in order to restrain the host residual T cells from rejecting the graft.[16]

Rejection or GVHD of kidney, liver, heart, or bone marrow allografts have been successfully reversed or prevented by *in vivo* serotherapy using mouse or rat MAbs.[17-20] *In vivo,* the main mechanisms involved are probably opsonization of the target cells and their elimination by the reticulo-endothelial system[21] or their elimination by killer cells through antibody-dependent cell-mediated cytolysis (ADCC).[22]

In this chapter, some of the more basic experiments concerning the choice of rat MAbs for complement-mediated cytolysis and antibody-dependent cell-mediated cytolysis are shown and discussed.

II. MATERIAL AND METHODS

Anti-T cell MAbs used in this study have been described in Chapter 20. Anti-leukemia and anti-small cell lung carcinoma (SCLC) cell MAbs have been characterized by Lebacq-Verheyden et al.[23-25]

Rabbit sera adsorbed on human red blood cells (A,B,O pool), baby rabbit sera or human serum have been used as a source of complement. A serum concentration yielding efficient (mostly >99%) lysis with specific MAbs and low aspecific lysis with irrelevant myeloma immonoglobulins was chosen.[26]

For microcytotoxicity assay, 2×10^3 cells were incubated in microplates[27] for 30 min at room temperature in 3 μl Eagle's medium containing 10% human AB serum and monoclonal ascites (20 ng MAb per assay). As a source of complement, 4 μl of rabbit or human serum were added and incubation was continued for 60 min at room temperature. After the addition of eosin (1 μl of a 5% solution), the percentage of dead cells was evaluated on an inverted microscope.

For a clonogenic cell assay, 1 million exponentially-growing Nalm-6[28] or NCI-H6929 cells in 100 μl HEPES-buffered Eagle's medium-containing 10% human AB serum were mixed with 0.8 to 8 μg MAb (in 20 μl) and 600 μl of rabbit/human complement if Nalm-6 cells were used, or 160 μl human complement if NCI-H69 cells were used. After a 75-min incubation at room temperature cells were washed, resuspended in 4 ml RPMI 1640 medium containing 2 m*M* L-glutamine, 20 m*M* HEPES, penicillin 100 U/ml, streptomycin (100 μg/ml), 10% heat-inactivated fetal calf serum and 10% filtered culture supernatant from exponentially growing Nalm-6 or NCI-H69 cells. Serial 10-fold dilutions in the same medium were prepared and distributed into 32 microwells (96-well culture plate). After 14 d at 37°C in a humid atmosphere with 5% CO_2, wells containing at least one clone (>100 cells) were counted using an inverted microscope. The number of clonogenic units (CFU-L) in the diluted cell suspension (m) was calculated, assuming a Poisson distribution for plated cells, i.e., $m = -\ln P(O)$, where P(O) is the proportion of negative wells.

Peripheral blood mononuclear cells from abnormal donors were separated by centrifugation on Ficoll-Paque and incubated overnight in plastic Petri dishes in HEPES-buffered RPMI 1640 medium containing 20% human AB serum (RPMI-H-ABS), and where stated, 10 U interleukin-2 per milliliter. Non adherent cells were harvested, washed, and used as effector cells. Lymphocytes, activated for 2 d with phytohemagglutinin (PHA), were labeled overnight by incubation with ^{51}Chromium (3×10^6 cells and 100 μCi/ml) and washed, then 10,000 ^{51}Cr-labeled target cells were incubated for 30 min at room temperature in 100 μl RPMI-H-ABS containing 0.1 to 0.5 μg MAb. Then, 3×10^4 (E:T = 3) to 1×10^6 (E: T = 100) effector cells in 100 μl of the same medium were added. In order to promote cellular contact, the plates (round-bottomed 96-well culture plates) were centrifuged for 5 min at 250 *g*, and placed for 6 h at 37°C. The plates were centrifuged for 10 min at 700 *g* and 100 μl of supernatant was for the counting of released radioactivity. Wells containing neither MAb, nor effector cells were used for estimation of spontaneous release. Maximal release was obtained by addition of 0.5% Triton X-100 to the wells. Results are expressed as (release in sample-spontaneous release)/(maximal release-spontaneous release).

III. RESULTS AND DISCUSSION

A. COMPLEMENT-MEDIATED CYTOLYSIS

1. Source and Concentration of Complement

Rabbit complement is more efficient than human complement. The disadvantage of rabbit complement is its frequent toxicity for human hematopoietic progenitor cells. Absorption of the rabbit sera on human red blood cells, or better, use of baby rabbit serum mostly eliminates this toxicity. For each batch of sera, lack of toxicity should be checked by *in vitro* culture of CFU-GEMM of a bone marrow before and after treatment with the complement. Both activity and toxicity of a batch of complement depend on the working dilution. Therefore, the range of concentrations for which the complement is 100% active and presents acceptable toxicity has to be found. The range of reliable concentrations is more extended with baby rabbit complement than with absorbed adult rabbit complement (not shown).

Human complement seems to lack toxicity for hematopoietic progenitor cells. Although human complement is less effective than rabbit complement, more than 99% killing can be achieved with some rat IgM MAbs or by combining different IgG2b MAbs. Among 72 sera samples of normal blood donors tested, 66 (91%) were found to kill >99% of SCLC cells in association with an IgM anti-SCLC MAb while 35 (48%) killed 40 to 70% cells in association with an IgG2b MAb. Hence, the activity of the human serum to be used should be checked, but is probably more dependent on the way of preparation than on the donor.

2. Temperature of Treatment

Complement is heat-labile. Even at 37°C, observed in vitro activity is progressively lost. We found cytolysis to be optimal between 16 and 30°C for IgG2b MAbs. For an IgM MAb, this temperature dependence was less pronounced (Figure 1). This important, but often neglected effect of temperature should be kept in mind, especially when dealing with human complement.

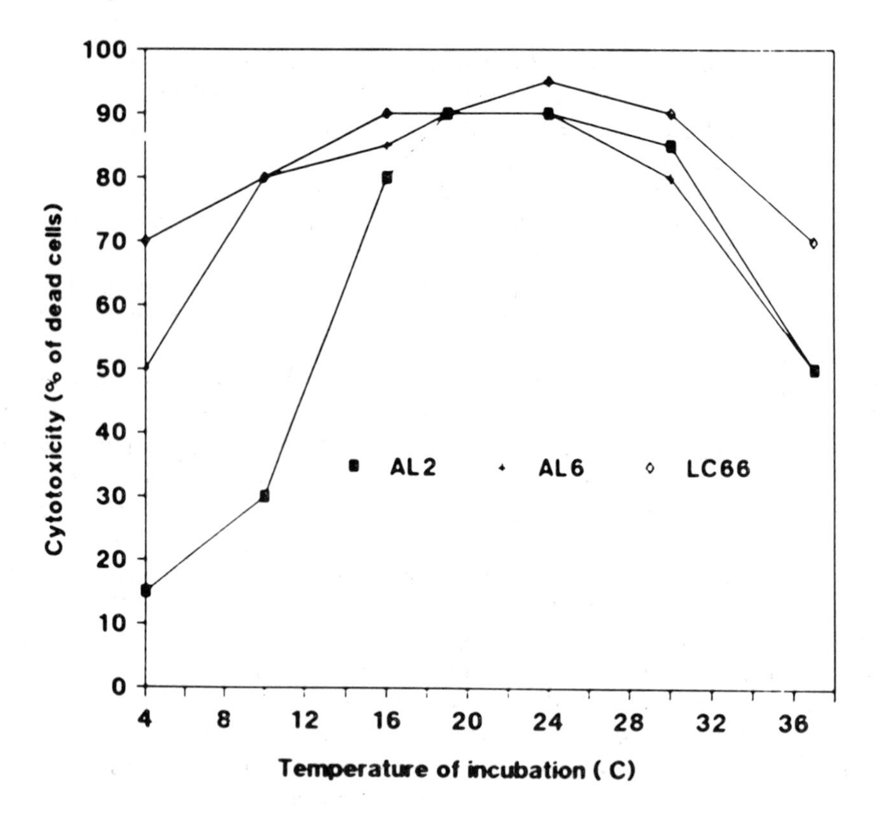

FIGURE 1. Effect of temperature on the lysis of Nalm-6 cells by rabbit complement mediated by AL2 (■), AL6 (+), or LC66 (◊).

3. Isotype of MAb

General agreement exists on the complement-fixing properties (rabbit, as well as human complements) of different rat Ig isotypes. IgM isotype is much more effective than IgG2b, which in turn is more effective than IgG2a, IgG2c, and IgG1.[30,31] This gradation holds for all cell types we have tested, although large quantitative differences in susceptibility to lysis between the different cell types were found (Table 1).

IgM fixes human as well as rabbit complements and triggers off substantial lysis. Our CD24 MAb LC66 kills on the average 3 to 4 logs of Nalm-6 cells or small cell lung carcinoma cells (Table 1). Until now, we have not produced any efficient pan-T MAb of the IgM isotype.

Killing in the presence of IgG2a or IgG2c remained very low, and high concentrations of MAb (10,000 times more than for an IgM MAb) were needed in order to achieve >50% killing.[24]

Lysis mediated by IgG2b MAbs varies widely, 0 to 2 log lysis with human complement and 0.5 to 5 log lysis with rabbit complement. Activity depends not only on the cell type involved (T lymphocytes were hardly lysed by human complement, even when using MAbs against major surface antigens), but also on the density of the antibody fixed on the cell surface (blends of 2 IgG2b MAbs can perform as well as an IgM MAb).

The group of H. Waldmann (Cambridge) studied lymphocyte lysis in detail: CAMPATH-1 mediated a 95 to >99% kill of mononuclear cells by human complement.[32] Bindon et al.[33] compared a series of anti-leukocyte common antigen (LCA) MAbs: lysis with IgG2a was invariably low (0 to 3%) and IgG2a never acted in synergy with IgG2a or IgG2b MAbs; lysis using a single IgG2b MAb was also low (0 to 10% with human complement and 10 to 20% with rabbit complement), but two IgG2b MAbs directed against different epitopes of LCA, or a different molecule on the cell surface, always acted in synergy (see below). IgG1 seemed less active than IgG2a.[31]

TABLE 1
Effect of MAb-Isotype on Lysis of Different Cell Types

	NCI-H69 (SCLC)	Nalm-6 (cALL)	T-lymphocytes (PHA-act)
Human c′			
IgM	Anti-LCA1: 95—>99% 3—4 log	LC66: >99% 3—4 log	LO-MN26: 40—60%
	LC66: 95->99% 2—3 log		
IgG2a	Anti-LCA3: 20—80%	NT	LO-CD5-a: 0%
			LO-CD6-a 0%
IgG2b	Anti-LCA2: 40—80%	AL2 : 80—95%	LO-Tact-1 0-5%
		AL6: 90—99%	LO-CD2-act: 0—5%
Irrelevant Ig	20%	0—10%	1—5%
Rabbit c′			
IgM	Anti-LCA1:	LC66: >99% ± 4 log	LO-MN26: 90—>99%
	LC66: 95—>99%		
IgG2b	Anti-LCA2: 40—90%	AL2: 80—>99% ± 4 log	LO-Tact-1: 40—50%
		AL6: >99% ± 5 log	LO-CD2-act: 70—80%
IgG2a	Anti-LCA3: 20%	NT	LO-CD5-a: 5—20%
			LO-CD6-a: 30%
IgG2c	LO-CD24: 60—>99%	AL1a: 50—90%	NT
Irrelevant Ig	10—20%	10—20%	5—20%

Note: Results are expressed as the percentage of dead cells, when vital dye-exclusion assays were used, or as the number of logs of tumor cell reduction when clonogenic assays were used.

Hence, it seems that IgG2b fixes rabbit as well as human complements, but is probably less efficient in bringing about lysis of the cell than IgM. The complement-fixing capacity of nonaggregated IgG2a, IgG2c, and IgG1 remains questionable.

4. Use of Blends of MAbs

The different reports on treatment with mouse MAbs have pointed out that blends of MAbs are more efficient than single MAbs.[3,4] We wanted to investigate if this is true for all non T-ALL cases, because in our first experiments on cALL, more than 99% killing was easily achieved by single MAbs and rabbit complement, and no beneficial effect resulting from the use of cocktails of MAbs could be demonstrated in these cases.[24]

In the first series of experiments, we therefore compared the responsiveness of different clinical samples to treatment with single or combined MAbs and rabbit complement.[34,35] Figure 2 shows the results obtained for four samples representative of each of four different phenotypes observed in cALL and null-ALL with phenotypes like 1 and 3 being more frequent than CD24— phenotypes (cases 2 and 4).

In case 1, more than 99% cell killing was achieved, using 40 ng of LC66 or of AL2, or 400 ng of AL6. For other samples of cALL, bearing moderate or high amounts of CD10, CD9, and CD24 antigens, similar results were obtained with sometimes higher cytotoxicity with AL6 or a somewhat lower cytotoxicity with AL2. On the contrary, for case 2, in which the CD24 antigen was hardly detected by immunofluorescence, single MAbs brought about 50 to 75% lysis and cocktails of 2 MAbs were required to obtain 90% tumor-cell elimination. For samples of null-ALL, bearing high densities of CD9 and CD24 on their surface, treatment with AL6 or LC66 resulted in >99% lysis and the combination of MAbs neither improved nor lowered these results (Figure 2). Null-ALL cell samples unreactive with CD24 MAbs were only slightly sensitive to cytolysis by AL6, AL2, or LC66. However, the combination of AL6 and LC66 resulted in the elimination of 95% of the tumor cells. Killing efficiency of the combined MAbs was even higher than the addition of the effects of the two MAbs used separately. HLA-DR leukemia cells were not efficiently lysed by any combination of AL2, AL6, or LC66 MAbs. Hence, complement

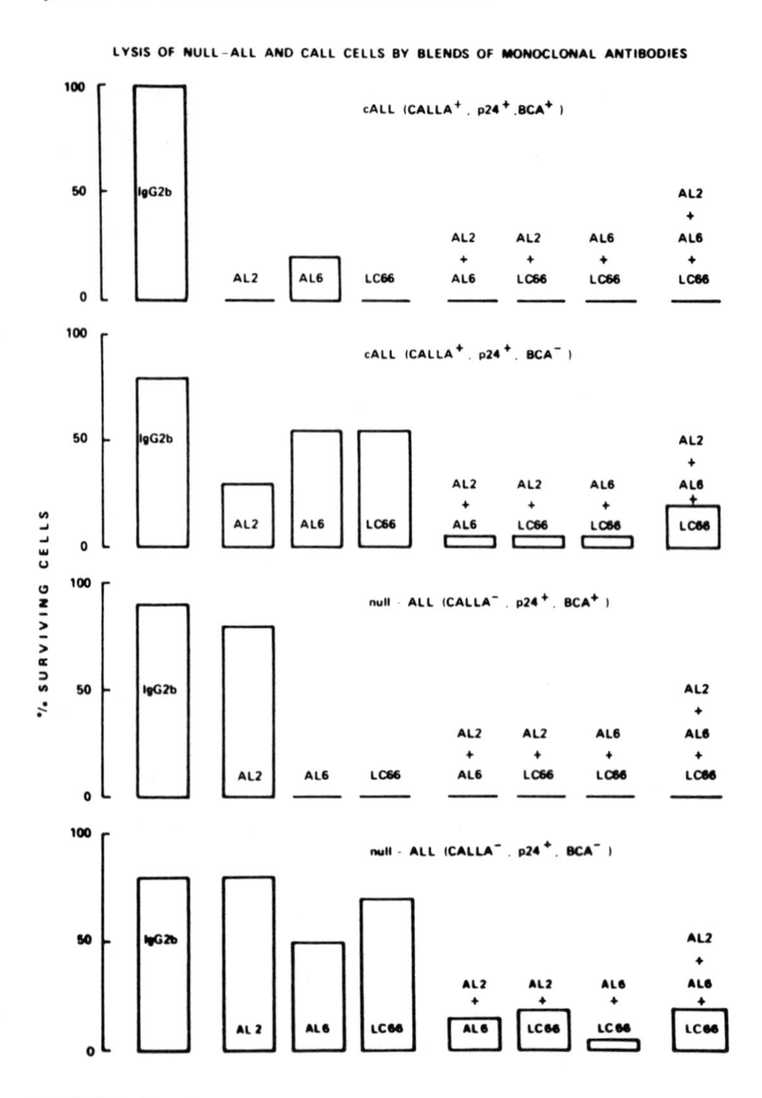

FIGURE 2. Complement-mediated cytolysis of malignant cells in null-ALL and cALL clinical samples by treatment with single or combined MAbs.

seems to lyse only those cells covered with enough MAb and complement. The threshold density level is obtained either with one MAb directed against a major antigen, or with a blend of two or three MAbs directed against different minor antigens.

These results are in agreement with those presented by Campana and Janossy,[36] who reported a higher than 4-log reduction of tumor cells in 48% of ALL cases using single MAbs and rabbit complement and in 77% of ALL cases when blends were used.

We were tempted to conclude from these first experiments that the choice of MAbs had to be adapted to each patient. However, the vital dye-exclusion technique used in the previous paragraph does not ensure detection of less than 1% surviving cells. These very few surviving cells could be those producing relapse. To study the effect of blend MAbs at this level, we set up a clonogenic assay allowing detection of 0.0001% surviving cells. When Nalm-6 cells were treated with increasing amounts of AL6 alone or LC66 alone, a plateau level of cytotoxicity was found for concentrations above 100 ng MAb per assay being, respectively, 5-log and 4-log cell elimination. Hence, whatever the amount of MAb used, a minor subpopulation escapes from treatment. When two MAbs were combined for treatment, 3 to 10 times less clonogenic units

TABLE 2
Recovery of Clonogenic Cells (CFU-L) after Treatment with Blends of MAbs

	CFU-L	
	Total amount	**Tumor cell elimination**
Rabbit complement		
Medium	3.5×10^6	—
LC66 0.8 µg	135	4 log
1.6 µg	141	4 log
AL2 8 µg	48	5 log
16 µg	204	4 log
AL6 2 µg	10	5.5 log
4 µg	22	5 log
LC66 0.8 µg + AL2 8 µg	9	5.5 log
LC66 0.8 µg + AL6 2 µg	0	>6 log
AL2 8 µg + AL6 2 µg	3	6 log
LC66 + AL6 + AL2	<13	>5 log
Human complement		
Medium	1.1×10^6	
AL2 8 µg	66,250	1.5 log
AL6 8 µg	20,000	2 log
LC66 8 µg	131	4 log
AL2 4 µg + AL6 4 µg	50,000	1.5 log
AL2 4 µg + LC66 4 µg	19	5 log
AL6 4 µg + LC66 4 µg	8	5 log
AL2 2.5 µg + AL6 2.5 µg + LC66 2.5 µg	13	5 log

Note: One million Nalm-6 cells were treated for 75 min at room temperature with medium alone or with the indicated combinations of MAb and 600 µl of rabbit or human complement.

were recovered than after treatment with single MAbs (Table 2). A blend of three MAbs did not enhance the killing efficiency. The same beneficial effect of blends was observed when using human complement: 1.5- or 4-log elimination with a single IgG2b or IgM MAb, respectively, and 5-log elimination for IgG2b + IgM combinations.

We suspect that the occasional cells that escape from lysis, when one MAb is used but that are killed when a blend of MAbs is used, escape because they lack or bear low amounts of the target antigen, even in a population assumed to be homogenous (here, the Nalm-6 cell line). LeBien et al.[4] came to the same conclusion, when studying cytolysis of the KM3 cell line, the cells surviving treatment with the BA1 + BA2 + BA3 cocktail expressed smaller amounts of the BA1, BA2, and BA3 antigens than the control KM3 cells.

Hence, a blend of MAbs should always be used not only to eradicate the occasional cells with "uncommon" phenotype, but also to enhance the probability of eliminating the true clonogenic leukemia cell, which has not necessarily the same phenotype as the predominant population.

As to the question of how many different MAbs can be used in a cocktail, Bindon et al.[33] have shown that a third MAb can interfere in the synergistic lysis obtained with two MAbs. In our experiment, the addition of a third MAb was rarely beneficial, and the reduction in efficiency, if any, was low (<1 log) and not significant. An alternative is to perform two cycles of treatment with two different blends of MAbs.

Another interesting point revealed by the work of Bindon et al.[33] is the advantage of using two MAbs directed against different epitopes on the same molecule, rather than on different molecules of the cell surface. They observed 50 to 66% lysis when two anti-leukocyte common antigen MAbs were combined, but only 5 to 27% when an anti-LCA was combined with an anti-T cell MAb.

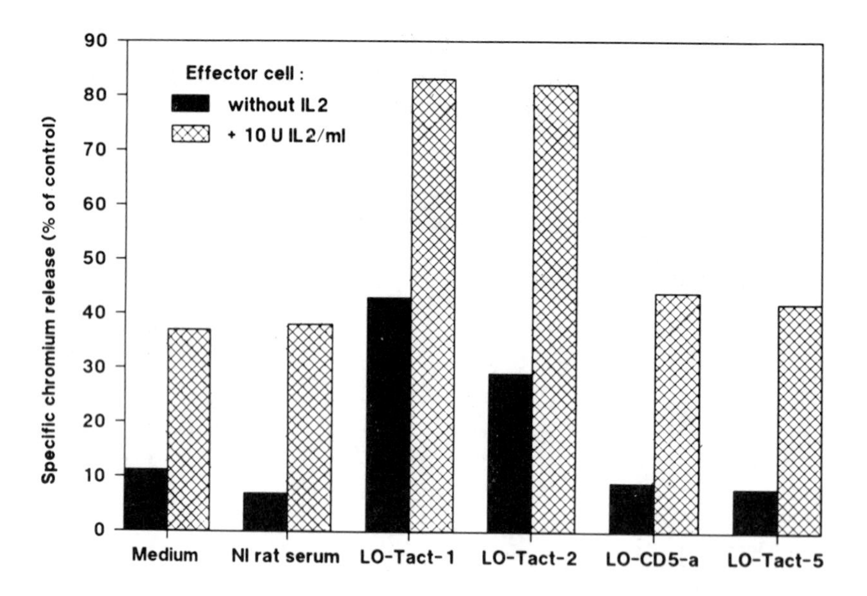

FIGURE 3. Effect of MAb isotype on ADCC. Cytolysis of PHA-activated lymphocytes by blood mononuclear cells (E/T:25/1) preincubated for 16 h in the absence (■) or the presence (❑) of 10 U of IL2 per milliliter was measured by ^{51}Cr-release.

From a practical point of use, it seems important to us to select the more efficient IgM MAbs, whatever their epitope specificity may be for bone marrow purging.

B. ANTIBODY-DEPENDENT CELL-MEDIATED CYTOTOXICITY

Complement-mediated cytolysis seems to play a minor role where serotherapy is concerned. In humans, mouse IgG2a and IgG3 are the most active isotypes for *in vivo* reduction of tumor[37] or for elimination of circulating T cells.[17,18] These subclasses, together with rat IgG2b, happen to be the most active *in vitro* in ADCC assays, mediated by killer cells, NK cells, or T-gamma lymphocytes.[22,31,38,39] IL2 has been shown to enhance ADCC *in vitro*[40] and to act in synergy with mouse anti-tumor IgG2a and IgG2b MAbs for the inhibition of tumor growth.[39]

We tested our rat MAbs for ADCC with PHA-activated T lymphocytes as targets (Figure 3). The results were clear-cut; all IgG2b MAbs mediated lysis, usually 40 to 50% ^{51}Cr-release was obtained and IgG2a and IgM MAbs were totally inefficient in this system. Only one IgG2c MAb was tested and shown to be inactive. The extent of cytolysis depended on the state of activation of the effector cells. After overnight incubation with interleukin-2 at low concentration (10 U/ml), lysis was markedly increased up to 80% ^{51}Cr-release with IgG2b MAbs, and corresponding to higher than 90% lysis by comparison with complement-mediated lysis estimated by vital dye-exclusion and ^{51}Cr-release (Figure 3). However, the specificity disappeared as lysis was observed with normal rat serum and with all MAbs tested. High levels of cytolysis are probably due to a superposition of true ADCC and short-term LAK effect. The same high levels of specific IgG2b MAb-dependent cellular cytolysis and high levels of aspecific cell cytolysis are obtained with effector cells of a donor during immunization/infection (Figure 4).

Rat MAbs certainly constitute a valuable clinical reagent complementary to mouse MAbs: first, rat IgM and IgG2b MAbs activate human and rabbit complements and IgG2b MAbs activate human killer cells for ADCC; second, they can be used in serotherapy as an alternative to mouse MAb, when the patient develops antibodies against the mouse Ig; and finally, rat and mouse having different immune repertoires and MAbs with slightly different specificities can be obtained. The fact that rat MAbs are easily produced in large quantities and separated from any contaminating antibody makes the system even more attractive.

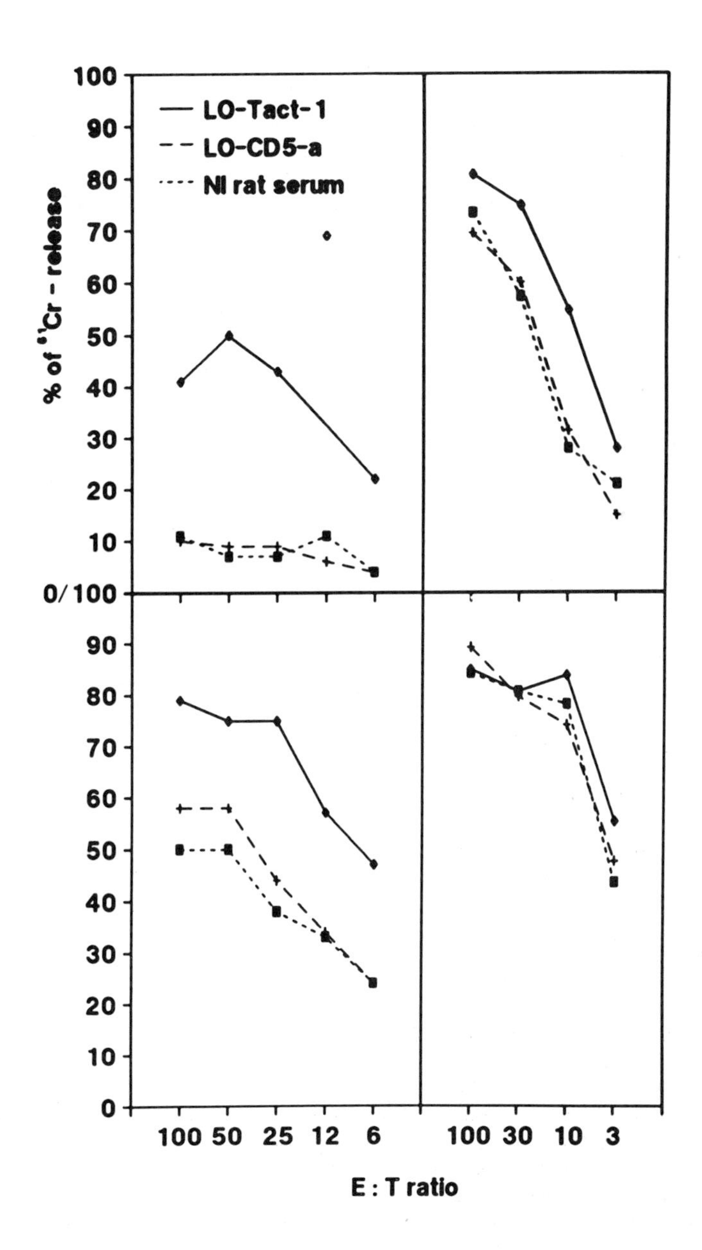

FIGURE 4. Effect on ADCC of the state of activation of effector cells. Cytolysis of PHA-activated lymphocytes was effected by nonadherent blood mononuclear cells from a normal blood donor (left) or a donor suffering from an infection (right). Effector cells had been preincubated for 18 h in the absence (upper) or presence of 10 U of IL2 per milliliter (lower).

REFERENCES

1. **Ritz, J., Bast, R. C., Jr., Clavell, L. A., Hercend, T., Sallan, S. E., Lipton, J. M., Feeney, M., Nathan, D. G., and Schlossman, S. F.,** Autologous bone-marrow transplantation in CALLA-positive acute lymphoblastic leukaemia after *in vitro* treatment with J5 monoclonal antibody and complement, *Lancet,* 2, 60, 1982.
2. **Ramsay, N., LeBien, T., Nesbit, M., McGlave, P., Weisdorf, D., Kenyon, P., Hurd, D., Goldman, A., Kim, T., and Kersey, J.,** Autologous bone marrow transplantation for patients with acute lymphoblastic leukemia in second or subsequent remission: results of bone marrow treated with monoclonal antibodies BA-1, BA-2, BA-3 plus complement, *Blood,* 66, 508, 1985.
3. **Bast, R. C., Jr., De Fabritiis, P., Lipton, J., Gelber, R., Mower, C., Nadler, L., Sallan, S., and Ritz, J.,** Elimination of malignant clonogenic cells from human bone marrow using multiple monoclonal antibodies and complement, *Cancer Res.,* 45, 499, 1985.
4. **Lebien, T. W., Stepan, D. E., Bartholomew, R. M., Stong, R. C., and Anderson, J. M.,** Utilization of a colony assay to assess the variables influencing elimination of leukemic cells from human bone marrow with monoclonal antibodies and complement, *Blood,* 65, 945, 1985.
5. **Poynton, C. H., Dicke, K. A., Culbert, S., Frankel, L. S., Jagannath, S., and Reading, C. L.,** Immunomagnetic removal of CALLA positive cells from human bone marrow, *Lancet,* 1, 524, 1983.
6. **Kemshead, J. T., Treleaven, J. G., Gibson, F. M., Ugelstad, J., Rembaum, A., and Philip, T.,** Monoclonal antibodies and magnetic microspheres used for depletion of malignant cells from bone marrow, in *Autologous Bone Marrow Transplantation,* Dicke, K. A., Spitzer, G., and Zander, A. R., Eds., University of Texas, M.D. Anderson Hospital and Tumor Institute at Houston, 1985, 409.
7. **Prentice, H. G., Janossy, G., Price-Jones, L., Trejdosiewicz, L. K., Panjwani, D., Graphakos, S., Ivory, K., Blacklock, H. A., Gilmore, M. J. M. L., Tidman, N., Skeggs, D. B. L., Ball, S., Patterson, J., and Hoffbrand, A. V.,** Depletion of T-lymphocytes in donor marrow prevents significant graft-versus-host disease in matched allogeneic leukaemic marrow transplant recipients, *Lancet,* March 3, 472, 1984.
8. **Waldmann, H., Hale, G., Cividalli, G., Weshler, Z., Manor, D., Rachmilewitz, E. A., Polliak, A., Or, R., Weiss, L., Samuel, S., Brautbar, C., and Slavin, S.,** Elimination of graft-versus-host disease by *in vitro* depletion of alloreactive lymphocytes with a monoclonal rat anti-human lymphocyte antibody (Campath-1), *Lancet,* Sept. 1, 483, 1984.
9. **Martin, P. J., Hansen, J. A., Buckner, C. D., Sanders, J. E., Deeg, H. J., Stewart, P., Appelbaum, F. R., Clift, R., Fefer, A., Witherspoon, R. P., Kennedy, M. S., Sullivan, K. M., Flournoy, N., Storb, R., and Thomas, E. D.,** Effects of *in vitro* depletion of T cells in HLA-identical allogeneic marrow grafts, *Blood,* 66, 664, 1985.
10. **Herve, P., Cahn, J. Y., and Flesch, M.,** Successful graft versus host disease prevention without graft failure in 32 HLA identical allogeneic bone marrow transplantation with marrow depleted of T cells by monoclonal antibodies and complement, *Blood,* 69, 388, 1987.
11. **Maraninchi, D., Blaise, D., Rio, B., Leblond, V., Dreyfus, F., Gluckman, E., Guyotat, D., Pico, J. L., Michallet, M., Ifrah, N., and Bordigoni, A.,** Impact of T-cell depletion on outcome of allogeneic bone-marrow transplantation for standard-risk leukaemias, *Lancet,* 25, 175, July 1987.
12. **Filipovich, A. H., Vallera, D. A., Youle, R. J., Haake, R., Blazar, B. R., Arthur, D., Neville, D. M., Ramsay, N. K. C., McGlave, P., and Kersey, J. H.,** Graft-versus-host disease prevention in allogeneic bone marrow transplantation from histocompatible siblings. A pilot study using immunotoxins for T cell depletion of donor bone marrow, *Transplantation,* 44, 62, 1987.
13. **Blazar, B. R., Quinones, R. R., Heinitz, K. J., Sevenich, E. A., and Filipovich, A. H.,** Comparison of three techniques for the ex vivo elimination of T cells from human bone marrow, *Exp. Hematol.,* 13, 123, 1985.
14. **de Witte, T., Hoogenhout, J., de Pauw, B., Holdrinet, R., Janssen, J., Wessels, J., van Daal, W., Hustinx, T., and Haanen, C.,** Depletion of donor lymphocytes by counterflow centrifugation successfully prevents acute graft-versus-host disease in matched allogeneic marrow transplantation, *Blood,* 67, 1302, 1986.
15. **Slavin, S., Or, R., Weshler, Z., Hale, G., and Waldmann, H.,** The use of total lymphoid irradiation for abrogation of host resistance to T-cell depleted marrow allografts, *Bone Marrow Transplant,* 1, 98, 1986.
16. **Hale, G. and Waldmann, H.,** Compath-1 for prevention of graft-versus-host disease and graft rejection. Summary of results from a mutli-centre study, *Proceedings of the EBMT Meeting,* Chamonix, France, April 10 to 13, 1988.
17. **Cosimi, A. B., Burton, B. C., Colvin, R. B., Goldstein, G., Delmonico, F. L., Laquaglia, M. P., Tolkoff-Rubin, N., Rubin, R. H., Herrin, J. T., and Russell, P. S.,** Treatment of acute allograft rejection with OKT3 monoclonal antibody, *Transplantation,* 32, 535, 1981.
18. Ortho MultiCenter Transplant Study Group, A randomized clinical trial of OKT3 monoclonal antibody for acute rejection of cadaveric renal transplants, *N. Engl. J. Med.,* 313, 337, 1985.
19. **Soulillou, J. P., Le Mauff, B., Olive, D., Delaage, M., Peyronnet, P., Hourmant, H., Mawas, C., Hirn, M., and Jacques, Y.,** Prevention of rejection of kidney transplants by monoclonal antibody directed against interleukin 2, *Lancet,* June 13, 1339, 1987.

20. **Beelen, D. W., Graeven, U., Schulz, G., Grosse-Wilde, H., Quabeck, K., Sayer, H., Schaefer, U. W., and Schmidt, C. G.,** Primary treatment of acute graft-versus-host disease with a monoclonal antibody (BMA 031) against the T-cell receptor, *Proceedings of the EBMT Meeting,* Chamonix, France, April 10 to 13, 1988.
21. **Rankin, E. M., Hekman, A., Somers, R., and ten Bokkel Huinink, W.,** Treatment of two patients with B cell lymphoma with monoclonal anti-idiotype antibodies, *Blood,* 65, 1373, 1985.
22. **Kipps, T. J., Parham, P., Punt, J., and Herzenberg, L. A.,** Importance of immunoglobulin isotype in human antibody-dependent, cell mediated cytotoxicity directed by murine monoclonal antibodies, *J. Exp. Med.,* 161, 1, 1985.
23. **Lebacq-Verheyden, A. M., Ravoet, A. M., Bazin, H., Sutherland, D. R., Tidman, N., and Greaves, M. F.,** Rat AL2, AL3, AL4 and AL5 monoclonal antibodies bind to the common acute lymphoblastic leukaemia antigen (CALLA gp 100), *Int. J. Cancer,* 32, 273, 1983.
24. **Lebacq-Verheyden, A. M., Humblet, Y., Neirynck, A., Ravoet, A., and Symann, M.,** Four rat cytotoxic monoclonal antibodies for the *in vitro* treatment of bone marrow autografts in non T non B acute lymphoblastic leukemias, in *Autologous Bone Marrow Transplantation,* Proc. 1st Int. Symp., Dicke, K. A., Spitzer, G., and Zander, A. R., Eds., University of Texas, M.P. Anderson Hospital and Tumor Institute at Houston, 1985, 419.
25. **Lebacq-Verheyden, A. M., Neirynck, A., Ravoet, A. M., Humblet, Y., Oie, H., Linnoila, I., Gazdar, A. F., Minna, J. D., and Symann, M.,** Monoclonal antibodies for the *in vitro* detection of small cell lung cancer metastases in human bone marrow, *Eur. J. Cancer Clin. Oncol.,* 24, 137, 1988.
26. **Lebacq-Verheyden, A. M., Humblet, Y., Ravoet, A. M., and Symann, M.,** Immunological removal of cancer cells in bone marrow autografts: setting the experimental conditions, in *Autologous Bone Marrow Transplantation and Solid Tumours,* McVie, J. G., Dalesio, O., and Smith, I. E., Eds., Raven Press, New York, 1984, 19.
27. **Terasaki, P. I., Bernoco, D., Park, M. S., Ozturk, G., and Iwaki, Y.,** Microdroplet testing for HLA A,B,C and D antigens, *Amer. J. Clin. Pathol.,* 69, 103, 1978.
28. **Minowada, J., Janossy, G., Greaves, M. F., Tsubota, T., Srivastava, B. I. S., Morikawa, S., and Tatsumi, E.,** Expression of an antigen associated with acute lymphoblastic leukemia in human leukemia-lymphoma cell lines, *J. Natl. Cancer Inst.,* 60, 1269, 1978.
29. **Gazdar, A. F., Carney, D. N., Russell, E. K., Sims, H. L., Baylin, S. B., Bunn, P. A., Guccion, J. G., and Minna, J. D.,** Establishment of continuous, clonable cultures of small-cell carcinoma of the lung which have amine precursor uptake and decarboxylation cell properties, *Cancer Res.,* 40, 3502, 1980.
30. **Füst, G., Medgyesi, G. A., Bazin, H., and Gergely, J.,** Differences in the ability of rat IgG subclasses to consume complement in homologous and heterologous serum, *Immunol. Lett.,* 1, 249, 1980.
31. **Hale, G., Cobbold, S. P., Waldmann, H., Easter, G., Matejtschuk, P., and Coombs, R. R. A.,** Isolation of low-frequency class-switch variants from rat hybrid myelomas, *J. Immunol. Methods,* 103, 59, 1987.
32. **Hale, G., Bright, S., Chumbley, G., Hoang, T., Metcalf, D., Munro, A. J., and Waldmann, H.,** Removal of T cells from bone marrow for transplantation: a monoclonal antilymphocyte antibody that fixes human complement, *Blood,* 62, 873, 1983.
33. **Bindon, C. I., Hale, G., Clark, M., and Waldmann, H.,** Therapeutic potential of monoclonal antibodies to the leukocyte-common antigen, *Transplantation,* 40, 538, 1985.
34. **Lebacq-Verheyden, A. M., Humblet, Y., Neirynck, A., Sekhavat, M., Ravoet, A. M., and Symann, M.,** Rat monoclonal antibodies for the *in vitro* treatment of bone marrow autografts in acute non-T non-B lymphoblastic leukemia, Proceedings of the Stohlman Jr. Memorial Symposium, Brussels, Oct. 26 to 27, 1984.
35. **Sekhavat, M., Ravoet, A. M., Humblet, Y., Neirynck, A., Lebacq, A. M., and Symann, M.,** *In vitro* elimination of leukaemic cells with cocktails of 2 or 3 rat monoclonal antibodies and complement as compared to single monoclonal antibodies, in *Proceedings of the Second International Symposium on Detection and Treatment of Minimal Residual Disease in Acute Leukemia,* Rotterdam, The Netherlands, Nov. 6 to 8, 1985.
36. **Campana, D. and Janossy, G.,** Leukemia diagnosis and testing of complement-fixing antibodies for bone marrow purging in acute lymphoid leukemia, *Blood,* 68, 1264, 1986.
37. **Houghton, A. N., Mintzer, D., Cordon-Cardo, C., Welt, S., Fliegel, B., Vadham, S., Carswell, E., Melamed, M. R., Oettgen, M. F., and Old, L. J.,** Mouse monoclonal IgG3 antibody detecting GD3 ganglioside: a phase I trial in patients with malignant melanoma, *Proc. Natl. Acad. Sci.,* 82, 1242, 1985.
38. **Hale, G., Clark, M., and Waldmann, H.,** Therapeutic potential of rat monoclonal antibodies: isotype specificity of antibody-dependent cell-mediated cytotoxicity with human lymphocytes, *J. Immunol.,* 134, 3056, 1985.
39. **Berinstein, N. and Levy, R.,** Treatment of a murine B cell lymphoma with monoclonal antibodies and IL2, *J. Immunol.,* 139, 971, 1987.
40. **Ortaldo, J. R., Woodhouse, C., Morgan, A. C., Herberman, R. B., Cheresh, D. A., and Reisfeld, R.,** Analysis of effector cells in human antibody-dependent cellular cytotoxicity with murine monoclonal antibodies, *J. Immunol.,* 138, 3566, 1987.

Chapter 30.II

RAT ANTIBODIES AND K CELLS

D. M. Chassoux, L. G. Linares-Cruz, and H. Bazin

TABLE OF CONTENTS

I. INTRODUCTION

K cell activity[1] has been described as the ability of certain mononuclear cells to kill target cells sensitized by antibody. They do so through their Fc receptor that recognizes the Fc part of bound, but not free, antibody.[2] Any nucleated cell capable of division may be destroyed by such an effector mechanism. Target structures may be histocompatiblity antigens, viral antigens, differentiation antigens, membrane determinants as receptors, or haptens artificially introduced at the membrane. Human K cell activity may be elicited by human antibodies, or classically by rabbit antisera. Usually a 10^{-4} to 10^{-5} dilution of an appropriate antiserum would sensitize relevant target cells for optimal K cell activity. The effector cell is found in peripheral blood, spleen, peritoneal cavity, and inflammatory exudates, but not in bone marrow or tonsils.

The origin of K cells is not known, but they do not belong to either T or B cells.[3] Their site of maturation is not known.[4] The role of K cells remains to be determined and this justifies further investigations.

An experimental system for studying K cells is offered by the rat species. In the rat, K cell activity is readily detected in spleen and peripheral blood. Spontaneous mutations, as the Rowett Nude mutation (athymic rat or rnu/rnu) and inbred strains with given characteristics may be used to study the development and the genetic control of K cell activity. Rat K cells are elicited by rat antibodies. The rat hybridoma technology thus offers the unique possibility to understand more on the role of antibody in this type of cytotoxicity with the obtainment of large quantities of purified antibodies against a given antigen. Moreover, the separation method using kappa allotype developed by Bazin et al.[5] allows the obtention of antibodies devoid of contamination by Ig of other isotypes.

II. RAT ANTIBODIES AND RAT K CELLS

The assessment of K cell activity in the rat is done by an *in vitro* cytotoxicity method that involves target cells with an isotopic label, a rat antiserum raised against the target, and splenic effector cells.

The target cells are chosen so that their susceptibility to lysis in absence of antibody is minimal (i.e., NK resistant). A cell line called CHANG, derived from a human hepatocarcinoma, has been largely used for that purpose. It is easily grown in standard tissue culture conditions, requires low amounts of serum and takes up ^{51}Cr readily. It is immunogenic in all rat strains tested. Also, it is possible to modify CHANG cells with an hapten.

A. IMMUNIZATION OF RATS

Viable CHANG cells were washed three times in phosphate buffered saline (PBS) at room temperature and 4 to 5 $\times$ 10^6 cells mixed with 0.5 ml of complete Freund's adjuvant. Subcutaneous injections were done in several points for each rat receiving 0.5 ml of the mixture. Five to seven weeks later, rats were challenged intraperitoneally with 10 to 12 $\times$ 10^6 live CHANG cells per individual. Sera were collected from blood drawn 5 d later, aliquoted and kept at –20°C. For hybridization, spleens were removed at day 4 after challenge and processed according to De Clercq et al.[6]

B. CYTOTOXICITY ASSAY

Target cells were incubated with ^{51}Cr for 1 h and washed. They were distributed at 10^4/100 microliter per well in round-bottom wells of culture microtiter plates. The wells contained dilutions of antisera or ascites to be tested, or hybridoma culture supernatants. Effector cells were then added. We found that Nude rat spleen cells have a higher relative activity compared with rnu/+ euthymic littermates, because K cells are not then diluted by T cells.[7] Such effector cells were routinely used to assess antibody activity, at a 50/1 effector/target ratio (E/T).

The assay is done in a 6 h incubation period at 37°C in a CO_2 incubator set at 5% CO_2 in air. Half of the supernatants are then collected and transferred into a gamma counter. The calculation of the percent radiolabel released by target cells compared with the total label, allows for the determination of cytotoxicity level. Specific cytotoxicity (SC) takes in account baseline ^{51}Cr released from target cells and maximum releasable value.

C. CELLULAR RADIO-IMMUNOASSAY

CHANG cells were distributed at 2×10^5/ml, 100 microliters per well in flat-bottom flexible assay plates, and left to adhere spontaneously on the plastic surface for 2 h at 37°C in RPMI culture medium supplemented with 10% fetal calf serum. Excess medium was removed and two washes with PBS at room temperature performed by flicking the plates. Cells were fixed with 0.25% glutaraldehyde. Incubation with samples to be tested was run for two h at room temperature and bound antibodies revealed by a mouse anti-rat kappa monoclonal antibody (MAb) MARK-1[8] labeled with ^{125}I. After extensive washing, wells were cut and retained label measured in a gamma counter.

D. ANTI-CHANG RAT MAb, INDUCING CYTOTOXICITY BY RAT K CELLS

Hybridomas were prepared by fusion of LOU/C splenocytes immunized against CHANG cells (as described above) and IR 983F non-secreting fusion cell line,[6] using PEG as fusing agent. Supernatants were screened on immunizing cells by cytotoxicity assay using spleen cells from rnu/rnu rats as effectors and counterscreened by cellular RIA.

Out of 288 wells seeded in a fusion between LOU/C immune spleen cells and IR 983F, hybrid cells grew in 24% of the wells. In cytotoxicity assay 9 clones were positive (11%) and 14 other clones were recorded in RIA. The reason for such a discrepancy of results given by these two assays has not been investigated. One may infer that antibodies recorded in RIA were of isotypes inoperative in cytotoxicity and that antibodies detected by a K cell assay were of a concentration and/or an affinity insufficient for detection in our cellular RIA. Only one clone called IIID10 was positive in both assays and was selected for further studies. Maximal level of cytotoxicity obtained with culture supernatants and ascites were of the same order of magnitude as that obtained with polyclonal rat antiserum. A 1:1000 dilution of IIID10 supernatant gave 50% SC with rnu/rnu effectors, at a 50/1 E/T; ascites may be diluted 1:12,000 to get the same activity. The antibody is of IgG2b isotype.

The specificity of the antibody is under study. In immunofluorescence, a bright perinuclear labeling is observed, in addition to a membrane staining of live CHANG cells after acetone fixation (Figure 1).

E. ISOTYPES AND RAT K CELLS

A unique possibility to study directly and not by inhibition,[10] the role of the isotype of antibodies in K cell cytotoxicity was offered by a series of rat monoclonals against the same antigen DNP (dinitrophenol) obtained by Bazin and colleagues[8] (see Chapter 19.XI.). They were used to sensitize hapten-modified CHANG cells. Seventeen IgG, one IgE, six IgA, and two IgM LO-DNP antibodies were tested in ascites form, some were also tested after purification.[5] We found that none of IgM or IgA LO-DNP antibodies could elicit rat K cell cytotoxicity. All IgGs worked. Concentrations of the antibodies in ascites fluids were assessed by cellular RIA. The maximum level of lysis (70%) was achieved by using IgG2a LO-DNP that could be diluted up to 1:32,000. IgG2b LO-DNP were less efficient at the same concentration and IgG1 LO-DNP antibodies displayed a maximum activity lower than IgG2b, that also required a higher antibody concentration. Results were confirmed when purified antibodies were used. The IgE LO-DNP was found to elicit rat K cell cytotoxicity.[24]

It is suggested that rat K cells may have distinct Fc receptors for these four isotypes. This recalls observations made on rat macrophages, where receptors for IgG2a, IgG2b/IgG1, and IgE

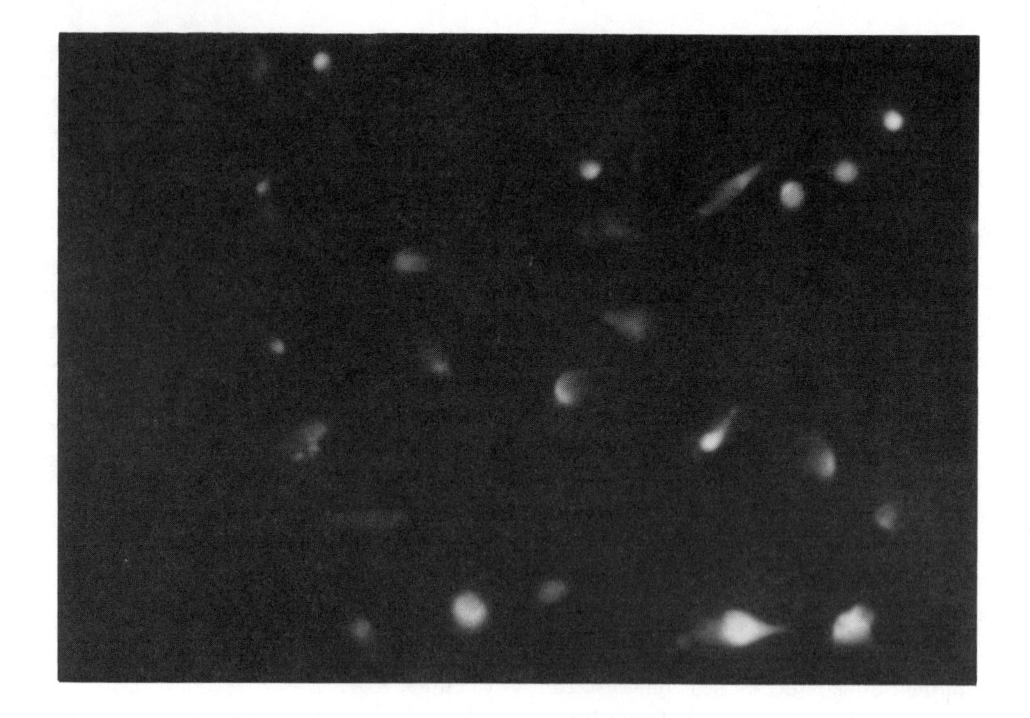

A

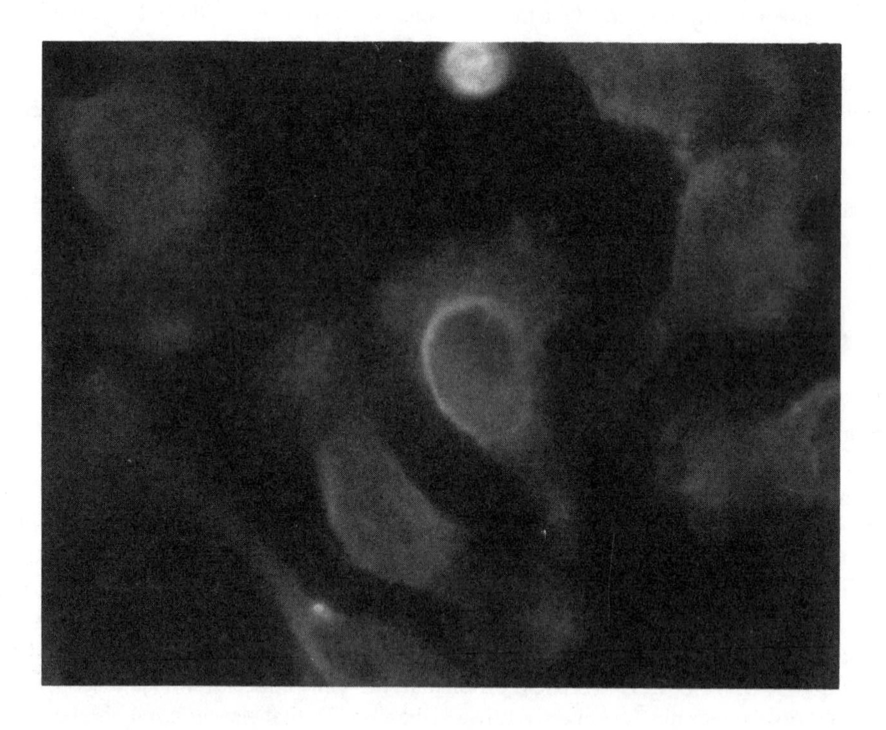

B

FIGURE 1. CHANG cells were allowed to adhere spontaneously onto a glass slide. They were fixed with acetone and then stained with the IIID10 rat monoclonal antibody revealed by a goat anti-rat reagent coupled to Rhodamine (A) or to Fluoresceine (B) (Nordic), magnification is ×500 (A) and ×1250 (B). An intense perinuclear staining is noted.

TABLE 1
Induction by Rat Monoclonal Antibodies of Human Peripheral Blood Mononuclear Cells K Cell Activity on Various Target Cells

Clone	Ref.	Isotype	Concentration	Activity (SC)	E/T	Target	Ag
30-H12	14	Rat IgG2b	0.1 μg/ml	80%	25/1	Mouse lymphoma cell line	Thy 1.2
YTH 54.12	13	Rat IgG2b	1 μg/ml	50%	20/1	PBL	LCA
YTS 154.7	13	Rat IgG2b	1 μg/ml	50%	50/1	Mouse thymocytes	Thy 1
LO-DNP-57	[a]	Rat IgG2b	0.O25 μg/ml	90%	20/1	CHANG-TNP	DNP
HL1-1	[b]	Rat IgG2b	1 μg/ml	90%	40/1	Human leukemic cell line	HLA-A-B-C
AL2	15	Rat IgG2b	ascites form 1/2 × 10^6	100%	40/1	Human leukemic cell line	CALLA

[a] See Chapter 19. XI.
[b] Ravoet, Ninane, De Bruyère, Sokal, and Bazin, unpublished results.

were found.[11] Another possibility is that a unique Fc receptor possesses different affinities for the Fc of reacted antibodies according to their isotype.

Thus, this rat model provides the possibility to study K cell Fc receptors.

III. RAT ANTIBODIES AND HUMAN K CELLS

The original description of K cells was done in the human, using human antibodies,[12] then rabbit antisera were routinely used. Rat antisera were usually weaker. A systematic analysis of rat MAbs was justified.

Human K cells are more efficient than rat K cells, and thus may be used at a smaller E/T ratio. They are obtained from defibrinated peripheral blood of healthy volunteers. Mononuclear cells are separated from granulocytes and red blood cells by density gradient and depleted from adherent cells by incubation on plastic at 37°C. The cell population recovered contains K cells.

Assays are performed as described for rat effectors, using NK resistant targets.

A. ISOTYPES AND HUMAN K CELLS

The main finding that arises from studies on rat MAbs and human K cells is the restriction to the one isotype IgG2b for obtaining cytotoxicity.

Work from Waldmann and collaborators[13] showed that a number of IgG2b rat MAbs against leucocyte common antigen (LCA), could elicit K cell cytotoxicity, whereas IgG2a and IgG2c were inactive. They did not obtain IgG1 antibodies.

We tested the LO-DNP series of MAbs directed against DNP described above, comparing isotypes IgG1, IgG2a, IgG2b, IgE, IgA, and IgM. We could not test antibodies of IgD, nor IgG2c isotypes. We found that all seven IgG2b antibodies tested were active, whereas none of the other isotypes worked. The maximum cytotoxicity level (80%) was similar for these reagents, and was of the same order of magnitude as that obtained with a polyclonal rabbit antiserum. A purified LO-DNP IgG2b antibody could sensitize target at an optimal dilution of 0.025 μg/ml. Cytotoxicity levels obtained were higher than those reached with rat effectors. Reports in the literature and our results indicate that IgG2b rat MAbs may be used at a concentration of 1 μg/ml or less for optimal cytotoxicity with human effectors, at a ratio of 20—50:1 target cell (Table 1).

It is interesting to compare this with results obtained using mouse MAbs which also elicit human K cells. There, isotype restriction is not so clear. Although most mouse IgG2a were active, IgG3 may also elicit strong cytotoxicity.[14] Occasionally, some MAbs of IgG1 or IgG2b

TABLE 2
Induction by Mouse Monoclonal Antibodies of Human Peripheral Blood Mononuclear Cells K Cell Activity on Various Target Cells

Clone	Ref.	Isotype	Concentration	Activity (SC)	E/T	Target	Ag
H 141-30	14	Mouse IgG3	50 μg/ml	85%	25/1	Murine lymphoma cell line	H2-D b
2B2	17	Mouse IgG3	10 μg/ml	85%	100/1	Human melanoma cell line	GD3
14.18	18	Mouse IgG3	100 μg/ml	30%	100/1	Lung carcinoma cell line	GD2
MD3-6	19	Mouse IgG3	10 μg/ml	20%	100/1	Human melanoma cell line	GD3
NE1-026	14	Mouse IgG2a	40 μg/ml	75%	25/1	Murine lymphoma	beta 2m
9H-1	16	Mouse IgG2a	50 μg/ml	75%	25/1	Human lympho-blastoid cell line	HU/beta 2m
ME-1	20	Mouse IgG2a	10 μg/ml	60%	100/1	Human lymphoblastoid cell line	HLA-B7
L6	21	Mouse IgG2a	20 μg/ml	58%	100/1	Lung adenocarcinoma	Human cancer antigen

isotypes have been found to elicit human K cells[16] to some extent. Some examples of activity reported in the literature are given in Table 2. We have only noted the best results achieved in each system. It is clear that either the antibody concentration or the number of effectors needs to be higher than in the rat system.

Confirming results with polyclonal antisera, cells most efficiently lysed by K cells are dividing cells of lymphoid origin. Tumor cells are adequate target cells, but dividing normal lymphocytes after a mitogen stimulation for instance, may be lysed efficiently with an appropriate antibody, much more so than unstimulated lymphocytes bearing the same antigen.[16]

B. IIID10 AND HUMAN K CELLS

We found that our MAb (IIID10) could sensitize CHANG cells for K cell cytotoxicity by human peripheral blood effectors. Culture supernatants of the hybridoma may be diluted up to 1:2000 for maximum cytotoxicity, which was higher than that obtained with rat effectors (70%). Lysis was abolished by a preincubation of effector cells with an anti-Fc receptor antibody.

This reagent will be proposed for the study of K cell activity. Serial dilutions of effector cells in the presence of a constant amount of sensitizing antibody may be tested in order to measure the cytotoxic activity of a cell population. A standard curve is obtained from tests done on a large number of healthy volunteers. Quantitation is done in terms of lytic units, a unit being the number of effector cells necessary to obtain a given level of lysis (for instance 50%).[1]

IV. CONCLUDING REMARKS

A better understanding of the mechanism of action of K cells is necessary, in particular for an eventual clinical application. Various factors, besides antibody isotypes, appear to play a role[22] and need to be analyzed.

The rat model offers the possibility to study parameters in a homologous situation that is difficult to obtain in the human, because human MAbs are not yet easily available.

However, mice and rat MAbs are being used as therapeutic agents in cancer patients or patients receiving organ transplantation. ADCC mechanisms are often considered a possible mechanism of action *in vivo*.

Genetically-engineered chimeric antibodies[23] with a mouse or rat antibody determinant linked to a human effector portion of immunoglobulin are already being produced in order to minimize *in vivo* host immunization against animal protein when repeated injections will be necessary.

REFERENCES

1. **MacLennan, I. C. M.,** Antibody in the induction and inhibition of lymphocyte cytotoxicity, *Transplant. Rev.*, 13, 67, 1972.
2. **MacLennan, I. C. M., Connell, G. E., and Gotch, F. M.,** Effector activating determinants on IgG. II. Differentiation of the combining site for C1q from those for cytotoxic K cells and neutrophils by plasmin digestion of rabbit IgG, *Immunology*, 26, 303, 1974.
3. **Chassoux, D., Kolb, J. P., Bazin, H., and MacLennan, I. C. M.,** Antibody-dependent cellular cytotoxicity (K) and natural killing (NK) in B-suppressed germ-free nude rats, *Immunology*, 50, 327, 1983.
4. **Chassoux, D., Kolb, J. P., and MacLennan, I. C. M.,** K-cell activity is independent of both thymus and spleen, *Adv. Exp. Med. Biol.*, 186, 331, 1985.
5. **Bazin, H., Cormont, F., and De Clercq, L.,** Rat monoclonal antibodies. II. A rapid and efficient method of purification from ascitic fluid or serum, *J. Immunol. Methods*, 71, 9, 1984.
6. **De Clercq, L., Cormont, F., and Bazin, H.,** Generation of rat-rat hybridomas with the use of the LOU IR983F non-secreting fusion cell line, *Methods Enzymol.*, 121, 234, 1986.
7. **Chassoux, D., Dokhelar, M. C., Tursz, T., and Salomon, J. C.,** Antibody-dependent cellular cytotoxicity and natural killing in nude rats: quantitative study according to age and sex, *Ann. Immunol. (Inst. Pasteur)*, 134D, 309, 1983.
8. **Bazin, H., Xhurdebise, L. M., Burtonboy, G., Lebacq, A. M., De Clercq, L., and Cormont, F.,** Rat monoclonal antibodies. I. Rapid purification from *in vitro* culture supernatants, *J. Immunol. Methods*, 66, 261, 1984.
9. **Bazin, H.,** Production of rat monoclonal antibodies with the LOU rat non-secreting IR983F myeloma cell line, in *Protitudes of the Biological Fluids*, Peeters, H., Ed., Pergamon Press, Oxford, 1982, 615.
10. **MacLennan, I. C. M., Howard, A., Gotch, F. M., and Quie, P. G.,** Effector activating determinants on IgG. I. The distribution and factors influencing the display of complement, neutrophil and cytotoxic B-cells determinants on human IgG subclasses, *Immunology*, 25, 459, 1973.
11. **Boltz-Nitulzscu, G., Bazin, H., and Speigelberg, H. L.,** Specificity of Fc receptors for IgG2a, IgG1/IgG2b and IgE on rat macrophages, *J. Exp. Med.*, 154, 374, 1981.
12. **MacLennan, I. C. M., Loewi, G., and Howard, A. L.,** A human serum immunoglobulin with specificity for certain homologous target cells, which induces target cell damage by normal human lymphocytes, *Immunology*, 17, 897, 1969.
13. **Hale, G., Clark, M., and Waldmann, H.,** Therapeutic potential of rat monoclonal antibodies: isotype specificity of antibody-dependent cell-mediated cytotoxicity with human lymphocytes, *J. Immunol.*, 134, 3056, 1985.
14. **Christiannsen, J. E. and Sears, D. W.,** Unusually efficient tumor cell lysis by human effectors of antibody-dependent cellular cytotoxicity mediated by monoclonal antibodies, *Cancer Res.*, 44, 3712, 1984.
15. **Lebacq-Verheyden, A. M., Ravoet, A. M., Bazin, H., Sutherland, D. R., Tidman, N., and Greaves, M. F.,** Rat AL2, AL3, AL4 and AL5 monoclonal antibodies bind to the common acute lymphoblastic leukaemia antigen (CALLA gp100), *Int. J. Cancer*, 32, 273, 1983.
16. **Sears, D. W. and Christiannsen, J. E.,** Mechanism of rapid tumor lysis by human ADCC: mediation by monoclonal antibodies and fragmentation of target cell DNA, *Adv. Exp. Med. Biol.*, 184, 509, 1985.
17. **Hellström, I., Brankovan, V., and Hellström, K. E.,** Strong antitumor activities of IgG3 antibodies to a human melanoma-associated ganglioside, *Proc. Natl. Acad. Sci.*, 82, 1499, 1985.
18. **Cheresh, D. A., Rosenberg, J., Mujoo, K., Hirschowitz, L., and Reisfeld, R. A.,** Biosynthesis and expression of the Disialoganglioside GD2 a relevant target antigen in small cell lung carcinoma for monoclonal antibody-mediated cytolysis, *Cancer Res.*, 46, 5112, 1986.
19. **Ortaldo, J. R. et al.,** Analysis of effector cells in human antibody-dependent cellular cytotoxicity with murine monoclonal antibodies, *J. Immunol.*, 138, 3566, 1987.
20. **Kipps, T. O., Parham, P., Punt, O., and Herzenberg, L. A.,** Importance of immunoglobulin isotype in human antibody-dependent, cell-mediated cytotoxicity directed by murine monoclonal antibodies, *J. Exp. Med.*, 161, 1, 1985.

21. **Hellström, I., Beaumier, P. L., and Hellström, K. E.,** Antitumor effect of L6, an IgG2a antibody that reacts with most human carcinomas, *Proc. Natl. Acad. Sci. U.S.A.*, 83, 7059, 1986.
22. **Chrisiaansen, J. E., Burnside, S. S., and Sears, D. W.,** Apparent sensitivity of human K lymphocytes to the spatial orientation and organisation of target cell-bound antibodies as measured by the efficiency of antibody-dependent cellular cytotoxicity (ADCC), *J. Immunol.*, 138, 2236, 1987.
23. **Bruggemann, M. et al.,** Comparison of the effector functions of human immunoglobulins using a matched set of chimeric antibodies, *J. Exp. Med.*, 166, 1351, 1987.
24. **Chassoux, D. M., Linares, L. G., Bazin, H.,** K cell mediated cytotoxicity induced with new monoclonal antibodies. I. Antibodies of various isotypes differ in their ability to induce cytotoxicity mediated by rat and human effectors, *Immunology*, 65, 623, 1988.

Chapter 31

ABSENCE OF VIRAL PARTICLES IN RAT HYBRIDOMAS. AN ULTRASTRUCTURAL STUDY

G. Burtonboy, N. Delferrière, M. Bodeus, and B. Mousset

By fusing rat spleen cells with 983 cells, a LOU/C rat immunocytoma cell line, it is possible to produce hybrid clones secreting monoclonal antibodies (MAbs).[1] This approach is to be compared with the mouse system originally described by Kölher and Milstein[2] and now widely used. As part of such a comparative study various rat hybridomas have been analyzed by electron microscopy mainly in order to detect the presence of possible viral particles. Indeed, mouse plasmacytomas and hybridomas have been shown to be infected by retroviruses: C-type virions are found in the culture supernatant and typical budding particles are observed on the surface of the cells.[3-7] Besides, intracisternal proliferation of what is referred to as type-A particles is frequently seen in their cytoplasm.[8-10]

Examples of these morphological features are given in Figures 1 and 2. The presence of the viruses can have practical implications and be a matter of concern if one considers the *in vivo* use of the MAb[11-14] for human.

In our laboratory, 12 rat hybridomas were examined by electron microscopy. Table 1 gives the characteristics of the hybrid clones. For this experiment, cells cultured in DMEM were taken during exponential growth and fixed in 2.5% glutaraldehyde at 4°C. Embedding in epon was made as described elsewhere;[15] thin sections were obtained with a diamond knife and stained with uranyl acetate and lead citrate.[16] This material was analyzed in a transmission electron microscope, Philips EM 300. The rat hybridoma cells were found to have a general morphology similar to what is observed in mouse hybridomas.[17] Occasional giant cells are present, but for most cells there is a single large nucleus often irregular in shape, slightly eccentric, with a finely dispersed chromatin, and prominent nucleolar structures. In none of the specimens examined, intranuclear particles or particle-like features which could suggest the presence of a virus were detected. Despite careful screening of all the nuclear areas in numerous sections of each specimen, no image of herpes, adeno, papova, nor parvovirus was found[18-21] (Figure 3).

The cytoplasm of the hybridoma cells is abundant. The rough endoplasmic reticulum is especially prominent in certain cells, whereas only a few cisterna are present in others (Figure 4). Only one fourth of the cells seem to be secreting. That is in agreement with the results of the study of the hybridomas by immunofluorescence using an antiserum against rat immunoglobulin, which shows that this protein is abundant only in part of the cell population; and it is indeed likely that the secretory process is dependent on the cell cycle. No viral particles were detected in the cytoplasm of the hybridoma cells, nor in the intercellular spaces and no image of budding was observed.

An electron microscopic study of IR983F cells,[22] the azaguanine resistant myeloma line used for the fusion, was performed and failed to reveal the presence of any virus: it was not possible to see any A- or C-type particle in these cells. If the absence of C-type particles was rather expected, the negative result obtained for A-type intracisternal particles was surprising. The myeloma cell line was derived from a tumor, namely IR983F, which was a rat immunocytoma.[23] Although IR983F itself was not examined from this point of view, an ultrastructural study of 15 such rat spontaneous ileocecal tumors had been made previously.[24] In this material no C particles were detected, but in all the immunocytomas analyzed the presence of intracisternal A-type particles was noticed (Figure 5). At the present time, there is no explanation for the disappearance of these A-type structures in the myeloma cell line and in the hybridomas derived from it.

This absence of viral particles is not unique to the hybrid clones obtained by fusion with

IR983F cells. Failure to demonstrate A and C particles has been reported for rat hybridoma cells produced by fusing rat spleen cells with the rat myeloma line Y3.Ag 1-2-329, a cell line which was derived from another LOU/C rat immunocytoma cell line, namely S21030-33.

Reverse transcriptase assay of the supernatant from the rat hybridoma cultures consistently gives negative results both with Mg++ and Mn++.[25-27] However, even in mouse hybridoma culture fluid, this enzyme activity was reported to be lower than expected from the amount of virions seen by electron microscopy,[28] and of course it is difficult to rule out the presence of a virus.[14] But in the rat hybridomas there is at least no morphological evidence of a viral infection. Retroviruses are not found to be released by the cells secreting rat MAb. This mere fact could be of interest: the viruses present in the mouse hybridomas have been shown to be possibly xenogenic, meaning that they could infect species other than mice[5,7] and it is well known that retroviruses from one animal species can be a pathogen for another.[34]

The point is probably worth being taken into account in the manufacture of injectable MAb products intended for human use *in vivo*;[13] this could be an advantage of the rat hybridoma technology.[1]

On the other hand, it has been suggested that the retroviruses present in the mouse myeloma cells could play an active role in the fusion process.[17] However, despite the absence of viral particles in the rat myeloma cells, the fusion efficiency is as good in the rat as in the mouse system. So the hypothesis that the retroviruses act synergistically with the fusing agent polyethylene glycol is difficult to support without further evidence.

ACKNOWLEDGMENTS

We thank H. Bazin for his enthusiastic advice, and suggestions. F. Bolle and P. Lambrecht are acknowledged for editing the manuscript.

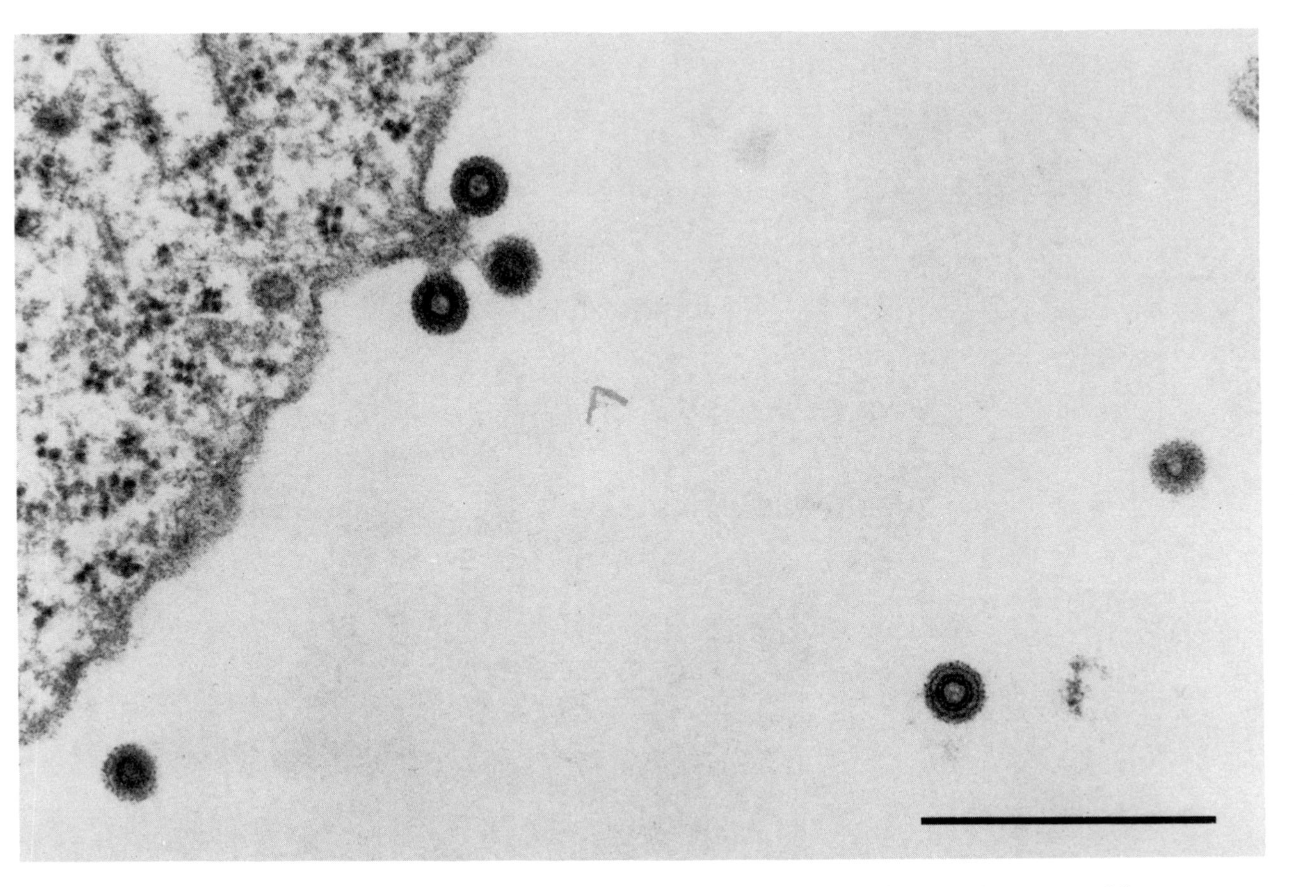

FIGURE 1. Budding of C-type particles on the surface of a mouse myeloma cell (MOPC 21). The bar represents 0.5 μm.

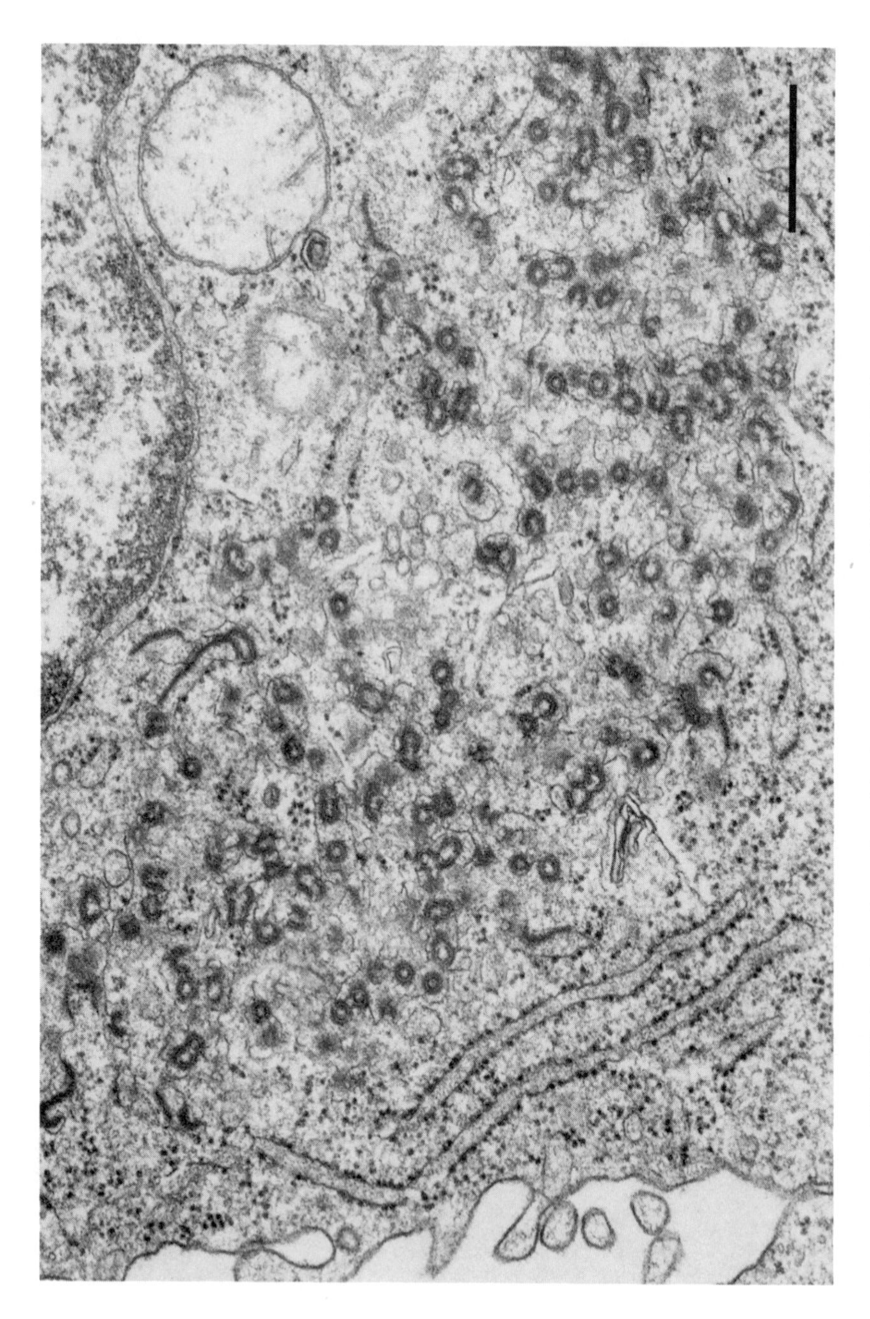

FIGURE 2. Intracisternal type A particles in a mouse myeloma cell (MOPC 21). The bar represents 0.5 μm.

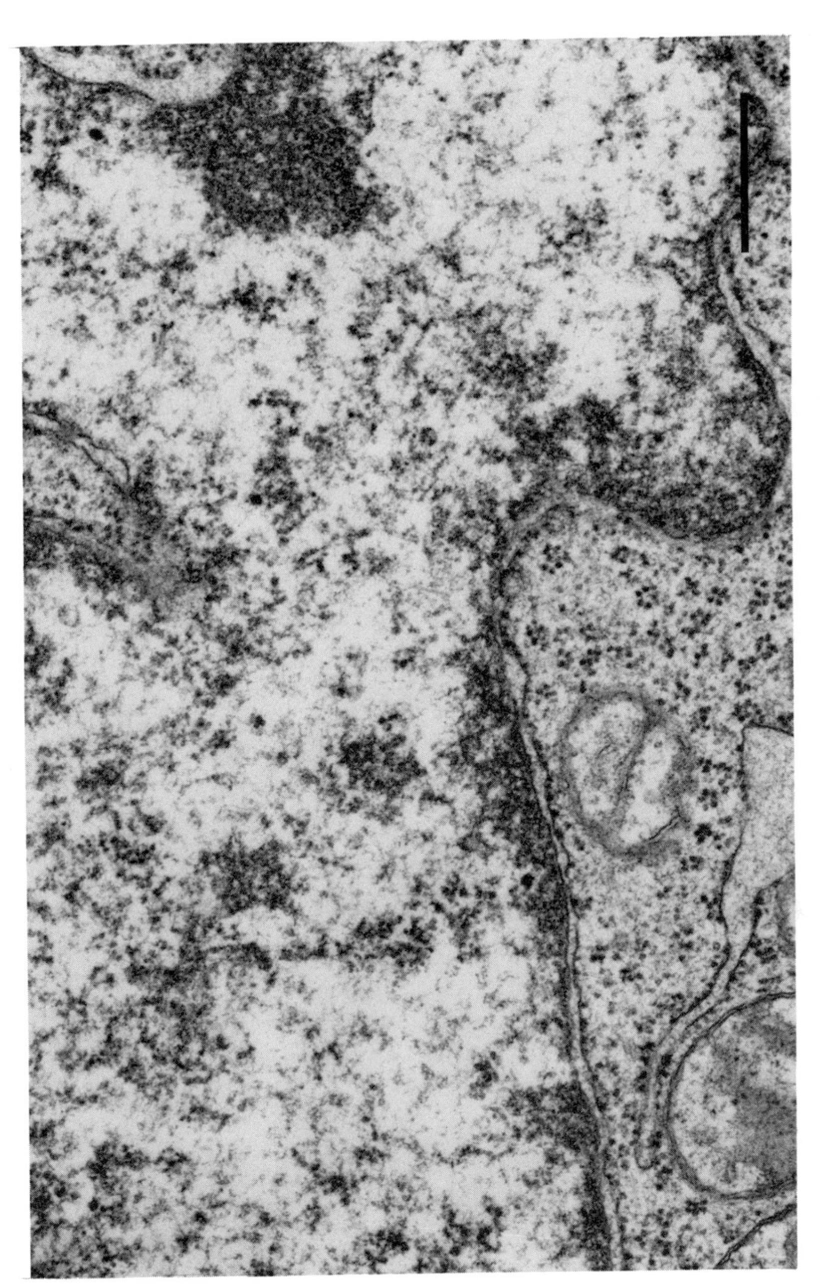

FIGURE 3. Nucleus of a rat hybridoma cell (RH1). The bar represents 0.5 μm.

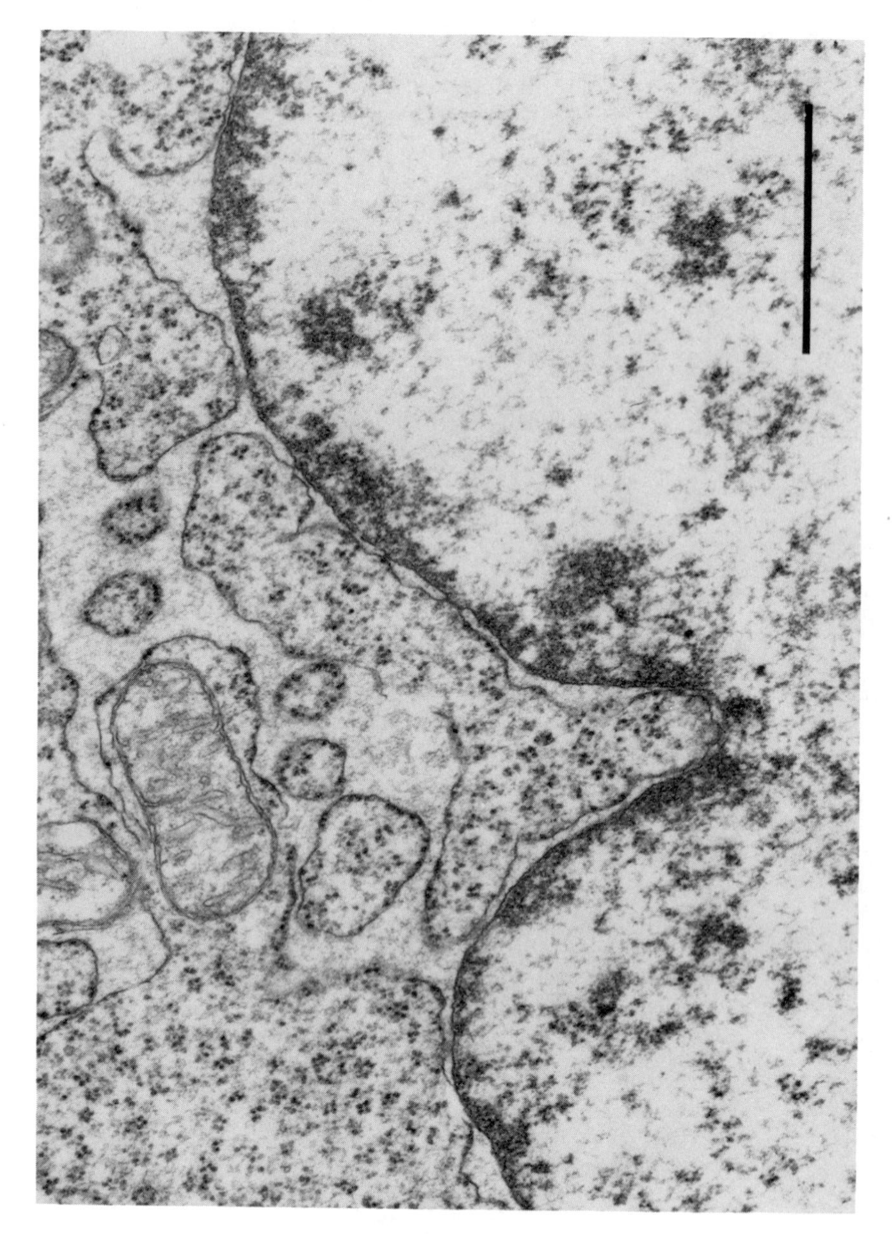

FIGURE 4. The cytoplasm of a cell in which the rough endoplasmic reticulum is abundant and dilated. Hybridoma cell 11C11. The bar represents 1 μm.

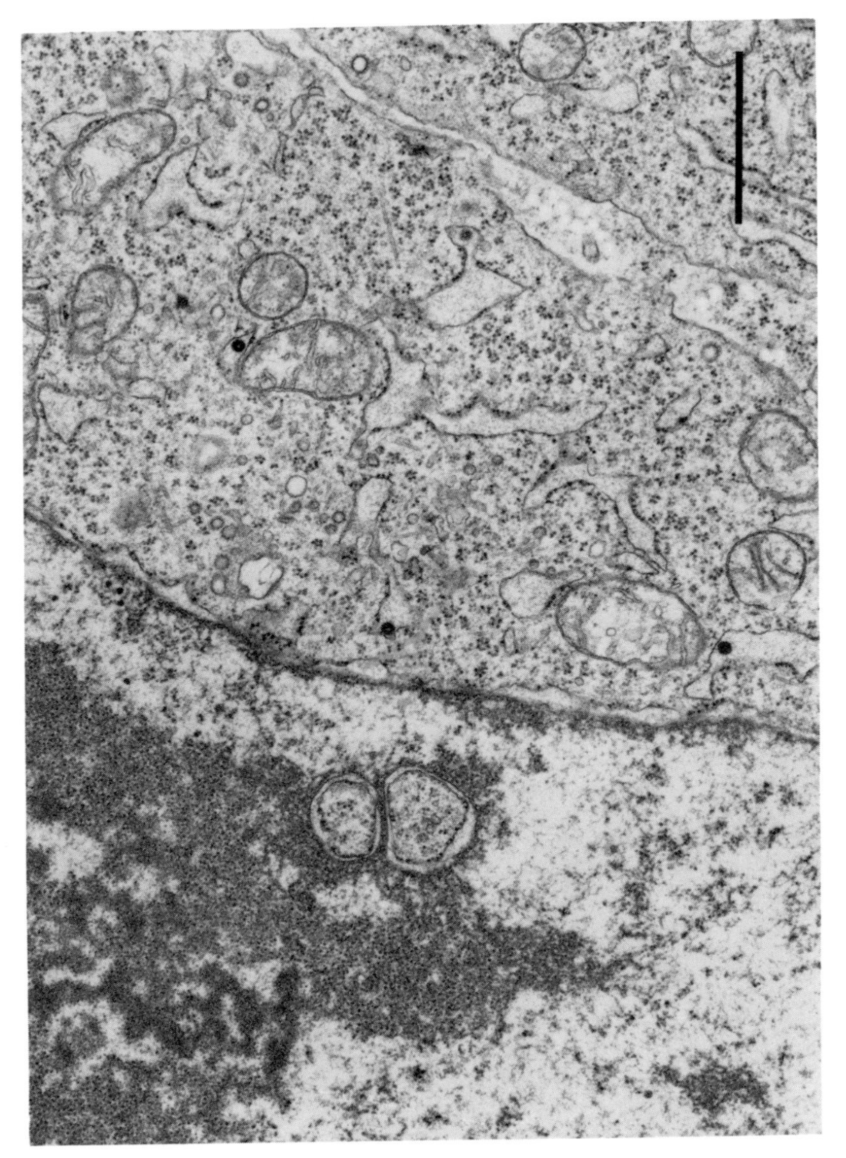

FIGURE 5. Intracisternal type A particles in the cytoplasm of a rat immunocytoma cell (IR656). The bar represents 1 μm.

TABLE 1
Rat Hybridoma Cell Lines Examined by Electron Microscopy

Hybridoma	Antigen	Isotype
RH 1	CPV	IgG2b
RH 2	CPV	IgG2b
RH 28	CPV	IgG2a
RH 57	CPV	IgG2b
G9E8	CEM cells	IgM
G4D11	CEM cells	IgG1
H8C10	CEM cells	IgG2c
H12F3	CEM cells	IgG2a
E5F4	HIV	IgG1
I1C11	HIV	IgG1
I8G11	HIV	IgG2a
I7G9	HIV	IgG1
I5B11	HIV	IgG2a

REFERENCES

1. **Bazin, H.,** Production of rat monoclonal antibodies with the LOU rat non secreting IR983F myeloma cell line, in *Protides of the Biological Fluids,* Peeters, H., Ed., Pergamon Press, Oxford, 1982, 615.
2. **Köhler, G. and Milstein, C.,** Continuous cultures of fused cells secreting antibody of predefined specificity, *Nature,* 256, 495, 1975.
3. **Galfré, G. and Milstein, C.,** Preparation of monoclonal antibodies: strategies and procedures, *Methods in Enzymology,* Vol. 73, Immunochemical Techniques, Part B, Langone, J. J. and Vunakis, H., Eds., Academic Press, New York, 1981.
4. **Bartal, A., Feit, C., Erlandson, R., and Hirshaut, Y.,** The presence of viral particles in hybridoma clones secreting monoclonal antibodies, *N. Engl. J. Med.,* 306, 1423, 1982.
5. **Bartal, A., Ludatscher, R., Robinson, E., Hirshaut, Y., and Lichtig, C.,** The presence of viral particles in hybridomas injected intraperitoneally to form ascitis, *Proc. Am. Soc. Clin. Oncol.,* 4, 220, 1985.
6. **Feliu, E., Rozman, C., Berga, L., Villela, R., Gallart, T., and Vives, J.,** Xenogenetic hybridoma cells contain and secrete virus particles, *Lancet,* 2, 1255, 1983.
7. **Rudolph, M., Karsten, U., and Michael, B.,** C type particles in hybridoma, *Immunol. Today,* 4, 183, 1983.
8. **Weiss, R.,** Hybridomas produce viruses as well as antibodies, *Immunol. Today,* 3, 292, 1982.
9. **Levy, J. A.,** Mouse plasmacytoma cells produce infectious type C virus, *Lancet,* 2, 522, 1983.
10. **Bartal, A. H., Feit, C., Erlandson, R., and Hirshaut, Y.,** The detection of retroviral particles in hybridomas, *Med. Microbiol. Immunol.,* 174, 325, 1986.
11. **Bernhard, W.,** The detection and study of tumor viruses with the electron microscope, *Cancer Res.,* 20, 712, 1960.
12. **Weimann Bernd-Jürgen,** Intra and extracellular particles from mouse plasmocytomas and hybridomas, in *Mechanisms of B Cell Neoplasma,* Melchers, F. and Potter, M., Eds., Roche, 1985, 217.
13. **Bussard, A.,** How pure are monoclonal antibodies, *Dev.. Biol. Standard,* 57, 13, 1984.
14. **Hoffman, T.,** Regulatory issues surrounding therapeutic use of monoclonal antibodies, in *Methods of Hybridoma Formation,* Bartal, A. and Hirshaut, Y., Eds., Humana Press, Clifton, NJ, 1987, 447.
15. **Miller, R. A. and Levy, R.,** Response of cutaneous B cell lymphoma to therapy with hybridoma monoclonal antibody, *Lancet,* 2, 226, 1981.
16. **Sears, H. F., Herlyn, D., Steplewski, Z., and Koprowski, H.,** Effects of monoclonal antibody immunotherapy on patients with gastrointestinal adenocarcinoma, *J. Biol. Response Mod.,* 3, 138, 1984.
17. **Merchant, E.,** The impact of hybridoma technology on the medical device and diagnosis product industry, *Proc. Ed. Sem. Hima Rep.,* 82, 83, 1982.
18. **Burtonboy, G., Delferrière, N., Meulemans, G., Vindevogel, H., Rodhain, J., and Lamy, M. E.,** Maladie de Gumboro, Etude au microscope électronique des cellules des bourses de Fabricius infectées, *Ann. Rech. Vet.,* 5, 291, 1974.

19. **Hayat, M. A.,** Multiple staining, in Principles and Techniques of Electron Microscopy, *Biological Applications,* 1, 139, 1970.
20. **Bartal, A. H. and Hirschaut, Y.,** Retroviruses and Hybridoma formation bystanders or active participants, in *Methods of Hybridoma Formation,* Bartal, A. and Hirshaut, Y., Eds., Humana Press, Clifton, NJ, 1987, 41.
21. **Burtonboy, G., Coignoul, F., Delferrière, N., and Pastoret, P. P.,** Canine hemorrhagic enteritis, detection of viral particles by electron microscopy, *Arch. Virol.,* 61, 1, 1979.
22. **Dalton, A. J. and Haguenau, F.,** *Ultrastructure of Animal Viruses and Bacteriophages,* Academic Press, New York, 1973.
23. **Doane, F. W. and Anderson, N.,** Electron and immunoelectron microscopy, in *Comparative Diagnosis of Viral Diseases,* Vol. II (Part B), Academic Press, New York, 1977, 505.
24. **Bazin, H.,** More on rat monoclonal antibodies, *Immunol. Today,* 4, 274, 1983.
25. **Bazin, H., Deckers, C., Beckers, A., and Heremans, J. F.,** Transplantable immunoglobulin secreting tumors in rats. I. General features of LOU/WSL strain rat immunocytomas and their monoclonal proteins, *Int. J. Cancer,* 10, 568, 1972.
26. **Burtonboy, G., Beckers, A., Rodhain, J., Bazin, H., and Lamy, M. E.,** Rat ileocecal immunocytoma. An ultrastructural study with special attention to the presence of viral particles, *J. Natl. Cancer Inst.,* 61, 477, 1978.
27. **Kacian, D. L. and Spiegelman, S.,** Purification and detection of reverse transcriptase in viruses and cells, *Meth. Enzymol,* 29, 150, 1974.
28. **Liu, W. T., Natori, T., Chang, K. S., and Wu, A. M.,** Reverse transcriptase of Foamy virus, *Arch. Virol.,* 55, 187, 1977.
29. **Lyon, M. and Huppert, J.,** Depression of reverse transcriptase activity by hybridoma supernatants. A potential problem in screening for retroviral contamination, *Biochem. Biophys. Res. Commun.,* 112, 265, 1983.
30. **Stavrou, D., Bilzer, T., Tsangaris, T., Durr, E., Steinecke, M., and Anzil, A. P.,** Presence and absence of virus particles in hybridomas secreting monoclonal antibodies against gliomas, *J. Cancer Res. Clin. Oncol.,* 186, 77, 1983.
31. **Burtonboy, G., Bazin, H., Deckers, C., Lamy, M., and Heremans, J. F.,** Transplantable immunoglobulin secreting tumors in rats. III. Establishment of immunoglobulin secreting cell lines from LOU/WSL strain rats, *Eur. J. Cancer,* 9, 259, 1973.
32. **Galfré, G., Milstein, C., and Wright, B.,** Rat x rat hybrid myelomas and a monoclonal anti-Fd portion of mouse IgG., *Nature,* 277, 131, 1979.
33. **Kilmartin, J. V., Wright, B., and Milstein, C.,** Rat monoclonal antibodies derived by using a new non secreting rat cell line, *J. Cell Biol.,* 93, 576, 1982.
34. **Munroe, J. S. and Windle, W. F.,** Tumor induced in primates by chicken sarcoma virus, *Science,* 140, 1415, 1963.

Chapter 32

ROLE OF CODATA-HDB IN THE CENTRALIZATION AND DISTRIBUTION OF INFORMATION CONCERNING THE RAT HYBRIDOMAS

B. Dutertre and Y. Bruinen

TABLE OF CONTENTS

I. RELATIONS BETWEEN CODATA AND HDB

Hybridoma Data Bank (HDB) was launched in 1983 by CODATA (Committee on Data for Science and Technology of the International Council of Scientific Unions), under the sponsorship of CODATA and IUIS (Internation Union of Immunological Societies).

HDB is an international organization aimed at collecting and distributing information on immunoclones and their products. Its center was established in Washington at the ATCC (American Type Culture Collection) and subsidized by various U.S. and European sources. Three branches have been created to facilitate contacts between producers and purchasers of MAbs: one in the U.S. at the ATCC in Rockville, MD, one in Nice, France (Centre Européen de Recherche et de Documentation sur les Immunoclones, CERDIC), and one in Japan, at the Riken Institute in Tokyo. They share all the data which are validated by a coordination center controlled by a technical committee, composed of the three nodes managers and two technical advisers.

The nodes and the technical committee are under the scientific supervision of the Task Group (TG) on a Hybridoma Data Bank, an official body of CODATA. It was created in 1983 and its chairman, currently Dr. B. Janicki, and members are chosen by the president of CODATA, with this choice being ratified by the General Assembly (the governing body of CODATA).

The data in the HDB and their copyright are property of CODATA. Under the statutes of CODATA, these data are public and must be provided to anyone who asks for it. The information can be disseminated on-line or off-line by any of the three nodes, in compliance with the contract between the nodes and CODATA, related to royalties payment.

II. CENTRALIZATION

A. COLLECTION

Collected data on rat hybridomas, like other hybridomas, come from several sources:

1. Commercial catalogs of biotechnology firms
2. Systematic search of the literature
3. Extracts of symposia and congress reports
4. Data reporting forms filled in by the scientists who produce monoclonal antibodies
5. Crude data from scientists

There are currently some 540 rat-hybridoma descriptions (including hybrid clones like mouse/rat, human/rat, etc.), i.e., 4.74% of the total amount of data in HDB.

1. Catalogs of Biotechnology Firms

In the bank, 20% of the data on rat hybridomas comes from the formating of catalogs of biotechnology firms.

HDB is being kept updated by the firms on their new products and/or distributors regularly. The first contacts by sending out form letters to the commercial companies came to a yield of over 50%, which shows the interest of the commercial field to our bank.

Of the whole data collected in commercial catalogs, 4.7% concern rat hybridomas.

2. Literature

The collection is made via an on-line access to bibliographic databases; essentially Medline for the information on rat (Molecular Immunology, European Journal of Immunology, Journal of Immunology, Journal of Experimental Medicine, etc.).

The extracts of congress reports are also very useful for the collection of data. Of the data on rat hybridomas, 37% originate in the literature, of which 41.6% from U.S., 29% from Belgium, 15.4% from U.K., 5.8% from Japan, 3.7% from The Netherlands, 2.6% from France, and 2.4% from the rest of the world.

3. Survey Forms

Scientists are asked to fill in specific questionnaires. Data Reporting Forms (DRF), with all the information that is needed to keep accurate data on their production. This is the main source for the Japanese node, whose essential aim is to collect all data available from all the Japanese research laboratories.

It is too scarce nevertheless, that information is sent by the scientists spontaneously. Most of the time, the DRF's to be filled in are sent out by HDB with the replies to information demands on the HDB activities and the form letters accompanying our demands for precisions on the descriptions contained in the abstracts or in the meeting reports.

On the other hand, we received almost 200 data from the Experimental Immunology Unit of the University of Louvain, already formatted on floppy disks. These are the first ones to be transmitted under a format which should be the main format used in the future.

Of the data on rat hybridomas, 43% are entered thanks to DRF's, i.e., this is the main source of data entry of rat hybridomas in the bank.

4. Special Contracts

The U.S. node already works on special contracts for specific searches, mainly with institutions or companies, and the European node is developing them, i.e., yearly contracts for specific searches with monthly updates, and eventually with the creation of private special files.

B. CENTRALIZATION

All data coded in our own format are sent to the validation center, which verifies that there are no duplicates and validates the information. Tapes are then sent back to each node so that data can be disseminated. Each record has a unique accession number.

C. DISSEMINATION

The database is available to all scientists over the world, either from basic research or from industry and can be queried by phone, telex, and electronic or ordinary mail.

It is obvious, that the fact that a MAb comes from a rat or a mouse or a human hybridoma is not a senior criterion for the scientists querying the data bank. Only a few queries stated precisely "rat hybridoma". Most of the queries refer to the antigen not to the nature of the hybridoma, and therefore we have disseminated a lot of data on rat hybridomas together with data on human or mouse hybridomas. These data are spread among the whole scientific community, at an increasing rate: the HDB is now known internationally in Japan, North America, and Europe, where the different nodes are located, but also in Israel, Egypt, South America, Australia, U.S.S.R., India, etc., and maybe in places that we are not aware of.

Nevertheless, the success of our endeavor depends highly on the number of data gathered and on its usefulness for the scientific community. Hence, we would be grateful if the scientists, whose laboratories produce MAbs, could send us information on their production. On the other hand, any request for information or any query on specific monoclonal antibodies can be addressed to:

In Europe: CERDIC, Faculté de Médecine, Avenue de Valombrose, F-06034 Nice Cedex, France, Telephone 33/93.20.01.81.., Telex 265451 Attn. DBI0044, Electronic mail 10075:DBI0044.

In Japan: RIKEN, Institute of Physical and Chemical Research, Wako, Saitama 351-01, Japan, Telephone 0484(62)1111, Telex 02362818 RIKEN J, Electronic mail 142:CDT0007.

In the U.S.: American Type Culture Collection, Bioinformatics Department, 12301 Parklawn Drive, Rockville, MD, 20852-1776, U.S., Telephone (301)231-5585, Electronic mail 142:CDT0043.

Chapter 33

LIST OF MONOCLONAL ANTIBODIES

J.M. Malache, P. Manouvriez, and H. Bazin

TABLE OF CONTENTS

I. SELECTED LIST OF MONOCLONAL ANTIBODIES ANTI-RAT IMMUNOGLOBULINS

A. FROM THE EXPERIMENTAL IMMUNOLOGY UNIT, UNIVERSITY OF LOUVAIN

From: Experimental Immunology Unit*
University of Louvain
Faculty of Medicine
Clos Chapelle aux Champs, 30-56
Brussels 1200 Belgium
Telephone: 32/2/7643430
Telex: UCL 23722 B
Telefax: 32/2/7643946

* **Code** M: mouse; A: anti; R: rat followed by a letter for the Ig class as M for IgM and a number.

MARM-4 HDB Number: 7027
Immunogen: IgM (Polyclonal rat IgM, IR968) from LOU/C Rat
Immunocyte Donor: BALB/c Mouse
Immortal Cell Partner: Sp2/O
Class of Antibody Produced: Mouse IgG1 kappa
Reactivity: Rat mu heavy chain of immunoglobulin (determined by immunodot)
Avidity: on IgM: 4×10^9
Applications: See reactivities; Second antibody in immunoassays; immunoaffinity chromatography for purification of rat IgM (solid phase Sepharose 4B CNBr act.); can be labeled with peroxydase, biotin, tritc; cannot be used for cell counting using a direct method of immunofluorescence.

MARM-7 HDB Number: 7030
Immunogen: IgM (IR202, IR968) from LOU/C Rat
Immunocyte Donor: BALB/c Mouse
Immortal Cell Partner: Sp2/O
Class of Antibody Produced: Mouse IgG1 kappa
Reactivity: Rat mu heavy chain of immunoglobulin (determined by immunodot)
Applications: See reactivities; second antibody in immunoassays; immunoaffinity chromatography for purification of rat IgM (Solid phase Sepharose 4B CNBr act.); can be labeled with peroxydase; biotin.

MARD-3 HDB Number: 7031
Immunogen: IgD (IR731) from LOU/C Rat
Immunocyte Donor: BALB/c Mouse

* Monoclonal antibodies available for scientific uses (limited quantities) upon request to Prof. H. Bazin. Monoclonal antibodies also available from commercial firms: Biosys, Quai du Clos des Roses 21, F-60200 Compiègne, France, Telephone: 33/4/4862275, Telex: 145171 F, Telefax: 33/4/4862297; Cosmo Bio, Mr. Y. Motohashi, Honcho Nihonbashi, Chuo-Ku, Tokyo, Japan, Telephone: 036630721, Telex: 2523541 COSBIO J, Telefax: 036636725; Janssen Biochemica, Turnhoutseweg 30, B-2340 Beerse, Belgium, Telephone: 014/602111, Telex: 34103 B; Oriental Yeast Co., Mr. M. Nishino, Biosciences Laboratory, Nihonbashi Fukawa Bldg. 6F, 10-11 Nihonbashi Kodenma-Cho, Chuo-Ku, Tokyo 103, Japan, Telephone: 03/6638218, Telex: 265-5069 OYEAST J, Telefax: 03/6638226 or 03/6638238; Sanbio B.V., Heinsbergenstraat 50, H-5402 EG-Uden, The Netherlands, Telephone: 04132-51115, Telex: 74827 SANBO NL, Telefax: 31413266605; Sera-Lab Ltd., Ms. Jennifer Murray, Crawley Down, Sussex RH10 4FF, England, Telephone: 0342/716366, Telex: 95317 SERLAB G; Serotec, Dr. T. M. Bernard, 22 Bankside, Station Approach, Kidlington, Oxford OX5 1JE, England, Telephone: 08675/79941, Telex: 838960 SERTEC G, Telefax: 08675/3899; Zymed Laboratories, Dr. A. Bella, 52 South Linden Avenue, Suite 5, South San Francisco, CA 94080, U.S., Telephone: 415/8714494, Telex: 4993181, Telefax: 415/871-4499.

Immortal Cell Partner: Sp2/O
Class of Antibody Produced: Mouse IgG1 kappa
Reactivity: Rat delta heavy chain of immunoglobulin (determined by immunodot)
Applications: See reactivities; can be used as a second antibody in immunoassays; can be labeled with peroxydase, FITC, TRITC, biotin. Can be used for immunopurification of rat IR731 (IgD) Protein (Solid phase Sepharose 4B CNBr act.); can be used for immunofluorescence.

MARA-1 HDB Number: 7023
Immunogen: IgA (IR1060, IR22, IR699) from LOU/C Rat
Immunocyte Donor: BALB/c Mouse
Immortal Cell Partner: Sp2/O
Class of Antibody Produced: Mouse IgG1 kappa
Reactivity: Rat alpha heavy chain of immunoglobulin (determined by immunodot)
Applications: See reactivities; immunoaffinity chromatography (Solid phase Sepharose 4B CNBr act.); can be labeled with biotin, peroxydase.

MARA-2 HDB Number: 7024
Immunogen: IgA (IR1060, IR22, IR699) from LOU/C Rat
Immunocyte Donor: BALB/c Mouse
Immortal Cell Partner: Sp2/O
Reactivity: Rat alpha heavy chain of immunoglobulin (determined by immunodot)
Avidity: on IgA: 4×10^8
Applications: See reactivities; second antibody in immunoassays; can be labeled with biotin, peroxydase; cannot be used for immunopurification.

MARE-1 HDB Number: 7021
Immunogen: IgE (IR162, IR410) from LOU/C Rat
Immunocyte Donor: BALB/c Mouse
Immortal Cell Partner: Sp2/O
Class of Antibody Produced: Mouse IgG1 kappa
Reactivity: Rat epsilon heavy chain of immunoglobulin (C epsilon 3) (determined by immunodot)
Avidity: On IgE: 4×10^9
Applications: See reactivities; can be used as a second antibody in immunoassays; can be used in immunoaffinity chromatography for purification of Rat IgE (Solid phase Sepharose 4B CNBr act.); can be labeled with iodine.

MARG1-1 HDB Number: 7000
Immunogen: IgG1 (IR27, IR31) from LOU/C Rat
Immunocyte Donor: BALB/c Mouse
Immortal Cell Partner: Sp2/O
Class of Antibody Produced: Mouse IgG1 kappa
Reactivity: Rat gamma 1 heavy chain of immunoglobulin (determined by immunodot)
Applications: See reactivities; second antibody in immunoassays; immunoaffinity chromatography for purification of rat IgG1 (Solid phase Sepharose 4B CNBr act.)

MARG1-2 HDB Number: still not attributed
Immunogen: IgG1 from LOU/C Rat
Immunocyte Donor: BALB/c Mouse
Immortal Cell Partner: Sp2/O
Class of Antibody Produced: Mouse IgG1 kappa
Reactivity: Rat gamma 1 heavy chain of immunoglobulin (determined by immunodot)
Avidity: on IgG1: 1×10^9
Applications: Cf Reactivities; second antibody in immunoassays; can be labeled with FITC.

MARG2a-8 HDB Number: 7008
Immunogen: IgG2a (Polyclonal antibodies anti horse spleen ferritin, IR33 from LOU/C Rat

Immunocyte Donor: BALB/c Mouse
Immortal Cell Partner: PAI-O
Class of Antibody Produced: Mouse IgG1 kappa
Reactivity: Rat gamma 2a heavy chain of immunoglobulin (determined by immunodot)
Avidity: on IgG2a: 6×10^9
Applications: See reactivities; second antibody in immunoassays; immunoaffinity chromatography for purification of Rat IgG2a (Solid phase Sepharose 4B CNBr act.); can be labeled with peroxydase.

MARG2b-3 HDB Number: 7010
Immunogen: IgG2b (IR863) from LOU/C Rat
Immunocyte Donor: BALB/c Mouse
Immortal Cell Partner: SP2/O
Class of Antibody Produced: Mouse IgG1 kappa
Reactivity: Rat gamma 2b heavy chain of immunoglobulin (determined by immunodot)
Avidity: on IgG2b: 3×10^9
Applications: See reactivities; second antibody in immunoassays; immunoaffinity chromatography for purification of rat IgG2b (Solid phase Sepharose 4B CNBr act.).

MARG2b-8 HDB Number: 7014
Immunogen: IgG2b (IR863) from LOU/C Rat
Immunocyte Donor: BALB/c Mouse
Immortal Cell Partner: Sp2/O
Class of Antibody Produced: Mouse IgG1 kappa
Reactivity: Rat gamma 2b heavy chain of immunoglobulin (determined by immunodot)
Applications: See reactivities; second antibody in immunoassays; immunoaffinity chromatography for purification of rat IgG2b (Solid phase Sepharose 4B CNBr act.); can be labeled with peroxydase.

MARG2c-2 HBD Number : 7019
Immunogen: IgG2c (IR304) from LOU/C Rat
Immunocyte Donor: BALB/c Mouse
Immortal Cell Partner: Sp2/O
Class of Antibody Produced: Mouse IgG1 kappa
Reactivity: Rat gamma 2c heavy chain of immunoglobulin (determined by immunodot)
Avidity: on IgG2c: 2×10^9
Applications: See reactivities; second antibody in immunoassays; immunoaffinity chromatography for purification of rat IgG2c (Solid phase Sepharose 4B CNBr act.); can be labeled with peroxydase.

MARG2c-3 HDB Number: 7020
Immunogen: IgG2c (IR304) from LOU/C Rat
Immunocyte Donor: BALB/c Mouse
Immortal Cell Partner: Sp2/O
Class of Antibody Produced: Mouse IgG2a kappa
Reactivity: Rat gamma 2c heavy chain of immunoglobulin (determined by immunodot)
Applications: See reactivities; second antibody in immunoassays; immunoaffinity chromatography for purification of rat IgG2c (Solid phase Sepharose 4B CNBr act.).

MARK-1* HDB Number: HDBS 230
Immunogen: kappa L-chain (IR202, IR968) from LOU/C Rat
Immunocyte Donor: BALB/c Mouse
Immortal Cell Partner: Sp2/O
Class of Antibody Produced: Mouse IgG1

* MARK-1 and Mark-3 hybridomas available upon request to Prof. H. Bazin or monoclonal antibodies available from commercial firms (see addresses listed in the footnote on page 480).

Reactivity: Rat kappa light chain of immunoglobulin (determined by immunodot)
Avidity: on IgD: 1×10^9
Applications: See reactivities; second antibody in immunoassays immunoaffinity chromatography for purification of rat Ig of the kappa type (Solid phase Sepharose 4B CNBr act.); can be labeled with peroxydase, tritc, iodine, alkaline, phosphatase, biotin.

MARK-2
Immunogen: IgD IR731 from LOU/C Rat
Immunocyte Donor: BALB/c Mouse
Immortal Cell Partner: SP2/O
Class of Antibody Produced: Mouse IgG1
Reactivity: Rat kappa light chain of immunoglobulin (allotype IgK-1a)

MARK-3* HDB Number: 7032
Immunogen: Kappa L-Chain of the IgKAPPA-1a allotype from LOU/C Rat
Immunocyte Donor: BALB/c Mouse
Immortal Cell Partner: Sp2/O
Class of Antibody Produced: Mouse IgG1
Reactivity: Rat kappa light chain of immunoglobulin (allotype IgK-1a) (determined by immunodot). Cross-react weakly in immunofluorescence, elisa RIA with rat kappa1b light chain of immunoglobulin. Can be employed for purifications in affinity chromatography.
Applications: See reactivities; second antibody in immunoassays immunoaffinity chromatography for purification of rat Ig of the kappa type, IgK-1a allotype (Solid phase Sepharose 4B CNBr act.); can be labeled with peroxidase, fitc, tritc, biotin.

MARL-4 HDB Number: 7049
Immunogen: Lambda L-chain (RH58, IR31) From LOU/C Rat
Immunocyte Donor: BALB/c Mouse
Immortal Cell Partner: PAI-O
Class of Antibody Produced: Mouse IgG2b kappa
Reactivity: Rat lambda light chain of immunoglobulin (determined by immunodot)
Avidity: on IgG1: 9×10^8
Applications: See reactivities; second antibody in immunoassays; immunoaffinity chromatography for purification of Rat Ig of the lambda type; can be labeled with peroxydase.

MARL-5 HDB Number: 7050
Immunogen: Lambda L-chain (RH58, IR31) from LOU/C Rat
Immunocyte Donor: BALB/c Mouse
Immortal Cell Partner: PAI-O
Class of Antibody Produced: Mouse IgM kappa
Reactivity: Rat lambda light chain of immunoglobulin (determined by immunodot)
Applications: See reactivities; second antibody in immunoassays; immunoaffinity chromatography for purification of rat Ig of the lambda type; can be labeled with peroxydase.

LO-RK1b-1 HDB Number: 7153
Immunogen: kappa 1b L-chain from OKA Rat
Immunocyte Donor: LOU/C Rat
Immortal Cell Partner: IR983F
Class of Antibody Produced: Rat IgG1 kappa
Reactivity: Rat kappa light chain of immunoglobulin (allotype IgK-1b) (determined by immunodot)
Applications: See reactivities; second antibody in immunoassays; immunoaffinity chromatography for purification of Rat IgK-1b kappa Ig (Solid phase Sepharose 4B CNBr act.); can be labeled with peroxidase, biotin, fitc.

* MARK-3 hybridoma available upon request to Prof. H. Bazin or monoclonal antibodies available from commercial firms (see addresses listed in the footnote on page 480).

LO-RK1b-2 HDB Number: 7154

Immunogen: kappa 1b L-chain from OKA Rat
Immunocyte Donor: LOU/C Rat
Immortal Cell Partner: IR983F
Class of Antibody Produced: Rat IgG1 kappa
Reactivity: Rat kappa light chain of immunoglobulin (allotype IgK-1b) (determined by immunodot)
Applications: See reactivities; second antibody in immunoassays; immunoaffinity chromatography for purification of Rat IgK-1b kappa Ig (Solid phase Sepharose 4B CNBr act.); can be labeled with biotin.

LO-RK1a-1 HDB Number: still not attributed

Immunogen: IgK-1a L-chain from LOU/C Rat
Immunocyte Donor: LOU/C.IgK-1b(OKA) Rat
Immortal Cell Partner: IR983F
Class of Antibody Produced: Rat IgG2a kappa
Reactivity: Rat kappa light chain of immunoglobulin (allotype IgK-1a) (determined by immunodot)
Applications: See reactivities

B. FROM OTHER LABORATORIES

M35/9F10

Immunogen: Rat Ig
Immunocyte Donor: BALB/c Mouse
Immortal Cell Partner: NSI
Class of Antibody Produced: Mouse IgG1 kappa
Reactivity: Rat kappa light chain of immunoglobulin (allotype IgK-1a)
Reference: Rat New Letter, 19, 4, 1987.

MAR80.2 Ref: HDB 2655

Immunogen: Rat Ig
Immunocyte Donor: SJL/J Mouse
Immortal Cell Partner: X63
Class of Antibody Produced: Mouse IgG2a kappa
Reactivity: Rat kappa light chain of immunoglobulin (allotype IgK-1a)

RG11/15 Ref: HDB 2665

Immunogen: Rat Ig
Immunocyte Donor: SJL Mouse
Immortal Cell Partner: NSI
Class of Antibody Produced: Mouse IgG2a
Reactivity: Rat kappa light chain of immunoglobulin (allotype IgK-1b)

RG9/6.13.HLK Ref: ATCC TIB 167; HDB 2670

Immunogen: Rat IgG
Immunocyte Donor: SJL Mouse
Immortal Cell Partner: NS-1
Class of Antibody Produced: Mouse IgG2b
Reactivity: Rat IgG2a Fab′

RG7/9.1 HLK Ref: ATCC TIB 169; HDB 2664

Immunogen: Rat IgG
Immunocyte Donor: SJL Mouse
Immortal Cell Partner: NS-1
Class of Antibody Produced: Mouse IgG2b kappa
Reactivity: Rat kappa light chain

RG11/39.4 Ref: ATCC TIB 170
Immunogen: Rat IgG
Immunocyte Donor: SJL Mouse
Immortal Cell Partner: NS-1
Class of Antibody Produced: Mouse IgG2b
Reactivity: Rat IgG1 Fc′
RG7/7.6.HL Ref: ATCC TIB 172; HDB 2666
Immunogen: Rat IgG
Immunocyte Donor: SJL Mouse
Immortal Cell Partner: NS-1
Class of Antibody Produced: Mouse IgG2a
Reactivity: Rat kappa light chain(K1b(LEW)
RG7/1.30 Ref: ATCC TIB 173; HDB 2668
Immunogen: Rat IgG
Immunocyte Donor: SJL Mouse
Immortal Cell Partner: NS-1
Class of Antibody Produced: Mouse IgG2b
Reactivity: Rat IgG2a Fc′
RG7/11.1 Ref: ATCC TIB 174; HDB 2669
Immunogen: Rat IgG
Immunocyte Donor: SJL Mouse
Immortal Cell Partner: NS-1
Class of Antibody Produced: Mouse IgG2b
Reactivity: Rat IgG2b Fc′
MAR18.5 Ref: ATCC TIB 216; HDB 2653
Immunogen: Rat IgG
Immunocyte Donor: SJL/J Mouse
Immortal Cell Partner: P3X63Ag8
Class of Antibody Produced: Mouse IgG2a
Reactivity: Rat kappa light chain
MAR 103.6 Ref: HDB 2654
Immunogen: Rat immunoglobulin
Immunocyte Donor: SJL/J Mouse
Immortal Cell Partner: P3-X63-Ag8
Class of Antibody Produced: IgG2a
Reactivity: Rat kappa Light Chain (Allotype IgK-1b)
G9/1 Ref: HDB 2661
Immunogen: Rat immunoglobulin
Immunocyte Donor: DA Rat
Immortal Cell Partner: 45.6.TG1.7
Class of Antibody Produced: IgG2b kappa
Reactivity: Rat kappa light chain of immunoglobulin (Allotype IgK-1b)
OX11 Ref: HDB 2663
Immunogen: Rat immunoglobulin
Immunocyte Donor: DA Rat
Immortal Cell Partner: NS-1
Class of Antibody Produced: IgG2b kappa
Reactivity: Rat kappa light chain of immunoglobulin (Allotype IgK-1b)
RG11/39.4 Ref: ATCC TIB 170; HDB 2667
Immunogen: Rat immunoglobulin
Immunocyte Donor: SJL Mouse

Immortal Cell Partner: NS-1
Class of Antibody Produced: IgG2b
Reactivity: Rat IgG1 Fc′

II. SELECTED LIST OF RAT MONOCLONAL ANTIBODIES USED IN THE PRESENT BOOK

A. SELECTED LIST OF RAT MAbS ANTI-MOUSE IMMUNOGLOBULINS FROM THE EXPERIMENTAL IMMUNOLOGY UNIT AT THE UNIVERSITY OF LOUVAIN

CODE LO: LOU rat; M: mouse followed by a letter for the Ig class as M for IgM and a number.

For all MAbs: Immunocyte donor: LOU/C rat; immortal cell partner: IR 983F.

LO-MM-3
Immunogen: Mouse polyclonal IgM
Class of Antibody Produced: Rat IgM kappa
Reactivity: Mouse mu heavy chain of immunoglobulin

LO-MM-9 Ref: HDB 7054
Immunogen: Mouse polyclonal IgM
Class of Antibody Produced: Rat IgG2a kappa
Reactivity: Mouse mu heavy chain of immunoglobulin

LO-ME-2
Immunogen: Mouse IgE
Class of Antibody Produced: Rat IgG2a kappa
Reactivity: Mouse epsilon heavy chain of immunoglobulin

LO-ME-3
Immunogen: Mouse IgE
Class of Antibody Produced: Rat IgG2a kappa
Reactivity: Mouse epsilon heavy chain of immunoglobulin

LO-MG1-2 Ref: HDB 7138
Immunogen: Mouse IgG1
Class of Antibody Produced: Rat IgG1 kappa
Reactivity: Mouse gamma1 heavy chain of immunoglobulin

LO-MG1-13
Immunogen: Mouse IgG1
Class of Antibody Produced: Rat IgG1 kappa
Reactivity: Mouse gamma1 heavy chain of immunoglobulin

LO-MG1-15
Immunogen: Mouse IgG1
Class of Antibody Produced: Rat IgG1 kappa
Reactivity: Mouse gamma1 heavy chain of immunoglobulin

LO-MG2a-2 Ref: HDB 7143
Immunogen: Mouse IgG
Class of Antibody Produced: Rat IgG2a kappa
Reactivity: Mouse gamma2a heavy chain of immunoglobulin

LO-MG2a-3 Ref: HDB 7144
Immunogen: Mouse IgG
Class of Antibody Produced: Rat IgG2a kappa
Reactivity: Mouse gamma2a heavy chain of immunoglobulin

LO-MG2a-7
Immunogen: Mouse IgG
Class of Antibody Produced: Rat IgG1 kappa
Reactivity: Mouse gamma2a heavy chain of immunoglobulin

LO-MG2b-1
Immunogen: Mouse IgG
Class of Antibody Produced: Rat IgG1 kappa
Reactivity: Mouse gamma2b heavy chain of immunoglobulin

LO-MG2b-2 Ref : HDB 7145
Immunogen: Mouse IgG
Class of Antibody Produced: Rat IgG1 kappa
Reactivity: Mouse gamma2b heavy chain of immunoglobulin

LO-MG3-7
Immunogen: Mouse IgG3
Class of Antibody Produced: Rat IgM kappa
Reactivity: Mouse gamma3 heavy chain of immunoglobulin

LO-MG-7 Ref: HDB 7147
Immunogen: Mouse IgG
Class of Antibody Produced: Rat IgM kappa
Reactivity: Mouse gamma heavy chain of immunoglobulin

LO-MK-1 Ref: HDB 7150
Immunogen: Mouse IgG
Class of Antibody Produced: Rat IgG2a kappa
Reactivity: Mouse kappa light chain of immunoglobulin

LO-MK-2 Ref: HDB 7151
Immunogen: Mouse IgG
Class of Antibody Produced: Rat IgG1 kappa
Reactivity: Mouse kappa light chain of immunoglobulin

LO-MK-3 Ref: HDB 7152
Immunogen: Mouse IgG
Class of Antibody Produced: Rat IgM kappa
Reactivity: Mouse kappa light chain of immunoglobulin

B. SELECTED LIST OF MAbs ANTI-PLANT PATHOGENS

From: Experimental Immunology Unit
University of Louvain
Faculty of Medicine
Clos Chapelle aux Champs 30-56
Brussels 1200 Belgium

and Phytophatology Unit
University of Louvain
Place Croix du Sud 3
Louvain-la-Neuve 1348 Belgium

1. Anti-Plant Bacterium: *Erwinia amylovora*

For all these antibodies: Immunogen: *Erwinia amylovora*
Immunocyte Donor: LOU/C Rat
Immortal Cell Partner: IR983F
Class of Antibody Produced: IgM

LO-Ea-4
Reactivity: *Erwinia amylovora, Erwinia uredovora, herbicola,* and *Pseudomonas*
LO-Ea-8
Reactivity: *Erwinia amylovora* and *Erwinia herbicola*
LO-Ea-10
Reactivity: *Erwinia amylovora*
LO-Ea-14
Reactivity: *Erwinia amylovora*
LO-Ea-15
Reactivity: *Erwinia amylovora*
LO-Ea-16
Reactivity: *Erwinia amylovora* and *Erwinia herbicola*
LO-Ea-17
Reactivity: *Erwinia amylovora, Erwinia carotodora* pv. *atroseptica,* and *Erwinia herbicola*
LO-Ea-18
Reactivity: *Erwinia amylovora*
LO-Ea-19
Reactivity: *Erwinia amylovora*
LO-Ea-21
Reactivity: *Erwinia amylovora*

2. Anti-Plant Virus: Potato Virus X
For all of this antibodies: Immunogen: Potatoes virus X
Immunocyte Donor: LOU/C Rat
Immortal Cell Partner: IR983F

LO-VX-1
Reactivity: Potatoes virus X
LO-VX-2
Reactivity: Potatoes virus X
LO-VX-3
Reactivity: Potatoes virus X
LO-VX-4
Reactivity: Potatoes virus X
LO-VX-5
Reactivity: Potatoes virus X
LO-VX-6
Reactivity: Potatoes virus X
LO-VX-7
Reactivity: Potatoes virus X
LO-VX-8
Reactivity: Potatoes virus X
LO-VX-9
Reactivity: Potatoes virus X
LO-VX-10
Reactivity: Potatoes virus X
LO-VX-11
Reactivity: Potatoes virus X
LO-VX-12
Reactivity: Potatoes virus X
LO-VX-13
Reactivity: Potatoes virus X

LO-VX-14
Reactivity: Potatoes virus X

C. SELECtED LIST OF MABS ANTI-PARVOVIRUS

From: Microbiology Unit
University of Louvain
Faculty of Medicine
Avenue Hippocrate 54
Brussels 1200 Belgium

For all monoclonal antibodies: Immunogen: Canine Parvovirus
Immunocyte Donor: LOU/C Rat
Immortal Cell Partner: IR983F

Class of Antibody Produced: Class
Reactivity: React
Immunofluorescens: IF
Immuno Electron Microscopy: IEM
Seroneutralization: SN
Inhibition Hemaglutination Test: IHA
Western Blot: WB

LO-RH-1
Class: IgG2b
React.: IF (+), IEM (+), SN: (+), IHA: (+)
WB: (85-70-67)
Specificity: Canine, mink, feline parvovirus

LO-RH-2
Class: IgG2b
React.: IF (+), IEM (+), SN: (+), IHA: (+)
WB: (–)
Specificity: Canine, mink, feline parvovirus

LO-RH-3
Class: IgG2b
React.: IF (+), IEM (–), SN: (–), IHA: (–)
WB: (NT)
Specificity: Canine, mink, feline parvovirus

LO-RH-4
Class: IgG2b
React.: IF (+), IEM (–), SN: (–), IHA: (–)
WB: (85-75)
Specificity: Canine, mink, feline parvovirus

LO-RH-5
Class: IgG2b
React.: IF (+), IEM (+), SN: (+), IHA: (+)
WB: (–)
Specificity: Canine, mink, feline parvovirus

LO-RH-6
Class: IgG2b
React.: IF (+), IEM (+), SN: (+), IHA: (+)
WB: (85-70-67)
Specificity: Canine, mink, feline parvovirus

LO-RH-7
Class: IgG2b
React.: IF (+), IEM (+), SN: (+), IHA: (+)
WB: (NT)
Specificity: Canine, mink, feline parvovirus

LO-RH-8
Class: IgM
React.: IF (+), IEM (+), SN: (+), IHA: (+)
WB: (85-70-67)
Specificity: Canine, mink, feline parvovirus

LO-RH-9
Class: IgG2b
React.: IF (+), IEM (+), SN: (+), IHA: (+)
WB: (85-70-67)
Specificity: Canine, mink, feline parvovirus

LO-RH-10
Class: IgG2b
React.: IF (+), IEM (+), SN: (+), IHA: (+)
WB: (85-70-67)
Specificity: Canine, mink, feline parvovirus

LO-RH-11
Class: IgG2b
React.: IF (+), IEM (+), SN: (+), IHA: (+)
WB: (–)
Specificity: Canine parvovirus

LO-RH-12
Class: IgG2b
React.: IF (+), IEM (+), SN: (+), IHA: (+)
WB: (–)
Specificity: Canine, mink, feline parvovirus

LO-RH-13
Class: IgG2b
React.: IF (+), IEM (+), SN: (+), IHA: (–)
WB: (85-70-67)
Specificity: Canine, mink, feline parvovirus

LO-RH-14
Class: IgG2b
React.: IF (+), IEM (+), SN: (+), IHA: (+)
WB: (–)
Specificity: Canine parvovirus

LO-RH-17
Class: IgG2b
React.: IF (+), IEM (+), SN: (+), IHA: (+)
WB: (–)
Specificity: Canine, mink, feline parvovirus

LO-RH-18
Class: IgG2b
React.: IF (+), IEM (+), SN: (+), IHA: (+)
WB: (–)
Specificity: Canine parvovirus

LO-RH-19
Class: IgG2b
React.: IF (+), IEM (+), SN: (+), IHA: (+)
WB: (–)
Specificity: Canine parvovirus

LO-RH-20
Class: IgM
React.: IF (+), IEM (+), SN: (+), IHA: (+)
WB: (–)
Specificity: Canine, mink, feline parvovirus

LO-RH-21
Class: IgG2b
React.: IF (+), IEM (+), SN: (+), IHA: (+)
WB: (–)
Specificity: Canine, mink, feline parvovirus

LO-RH-22
Class: IgG2b
React.: IF (+), IEM (+), SN: (+), IHA: (+)
WB: (NT)
Specificity: Canine, mink, feline parvovirus

LO-RH-23
Class: ——
React.: IF (+), IEM (–), SN: (–), IHA: (–)
WB: (NT)
Specificity: Canine parvovirus

LO-RH-24
Class: IgG2b
React.: IF (+), IEM (+), SN: (+), IHA: (+)
WB: (–)
Specificity: Canine, mink, feline parvovirus

LO-RH-25
Class: IgG2b
React.: IF (+), IEM (+), SN: (+), IHA: (+)
WB: (–)
Specificity: Canine, mink parvovirus

LO-RH-27
Class: IgG2b
React.: IF (+), IEM (+), SN: (+), IHA: (+)
WB: (–)
Specificity: Canine, mink, feline parvovirus

LO-RH-28
Class: IgG2a
React.: IF (+), IEM (–), SN: (–), IHA: (–)
WB: (85-70-67)
Specificity: Canine, mink, feline parvovirus

LO-RH-29
Class: IgG2b
React.: IF (+), IEM (+), SN: (+), IHA: (+)
WB: (–)
Specificity: Canine, mink, feline parvovirus

LO-RH-30
Class: IgG2b
React.: IF (+), IEM (+), SN: (–), IHA: (–)
WB: (–)
Specificity: Canine, mink, feline parvovirus

LO-RH-34
Class: IgG2b
React.: IF (+), IEM (–), SN: (+), IHA: (–)
WB: (–)
Specificity: Canine, mink, feline parvovirus

LO-RH-35
Class: IgG2b
React.: IF (+), IEM (+), SN: (+), IHA: (+)
WB: (–)
Specificity: Canine, mink, feline parvovirus

LO-RH-37
Class: IgG1
React.: IF (+), IEM (–), SN: (–), IHA: (–)
WB: (NT)
Specificity: Canine, mink, feline parvovirus

LO-RH-38
Class: IgG2b
React.: IF (+), IEM (–), SN: (–), IHA: (–)
WB: (–)
Specificity: Canine, mink, feline parvovirus

LO-RH-40
Class: IgG2b
React.: IF (+), IEM (+), SN: (+), IHA: (–)
WB: (85)
Specificity: Canine, mink, feline parvovirus

LO-RH-41
Class: IgM
React.: IF (+), IEM (–), SN: (–), IHA: (–)
WB: (–)
Specificity: Canine, mink, feline parvovirus

LO-RH-42
Class: IgG2b
React.: IF (+), IEM (–), SN: (–), IHA: (–)
WB: (NT)
Specificity: Canine, mink, feline parvovirus

LO-RH-43
Class: IgG2b
React.: IF (+), IEM (+), SN: (+), IHA: (+)
WB: (NT)
Specificity: Canine, mink, feline parvovirus

LO-RH-45
Class: NT
React.: IF (+), IEM (+), SN: (+), IHA: (+)
WB: (–)
Specificity: Canine, mink, feline parvovirus

LO-RH-46
Class: IgG2b
React.: IF (+), IEM (+), SN: (+), IHA: (+)
WB: (–)
Specificity: Canine, mink, feline parvovirus

LO-RH-50
Class: IgM
React.: IF (+), IEM (–), SN: (–), IHA: (–)
WB: (–)
Specificity: Canine, mink, feline parvovirus

LO-RH-51
Class: IgG1
React.: IF (+), IEM (–), SN: (–), IHA: (–)
WB: (–)
Specificity: Canine, mink, feline parvovirus

LO-RH-54
Class: IgG2b
React.: IF (+), IEM (+), SN: (+), IHA: (+)
WB: (–)
Specificity: Canine, mink, feline parvovirus

LO-RH-55
Class: IgG2b
React.: IF (+), IEM (+), SN: (+), IHA: (+)
WB: (–)
Specificity: Canine, mink, feline parvovirus

LO-RH-57
Class: IgG2b
React.: IF (+), IEM (+), SN: (+), IHA: (+)
WB: (85-70-67)
Specificity: Canine, mink, feline parvovirus

LO-RH-58
Class: IgG2b
React.: IF (+), IEM (+), SN: (+), IHA: (+)
WB: (NT)
Specificity: Canine parvovirus

LO-RH-59
Class: IgG2a
React.: IF (+), IEM (+), SN: (+), IHA: (+)
WB: (–)
Specificity: Canine, mink, feline parvovirus

LO-RH-60
Class: IgG2b
React.: IF (+), IEM (–), SN: (–), IHA: (–)
WB: (–)
Specificity: Canine, mink, feline parvovirus

LO-RH-61
Class: IgG2b
React.: IF (+), IEM (+), SN: (+), IHA: (–)
WB: (85-70-67)
Specificity: Canine, mink, feline parvovirus

LO-RH-62
Class: IgG2b
React.: IF (+), IEM (+), SN: (+), IHA: (+)
WB: (–)
Specificity: Canine, mink, feline parvovirus

LO-RH-63
Class: IgG2b
React.: IF (+), IEM (+), SN: (+), IHA: (–)
WB: (85-70-67)
Specificity: Canine, mink, feline parvovirus

LO-RH-64
Class: IgG2b
React.: IF (+), IEM (+), SN: (+), IHA: (+)
WB: (85-70-67)
Specificity: Canine, mink, feline parvovirus

LO-RH-65
Class: NT
React.: IF (+), IEM (+), SN: (+), IHA: (+)
WB: (–)
Specificity: Canine parvovirus

LO-RH-67
Class: IgG2b
React.: IF (+), IEM (–), SN: (–), IHA: (–)
WB: (–)
Specificity: Canine, mink, feline parvovirus

LO-RH-68
Class: IgG2b
React.: IF (+), IEM (+), SN: (+), IHA: (–)
WB: (NT)
Specificity: Canine, mink, feline parvovirus

LO-RH-70
Class: IgG2b
React.: IF (+), IEM (+), SN: (+), IHA: (+)
WB: (–)
Specificity: Canine, mink, feline parvovirus

LO-RH-71
Class: IgG2b
React.: IF (+), IEM (+), SN: (+), IHA: (+)
WB: (85-70-67)
Specificity: Canine, mink, feline parvovirus

D. SELECTED LIST OF MAbs ANTI-HUMAN IMMUNODEFICIENCY VIRUS

From: Microbiology Unit
University of Louvain

For all monoclonal antibodies: Immunogen: Human Immunodeficiency Virus
Immunocyte Donor: LOU/C Rat
Immortal Cell Partner: IR983F

LO-HIV-1
Class of Antibody Produced: IgG1
Specificity: LAV1, HTLV3, ARV

Detected Protein (Western Blot): 24-55-40

LO-HIV-2

Class of Antibody Produced: IgG1
Specificity: LAV1, HTLV3, ARV
Detected Protein (Western Blot): 24-55-40

LO-HIV-3

Class of Antibody Produced: IgG2a
Specificity: LAV1, HTLV3
Detected Protein (Western Blot): 55-13

LO-HIV-4

Class of Antibody Produced: IgG1
Specificity: ARV
Detected Protein (Western Blot): 34

LO-HIV-5

Class of Antibody Produced: IgG2a
Specificity: ARV
Detected Protein (Western Blot): 34

E. LIST OF MAbs ANTI-DNP-HAPTEN*

From: Experimental Immunology Unit,
University of Louvain

Immunogen: DNP-OVA: Dinitrophenyl hapten - Ovalbumin
DNP-As: Dinitrophenyl hapten - Ascaris extract
DNP-NP: Dinitrophenyl hapten - Nippostrongylus brasiliensis extract
DNP-SAL: Dinitrophenyl hapten - Salmonella

Immunocyte Donor: LOU/C rat

Immortal Cell Partner: IR983F

MAB name	HDB number	Class of antibody produced	Immunogen
LO-DNP-1	232	IgG1 kappa	DNP-OVA
LO-DNP-2	7055	IgG1 kappa	DNP-As, DNP-NP
LO-DNP-3	7056	IgG1 kappa	DNP-As, DNP-NP
LO-DNP-4	7057	IgG1 kappa	DNP-As, DNP-NP
LO-DNP-5	7058	IgG1 kappa	DNP-As, DNP-NP
LO-DNP-6	7059	IgG1 kappa	DNP-As, DNP-NP
LO-DNP-7	7060	IgG2a kappa	DNP-As, DNP-NP
LO-DNP-8	7061	IgG2a kappa	DNP-As, DNP-NP
LO-DNP-9	7062	IgG1 kappa	DNP-As, DNP-NP
LO-DNP-10	7063	IgE kappa	DNP-OVA
LO-DNP-11	7064	IgG2b kappa	DNP-OVA
LO-DNP-12	7065	IgE kappa	DNP-OVA
LO-DNP-13	7066	IgE kappa	DNP-OVA
LO-DNP-14	7067	IgE kappa	DNP-OVA
LO-DNP-15	7068	IgG1 kappa	DNP-OVA
LO-DNP-16	7069	IgG2a kappa	DNP-OVA
LO-DNP-17	7070	IgG1 kappa	DNP-OVA
LO-DNP-18	7071	IgG1 kappa	DNP-OVA
LO-DNP-19	7072	IgG1 kappa	DNP-OVA
LO-DNP-20	7073	IgG2a kappa	DNP-OVA

* Monoclonal antibodies available upon request to Prof. H. Bazin for scientific uses (limited quantities). For large quantities, conditions to be discussed. Availability of the corresponding hybridomas to be discussed upon request.

MAB name	HDB number	Class of antibody produced	Immunogen
LO-DNP-21	7074	IgG2b ND	DNP-NP
LO-DNP-22	7075	IgG2a kappa	DNP-NP
LO-DNP-23	7076	IgG1 kappa	DNP-NP
LO-DNP-24	7077	IgG1 kappa	DNP-NP
LO-DNP-25	7078	IgG2a kappa	DNP-NP
LO-DNP-26	7079	IgG2b kappa	DNP-NP
LO-DNP-27	7080	IgG2c kappa	DNP-NP
LO-DNP-28	7081	IgG2c kappa	DNP-NP
LO-DNP-29	7082	IgE kappa	DNP-NP
LO-DNP-30	7083	IgE kappa	DNP-NP
LO-DNP-31	7084	IgE kappa	DNP-NP
LO-DNP-32	7085	IgG2a	DNP-NP
LO-DNP-33	7086	IgG2a	DNP-NP
LO-DNP-34	7087	IgM kappa	DNP-NP
LO-DNP-35	7088	IgG1	DNP-As, DNP-NP
LO-DNP-36	7089	IgE	DNP-OVA
LO-DNP-37	7090	IgG2b	DNP-NP
LO-DNP-38	7091	IgM	DNP-NP
LO-DNP-39	7092	IgM kappa	DNP-NP
LO-DNP-40	7093	IgM	DNP-NP
LO-DNP-41	7094	IgM	DNP-NP
LO-DNP-42	7095	ND	DNP-NP
LO-DNP-43	7096	ND	DNP-NP
LO-DNP-44	7097	IgA	DNP-NP
LO-DNP-45	7098	IgA kappa	DNP-SAL
LO-DNP-46	7099	IgA	DNP-SAL
LO-DNP-47	7100	IgA	DNP-SAL
LO-DNP-48	7101	IgG2b	DNP-SAL
LO-DNP-49	7102	IgA	DNP-SAL
LO-DNP-50	7103	IgA	DNP-SAL
LO-DNP-51	7104	IgA	DNP-SAL
LO-DNP-52	7105	IgA	DNP-SAL
LO-DNP-53	7106	IgA	DNP-SAL
LO-DNP-54	7107	IgG2b kappa	DNP-SAL
LO-DNP-55	7108	IgG2b	DNP-SAL
LO-DNP-56	7109	IgG2b	DNP-SAL
LO-DNP-57	7110	IgG2b kappa	DNP-SAL
LO-DNP-58	7111	IgA	DNP-SAL
LO-DNP-59	7112	IgA	DNP-SAL
LO-DNP-60	7113	IgA	DNP-SAL
LO-DNP-61	7114	IgG2a kappa	DNP-SAL
LO-DNP-62	7115	IgG2b	DNP-SAL
LO-DNP-63	7116	IgA	DNP-SAL
LO-DNP-64	7117	IgA kappa	DNP-SAL
LO-DNP-65	7118	IgA	DNP-SAL
LO-DNP-66	7119	IgG2b	DNP-SAL
LO-DNP-67	7120	IgA kappa	DNP-SAL
LO-DNP-68	7121	IgA	DNP-SAL

F. SELECTED LIST OF MAbs ANTI-HUMAN LIPOPROTEIN

From: Experimental Immunology Unit,
University of Louvain

1. Anti-Human High Density Lipoprotein

For all of these antibodies: Immunogen: Human high density lipoprotein
Immunocyte Donor: LOU/C Rat
Immortal Cell Partner: IR983F
Class of Antibody Produced: rat IgG2a Kappa

LO-HApoAI-1
Reactivity: Human high density lipoprotein

LO-HApoAI-2
Reactivity: Human high density lipoprotein

LO-HApoAI-3
Reactivity: Human high density lipoprotein

LO-HApoAI-4
Reactivity: Human high density lipoprotein

LO-HApoAI-5
Reactivity: Human high density lipoprotein

2. Anti-Human Low Density Lipoprotein

For all of these antibodies: Immunogen: Human low density lipoprotein
Immunocyte Donor: LOU/C Rat
Immortal Cell Partner: IR983F
Reactivity: Human low density lipoprotein

LO-HApoB-1
Class of Antibody Produced:Rat IgG1 kappa

LO-HApoB-2
Class of Antibody Produced: Rat IgG1 kappa

LO-HApoB-3
Class of Antibody Produced: Rat IgG kappa

LO-HApoB-4
Class of Antibody Produced: Rat IgG1 kappa

LO-HApoB-5
Class of Antibody Produced: Rat IgG kappa

LO-HApoB-6
Class of Antibody Produced: Rat IgG2a kappa

G. SELECTED LIST OF MAbS ANTI-HUMAN LEUCOCYTES

From: Haematology and Experimental Immunology Units
University of Louvain
Faculty of Medicine
Clos Chapelle aux Champs, 30
Brussels 1200 Belgium

For all of these antibodies: Immunocyte Donor: LOU/C Rat
Immortal Cell Partner: IR983F

LO-CD6-a
Immunogen: HPB-All
Class of Antibody Produced: Rat IgG2a
Reactivity: CD6, p129, anti mature

LO-CD4-a
Immunogen: HPB-All
Class of Antibody Produced: Rat IgG2a
Reactivity: CD4, Sub-population of T lymphocytes

LO-CD4-b

Immunogen: Peripheal blood lymphocytes
Class of Antibody Produced: Rat IgG2a
Reactivity: CD4, Sub-population of T lymphocytes

LO-CD2-out

Immunogen: Peripheal blood lymphocytes
Class of Antibody Produced: Rat IgG2b
Reactivity: T112, weak on resting T, bright on immature and activated T

LO-Tmat

Immunogen: Peripheal blood lymphocytes
Class of Antibody Produced: Rat IgG2a
Reactivity: anti mature T cells

LO-CD1-a

Immunogen: Thymocytes
Class of Antibody Produced: Rat IgG2a
Reactivity: CD1, cortical thymocytes

LO-CD5-a

Immunogen: Thymocytes
Class of Antibody Produced: Rat IgG2a kappa
Reactivity: CD5, T lineage + B-CLL

LO-Tact-1

Immunogen: PHA-activated lymphocytes
Class of Antibody Produced: Rat IgG2b kappa
Reactivity: CD25, T activated cells

LO-Tact-2

Immunogen: PHA-activated lymphocytes
Class of Antibody Produced: Rat IgG2b kappa
Reactivity: CD25, T activated cells

LO-panB

Immunogen: B-CLL
Class of Antibody Produced: Rat IgG2b
Reactivity: anti B lineage

LO-DR-a

Immunogen: PHA-activated lymphocytes
Class of Antibody Produced: Rat IgG2b kappa
Reactivity: anti HLA DR framework

LO-MO1

Immunogen: Peripheal blood mononuclear cells
Class of Antibody Produced: Rat IgG2b kappa
Reactivity: anti mature monocytes

LO-PL3-a

Immunogen: Peripheal blood mononuclear cells
Class of Antibody Produced: Rat IgG2a kappa
Reactivity: anti platelet, megakaryocyte lineage

LO-PL3-b

Immunogen: Peripheal blood mononuclear cells
Class of Antibody Produced: Rat IgG2a kappa
Reactivity: anti platelet, megakaryocyte lineage

LO-PL4

Immunogen: Peripheal blood mononuclear cells
Class of Antibody Produced: ND
Reactivity: CD9

H. SELECTED LIST OF MAbs ANTI-HUMAN IMMUNOGLOBULINS

From: Experimental Immunology Unit,
University of Louvain

See Chapter 19. VI., Table 1.

I. SELECTED LIST OF MAbs ANTI-HORSERADISH PEROXIDASE

From: Experimental Immunology Unit,
University of Louvain

See Chapter 19.IX., Table 2.

J. SELECTED LIST OF MAbs ANTI-*STREPTOCOCCUS MUTANS*

From: Experimental Immunology Unit,
University of Louvain

For all monoclonal antibodies: Immunogen: Wall extracted antigens of *Streptococcus mutans* serotype f (OML 175)
Immunocyte Donor: WISTAR R Rat
Immortal Cell Partner: IR983F

LO-SM-2

Class of Antibody Produced: Rat IgM kappa
Reactivity: Wall extracted antigens of *Streptococcus mutans* serotype a,b,c,d,e,f

LO-SM-11

Class of Antibody Produced: Rat IgG2a kappa
Reactivity: Wall extracted antigens of *Streptococcus mutans* serotype a,b,c,d,e,f

K. SELECTED LIST OF MAbs ANTI-*BACTEROIDES GINGIVALIS*

From: Experimental Immunology Unit,
University of Louvain

LO-BG-1

Immunogen: Whole cells of *Bacteroides gingivalis* ATCC 33277
Immunocyte Donor: LOU/C Rat
Immortal Cell Partner: IR983F
Class of Antibody Produced: Rat IgG2aKappa
Reactivity: Whole cells of *Bacteroides gingivalis* ATCC 33277 and antigen preparation of outer membrane and cell surface componants of *B. gingivalis*

III. IR 983F CELL LINE

IR983F cell line available upon request to: Prof. H. Bazin
Experimental Immunology Unit
Faculty of Medicine
University of Louvain
Clos Chapelle aux Champs 30.56
B-1200 Brussels
Belgium
Telephone: 32/2/7643430
Telex: UCL 23722 B
Telefax: 32/2/7643946

Index

INDEX

A

B

C

D

E

F

G

H

I

J

K

L

M

N

O

P

Q

R

S

T

U

V

W

X

Y

Z